Adventures in
Stochastic Processes

T0320711

WELCOME TO THE RANDOM WORLD OF HAPPY HARRY—famed restaurateur, happy hour host, community figure, former semi–pro basketball player, occasional software engineer, talent agent, budding television star, world traveller, nemesis of the street gang called the Mutant Creepazoids, theatre patron, supporter of precise and elegant use of the English language, supporter of the war on drugs, unsung hero of the fairy tale *Sleeping Beauty*, and the target of a vendetta by the local chapter of the Young Republicans. Harry and his restaurant are well known around his Optima Street neighborhood both to the lovers of fine food and the public health service. Obviously this is a man of many talents and experiences who deserves to have a book written about his life.

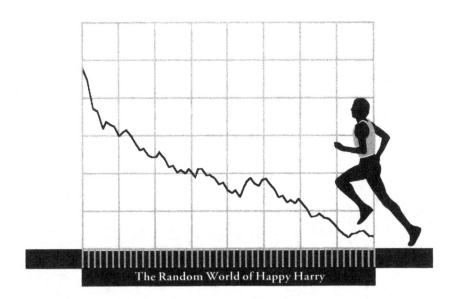

The Random World of Happy Harry

Sidney Resnick

Adventures in Stochastic Processes

with Illustrations

Birkhäuser

Boston · Basel · Berlin

Sidney I. Resnick
School of Operations Research and
  Industrial Engineering
Cornell University
Ithaca, NY 14853
USA

Library of Congress Cataloging-in-Publication Data
Resnick, Sidney I.
    Adventures in stochastic processes / Sidney Resnick.
       p.   cm.
    Includes bibliographical references and index.
    ISBN 0-8176-3591-2   (hard : U.S. acid-free paper). –ISBN
  3-7643-3591-2 (hard : Switzerland : acid-free paper)
    1. Stochastic processes.   I. Title.
  QA274.R46   1992   92-4431
  519'.2–dc20   CIP

Printed on acid-free paper
©1992 Birkhäuser Boston
©1994 Birkhäuser Boston, 2nd printing
©2002 Birkhäuser Boston, 3rd printing

Birkhäuser

ISBN 0-8176-3591-2   SPIN 10854451
ISBN 3-7643-3591-2

Cover design by Minna Resnick
Typeset by the author in A$\mathcal{M}$S TEX
Printed and bound by Hamilton Printing Co., Rennselaer, NY
Printed in the United States of America

9 8 7 6 5 4 3

Table of Contents

*This section contains advanced material which may be skipped on first reading by beginning readers.

CHAPTER 3. RENEWAL THEORY

CHAPTER 4. POINT PROCESSES

---

*This section contains advanced material which may be skipped on first reading by beginning readers.

CHAPTER 5. CONTINUOUS TIME MARKOV CHAINS

CHAPTER 6. BROWNIAN MOTION

---

*This section contains advanced material which may be skipped on first reading by beginning readers.

CHAPTER 7. THE GENERAL RANDOM WALK*

---

*This section contains advanced material which may be skipped on first reading by beginning readers.

Preface

While this is a book about Harry and his adventurous life, it is primarily a serious text about stochastic processes. It features the basic stochastic processes that are necessary ingredients for building models of a wide variety of phenomena exhibiting time varying randomness.

The book is intended as a first year graduate text for courses usually called *Stochastic Processes* (perhaps amended by the words "Applied" or "Introduction to ... ") or *Applied Probability*, or sometimes *Stochastic Modelling*. It is meant to be very accessible to beginners, and at the same time, to serve those who come to the course with strong backgrounds. This flexiblity also permits the instructor to push the sophistication level up or down. For the novice, discussions and motivation are given carefully and in great detail. In some sections beginners are advised to skip certain developments, while in others, they can read the words and skip the symbols in order to get the content without more technical detail than they are ready to assimilate. In fact, with the numerous readings and variety of problems, it is easy to carve a path so that the book challenges more advanced students, but remains instructive and manageable for beginners. Some sections are starred and come with a warning that they contain material which is more mathematically demanding. Several discussions have been modularized to facilitate flexible adaptation to the needs of students with differing backgrounds. The text makes crystal clear distinctions between the following: proofs, partial proofs, motivations, plausibility arguments and good old fashioned hand-waving.

Where did Harry, Zeke and the rest of the gang come from? Courses in Stochastic Processes tend to contain overstuffed curricula. It is, therefore, useful to have quick illustrations of how the theory leads to techniques for calculating numbers. With the Harry vignettes, the student can get in and out of numerical illustrations quickly. Of course, the vignettes are not meant to replace often stimulating but time consuming real applications. A variety of examples with applied appeal are sprinkled throughout the exposition and exercises. Our students are quite fond of Harry and enjoy psychoanalyzing him, debating whether he is "a polyester sort of guy" or the "jeans and running shoes type." They seem to have no trouble discerning the didactic intent of the Harry stories and accept the need for some easy numerical problems before graduating to more serious ones. Student culture has become so ubiquitous that foreign students who are not native English speakers can quickly get into the swing. I think Harry is a useful and entertaining guy but if you find that you loathe him, he is easy to avoid in the text.

Where did they come?
I can't say.
But I bet they have come a long long way.[1]

To the instructor: The discipline imposed during the writing was that the first six chapters should not use advanced notions of conditioning which involve relatively sophisticated ideas of integration. Only the elementary definition is used: $P(A|B) = P(A \cap B)/P(B)$. Instead of conditioning arguments we find independence where we need it and apply some form of the product rule: $P(A \cap B) = P(A)P(B)$ if A and B are independent. This maintains rigor and keeps the sophistication level down.

No knowledge of measure theory is assumed but it is assumed that the student has already digested a good graduate level pre–measure theoretic probability course. A bit of measure theory is discussed here and there in starred portions of the text. In most cases it is simple and intuitive but if it scares you, skip it and you will not be disadvantaged as you journey through the book. If, however, you know some measure theory, you will understand things in more depth. There is a sprinkling of references throughout the book to *Fubini's theorem, the monotone convergence theorem* and *the dominated convergence theorem*. These are used to justify the interchange of operations such as summation and integration. A relatively unsophisticated student would not and should not worry about justifications for these interchanges of operations; these three theorems should merely remind such students that somebody knows how to check the correctness of these interchanges.

Analysts who build models are supposed to know how to build models. So for each class of process studied, a construction of that process is included. Independent, identically distributed sequences are usually assumed as primitives in the constructions. Once a concrete version of the process is at hand, many properties are fairly transparent. Another benefit is that if you know how to construct a stochastic process, you know how to *simulate* the process. While no specific discussion of simulation is included here, I have tried to avoid pretending the computer does not exist. For instance, in the Markov chain chapters, formulas are frequently put in matrix form to make them suitable for solution by machine rather than by hand. Packages such as Minitab, Mathematica, Gauss, Matlab, etc., have been used successfully as valuable aids in the solution of problems but local availability of computing resources and the rapidly changing world of hardware and software make specific suggestions unwise. Ask your local guru

---

[1]Dr. Seuss, *One Fish, Two Fish, Red Fish, Blue Fish*

for suggestions. You need to manipulate some matrices, and find roots of polynomials; but nothing too fancy. If you have access to a package that does symbolic calculations, so much the better. A companion disk to this book is being prepared by Douglas McBeth which will allow easy solutions to many numerical problems.

There is much more material here than can be covered in one semester. Some selection according to the needs of the students is required. Here is the core of the material: Chapter 1: 1.1–1.6. Skip the proof of the continuity theorem in 1.5 if necessary but mention Wald's identity. Some instructors may prefer to skip Chapter 1 and return later to these topics, as needed. If you are tempted by this strategy, keep in mind that Chapter 1 discusses the interesting and basic random walk and branching processes and that facility with transforms is worthwhile. Chapter 2: 2.1–2.12,2.12.1. In Section 2.13, a skilled lecturer is advised to skip most of the proof of Theorem 2.13.2, explain coupling in 15 minutes, and let it go at that. This is one place where hand-waving really conveys something. The material from Section 2.13.1 should be left to the curious. If time permits, try to cover Sections 2.14 and 2.15 but you will have to move at a brisk pace. Chapter 3: In renewal theory stick to basics. After all the discrete state space theory in Chapters 1 and 2, the switch to the continuous state space world leaves many students uneasy. The core is Sections 3.1–3.5, 3.6, 3.7, and 3.7.1. Sections 3.8 and 3.12.3 are accessible if there is time but 3.9–3.12.2 are only for supplemental reading by advanced students. Chapter 4: The jewels are in Sections 4.1 to 4.7. You can skip 4.3.1. If you have a group that can cope with a bit more sophistication, try 4.7.1, 4.8 and 4.9. Once you come to know and love the Laplace functional, the rest is incredibly easy and short. Chapter 5: The basics are 5.1–5.8. If you are pressed for time, skip possibly 5.6 and 5.8; beginners may avoid 5.2.1, 5.3.1 and 5.5.1. Section 5.7.1 is on queueing networks and is a significant application of standard techniques, so try to reserve some time for it. Section 5.9 is nice if there is time. Despite its beauty, leave 5.11 for supplemental reading by advanced students. Chapter 6: Stick to some easy path properties, strong independent increments, reflection, and some explicit calculations. I recommend 6.1, 6.2, 6.4, 6.5, 6.6, 6.7, and 6.8. For beginners, a quick survey of 6.11–6.13 may be adequate. If there is time and strong interest in queueing, try 6.9. If there is strong interest in statistics, try 6.10. I like Chapter 7, but it is unlikely it can be covered in a first course. Parts of it require advanced material.

In the course of teaching, I have collected problems which have been inserted into the examples and problem sections; there should be a good supply. These illustrate a variety of applied contexts where the skills mastered in the chapter can be used. Queueing theory is a frequent context for many exercises. Many problems emphasize calculating numbers which seems to be a skill most students need these days, especially considering the wide clientele who enroll for courses in stochastic processes. There is a big payoff for the student who will spend serious time working out the problems. Failure to do do will relegate the novice reader to the status of voyeur.

Some acknowledgements and thank you's: The staff at Birkhäuser has been very supportive, efficient and colleagal, and the working relationship could not have been better. Minna Resnick designed a stunning cover and logo. Cornell's Kathy King probably does not realize how much cumulative help she intermittently provided in turning scribbled lecture notes into something I could feed the TEX machine. Richard Davis (Colorado State University), Gennady Sammorodnitsky (Cornell) and Richard Serfozo (Georgia Institute of Technology) used the manuscript in classroom settings and provided extensive lists of corrections and perceptive suggestions. A mature student perspective was provided by David Lando (Cornell) who read almost the whole manuscript and made an uncountable number of amazingly wise suggestions about organization and presentation, as well as finding his quota of mistakes. Douglas McBeth made useful comments about appropriate levels of presentation and numerical issues. David Lando and Eleftherios Iakavou helped convince me that Harry could become friends with students whose mother tongue was different from English. Joan Lieberman convinced me even a lawyer could appreciate Harry. Minna, Rachel and Nathan Resnick provided a warm, loving family life and generously shared the home computer with me. They were also very consoling as I coped with two hard disk crashes and a monitor melt-down.

While writing a previous book in 1985, I wore out two mechanical pencils. The writing of this book took place on four different computers. Financial support for modernizing the computer equipment came from the National Science Foundation, Cornell's Mathematical Sciences Institute and Cornell's School of Operations Research and Industrial Engineering. Having new equipment postponed the arrival of bifocals and made that marvellous tool called TEX almost fun to use.

CHAPTER 1

Preliminaries
Discrete Index Sets
and/or Discrete State Spaces

T HIS CHAPTER eases us into the subject with a review of some useful techniques for handling non-negative integer valued random variables and their distributions. These techniques are applied to some significant examples, namely, the simple random walk and the simple branching process. Towards the end of the chapter stopping times are introduced and applied to obtain Wald's identity and some facts about the random walk. The beginning student can skip the advanced discussion on sigma-fields and needs only a primitive understanding that sigma fields organize information within probability spaces.

Section 1.7, intended for somewhat advanced students, discusses the *distribution of a process* and leads to a more mature and mathematically useful understanding of what a stochastic process is rather than what is provided by the elementary definition: *A stochastic process is a collection of random variables* $\{X(t), t \in T\}$ *defined on a common probability space indexed by the index set T which describes the evolution of some system.* Often $T = [0, \infty)$ if the system evolves in continuous time. For example, $X(t)$ might be the number of people in a queue at time t, or the accumulated claims paid by an insurance company in $[0, t]$. Alternatively, we could have $T = \{0, 1, \dots\}$ if the system evolves in discrete time. Then $X(n)$ might represent the number of arrivals to a queue during the service interval of the nth customer, or the socio-economic status of a family after n generations. When considering stationary processes, $T = \{\dots, -1, 0, 1, \dots\}$ is a common index set. In more exotic processes, T might be a collection of regions, and $X(A)$, the number of points in region A.

1.1. NON-NEGATIVE INTEGER VALUED RANDOM VARIABLES.

Suppose X is a random variable whose range is $\{0, 1, \dots, \infty\}$. (Allowing a possible value of ∞ is a convenience. For instance, if X is the waiting time for a random event to occur and if this event never occurs, it is natural to think of the value of X as ∞.) Set

$$P[X = k] = p_k, \quad k = 0, 1, \dots,$$

so that

$$P[X < \infty] = \sum_{k=0}^{\infty} p_k, \quad P[X = \infty] = 1 - \sum_{k=0}^{\infty} p_k =: p_\infty.$$

(Note that the notation ":=" means that a definition is being made. Thus $1 - \sum_{k=0}^{\infty} p_k =: p_\infty$ means that p_∞ is defined as $1 - \sum_{k=0}^{\infty} p_k$. In general $A =: B$ or equivalently $B := A$ means B is defined as A.) If $P[X = \infty] > 0$, define $E(X) = \infty$; otherwise

$$E(X) = \sum_{k=0}^{\infty} k p_k.$$

If $f : \{0, 1, \dots, \infty\} \mapsto [0, \infty]$ then in an elementary course you probably saw the derivation of the fact that

$$Ef(X) = \sum_{0 \le k \le \infty} f(k) p_k.$$

If $f : \{0, 1, \dots, \infty\} \mapsto [-\infty, \infty]$ then define two positive functions f^+ and f^- by

$$f^+ = \max\{f, 0\}, \quad f^- = -\min\{f, 0\}$$

so that $Ef^+(X)$ and $Ef^-(X)$ are both well defined and

$$Ef^\pm(X) = \sum_{0 \le k \le \infty} f^\pm(k) p_k.$$

Now define

$$Ef(X) = Ef^+(X) - Ef^-(X)$$

provided at least one of $Ef^+(X)$ and $Ef^-(X)$ is finite. In the contrary case, where both are infinite, the expectation does not exist. The expectation is finite if $\sum_{0 \le k \le \infty} |f(k)| p_k < \infty$.

If $p_\infty = 0$ and

$$f(k) = k^n, \text{then} Ef(X) = EX^n = \ n\text{th moment};$$
$$f(k) = (k - E(X))^n, \text{ then } Ef(X) = E(X - E(X))^n$$
$$= n\text{th central moment}.$$

In particular, when $n = 2$ in the second case we get

$$\text{Var}(X) = E(X - E(X))^2 = EX^2 - (E(X))^2.$$

Some examples of distributions $\{p_k\}$ that you should review and that will be particularly relevant are

1. Binomial, denoted $b(k; n, p)$, which is the distribution of the number of successes in n Bernoulli trials when the success probability is p. Then

$$P[X = k] = b(k; n, p) := \binom{n}{k} p^k (1 - p)^{n-k}, \quad 0 \le k \le n, 0 \le p \le 1,$$

and $E(X) = np$, $\mathrm{Var}(X) = np(1 - p)$.

2. Poisson, denoted $p(k; \lambda)$. Then for $k = 0, 1, \ldots, \lambda > 0$

$$P[X = k] = p(k, \lambda) := e^{-\lambda} \lambda^k / k!$$

and $E(X) = \lambda$, $\mathrm{Var}(X) = \lambda$.

3. Geometric, denoted $g(k; p)$, so that for $k = 0, 1, \ldots$

$$P[X = k] = g(k, p) := (1 - p)^k p, \quad 0 \le p \le 1,$$

which is the distribution of the number of failures before the first success in repeated Bernoulli trials. The usual notation is to set $q = 1 - p$. Then

$$E(X) = \sum_{k=0}^{\infty} kq^k p = p \sum_{k=1}^{\infty} kq^k$$

$$= p \sum_{k=1}^{\infty} \left(\sum_{j=1}^{k} 1 \right) q^k,$$

and reversing the order of summation yields

$$= p \sum_{j=1}^{\infty} \sum_{k=j}^{\infty} q^k = p \sum_{j=1}^{\infty} q^j / (1 - q)$$

(1.1.1)
$$= \sum_{j=1}^{\infty} q^j = q/(1 - q) = q/p.$$

Alternatively, we could have computed this by summing tail probabilities:

Lemma 1.1.1. *If X is non-negative integer valued then*

(1.1.2)
$$E(X) = \sum_{k=0}^{\infty} P[X > k].$$

Proof. To verify this formula involves reversing the steps of the previous computation:

$$\sum_{k=0}^{\infty} P[X > k] = \sum_{k=0}^{\infty} \sum_{j=k+1}^{\infty} p_j = \sum_{j=1}^{\infty} \left(\sum_{k=0}^{j-1} 1 \right) p_j$$

$$= \sum_{j=1}^{\infty} j p_j = E(X). \quad \blacksquare$$

In the multivariate case we have a random vector with non-negative integer valued components $\mathbf{X}' = (X_1, \dots, X_k)$ with a mass function

$$P[X_1 = j_1, \dots, X_k = j_k] = p_{j_1, \dots, j_k}$$

for non-negative integers $j_1, \dots, j_k$. If

$$f : \{0, 1, \dots, \infty\}^k \mapsto [0, \infty]$$

then

$$Ef(X_1, \dots, X_k) = \sum_{(j_1, \dots, j_k)} f(j_1, \dots, j_k) p_{j_1, \dots, j_k}.$$

If $f : \{0, 1, \dots, \infty\}^k \mapsto R$ then

$$Ef(X_1, \dots, X_k) = Ef^+(X_1, \dots, X_k) - Ef^-(X_1, \dots, X_k)$$

as in the case $k = 1$ if at least one of the expectations on the right is finite. Now recall the following *properties:*

1. For $a_1, \dots, a_k \in R$

$$E \left(\sum_{i=1}^{k} a_i X_i \right) = \sum_{i=1}^{k} a_i E(X_i)$$

(provided the right side makes sense; no $\infty - \infty$ please!).

2. If $X_1, \dots, X_k$ are independent so that the joint mass function of $X_1, \dots, X_k$ factors into a product of marginal mass functions, then for any bounded functions $f_1, \dots, f_k$ with domain $\{0, 1, \dots, \infty\}$ we have

(1.1.3) $$E \prod_{i=1}^{k} f_i(X_i) = \prod_{i=1}^{k} Ef_i(X_i).$$

The proof is easy for non-negative integer valued random variables based on the factorization of the joint mass function.

3. If $EX_i^2 < \infty, i = 1, \ldots, k$ and

$$\text{Cov}(X_i, X_j) = 0, 1 \le i < j \le k$$

then

$$\text{Var}\left(\sum_{i=1}^{k} a_i X_i\right) = \sum_{i=1}^{k} a_i^2 \text{Var}(X_i)$$

for $a_i \in R, i = 1, \ldots, k$.

1.2. CONVOLUTION.

Suppose X and Y are independent, non-negative, integer valued random variables with

$$P[X = k] = a_k, \quad P[Y = k] = b_k, \quad k = 0, 1, \ldots.$$

Since for $n \ge 0$

$$[X + Y = n] = \bigcup_{i=0}^{n} [X = i, Y = n - i]$$

we get

$$P[X + Y = n] = P\left\{\bigcup_{i=0}^{n} [X = i, Y = n - i]\right\}$$

$$= \sum_{i=0}^{n} P[X = i, Y = n - i]$$

$$= \sum_{i=0}^{n} a_i b_{n-i} =: p_n.$$

This operation on sequences arises frequently and is called *convolution*:

Definition. *The convolution of the two sequences $\{a_n, n \ge 0\}$ and $\{b_n, n \ge 0\}$ is the new sequence $\{c_n, n \ge 0\}$ whose nth element c_n is defined by*

$$c_n = \sum_{i=0}^{n} a_i b_{n-i}.$$

We write
$$\{c_n\} = \{a_n\} * \{b_n\}.$$

Although this definition applies to any two sequences defined on the non-negative integers, it is most interesting to us because it gives the distribution of a sum of independent non-negative integer valued random variables. The calculation before the definition shows that

$$\{P[X + Y = n]\} = \{P[X = n]\} * \{P[Y = n]\}.$$

For the examples, the following notational convention is convenient: Write $X \sim \{p_k\}$ to indicate the mass function of X is $\{p_k\}$. The notation $X \overset{d}{=} Y$ will be used to mean that X and Y have the same distribution. In the case of non-negative integer valued random variables, the mass functions or discrete densities are the same.

Example 1.2.1. If $X \sim p(k; \lambda)$ and $Y \sim p(k; \mu)$ and X and Y are independent then $X + Y \sim p(k; \lambda + \mu)$.

Example 1.2.2. If $X \sim b(k; n, p)$ and $Y \sim b(k; m, p)$ and X and Y are independent then $X + Y \sim b(k; n + m, p)$ since the number of successes in n Bernoulli trials coupled with the number of successes in m independent Bernoulli trials yields the number of successes in $n + m$ Bernoulli trials.

The results of both examples have easy analytic proofs using just the definition and a bit of manipulation. Easier proofs using generating functions are soon forthcoming.

Some obvious properties of convolution are:

1. The convolution of two probability mass functions on the non-negative integers is a *probability mass function*.

2. Convolution is a *commutative operation* (which corresponds to the probability statement that $X + Y \overset{d}{=} Y + X$). Thus

$$\{a_n\} * \{b_n\} = \{b_n\} * \{a_n\}.$$

3. Convolution is an *associative operation* so that the order in which three or more sequences are convolved is immaterial. Thus

$$\{a_n\} * (\{b_n\} * \{c_n\}) = (\{a_n\} * \{b_n\}) * \{c_n\}.$$

The last assertion corresponds to the probability statement that

$$X + (Y + Z) \overset{d}{=} (X + Y) + Z$$

where $X, Y,$ and Z are independent non-negative integer valued random variables.

It is convenient to have a power notation when we convolve a sequence with itself repeatedly; thus we define

$$\{p_n\}^{2*} := \{p_n\} * \{p_n\}$$

so that if X_1, X_2 are independent, identically distributed (iid) with common density $\{p_k\}$ we get

$$X_1 + X_2 \sim \{p_n\}^{2*}.$$

Similarly for $X_1, \ldots, X_k$ iid we get

$$X_1 + \cdots + X_k \sim \{p_n\}^{k*} := \{p_n\} * \cdots * \{p_n\}$$

and recall that it does not matter in what order we perform the convolutions on the right side since the operation is associative.

1.3. Generating Functions.

Let $a_0, a_1, a_2, \ldots$ be a numerical sequence. If there exists $s_0 > 0$ such that

$$A(s) := \sum_{j=0}^{\infty} a_j s^j$$

converges in $|s| < s_0$, then we call $A(s)$ the *generating function* (gf) of the sequence $\{a_i\}$.

We are most interested in generating functions of probability densities. Let X be a non-negative integer valued random variable with density $\{p_k, k \geq 0\}$. The generating function of $\{p_k\}$ is

$$P(s) = \sum_{k=0}^{\infty} p_k s^k$$

and by an abuse of language this is also called the generating function of X. Note that

$$P(s) = Es^X$$

and that $P(1) = \sum_{k=0}^{\infty} p_k \leq 1$ so the radius of convergence of $P(s)$ is at least 1 (and may be greater than 1). Note $P(1) = 1$ iff $P[X < \infty] = 1$.

We will see that a generating function, when it exists, uniquely determines its sequence (and in fact we will give a differentiation scheme which *generates* the sequences from the gf). There are five main uses of generating functions:

(1) Generating functions aid in the computation of the mass function of a sum of independent non-negative integer valued random variables.

(2) Generating functions aid in the calculation of moments. Moments are frequently of interest in stochastic models because they provide easy (but rough) methods for statistical estimation of the parameters of the model.

(3) Using the continuity theorem (see Section 1.5), generating functions aid in the calculation of limit distributions.

(4) Generating functions aid in the solution of difference equations or recursions. Generating function techniques convert the problem of solving a recursion into the problem of solving a differential equation.

(5) Generating functions aid in the solution of linked systems of differential-difference equations. The generating function technique necessitates the solution of a partial differential equation. This technique is applied frequently in continuous time Markov chain theory and is discussed in Section 5.8.

Example 1.3.1. $X \sim p(k; \lambda)$. Then

$$P(s) = \sum_{k=0}^{\infty} \frac{e^{-\lambda}\lambda^k}{k!} s^k = e^{-\lambda} \sum_{k=0}^{\infty} \frac{(\lambda s)^k}{k!}$$

$$= e^{\lambda(s-1)}$$

for all $s > 0$.

Example 1.3.2. $X \sim b(k; n, p)$. Then

$$P(s) = \sum_{k=0}^{n} \left(\binom{n}{k} p^k q^{n-k} \right) s^k$$

$$= \sum_{k=0}^{n} \binom{n}{k} (ps)^k q^{n-k} = (q + ps)^n$$

for all $s > 0$.

Example 1.3.3. $X \sim g(k; p)$. Then

$$P(s) = \sum_{k=0}^{\infty} (q^k p) s^k = p \sum_{k=0}^{\infty} (qs)^k$$

$$= p/(1 - qs)$$

for $0 < s < q^{-1}$.

1.3.1. DIFFERENTIATION OF GENERATING FUNCTIONS.

A generating function $P(s)$ is a power series with radius of convergence at least equal to 1. P can be differentiated as many times as desired by the interchange of summation and differentiation to obtain

$$\frac{d^n}{ds^n}P(s) = \sum_{k=n}^{\infty} k(k-1)\ldots(k-n+1)p_k s^{k-n}$$

$$(1.3.1.1) \qquad = \sum_{k=n}^{\infty} \frac{k!}{(k-n)!}p_k s^{k-n}$$

for $0 \leq s < 1$. For instance we may readily verify formula (1.3.1.1) for the case $n = 1$ with the following argument. (Novices may skip this argument and resume reading at the next paragraph.) We have that

$$P'(s) = \lim_{h\to 0}(P(s+h) - P(s))/h$$

$$= \lim_{h\to 0}\sum_{k=1}^{\infty} p_k((s+h)^k - s^k)/h.$$

For $s \in (0,1)$, there exist $\eta < 1, h_0 > 0$ such that for $|h| \leq h_0$

$$|s+h| \vee s \leq \eta.$$

Now

$$|((s+h)^k - s^k)/h| = |h^{-1}\int_s^{s+h} ku^{k-1}du|$$

$$\leq |h|^{-1}k\eta^{k-1}|s+h-s| = k\eta^{k-1}$$

independent of h. Since

$$\sum_{k=1}^{\infty} p_k k\eta^{k-1} \leq \sum_{k=1}^{\infty} k\eta^{k-1} < \infty$$

by dominated convergence we get,

$$\lim_{h\to 0}\sum_{k=1}^{\infty} p_k((s+h)^k - s^k)/h = \sum_{k=1}^{\infty} p_k \lim_{h\to 0}((s+h)^k - s^k)/h$$

$$= \sum_{k=1}^{\infty} p_k k s^{k-1},$$

which gives (1.3.1.1) for $n = 1$. The procedure just outlined and induction yield the general case.

If we evaluate (1.3.1.1) at $s = 0$ we get

$$(1.3.1.2) \qquad \frac{d^n}{ds^n}P(s)|_{s=0} = n!p_n, \quad n = 0, 1, 2, \ldots,$$

and we conclude that the following result must be true.

Proposition 1.3.1. *A generating function uniquely determines (or generates) its sequence.*

1.3.2. GENERATING FUNCTIONS AND MOMENTS.

As a lead in to the subject of finding moments of non-negative integer valued random variables, consider the following relation between the gf of $\{p_k\}$ and the gf of the tail probabilities $\{P[X > k]\}$.

Proposition 1.3.2. *Let X have mass function $\{p_k\} = \{P[X = k], k \geq 0\}$ satisfying $\sum_{k=0}^{\infty} p_k = 1$. Define*

$$P(s) = Es^X,$$
$$q_k = P[X > k], \quad k = 0, 1, \dots$$

and

$$Q(s) = \sum_{k=0}^{\infty} q_k s^k.$$

Then

(1.3.2.1) $$Q(s) = \frac{1 - P(s)}{1 - s}, \quad 0 \leq s < 1.$$

Proof. Follow your nose: Since $q_k = \sum_{i=k+1}^{\infty} p_i$ we have

$$Q(s) = \sum_{k=0}^{\infty} \left(\sum_{i=k+1}^{\infty} p_i \right) s^k = \sum_{i=1}^{\infty} \left(\sum_{k=0}^{i-1} s^k \right) p_i,$$

and summing the geometric terms shows that $Q(s)$ equals

$$\sum_{i=1}^{\infty} \left(\frac{1 - s^i}{1 - s} \right) p_i$$
$$= \sum_{i=0}^{\infty} \left(\frac{1 - s^i}{1 - s} \right) p_i$$
$$= (1 - s)^{-1}(1 - P(s)). \blacksquare$$

In (1.3.2.1) let $s \uparrow 1$. On the one hand, by monotone convergence, we get

$$\lim_{s\uparrow 1} Q(s) = \lim_{s\uparrow 1} \sum_{k=0}^{\infty} q_k s^k = \sum_{k=0}^{\infty} q_k$$
$$= \sum_{k=0}^{\infty} P[X > k] = E(X)$$

the last step following from Lemma 1.1.1. On the other hand

$$\lim_{s\uparrow 1} Q(s) = \lim_{s\uparrow 1}(1 - P(s))/(1 - s)$$
$$= \lim_{s\uparrow 1}(P(1) - P(s))/(1 - s) = P'(1).$$

(Strictly speaking, $P'(1)$ is the left derivative of P at 1.) We conclude that

(1.3.2.1) $E(X) = P'(1).$

Of course, this would also follow from (1.3.1.1).
    For example, if $X \sim g(k; p)$ then

$$P(s) = p/(1 - qs)$$

so

$$P'(s) = pq/(1 - qs)^2$$

and

$$E(X) = P'(1) = pq/(1 - q)^2 = pq/p^2 = q/p$$

in agreement with (1.1.1).
    Higher order derivatives of $P(s)$ may be used to calculate higher order moments. From (1.3.1.1) we have

$$\frac{d^n}{ds^n} P(s) = \sum_{k=n}^{\infty} (k(k-1)(k-2)\ldots(k-n+1))p_k s^{k-n}$$
$$= \sum_{k=0}^{\infty} (k(k-1)(k-2)\ldots(k-n+1))\, p_k s^{k-n}$$

so that letting $s \uparrow 1$, from monotone convergence, we get

$$\lim_{s\uparrow 1} \frac{d^n}{ds^n} P(s) =: P^{(n)}(1) = \sum_{k=0}^{\infty} k(k-1)(k-2)\ldots(k-n+1)p_k$$
$$= E\left(X(X-1)\ldots(X-n+1)\right).$$

In particular, when $n = 2$ we get

$$P''(1) = \sum_{k=0}^{\infty} k(k-1)p_k = \sum_{k=0}^{\infty} k^2 p_k - \sum_{k=0}^{\infty} kp_k$$

so that

$$P''(1) = E(X)^2 - EX$$

and therefore

(1.3.2.3)
$$\text{Var}(X) = P''(1) + P'(1) - (P'(1))^2.$$

For example, continuing with the case that $X \sim g(k;p)$, we found $P(s) = p/(1 - qs)$ and $P'(s) = qp/(1 - qs)^2$ so that

$$P''(s) = 2q^2 p(1 - qs)/(1 - qs)^4.$$

Thus

$$P'(1) = q/p$$
$$P''(1) = 2q^2/p^2$$

and from (1.3.2.3)

$$\begin{aligned}
\text{Var}(X) &= P''(1) + P'(1) - (P'(1))^2 \\
&= 2q^2/p^2 + q/p - q^2/p^2 \\
&= q^2/p^2 + q/p = \frac{q}{p}\left(\frac{q}{p} + 1\right) \\
&= q/p^2.
\end{aligned}$$

1.3.3. GENERATING FUNCTIONS AND CONVOLUTION.

Convolution is an awkward operation to perform on sequences. As shown in Proposition 1.3.3 below, convolution is converted into an ordinary product by the process of taking generating functions. The price paid for this gain in simplicity is the work that must be expended in recovering the convolution sequence from the product of gf's. For emphasis, the statement and proof contain redundancies.

Proposition 1.3.3. *The gf of a convolution is the product of the gf's.*

(1) *If $X_i, i = 1, 2$ are independent non-negative integer valued random variables with gf's $(i = 1, 2)$*

$$P_{X_i}(s) = Es^{X_i}, \quad 0 \le s \le 1,$$

then

$$P_{X_1+X_2}(s) = P_{X_1}(s)P_{X_2}(s).$$

(2) *If $\{a_j\}$ and $\{b_j\}$ are two sequences with gf's $A(s), B(s)$, then the gf of $\{a_n\} * \{b_n\}$ is $A(s)B(s)$.*

Note that if in (1), $X_1 \overset{d}{=} X_2$ in addition to X_1, X_2 being independent, we have

$$P_{X_1+X_2}(s) = (P_{X_1}(s))^2.$$

Proof. Although (2) implies (1), there is something to be learned by proving each separately.

(1) We have

$$
\begin{aligned}
P_{X_1+X_2}(s) &= Es^{X_1+X_2} = Es^{X_1}s^{X_2} \\
&= Es^{X_1}Es^{X_2} = P_{X_1}(s)P_{X_2}(s)
\end{aligned}
$$

since s^{X_1}, s^{X_2} are independent random variables.

(2) Suppose for concreteness that the radius of convergence of both $A(s)$ and $B(s)$ is s_0. The gf of the convolution is

$$\sum_{n=0}^{\infty} \left(\sum_{k=0}^{n} a_k b_{n-k} \right) s^n, \quad |s| < s_0$$

and, since Fubini's Theorem allows us to reverse the order of summation, we get

$$\sum_{k=0}^{\infty} \sum_{n=k}^{\infty} a_k b_{n-k} s^n = \sum_{k=0}^{\infty} a_k s^k \sum_{n=k}^{\infty} b_{n-k} s^{n-k} = A(s)B(s). \quad \blacksquare$$

We now consider some examples illustrating the techniques involved.

Example 1.3.4. If $X_1 \sim p(k; \lambda), X_2 \sim p(k; \mu)$ and X_1, X_2 are independent then

$$
\begin{aligned}
P_{X_1+X_2}(s) &= P_{X_1}(s)P_{X_2}(s) = e^{\lambda(s-1)}e^{\mu(s-1)} \\
&= e^{(\lambda+\mu)(s-1)},
\end{aligned}
$$

so we conclude

$$X_1 + X_2 \sim p(k; \lambda + \mu).$$

Example 1.3.5. If X_1 has range $\{0, 1\}$ with

$$P[X_1 = 1] = p = 1 - P[X = 0]$$

then

$$P_{X_1}(s) = Es^{X_1} = qs^0 + ps^1 = q + ps.$$

If $X \sim b(k; n, p)$, think of X as the number of successes in n Bernoulli trials and let X_i be 1 if the ith trial is a success and 0 otherwise. Thus

$$X = X_1 + \cdots + X_n$$

and

$$P_X(s) = \prod_{i=1}^{n} P_{X_i}(s) = (P_{X_1}(s))^n = (q + ps)^n$$

in agreement with Example 1.3.2.

Example 1.3.6. Let $X_1, \ldots, X_r$ be iid, $\sim g(k; p)$ and set

$$X = X_1 + \cdots + X_r,$$

so that X is the number of failures necessary to obtain r successes in repeated independent Bernoulli trials. The density of X is called the negative binomial distribution, and we now derive its form using gf's. Since we know

$$P_{X_1}(s) = p/(1 - qs)$$

we have

$$P_X(s) = \prod_{i=1}^{r} P_{X_i}(s) = (P_{X_1}(s))^r$$

$$= (p/(1 - qs))^r = \sum_{k=0}^{\infty} P[X = k] s^k.$$

The plan is to expand $(p/(1 - qs))^r$ as a power series in s, and then the coefficient of s^k is $P[X = k]$. We require the *binomial theorem*: For $a \in R$,

$$(1 + t)^a = \sum_{k=0}^{\infty} \binom{a}{k} t^k$$

for $|t| < 1$. Recall that the definition of $\binom{a}{k}$ is

$$\binom{a}{k} = (a)_k / k! = a(a - 1)...(a - k + 1)/k!.$$

Using this with $a = -r$, we get

$$(p/(1 - qs))^r = p^r (1 - qs)^{-r}$$

$$= p^r \sum_{k=0}^{\infty} \binom{-r}{k} (-1)^k q^k s^k$$

from which

$$P[X = k] = (-1)^k \binom{-r}{k} p^r q^k.$$

1.3.4. GENERATING FUNCTIONS, COMPOUNDING AND RANDOM SUMS.

There is a useful and simple generalization of Proposition 1.3.3 which will prove basic to our study of branching processes. Let $\{X_n, n \geq 1\}$ be independent, identically distributed (iid) non-negative integer valued random variables and suppose $X_1 \sim \{p_k\}$ and

$$Es^{X_1} = P_{X_1}(s), \quad 0 \leq s \leq 1.$$

Let N be independent of $\{X_n, n \geq 1\}$ and suppose N is non-negative integer valued with

$$P[N = j] = \alpha_j, j \geq 0; \quad Es^N = P_N(s), 0 \leq s \leq 1.$$

Define

$$S_0 = 0$$
$$S_n = X_1 + \cdots + X_n, \quad n \geq 1.$$

Then S_N is a random sum with a *compound distribution*: For $j \geq 0$

$$P[S_N = j] = \sum_{k=0}^{\infty} P[S_N = j, N = k]$$

$$= \sum_{k=0}^{\infty} P[S_k = j, N = k] = \sum_{k=0}^{\infty} P[S_k = j]P[N = k]$$

$$= \sum_{k=0}^{\infty} p_j^{k*} \alpha_k$$

where $P[S_k = j] = p_j^{k*}$ is the jth element of the kth convolution power of the sequence $\{p_n\}$. Thus the gf of S_N is

$$P_{S_N}(s) = \sum_{j=0}^{\infty} P[S_N = j]s^j$$

$$= \sum_{j=0}^{\infty} \left(\sum_{k=0}^{\infty} p_j^{k*} \alpha_k \right) s^j = \sum_{k=0}^{\infty} \alpha_k \left(\sum_{j=0}^{\infty} p_j^{k*} s^j \right)$$

$$= \sum_{k=0}^{\infty} \alpha_k \left(\sum_{j=0}^{\infty} P[S_k = j]s^j \right) = \sum_{k=0}^{\infty} \alpha_k \left(P_{X_1}(s) \right)^k$$

(from Proposition 1.3.3)

$$= P_N(P_{X_1}(s)),$$

and we conclude

(1.3.4.1) $$P_{S_N}(s) = P_N(P_{X_1}(s)).$$

Note the gf of the random index is on the outside of the functional composition.

In the special case where $N \sim p(k; \lambda)$ we get a compound-Poisson distribution with gf given by (1.3.4.1): Since $P_N(s) = \exp\{\lambda(s - 1)\}$

(1.3.4.2) $$P_{S_N}(s) = \exp\{\lambda(P_{X_1}(s) - 1)\}.$$

Example. Harry Delivers Pizzas. Harry's restaurant has a delivery service for pizzas. Friday night Harry goes on a drinking binge, which causes him to be muddled all day Saturday while answering the phone at the restaurant. A Poisson(λ) number of orders are phoned in on Saturday but there is only probability p that Harry notes the address of a caller correctly. What is the distribution of the number of pizzas successfully delivered?

Let

$$X_i = \begin{cases} 1, & \text{if the address of the } i\text{th caller is correctly noted;} \\ 0, & \text{otherwise.} \end{cases}$$

Let $N \sim p(k, \lambda)$. The number of successful deliveries is S_N with gf

$$P_{S_N}(s) = P_N(P_{X_1}(s)) = P_N(q + ps)$$

and from (1.3.4.2) this is

$$= \exp\{\lambda(q + ps - 1)\}$$
$$= \exp\{\lambda(ps - p)\} = \exp\{\lambda p(s - 1)\}$$

so we recognize that

$$S_N \sim p(k; \lambda p). \quad \blacksquare$$

The effect of compounding has been to reduce the parameter from λ to λp, a phenomenon which is called *thinning* in Poisson process theory. The Poisson compounding of Bernoulli variables occurs often, and the previous simple example serves as a paradigm. Other examples: Imagine a Poisson number of customers have arrived at a service facility to be served in turn. Each customer has probability p of being satisfied with the service received so that S_N is the number of satisfied customers. Alternatively, imagine telephone traffic arriving at a gateway are of two types, I and II. Type I traffic gets routed through trunk line I and type II traffic gets routed through trunk line II. If $100p\%$ of the calls are of type II and if the net number of calls to arrive in an hour is Poisson distributed with parameter λ then the number of calls routed to trunk line I is Poisson with parameter $p\lambda$. This thinning procedure is further discussed in Chapter 4.

Example. Harry Drives Cross-Country. For a vacation Harry drives cross-crountry. Because of his lead foot he encounters a seemingly infinite sequence of patrolling police cars which stop him. Half of the time he is stopped, he has to pay a fine of $50 and the other half he pays a fine of $100. However when it comes to dealing with the police, Harry is a smart-aleck with an uncontrollable mouth. So whenever he is stopped, there is probability p that not only will he have to pay a fine but he will foolishly make some sarcastic and rude comment which will result in his license being taken away. What is the distribution of the total fines assessed until his license is revoked?

Let $\{X_n, n \geq 1\}$ be iid with

$$P[X_1 = 50] = 1/2,$$
$$P[X_1 = 100] = 1/2$$

and let N be independent of $\{X_n, n \geq 1\}$ with

$$P[N = k] = q^{k-1}p, \quad k \geq 1.$$

(Note this N has range $\{1, 2, \ldots\}$. Earlier we considered the geometric distribution which concentrates on $\{0, 1, 2, \ldots\}$.) The total fines paid is S_N and since

$$P_N(s) = \sum_{k=1}^{\infty} q^{k-1}ps^k = ps \sum_{k=1}^{\infty} (qs)^{k-1}$$
$$= ps/(1 - qs)$$

we get

$$P_{S_N}(s) = P_N(P_{X_1}(s)) = P_N(\frac{1}{2}s^{50} + \frac{1}{2}s^{100}). \quad \blacksquare$$

Note from (1.3.4.1)

$$ES_N = P'_{S_N}(s)|_{s=1} = P'_N(P_{X_1}(s))P'_{X_1}(s)|_{s=1}$$

(remember how to differentiate the composition of two functions?) and assuming $P_{X_1}(1) = 1$ this gives

$$= P'_N(P_{X_1}(1))P'_{X_1}(1) = P'_N(1)P'_{X_1}(1)$$

so

(1.3.4.3) $$ES_N = E(N)E(X_1).$$

The pattern of this expectation will repeat itself when we discuss Wald's Identity in Section 1.8.1.

In the second example, $EX_1 = \frac{1}{2}(50 + 100) = 75$ and

$$P'_N(s)|_{s=1} = \frac{(1-qs)p + pqs}{(1-qs)^2}|_{s=1} = \frac{(1-q)p + pq}{(1-q)^2}$$
$$= (p - pq + pq)/p^2 = p/p^2 = 1/p = E(N)$$

therefore

$$E(S_N) = 75/p.$$

1.4. THE SIMPLE BRANCHING PROCESS.

We now discuss a significant application of generating functions. The simple branching process (sometimes called the Galton-Watson-Bienymé Process) uses generating functions in an essential manner.

Informally the process is described as follows: The basic ingredient is a density $\{p_k\}$ on the non-negative integers. A population starts with a progenitor who forms generation number 0. This initial progenitor splits into k offspring with probability p_k. These offspring constitute the first generation. Each of the first generation offspring independently split into a random number of offspring; the number for each is determined by the density $\{p_k\}$. This process continues until extinction, which occurs whenever all the members of a generation fail to produce offspring.

Such an idealized model can be thought of as the model for population growth in the absence of environmental pressures. In nuclear fission experiments, it was an early model for the cascading of neutrons. Historically, it first arose from a study of the likelihood of survival of family names:

How fertile must a family be to insure that in no future generation will the family name die out? The branching process formalism has also been used in the study of queues; the offspring of a customer in the system are those who arrive while the customer is in service.

Here is a formal definition of the model: Let $\{Z_{n,j}, n \geq 1, j \geq 1\}$ be iid non-negative integer valued random variables, with each variable having the common distribution $\{p_k\}$. For what follows, interpret a random sum as 0 when the number of summands is 0. Define the branching process $\{Z_n, n \geq 0\}$ by

$$Z_0 = 1$$
$$Z_1 = Z_{1,1}$$
$$Z_2 = Z_{2,1} + \cdots + Z_{2,Z_1}$$

$$\vdots \qquad \vdots$$

$$Z_n = Z_{n,1} + \cdots + Z_{n,Z_{n-1}},$$

so that $Z_{n,j}$ can be thought of as the number of members of the nth generation which are offspring of the jth member of the $(n-1)$st generation.

Note that $Z_n = 0$ implies $Z_{n+1} = 0$ so that once the process hits 0 it stays at 0. Also observe that Z_{n-1} is independent of $\{Z_{n,j}, j \geq 1\}$ which is crucial to what follows, since we will need this independence to apply (1.3.4.1).

For $n \geq 0$ define $P_n(s) = Es^{Z_n}$ and set

$$P(s) = Es^{Z_1} = \sum_{k=0}^{\infty} p_k s^k, \qquad 0 \leq s \leq 1.$$

Thus $P_0(s) = s, P_1(s) = P(s)$, and from (1.3.4.1) we get

$$P_n(s) = P_{n-1}(P(s)).$$

And therefore

$$P_2(s) = P(P(s))$$
$$P_3(s) = P_2(P(s)) = P(P(P(s))) = P(P_2(s))$$

$$\vdots \qquad \vdots$$

$$P_n(s) = P_{n-1}(P(s)) = P(P_{n-1}(s)).$$

Thus, the analytic equivalent of the branching effect is functional composition. In general, explicit calculations are hard, but in principle this determines the distribution of Z_n for any $n \geq 0$. One case where some explicit calculations are possible is the case of binomial replacement.

Example. Binomial replacement: If $P(s) = q + ps$ then

$$P_2(s) = q + p(q + ps) = q + pq + p^2 s$$
$$P_3(s) = q + pq + p^2(q + ps) = q + pq + p^2 q + p^3 s$$
$$\vdots \qquad \vdots$$
$$P_{n+1}(s) = q + pq + p^2 q + \cdots + p^n q + p^{n+1} s.$$

For later purposes note that for $0 \le s \le 1$

$$\lim_{n \to \infty} P_{n+1}(s) = q \sum_{j=0}^{\infty} p^j = q/(1 - p) = 1.$$

Moments: Although the distribution of Z_n is determined by the awkward process of functional iteration of generating functions, certain explicit expressions are possible for moments. These are generally obtained by solving difference equations. Set

$$m = E(Z_1) = \sum_{k=0}^{\infty} k p_k$$
$$\sigma^2 = \mathrm{Var}(Z_1)$$

and assume $m < \infty$ and $\sigma^2 < \infty$. To compute $E(Z_n) =: m_n$ we note $m_n = P_n'(1)$, and since $P_n(s) = P_{n-1}(P(s))$ we get

$$P_n'(s) = P_{n-1}'(P(s)) P'(s).$$

Letting $s \uparrow 1$ yields

$$m_n = m_{n-1} m$$

and iterating back gives

$$m_n = m_{n-2} m^2 = m_{n-3} m^3 = \cdots = m_1 m^{n-1} = m^n$$

since $m_1 = m$.

In the special case of binomial replacement where $P(s) = q + ps$, we have $m = p$ and $E(Z_n) = p^n$.

For the general case, we may calculate $\mathrm{Var}(Z_n)$ by a similar but more tedious procedure starting with the relation

$$P_n(s) = P(P_{n-1}(s)).$$

The procedure yields (if you do not make an algebraic error)

$$\mathrm{Var}(Z_{n+1}) = \begin{cases} \sigma^2 m^n \left(\frac{1-m^{n+1}}{1-m} \right), & \text{if } m \neq 1 \\ \sigma^2(n+1), & \text{if } m = 1. \end{cases}$$

Extinction Probability: Define the event

$$[\text{ extinction }] = \bigcup_{n=1}^{\infty} [Z_n = 0].$$

We seek to compute

$$\pi = P[\text{ extinction }].$$

Since

$$[Z_n = 0] \subset [Z_{n+1} = 0]$$

we have

$$\pi = P\{\bigcup_{k=1}^{\infty} [Z_k = 0]\} = \lim_{n \to \infty} P\{\bigcup_{k=1}^{n} [Z_k = 0]\}$$
$$= \lim_{n \to \infty} P[Z_n = 0] = \lim_{n \to \infty} P_n(0)$$
$$=: \lim_{n \to \infty} \pi_n$$
$$= \lim_{n \to \infty} P[\text{ extinction on or before generation } n].$$

This yields π in principle, but the goal is to be able to calculate it without having to compute all the functional iterates $P_n(s)$. The method to do this is given in the next result. To prevent degeneracies, assume that $0 < p_0 < 1$. (If $p_0 = 0, \pi = 0$; if $p_0 = 1, \pi = 1$.) We will need to use the fact that $P(s)$ is convex on $[0,1]$.

Theorem 1.4.1. *If $m = E(Z_1) \leq 1$ then $\pi = 1$. If $m > 1$, then $\pi < 1$ and is the unique non-negative solution to the equation*

$$s = P(s)$$

which is less than 1.

Proof. STEP 1: We first show π is a solution of the equation $s = P(s)$. Since the events $\{[Z_n = 0]\}$ are non-decreasing,

$$[Z_n = 0] \subset [Z_{n+1} = 0];$$

we have

$$\pi_n := P[Z_n = 0]$$

is a non-decreasing sequence converging to π. Since

$$P_{n+1}(s) = P(P_n(s)),$$

we get, by setting $s = 0$, that

$$\pi_{n+1} = P(\pi_n).$$

Letting $n \to \infty$ and using the continuity of $P(s)$ yields

$$\pi = P(\pi).$$

STEP 2: We show π is the smallest solution of $s = P(s)$ in $[0,1]$. Suppose q is some solution of the equation $s = P(s)$ satisfying $0 \le q \le 1$. Then, since $0 \le q$,

$$\pi_1 = P(0) \le P(q) = q$$

and therefore

$$\pi_2 = P_2(0) = P(\pi_1) \le P(q) = q$$

and, continuing in this manner one more step,

$$\pi_3 = P_3(0) = P(\pi_2) \le P(q) = q.$$

In general we obtain

$$\pi_n \le q.$$

Letting $n \to \infty$ yields $\pi \le q$ showing π is minimal among solutions in $[0,1]$.

STEP 3: Note $P(s)$ is convex since

$$P''(s) = \sum_{k=2}^{\infty} k(k-1)p_k s^{k-2} \ge 0.$$

Because of convexity and the fact that $P(0) = p_0 > 0$, the graphs

$$y = P(s), \qquad y = s$$

for $0 \le s \le 1$ have at most two points in common. One of these is $s = 1$. If $P'(1) = m \le 1$, then in a left neighborhood of 1 the graph of $y = P(s)$ cannot be below that of $y = s$ and hence by convexity of $P(s)$ the only intersection is $s = 1$. In the contrary case, if $P'(1) = m > 1$ then in a left neighborhood of 1 the graph of $y = P(s)$ is below the diagonal and there must be an additional intersection to the left of 1 of the two graphs. See Figure 1.4.1. ∎

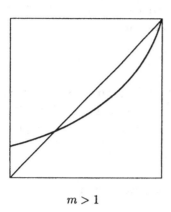
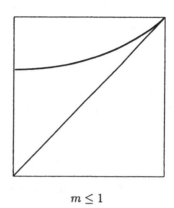

$$m > 1 \qquad\qquad\qquad\qquad m \leq 1$$

Figure 1.4.1

Note in the binomial replacement example that $s = P(s)$ yields the equation

$$s = q + ps$$

whose only solution is $s = 1$ which agrees with the fact that $m = p \leq 1$.

We now give the following complement, which ties in with the continuity theorem of the next section.

Complement. *For $0 \leq s < 1$*

(1.4.1) $$\lim_{n \to \infty} P_n(s) = \pi.$$

Verification. First suppose that $s \leq \pi$. Then $P(s) \leq P(\pi) = \pi$ and

$$P(s) \leq \pi.$$

Therefore $P_2(s) = P(P(s)) \leq P(\pi) = \pi$; thus

$$P_2(s) \leq \pi.$$

In general, by continuing this procedure we get

$$P_n(s) \leq \pi$$

from which, for $0 \leq s \leq \pi$,

$$\pi_n = P_n(0) \leq P_n(s) \leq \pi.$$

And, since $\pi_n \to \pi$, by letting $n \to \infty$ we get that $P_n(s) \to \pi$.

To finish the complement we must suppose $\pi < 1$ and investigate what happens on the domain $\pi \leq s < 1$. We have for this range

$$\pi = P(\pi) \leq P(s) \leq s,$$

the last inequality following from the fact that the graph of $y = P(s)$ must be under the diagonal on the set $\{s : \pi \leq s \leq 1\}$ (See Figure 1.4.1 for the case $m > 1$.) Since $P(s)$ is non-decreasing, from the previous inequalities we get

$$P(\pi) = \pi \leq P_2(s) \leq P(s) \leq s.$$

Continuing in this manner, for the general case we obtain

$$\pi \leq P_n(s) \leq P_{n-1}(s) \leq \cdots \leq P(s) \leq s.$$

Thus we conclude that $\{P_n(s)\}$ is non-increasing for $\pi \leq s \leq 1$. Let $P_\infty(s) = \lim_{n \to \infty} P_n(s)$. Suppose for some $s_0 \in (\pi, 1)$ we have $P_\infty(s_0) =: \alpha > \pi$. Then

$$P(\alpha) = \lim_{n \to \infty} P(P_n(s_0)) = \lim_{n \to \infty} P_{n+1}(s_0) = \alpha$$

and on the domain $(\pi, 1)$ we have $P(s) \leq s$. If $P(s)$ is linear then $P(s) = s$ on $(\pi, 1)$. But then we would have $p_k = 0$ for $k \geq 2$ and $m \leq 1$ so $\pi = 1$ against our supposition that $\pi < 1$. If $P(s)$ is not linear, then by convexity, $P(s) < s$ for $s \in (\pi, 1)$. Thus we have

$$\pi < \alpha = P_\infty(s_0) \leq P(s_0) < s_0 < 1$$

but also

$$\alpha = P(\alpha),$$

which contradicts the fact that there are no solutions of $s = P(s)$ in $(\pi, 1)$. The contradiction arose because of the supposition $\alpha > \pi$. ∎

What $P_n(s) \to \pi$ says is that

$$\sum_{k=0}^\infty P[Z_n = k] s^k \to \pi = \pi s^0 + \sum_{k=1}^\infty 0 s^k.$$

Anticipating the *Continuity Theorem* for generating functions presented in the next section, this implies that

$$P[Z_n = 0] \to \pi,$$
$$P[Z_n = k] \to 0, \quad \text{for } k \geq 1.$$

In fact, using Markov Chain or martingale methods, we can get a stronger result, namely that

$$P[Z_n \to 0 \quad \text{or} \quad Z_n \to \infty] = 1$$

and

$$\pi = P[Z_n \to 0] = 1 - P[Z_n \to \infty].$$

The simple branching process exhibits an instability: Either extinction occurs or the process explodes.

Example. Harry Yearns for a Coffee Break. In order to help some friends, Harry becomes the east coast sales representative of B & D Software. The software has been favorably reviewed and demand is heavy. Harry sets up a sales booth at the local computer show and takes orders. Each order takes three minutes to fill. While each order is being filled there is probability p_j that j more customers will arrive and join the line. Assume $p_0 = .2, p_1 = .2$ and $p_2 = .6$. Harry cannot take a coffee break until a service is completed and no one is waiting in line to order the software. If present conditions persist, what is the probability that Harry will ever take a coffee break?

Consider a branching process with offspring distribution given by (p_0, p_1, p_2). Harry can take a coffee break if and only if extinction occurs in the branching process. We have

$$P(s) = .2 + .2s + .6s^2$$

and

$$m = (.2)1 + (.6)(2) = 1.4 > 1,$$

so $s = P(s)$ yields the equation

$$s = .2 + .2s + .6s^2.$$

Therefore we must solve the quadratic equation

$$.6s^2 - .8s + .2 = 0,$$

and the two roots

$$\frac{.8 + \sqrt{(.8)^2 - 4(.6)(.2)}}{2(.6)}, \frac{.8 - \sqrt{(.8)^2 - 4(.6)(.2)}}{2(.6)}$$

yield the numbers $1, \frac{1}{3}$, and thus $\pi = \frac{1}{3}$. Thus the probability that Harry can ever take a coffee break if present conditions persist is $1/3$. ∎

When $P(s)$ is of degree higher than two, solution by hand of the equation $s = P(s)$ may be difficult, while a numerical solution is easy. The procedure is first to compute $m = \sum_{k=0}^{\infty} kp_k$. If $m \le 1$, then $\pi = 1$ and we are done. Otherwise we must solve numerically. Root finding packages are common. A program such as *Mathematica* makes short work of finding the solution. Typing

$$\text{Solve}[P(s) - s = 0, s]$$

will immediately yield all the roots of the equation and the smallest root in the unit interval can easily be identified. Alternately, π can be found by computing $\pi = \lim_{n \to \infty} P_n(0)$. The recursion

$$\pi_0 = P(0), \qquad \pi_{n+1} = P(\pi_n)$$

can be easily programmed on a computer and the solution will converge quickly. In fact the convergence will be geometrically fast since for some $\rho \in (0, 1)$

$$0 \le \pi - \pi_n \le \rho^n \to 0.$$

The reason for this inequality is that by the mean value theorem and the monotonicity of P'

$$\pi - \pi_n = P_n(\pi) - P_n(0) \le P'_n(\pi).$$

We need to check that

$$P'_n(\pi) = (P'(\pi))^n \text{ and } P'(\pi) < 1.$$

Since

$$P'_{n+1}(s) = P'_n(P(s))P'(s)$$

we get

$$P'_{n+1}(\pi) = P'_n(\pi)P'(\pi).$$

The difference equation when iterated shows the desired power solution. It remains to check that $P'(\pi) < 1$. If this were false and $P'(\pi) \ge 1$, then for $s \ge \pi$, by monotonicity and the mean value theorem, we get

$$P(s) - P(\pi) \ge P'(\pi)(s - \pi) \ge (s - \pi),$$

i.e.,

$$P(s) \ge \pi + s - \pi = s$$

which for $s \ge \pi$ is a contradiction since on $(\pi, 1)$ we have $P(s) < s$ (assuming $P(s)$ is not linear).

1.5. LIMIT DISTRIBUTIONS AND THE CONTINUITY THEOREM.

Let $\{X_n, n \geq 0\}$ be non-negative, integer valued random variables with $(n \geq 0, k \geq 0)$

$$(1.5.1) \qquad P[X_n = k] = p_k^{(n)}, \quad P_n(s) = Es^{X_n}.$$

Then X_n converges in distribution to X_0, written $X_n \Rightarrow X_0$, if

$$(1.5.2) \qquad \lim_{n \to \infty} p_k^{(n)} = p_k^{(0)}$$

for $k = 0, 1, 2, \ldots$. As the next result shows, this is equivalent to

$$(1.5.3) \qquad P_n(s) \to P_0(s)$$

for $0 \leq s \leq 1$ as $n \to \infty$.

Theorem 1.5.1. The Continuity Theorem. *Suppose for each* $n = 1, 2, \ldots$ *that* $\{p_k^{(n)}, k \geq 0\}$ *is a probability mass function on* $\{0, 1, 2, \ldots\}$ *so that*

$$p_k^{(n)} \geq 0, \quad \sum_{k=0}^{\infty} p_k^{(n)} = 1.$$

Then there exists a sequence $\{p_k^{(0)}, k \geq 0\}$ *such that*

$$(1.5.4) \qquad \lim_{n \to \infty} p_k^{(n)} = p_k^{(0)}, \quad k \geq 0,$$

iff there exists a function $P_0(s), 0 < s < 1$ *such that*

$$(1.5.5) \qquad \lim_{n \to \infty} P_n(s) = \lim_{n \to \infty} \sum_{k=0}^{\infty} p_k^{(n)} s^k = P_0(s).$$

for $0 < s < 1$. *In this case* $P_0(s) = \sum_{k=0}^{\infty} p_k^{(0)} s^k$ *and* $\sum_{k=0}^{\infty} p_k^{(0)} = 1$ *iff* $\lim_{s \uparrow 1} P_0(s) =: P_0(1) = 1$.

Remarks. As we will see, this provides an alternative to brute force when proving convergence in distribution. Frequently it is easier to prove that the generating functions converge rather than trying to show the convergence of a sequence of mass functions.

From (1.5.4) we have

$$(1.5.6) \qquad 0 \leq p_k^{(0)} \leq 1$$

since the same is true for $p_k^{(n)}$ and $\lim_{n \to \infty} p_k^{(n)} = p_k^{(0)}$. But it does not follow that $\sum_{k=0}^{\infty} p_k^{(0)} = 1$ since mass can escape to infinity. As a graphic example suppose

$$p_k^{(n)} = \delta_{k,n} = \begin{cases} 1, & \text{if } k = n \\ 0, & \text{if } k \neq n. \end{cases}$$

For any fixed k,

(1.5.7) $$\lim_{n \to \infty} p_k^{(n)} = 0,$$

from which

$$(p_0^{(0)}, p_1^{(0)}, ...) = (0, 0, ...).$$

This phenomenon arises because we consider the state space $\{0, 1, 2, \dots\}$. If we enlarge the state space to $\{0, 1, 2, \dots, \infty\}$ then $X_n \Rightarrow \infty$ and the limit distribution concentrates all mass at ∞.

Proof. Suppose (1.5.2). Fix $s \in (0, 1)$ and for any $\epsilon > 0$ we may pick m so large that

$$\sum_{i=m+1}^{\infty} s^i < \epsilon.$$

We have

$$|P_n(s) - P_0(s)| \leq \sum_{k=1}^{\infty} |p_k^{(n)} - p_k^{(0)}| s^k$$

$$\leq \sum_{k=1}^{m} |p_k^{(n)} - p_k^{(0)}| + \sum_{k=m+1}^{\infty} s^k$$

$$\leq \sum_{k=1}^{m} |p_k^{(n)} - p_k^{(0)}| + \epsilon.$$

Letting $n \to \infty$, we get

$$\limsup_{n \to \infty} |P_n(s) - P_0(s)| \leq \epsilon$$

and because ϵ is arbitrary we obtain (1.5.5).

The proof of the converse is somewhat more involved and is deferred to the appendix at the end of this section, which can be read by the interested student or skipped by a beginner.

Example. Harry and the Mushroom Staples.* Each morning, Harry buys enough salad ingredients to prepare 200 salads for the lunch crowd at his restaurant. Included in the salad are mushrooms which come in small boxes held shut by relatively large staples. For each salad, there is probability .005 that the person preparing the salad will sloppily drop a staple into it. During a three week period, Harry's precocious twelfth grade niece, who has just completed a statistics unit in high school, keeps track of the number of staples dropped in salads. (Harry's customers are not reticent about complaining about such things so detection of the sin and collection of the data pose no problem.) After drawing a histogram, the niece decides that the number of salads per day containing a staple is Poisson distributed with parameter $(200)(.005) = 1$. ∎

Harry's niece has empirically rediscovered the Poisson approximation to the binomial distribution: If $X_n \sim b(k; n, p(n))$ and

$$(1.5.8) \qquad \lim_{n\to\infty} np(n) = \lim_{n\to\infty} EX_n = \lambda \in (0, \infty),$$

then

$$X_n \Rightarrow X_0$$

as $n \to \infty$ where $X_0 \sim p(k; \lambda)$.

The verification is easy using generating functions. We have

$$\lim_{n\to\infty} P_n(s) = \lim_{n\to\infty} Es^{X_n} = \lim_{n\to\infty} (1 - p(n) + p(n)s)^n$$

$$= \lim_{n\to\infty} \left(1 + \frac{(s-1)np(n)}{n}\right)^n = e^{\lambda(s-1)}$$

using (1.5.8).

Appendix: Continuation of the Proof of Theorem 1.5.1.

We now return to the proof of Theorem 1.5.1 and show why convergence of the generating functions implies convergence of the sequences.

Assume we know the following fact: Any sequence of mass functions $\{\{f_j^{(n)}, j \geq 0\}, n \geq 1\}$ has a convergent subsequence $\{\{f_j^{(n')}, j \geq 0\}\}$ meaning that for all j

$$\lim_{n'\to\infty} f_j^{(n')}$$

exists. If $\{p_k^{(n)}\}$ has two different subsequential limits along $\{n'\}$ and $\{n''\}$, by the first half of Theorem 1.5.1 and hypothesis (1.5.3), we would have

$$\lim_{n'\to\infty} \sum_{k=0}^{\infty} p_k^{(n')} s^k = \lim_{n'\to\infty} P_{n'}(s) = P_0(s)$$

---

*A semi-true story.

and also

$$\lim_{n''\to\infty}\sum_{k=0}^{\infty}p_k^{(n'')}s^k = \lim_{n''\to\infty}P_{n''}(s) = P_0(s).$$

Thus any two subsequential limits of $\{p_k^{(n)}\}$ have the same generating function. Since generating functions uniquely determine the sequence, all subsequential limits are equal and thus $\lim_{n\to\infty}p_k^{(n)}$ exists for all k. The limit has a generating function $P_0(s)$.

It remains to verify the claim that a sequence of mass functions $\{\{f_j^{(n)}, j \geq 0\}, n \geq 1\}$ has a subsequential limit. Since for each n we have

$$\{f_j^{(n)}, j \geq 0\} \subset [0,1]^{\infty},$$

and $[0,1]^{\infty}$ is a compact set (being a product of the compact sets $[0,1]$), we have an infinite sequence of elements in a compact set. Hence a subsequential limit must exist.

If the compactness argument is not satisfying, a subsequential limit can be manufactured by a diagonalization procedure. (See Billingsley, 1986, page 566.)

1.5.1. THE LAW OF RARE EVENTS.

A more sophisticated version of the Poisson approximation, sometimes called the Law of Rare Events, is discussed next.

Proposition 1.5.2. *Suppose we have a doubly indexed array of random variables such that for each $n = 1, 2, \ldots$, $\{X_{n,k}, k \geq 1\}$, is a sequence of independent (but not necessarily identically distributed) Bernoulli random variables satisfying*

$$(1.5.1.1) \qquad P[X_{n,k} = 1] = p_k(n) = 1 - P[X_{n,k} = 0],$$

$$(1.5.1.2) \qquad \bigvee_{1\leq k\leq n} p_k(n) =: \delta(n) \to 0, \quad n \to \infty,$$

$$(1.5.1.3) \qquad \sum_{k=1}^{n} p_k(n) = E\sum_{k=1}^{n} X_{n,k} \to \lambda \in (0,\infty), \quad n \to \infty.$$

If $PO(\lambda)$ is a Poisson distributed random variable with mean λ then

$$\sum_{k=1}^{n} X_{n,k} \Rightarrow PO(\lambda).$$

Remarks. The effect of (1.5.1.2) is that each $X_{n,k}, k = 1, \ldots, n$ has a uniformly small probability of being 1. Think of $X_{n,k}$ as being the indicator of the event $A_{n,k}$, viz $X_{n,k} = 1_{A_{n,k}}$ in which case

$$\sum_{k=1}^{n} X_{n,k} = \sum_{k=1}^{n} 1_{A_{n,k}} = \text{the number of } A_{n,k}, 1 \leq k \leq n \text{ which occur.}$$

So when each of a large number of independent events has a small probability of occurring (i.e. events are rare), the number of events which occur is approximately Poisson distributed.

A stochastic process version of this result is frequently used as one of the justifications for assuming a queueing model has input determined by a Poisson process: Imagine several different streams of customers from independent sources converging on a service facility. Each stream is fairly sparse. If the different inputs are combined, then the arrival pattern observed by the service facility is approximately that of a Poisson process. The Poisson process will be defined in Chapter 3 and studied carefully in Chapter 4, but for now you should understand by this that the total number of arrivals in time $[0, t]$ to the service facility from all the different sources is approximately Poisson distributed.

Example. Traffic to Harry's Restaurant. Harry's restaurant is surrounded by 25 small Mom and Pop shops (i.e., small family owned and operated businesses). Between noon and 1 p.m., each Pop has his lunch break, but for the ith Pop there is only the small probability p_i that he will eat at Happy Harry's. (Alternatives include the sprouts bar across the street or a low cholesterol lunch of jogging.) The number of Pops to come to Happy Harry's between noon and 1 PM is approximately $p(k; \lambda) = \sum_{i=1}^{25} p_i$. ∎

Proof. The gf of $\sum_{k=1}^{n} X_{n,k}$ is

$$\prod_{k=1}^{n} P_{X_{n,k}}(s) = \prod_{k=1}^{n} (1 - p_k(n) + p_k(n)s).$$

So it suffices to show (take logarithms)

(1.5.1.4) $$\lim_{n \to \infty} \sum_{k=1}^{n} \log(1 - p_k(n)(1 - s)) = \lambda(s - 1)$$

for $0 < s < 1$ since this would imply after exponentiating that

$$P_{\sum_{k=1}^{n} X_{n,k}}(s) \to e^{\lambda(s-1)}.$$

To prove this we need the following estimate. For $0 < x < 1$

(1.5.1.5) $-\log(1-x) = x + R(x),$

where $R(x)$ is non-decreasing and

(1.5.1.6) $R(x) \le 2x^2, \quad 0 < x \le 1/2.$

To check these inequalities, use the infinite series representation: For $0 < x < 1$

$$-\log(1-x) = \sum_{n=1}^{\infty} \frac{x^n}{n} = x + \sum_{n=2}^{\infty} \frac{x^n}{n}$$

and set $R(x) = \sum_{n=2}^{\infty} x^n/n$. Since $x^n/n \le x^n$ we have

$$R(x) \le \sum_{n=2}^{\infty} x^n = x^2/(1-x).$$

Next, if $0 < x \le 1/2$, $(1-x)^{-1} \le 2$, giving (1.5.1.6).

So multiplying the left side of (1.5.1.4) by -1, and using (1.5.1.6) yields

$$\lim_{n\to\infty} \sum_{k=1}^{\infty} -\log(1 - p_k(n)(1-s)) = \sum_{k=1}^{n} p_k(n)(1-s) + \sum_{k=1}^{n} R(p_k(n)(1-s)).$$

Since

$$\lim_{n\to\infty} \sum_{k=1}^{n} p_k(n)(1-s) = \lambda(1-s),$$

the desired result will be proved if we show

$$\lim_{n\to\infty} \sum_{k=1}^{n} R(p_k(n)(1-s)) = 0.$$

For $0 < s < 1$ and n so large that

$$\bigvee_{1\le k\le n} p_k(n) \le \delta(n) \le 1/2,$$

we get from the monotonicity of R and (1.5.1.6)

$$\sum_{k=1}^{n} R(p_k(n)(1-s)) \le \sum_{k=1}^{n} R(p_k(n)) \le 2 \sum_{k=1}^{n} (p_k(n))^2$$

$$\le 2 \bigvee_{k=1}^{n} p_k(n) \sum_{k=1}^{n} p_k(n)$$

$$\le 2\delta(n) \sum_{k=1}^{n} p_k(n) \to 2 \cdot 0 \cdot \lambda = 0.$$

Thus (1.5.1.4) is true. ∎

1.6. THE SIMPLE RANDOM WALK.

Let $\{X_n, n \geq 1\}$ be independent, identically distributed random variables with only two possible values $\{-1, 1\}$ and

$$P[X_1 = 1] = p = 1 - P[X_1 = -1] = 1 - q,$$

$(0 \leq p, q \leq 1, p + q = 1)$ and define the simple random walk process $\{S_n, n \geq 0\}$ by

$$S_0 = 0, S_n = X_1 + \cdots + X_n, n \geq 1.$$

Such a process is important didactically and as a source of examples. It is often used in the following primitive gambling model: Toss a coin. If the outcome is heads, you win one dollar; if the outcome is tails you lose one dollar. S_n is the fortune after n plays. Many of the problems encountered in the study of $\{S_n\}$ are typical of those with more sophisticated models.

There are powerful martingale and Markov chain methods for the study of $\{S_n\}$. Here we emphasize generating function methods.

Define

$$N = \inf\{n \geq 1 : S_n = 1\}$$

to be the first time the random walk hits 1, i.e., the first time the gambler is ahead. We wish to derive the distribution of N. The plan is to write a difference equation and solve it using generating functions. Set

$$\phi_n = P[N = n], n \geq 0$$

so that $\phi_0 = 0, \phi_1 = p$. What equations do we expect for $\{\phi_n\}$? If $n \geq 2$, then in order for the random walk to go from 0 to 1 in n steps the first step must be to -1 (which has probability q). From -1 the walk must make its way back up to 0. Say this takes j steps. Then it seems reasonable that the probability of the walk going from -1 to 0 in j steps is ϕ_j. From 0 the random walk still must get up to 1. Say this takes k steps. Then this probability should be ϕ_k and the constraint on j and k is that $1 + j + k = n$ where the 1 is used for the initial step to -1. Thus the equation should be

$$\phi_n = \sum_{j=1}^{n-2} q\phi_j\phi_{n-j-1}, \quad n \geq 2.$$

The argument just given seems plausible and we now make it precise. (Those who found the argument convincing can skip to (1.6.2).)

For $n \geq 2$ we have

(1.6.1) $$[N = n] = \bigcup_{j=1}^{n-2} [X_1 = -1] \cap A_j \cap B_{n-j-1}$$

where

$$A_j = [\text{the random walk makes a first return from } -1 \text{ to } 0 \text{ in } j \text{ steps}]$$

$$= [\inf\{n : \sum_{i=1}^{n} X_{i+1} = 1\} = j]$$

and

$$B_{n-j-1} = [\text{the random walk makes a first passage from } 0 \text{ to } 1 \text{ in}$$
$$n - j - 1 \text{ steps}]$$

$$= [\inf\{n : \sum_{i=1}^{n} X_{j+1+i} = 1\} = n - j - 1].$$

When $n = 2$, interpret the right side of (1.6.1) as $\emptyset$, the empty set. Note A_j is determined by $X_2, \ldots, X_{j+1}$, and similarly B_{n-j-1} is determined by $X_{j+2}, \ldots, X_n$ Thus, the three events

$$[X_1 = -1], \quad A_j, \quad B_{n-j-1}$$

are independent because they depend on disjoint blocks of the X's. Since the union in (1.6.1) is a union of disjoint events we have

$$P[N = n] = \phi_n = \sum_{j=1}^{n-2} qP(A_j)P(B_{n-j-1}).$$

Now

$$\{X_1, X_2, \ldots\} \stackrel{d}{=} \{X_2, X_3, X_4, \ldots\}$$

meaning the finite dimensional distributions of both sequences are identical; i.e., for any m and sequence $k_1, \ldots, k_m$ of elements chosen from $\{-1, 1\}$ we have

$$P[X_1 = k_1, \ldots, X_m = k_m] = P[X_2 = k_1, \ldots, X_{m+1} = k_m]$$

since both sequences are just independent, identically distributed. Therefore,

$$P(A_j) = P[\inf\{n : \sum_{i=1}^{n} X_i = 1\} = j] = \phi_j$$

and similarly

$$PB_{n-j-1} = \phi_{n-j-1},$$

from which we get the recursion

$$\phi_0 = 0, \phi_1 = p$$

(1.6.2)
$$\phi_n = \sum_{j=1}^{n-2} q\phi_j\phi_{n-j-1}, \quad n \geq 2.$$

This difference equation summarizes the probability structure.

To solve, multiply (1.6.2) by s^n and sum over n. Set $\Phi(s) = \sum_{n=0}^{\infty} \phi_n s^n$. We have

(1.6.3)
$$\sum_{n=2}^{\infty} \phi_n s^n = \sum_{n=2}^{\infty} \left(\sum_{j=1}^{n-2} q\phi_j\phi_{n-j-1} \right) s^n$$

$$= \sum_{n=2}^{\infty} \left(\sum_{j=0}^{n-2} q\phi_j\phi_{n-j-1} \right) s^n.$$

Reversing the summation order (note $n - 2 \geq j \geq 0$ implies $n \geq j + 2$), we get the above equal to

$$= \sum_{j=0}^{\infty} \left(\sum_{n=j+2}^{\infty} \phi_{n-j-1} s^{n-j-1} \right) \phi_j s^j q s.$$

Setting $m = n - j - 1$ yields

$$= \sum_{j=0}^{\infty} \left(\sum_{m=1}^{\infty} \phi_m s^m \right) \phi_j s^j q s$$

$$= \sum_{j=0}^{\infty} \Phi(s)\phi_j s^j q s = q s \Phi(s) \sum_{j=0}^{\infty} \phi_j s^j$$

$$= q s \Phi^2(s).$$

The left side of (1.6.3) is

$$\sum_{n=1}^{\infty} \phi_n s^n - \phi_1 s = \Phi(s) - ps,$$

and we conclude

$$\Phi(s) - ps = q s \Phi^2(s).$$

Solve the quadratic for the unknown $\Phi(s)$. We get

$$\Phi(s) = \left(1 \pm \sqrt{1 - 4pqs^2}\right)/2qs.$$

The solution with the "+" sign is probabilistically inadmissible. We know $\Phi(0) \leq 1$ but the solution with the "+" sign has the property

$$\frac{1 + \sqrt{1 - 4pqs^2}}{2qs} \sim \frac{1+1}{2qs} = \frac{2}{2qs} \to \infty$$

as $s \to 0$ (where $a(s) \sim b(s)$ as $s \to 0$ means $\lim_{s \to 0} a(s)/b(s) = 1$). So we conclude

(1.6.4) $$\Phi(s) = \frac{1 - \sqrt{1 - 4pqs^2}}{2qs}, \quad 0 \leq s \leq 1.$$

We can expand this to get an explicit solution for $\{\phi_n\}$ using the Binomial Theorem. On the one hand we know

$$\Phi(s) = \sum_{n=1}^{\infty} \phi_n s^n$$

and on the other, by expanding 1.6.4, we also have

$$\Phi(s) = \left(1 - \sum_{j=0}^{\infty} \binom{\frac{1}{2}}{j}(-1)^j (4pqs^2)^j\right)/2qs.$$

The "1" and the "$j = 0$" terms cancel; taking the minus sign in front of the sum inside the sum yields

$$\Phi(s) = \sum_{j=1}^{\infty} \binom{1/2}{j}(-1)^{j+1}(4pq)^j s^{2j}/2qs$$

$$= \sum_{j=1}^{\infty} \binom{1/2}{j}(-1)^{j+1}\frac{(4pq)^j}{2q} s^{2j-1}$$

$$= (\cdot)s + (\cdot)s^3 + (\cdot)s^5 + \dots.$$

We conclude

$$\phi_{2j-1} = \binom{1/2}{j}(-1)^{j+1}(4pq)^j/2q, \quad j \geq 1,$$

and, for the even indices, we have for $j \geq 1$

$$\phi_{2j} = 0.$$

For obtaining qualitative conclusions, it may be easier to extract information from the generating function. For instance, from (1.6.4)

$$P[N < \infty] = \Phi(1) = \left(1 - \sqrt{1 - 4p(1-p)}\right)/2q$$
$$= \left(1 - \sqrt{1 - 4p + 4p^2}\right)/2q = \left(1 - \sqrt{(2p-1)^2}\right)/2q$$
$$= (1 - |2p - 1|)/2q = (1 - |2p - p - q|)/2q$$
$$= (1 - |p - q|)/2q,$$

and we summarize

$$P[N < \infty] = (1 - |p - q|)/2q$$
$$= \begin{cases} 1, & \text{if } p \geq q \\ p/q, & \text{if } p < q. \end{cases}$$

Note that if $p < q$, i.e., the pressure pushing the random walk in the positive direction is weak so that

$$P[N = \infty] = 1 - p/q > 0.$$

In this case $P[S_n \leq 0,$ for all $n \geq 0] > 0$ and on the set of positive probability

$$\cap_{n=0}^{\infty}[S_n \leq 0]$$

the gambler is never ahead.

When $P[N = \infty] > 0$ we have by definition $EN = \infty$. When $p \geq q$ we compute $EN = \Phi'(1)$ by differentiating in (1.6.4)

$$\Phi'(s) = \frac{2qs(-\frac{1}{2}(1 - 4pqs^2)^{-1/2}(-8pqs) - (1 - \sqrt{1 - 4pqs^2})2q}{4q^2 s^2}$$

(ugly but correct), and letting $s \uparrow 1$ yields

$$EN = \left(2q\left(\frac{4pq}{\sqrt{1 - 4pq}}\right) - 2q(1 - \sqrt{1 - 4pq})\right)/4q^2.$$

Dividing numerator and denominator by $2q$ we get

$$EN = \left(\frac{4pq}{|2p - 1|} - (1 - |2p - 1|)\right)/2q$$
$$= \frac{2p}{|p - q|} - \frac{(1 - |p - q|)}{2q},$$

and so

$$EN = \begin{cases} \infty, & \text{if } p = q = 1/2 \\ (p-q)^{-1}, & \text{if } p > q. \end{cases}$$

An extension of these techniques can be used to analyze the distribution of the first return time to 0. Define

$$N_0 = \inf\{n \geq 1 : S_n = 0\},$$

set $f_0 = 0$ and

$$f_{2n} = P[N_0 = 2n], \ n \geq 1.$$

Also

$$F(s) = \sum_{n=0}^{\infty} f_{2n} s^{2n}, \quad 0 \leq s \leq 1.$$

Now we have

$$N_0 = \begin{cases} 1 + \inf\{n : \sum_{i=1}^{n} X_{i+1} = 1\} & \text{on } [X_1 = -1] \\ 1 + \inf\{n : \sum_{i=1}^{n} X_{i+1} = -1\} & \text{on } [X_1 = 1]. \end{cases}$$

Set

$$N^+ = \inf\{n : \sum_{i=1}^{n} X_{i+1} = 1\}$$

$$N^- = \inf\{n : \sum_{i=1}^{n} X_{i+1} = -1\}$$

and observe that because $\{X_i, i \geq 1\} \overset{d}{=} \{X_i, i \geq 2\}$ we have $N \overset{d}{=} N^+$. Also N^+ is determined by $\{X_{i+1}, i \geq 1\}$ and is therefore independent of X_1. Similarly N^- is independent of X_1. We have

$$F(s) = Es^{N_0} = Es^{N_0} 1_{[X_1 = -1]} + Es^{N_0} 1_{[X_1 = 1]}$$

$$= Es^{1+N^+} 1_{[X_1 = -1]} + Es^{1+N^-} 1_{[X_1 = 1]}.$$

By independence this is

$$= sEs^{N^+} P[X_1 = -1] + sEs^{N^-} P[X_1 = 1]$$

(1.6.7) $$= s\Phi(s)q + spEs^{N^-}.$$

Note

$$N^- = \inf\{n : \sum_{i=1}^{n} X_{i+1} = -1\} \stackrel{d}{=} \inf\{n : \sum_{i=1}^{n} X_i = -1\}$$

$$= \inf\{n : \sum_{i=1}^{n}(-X_i) = 1\} = \inf\{n : \sum_{i=1}^{n} X_i^{\#} = 1\}.$$

Moreover, the process $\{\sum_{i=1}^{n} X_i^{\#}, n \geq 1\}$ is a simple random walk with step distribution

$$P[X_1^{\#} = 1] = P[-X_1 = 1] = P[X_1 = -1] = q$$
$$P[X_1^{\#} = -1] = p.$$

To get

$$\Phi^-(s) = Es^{N^-},$$

we simply use the formula (1.6.4) with p and q reversed. Consequently, from (1.6.7),

$$F(s) = sq\left(\frac{1 - \sqrt{1 - 4pqs^2}}{2qs}\right) + sp\left(\frac{1 - \sqrt{1 - 4pqs^2}}{2ps}\right)$$

(1.6.8) $$= 1 - \sqrt{1 - 4pqs^2}.$$

Further,

$$F(1) = P[N_0 < \infty] = 1 - \sqrt{1 - 4pq} = 1 - |p - q|$$

so

$$P[N_0 < \infty] = \begin{cases} 1, & \text{if } p = q \\ 2q, & \text{if } p > q \\ 2p, & \text{if } p < q \end{cases}$$

and only in the balanced case $p = q = 1/2$ does the random walk return to 0 with probability 1. However, even in the balanced case when $p = q = 1/2$ and $P[N_0 < \infty] = 1$ we have

$$F(s) = 1 - \sqrt{1 - s^2}$$

from which

$$F'(s) = -\frac{1}{2}(1 - s^2)^{-1/2}2s \to \infty$$

as $s \to 1$ so that

$$EN_0 = F'(s)|_{s=1} = \infty.$$

Thus, a return to the origin is certain but only after a random amount of time whose expectation is infinite.

1.7. THE DISTRIBUTION OF A PROCESS*.

Given a probability space $(\Omega, \mathcal{A}, P)$, a random variable is a measurable mapping $X : (\Omega, \mathcal{A}) \mapsto (R, \mathcal{R})$ where $\mathcal{R}$ are the Borel subsets of R; i.e., the sigma field generated by open sets (or open intervals). This means that for every Borel subset $B \in \mathcal{R}$ we have

$$X^{-1}(B) := \{\omega : X(\omega) \in B\} \in \mathcal{A}.$$

Define

$$\sigma(X) := X^{-1}(\mathcal{R}) = \{X^{-1}(B) : B \in \mathcal{R}\} \subset \mathcal{A}.$$

This is the σ-algebra generated by X. It is the smallest σ-algebra containing all sets of the form

$$\{\omega : X(\omega) \in B\} = [X \in B], B \in \mathcal{R}.$$

Think of $\sigma(X)$ as that portion of the information in the probability space obtainable from a knowledge of X.

The *distribution* of X is the probability measure on $(R, \mathcal{R})$ defined by $P \circ X^{-1}$ so that for $B \in \mathcal{R}$ we have

$$P \circ X^{-1}(B) = P[X \in B].$$

The sets $\{(-\infty, x], x \in R\}$ are closed under finite intersection and generate $\mathcal{R}$; hence by a set induction result called Dynkin's Theorem (Billingsley, 1986, p. 37, 38) $P \circ X^{-1}$ is determined by its values on semi-infinite intervals. This is the familiar statement that the distribution of a random variable is determined by

$$F(x) = P \circ X^{-1}(-\infty, x] = P[X \le x]$$

$x \in R$; i.e., the distribution is determined by the distribution function. The distribution of X is important because it determines all probabilities pertaining to X since the distribution of X determines probabilities of all sets in $X^{-1}(\mathcal{R}) = \sigma(X) \subset \mathcal{A}$.

Now we generalize the previous discussion. Let $(\Omega, \mathcal{A}, P)$ be a probability space as before and let $(S, \mathcal{S})$ be a measurable space where S is a complete, separable metric space and $\mathcal{S}$ is the σ-algebra generated by open subsets of S. A measurable mapping

$$X : (\Omega, \mathcal{A}) \mapsto (S, \mathcal{S})$$

is called a random element of S. Reaching into our Ω-bag, we pull out an ω, put it into the map X and out comes an element of S. Important examples are:

---

* This section contains advanced material which may be skipped on first reading by beginning readers.

(1) If $S = R$ then X is a random variable.

(2) If $S = R^d$ for $d \geq 2$, then X is a random vector.

(3) If $S = R^\infty$, then X is a stochastic sequence; i.e., a stochastic process with a discrete index set. So a stochastic process with index set $\{0, 1, \dots\}$ is a random element of R^∞.

(4) If $S = C$, the space of continuous functions on $[0, \infty)$, then X is a continuous path stochastic process. A prominent example of such a process is Brownian motion.

(5) If $S = M_p(E)$, the space of all point measures on some nice space E, then X is a stochastic point process. Some prominent examples when $E = [0, \infty)$ are the Poisson processes and renewal processes.

Other examples abound.

The *distribution* of the random element X is the probability measure on $(S, \mathcal{S})$ induced by X, namely $P \circ X^{-1}$, so that for $B \in \mathcal{S}$

$$P \circ X^{-1}(B) = P[X \in B].$$

As before, the distribution of X determines the probabilities of all events determined by X, namely $X^{-1}(\mathcal{S})$.

Usually it is convenient to find a small class of sets, as we did in the case of random variables, so that $P \circ X^{-1}$ is determined by its values on this small class. Recall that the relevant technique is the consequence of Dynkin's Theorem (Billingsley, 1986, p. 38) given next.

Proposition 1.7.1. *If $\mathcal{C}$ is a class of subsets of S which is closed under finite intersections and generates $\mathcal{S}$, i.e., if $\sigma(\mathcal{C}) = \mathcal{S}$, and if two probability measures P_1, P_2 agree on $\mathcal{C}$, then $P_1 = P_2$ on $\mathcal{S}$.*

Sequence Space: We now concentrate attention on

$$R^\infty = \{\mathbf{x} : \mathbf{x} = (x_1, x_2, \dots) \text{ and } x_i \in R, \ i \geq 1\}$$

since this class is most appropriate for discrete time stochastic processes. Let $\mathcal{C}$ be the class of finite dimensional rectangles in R^∞; i.e., $\Lambda \in \mathcal{C}$ if there exist real intervals $I_1, \dots, I_k$ for some k and

$$\Lambda = \{\mathbf{x} \in R^\infty : x_i \in I_i, \ i = 1, \dots, k\}.$$

Note that $\mathcal{C}$ is closed under intersections, and it is also true that $\mathcal{R}^\infty$, the σ-algebra generated by the open sets in R^∞, is generated by the finite dimensional rectangles in $\mathcal{C}$ (cf. for example, Billingsley, 1968). So we have the important conclusion that any measure on $(R^\infty, \mathcal{R}^\infty)$ is uniquely determined by its values on the finite dimensional rectangles $\mathcal{C}$.

Suppose $\mathbf{X} = (X_1, X_2, \dots)$ is a random element of R^∞ defined on the probability space $(\Omega, \mathcal{A}, P)$. Its distribution $P \circ \mathbf{X}^{-1}$ is determined by its values on the finite dimensional rectangles $\mathcal{C}$. This can be expressed another way. We say two random elements $\mathbf{X}$ and $\mathbf{X}'$ are equal in distribution (written $\mathbf{X} \overset{d}{=} \mathbf{X}'$) if $P \circ \mathbf{X}^{-1} = P \circ (X')^{-1}$ on $\mathcal{R}^\infty$. $\mathbf{X}'$ is then called a *version* of $\mathbf{X}$.

Proposition 1.7.2. *If* $\mathbf{X}$ *and* $\mathbf{X}'$ *are two random elements of* R^∞ *then*

$$\mathbf{X} \overset{d}{=} \mathbf{X}'$$

iff

$$\text{for every } k \geq 1 : (X_1, \dots, X_k) \overset{d}{=} (X_1', \dots, X_k') \in R^k.$$

Proof. Define the projections $\Pi_k : R^\infty \mapsto R^k$ by

$$\Pi_k(x_1, x_2, \dots) = (x_1, \dots, x_k).$$

Each Π_k is continuous and hence measurable. If $\mathbf{X} \overset{d}{=} \mathbf{X}'$ then also

$$\Pi_k(\mathbf{X}) = (X_1, \dots, X_k) \overset{d}{=} \Pi_k(\mathbf{X}') = (X_1', \dots, X_k')$$

as desired. Conversely if

$$\text{for every } k \geq 1 : (X_1, \dots, X_k) \overset{d}{=} (X_1', \dots, X_k') \in R^k$$

then the distributions of $\mathbf{X}$ and $\mathbf{X}'$ agree on $\mathcal{C}$ and hence everywhere. ∎

Call the collection of distributions

$$P \circ \mathbf{X}^{-1} \circ \Pi_k(\cdot) = P[(X_1, \dots, X_k) \in \cdot]$$

on R^k ($k \geq 1$) the *finite dimensional distributions* of the process $\mathbf{X}$ and our proposition may be phrased as *the distribution of a process is determined by the finite dimensional distributions.*

Define a new class $\mathcal{C}'$ as follows: A set Λ' is in $\mathcal{C}'$ if it is of the form

$$\Lambda' = \{ y \in R^\infty : y_i \leq x_i, i = 1, \dots, k \}$$

for some $k \geq 1, (x_1, \dots, x_k) \in R^k$. Note that $\mathcal{C}'$ is still closed under intersections and still generates $\mathcal{R}^\infty$; also

$$P \circ \mathbf{X}^{-1}(\Lambda') = P[X_1 \leq x_1, \dots, X_k \leq x_k],$$

which is a k-dimensional distribution function. The analogue of Proposition 1.7.1 is that the distribution of the process is determined by the finite dimensional distribution *functions*.

Two random elements $\mathbf{X}, \mathbf{X}'$ in R^∞ which are equal in distribution will be probabilistically indistinguishable. This last statement is somewhat vague. What is meant is that any probability calculation done for $\mathbf{X}$ yields the same answer when done for $\mathbf{X}'$. (This rephrases the statement

$$P[\mathbf{X} \in B] = P[\mathbf{X}' \in B], \quad \forall B \in \mathcal{R}^\infty.)$$

In succeeding chapters we frequently will construct a convenient representation of the stochastic process $\mathbf{X} = \{X_n\}$. (This was already done with the branching process.) We are assured that any other version $\mathbf{X}' = \{X_n'\}$ will have the same properties as the constructed $\mathbf{X}$.

Here is one last bit of information: Define the coordinate map $\pi_k : R^\infty \mapsto R$ by

$$\pi_k(x_1, x_2, \dots) = x_k$$

for $k \geq 1$. The following shows there is nothing mysterious about a *measurable map* from Ω to R^∞.

Proposition 1.7.3. *If $\mathbf{X}$ is a random element of R^∞ then for each $k \geq 1$ we have that $\pi_k(\mathbf{X})$ is a random variable. Conversely, if $X_1, X_2, \dots$ are random variables defined on Ω, then defining $\mathbf{X}$ by*

$$\mathbf{X}(\omega) = (X_1(\omega), X_2(\omega), \dots)$$

yields a random element of R^∞.

A random element then is just a sequence of random variables.

Proof. π_k is continuous and hence measurable. Therefore if $\mathbf{X}$ is a random element of R^∞,

$$\pi_k \circ \mathbf{X} : \Omega \mapsto R$$

being the composition of two measurable maps is measurable and hence is a random variable.

For the converse, we must show

$$\mathbf{X}^{-1}(\mathcal{R}^\infty) \subset \mathcal{A}.$$

However, $R^\infty = \sigma(\mathcal{C})$ and

$$\mathbf{X}^{-1}(\sigma(\mathcal{C})) = \sigma(\mathbf{X}^{-1}(\mathcal{C})).$$

But for a typical $\Lambda \in \mathcal{C}$,

$$\mathbf{X}^{-1}(\Lambda) = \bigcap_{j=1}^{k} [X_j \in I_j] \in \mathcal{A}$$

since $X_1, \ldots, X_k$ are random variables. Hence

$$\mathbf{X}^{-1}(\mathcal{C}) \subset \mathcal{A}$$

and

$$\sigma(\mathbf{X}^{-1}(\mathcal{C})) \subset \mathcal{A}$$

as desired. ∎

1.8. STOPPING TIMES. *

Information in a probability space is organized with the help of σ-fields. If we have a stochastic process $\{X_n, n \geq 1\}$ we frequently have to know what information is available if we hypothetically observe the process for n time units. Imagine that you will observe the process next week for n time units. The information at our disposal today from this hypothetical future experiment is the σ-algebra generated by $X_1, \ldots, X_n$ which we denote as $\sigma(X_1, \ldots, X_n)$. Another way to think about this is that $\sigma(X_1, \ldots, X_n)$ consists of those events such that when we know the value of $X_1, \ldots, X_n$, we can decide whether or not the events occurred. Note for $n \geq 0$

$$\sigma(X_1, \ldots, X_n) \subset \sigma(X_1, \ldots, X_{n+1})$$

and the information from hypothetically observing the whole process is

$$\sigma(X_j, j \geq 1) = \sigma\{\bigcup_{n=0}^{\infty} \sigma(X_1, \ldots, X_n)\}.$$

In general, suppose we have a probability space $(\Omega, \mathcal{A}, P)$ and an increasing family of σ-fields $\mathcal{F}_n, n \geq 0$; i.e., $\mathcal{F}_n \subset \mathcal{F}_{n+1} \subset \mathcal{A}$. Define

$$\mathcal{F}_\infty = \bigvee_{n \geq 0} \mathcal{F}_n := \sigma(\bigcup_{n=0}^{\infty} \mathcal{F}_n)$$

---

* This section contains advanced material which may be skipped on first reading by beginning readers.

so that $\mathcal{F}_n \subset \mathcal{F}_\infty \subset \mathcal{A}$. Think of $\{\mathcal{F}_n, 0 \leq n \leq \infty\}$ as a history; as time goes on, more information accumulates. Note if $\mathcal{F}_n = \sigma(X_1, \ldots, X_n)$ then $\mathcal{F}_\infty = \sigma(X_j, j \geq 1)$.

An important technical and conceptual concept is that of a *stopping time*. A random variable $\alpha : \Omega \mapsto \{0, 1, \ldots, \infty\}$ is a stopping time with respect to $\{\mathcal{F}_n, n \geq 0\}$ if for every $n \geq 0$ we have

(1.8.1)
$$[\alpha = n] \in \mathcal{F}_n,$$

or equivalently

(1.8.2)
$$[\alpha \leq n] \in \mathcal{F}_n.$$

In the case $\mathcal{F}_n = \sigma(X_1, \ldots, X_n)$, the event $[\alpha = n]$ is determined by $X_1, \ldots, X_n$. It is necessary to allow ∞ as a possible value of α since when α is the waiting time for an event to occur, it may be possible that the event will never occur, in which case the waiting time is infinite. The terminology "stopping time" suggests a gambler whose decision when to stop playing must be based on past events and not on future ones.

Example 1.8.1. Let $\{X_j, j \geq 1\}$ be iid,

$$P[X_1 = 1] = p = 1 - P[X_1 = -1],$$

$0 < p < 1$ and let $\{S_n\}$ be the associated random walk. Then

$$N = \inf\{n \geq 1 : S_n = 1\}$$

is a stopping time. Note the *convention*: The infimum of an empty set is $+\infty$. So

$$[S_n < 1, \text{ for all n }] = [\{n \geq 1 : S_n = 1\} = \emptyset] = [N = \infty].$$

We have that N is a stopping time with respect to $\{\mathcal{F}_n, n \geq 0\}$ where

$$\mathcal{F}_0 = \{\emptyset, \Omega\}, \quad \mathcal{F}_n = \sigma(X_1, \ldots, X_n), n \geq 1.$$

The reason is that for $n \geq 1$

$$[N = n] = [S_1 < 1, \ldots, S_{n-1} < 1, S_n = 1] \in \mathcal{F}_n.$$

Example 1.8.2. A slight generalization shows hitting times are stopping times. Let $\{\xi_j, j \geq 0\}$ be a process on $(\Omega, \mathcal{A}, P)$ and define

$$\mathcal{F}_n = \sigma(\xi_0, \ldots, \xi_n), n \geq 0.$$

Suppose the state space of the process is $(S, \mathcal{S})$ and let $B \in \mathcal{S}$ be a subset of the state space. Define the hitting time of B to be

$$\tau_B = \inf\{n \geq 0 : \xi_n \in B\}.$$

Then τ_B is a stopping time with respect to $\{\mathcal{F}_n\}$ since for $n \geq 1$

$$[\tau_B = n] = [\xi_0 \in B^c, \dots, \xi_{n-1} \in B^c, \xi_n \in B] \in \mathcal{F}_n.$$

In Markov chain models, the state space may be $\{0, 1, \dots\}$. A typical case is that $B = \{0\}$. We are interested in τ_0 where

$$\tau_0 = \inf\{k \geq 0 : \xi_k = 0\}.$$

Back to the general discussion. If α is a stopping time with respect to $\{\mathcal{F}_n\}$, the information up to time α is contained in the σ-field

$$\mathcal{F}_\alpha = \{\Lambda \in \mathcal{F}_\infty : \Lambda \cap [\alpha = n] \in \mathcal{F}_n \text{ for all } 1 \leq n \leq \infty\}.$$

Thus $\Lambda \in \mathcal{F}_\alpha$ if, when imposing the additional restriction that $[\alpha = n]$, the resulting set is placed in $\mathcal{F}_n$. Check that $\mathcal{F}_\alpha$ is a σ-field and that $\mathcal{F}_\alpha$ consists of sets of the form

$$\bigcup_{0 \leq n \leq \infty} [\alpha = n] \cap \Lambda_n$$

where $\Lambda_n \in \mathcal{F}_n, 0 \leq n \leq \infty$.

Suppose $\{\xi_n, n \geq 0\}$ is a process on $(\Omega, \mathcal{A}, P)$ with state space $(S, \mathcal{S})$ so that each ξ has domain Ω and range S. Suppose

$$\sigma(\xi_0, \dots, \xi_n) \subset \mathcal{F}_n.$$

The *post-α process* is the process

$$\{\xi_{\alpha+n}, n \geq 1\}$$

which is only defined on $[\alpha < \infty]$. The definition of $\xi_{\alpha+n}$ is that, at $\omega \in [\alpha < \infty]$, the value of the function $\xi_{\alpha+n}$ is $\xi_{\alpha(\omega)+n}(\omega)$. If $[\alpha < \infty] = \Omega$ then ξ_α is $\mathcal{F}_\alpha$ measurable, written $\xi_\alpha \in \mathcal{F}_\alpha$, since for $B \in \mathcal{S}$ and $0 \leq n < \infty$

$$[\xi_\alpha \in B] \cap [\alpha = n] = [\xi_n \in B] \cap [\alpha = n] \in \mathcal{F}_n.$$

If $[\alpha < \infty] \subset \Omega$ but $[\alpha < \infty] \neq \Omega$, then one can check that either

$$\xi_\alpha 1_{[\alpha < \infty]} \in \mathcal{F}_\alpha \text{ or } \xi_\alpha \in [\alpha < \infty] \cap \mathcal{F}_\alpha,$$

the latter considered in the trace space $(\Omega \cap [\alpha < \infty], [\alpha < \infty] \cap \mathcal{A})$. The post-$\alpha$ field $\mathcal{F}'_\alpha$ is the subfield of $[\alpha < \infty] \cap \mathcal{F}_\infty$ generated by the post-α process $\{\xi_{\alpha+n}, n \geq 1\}$. It is the information available after time α.

1.8.1. Wald's Identity.

Wald's identity and its generalizations are special cases of martingale stopping theorems. They are useful for computing moments of randomly stopped sums, although checking the validity of moment assumptions necessary for the identities to hold can be tricky. We have already seen an identity like Wald's in (1.3.4.3).

If you did not read Section 1.7, think of a stopping time α with respect to the sequence $\{X_n, n \geq 1\}$ as a random variable such that the set $[\alpha = m]$ is determined only by $X_1, \ldots, X_m$, for any $m \geq 1$. Thus α takes on the value m regardless of future (beyond time m) values of the process.

Proposition 1.8.1. *Suppose $\{X_n, n \geq 1\}$ are independent, identically distributed with $E|X_1| < \infty$. Suppose α is a stopping time with respect to $\{X_n, n \geq 1\}$ and*

$$E\alpha < \infty.$$

Then

$$E(\sum_{i=1}^{\alpha} X_i) = E(X_i)E\alpha.$$

Proof. We have

$$E(\sum_{i=1}^{\alpha} X_i) = E \sum_{i=1}^{\infty} X_i 1_{[i \leq \alpha]}.$$

If we can interchange expectation and the infinite sum, then the above is

$$\sum_{i=1}^{\infty} E X_i 1_{[i \leq \alpha]}.$$

Since $[i \leq \alpha] = [\alpha < i]^c = [\alpha \leq i - 1]^c$ depends only on $X_1, \ldots, X_{i-1}$, we have that X_i and $1_{[i \leq \alpha]}$ are independent. The above then equals

$$\sum_{i=1}^{\infty} E X_1 E 1_{[i \leq \alpha]} = E(X_1) \sum_{i=1}^{\infty} P[\alpha \geq i] = E(X_1) \sum_{i=0}^{\infty} P[\alpha > i] = E(X_1)E\alpha$$

from Lemma 1.1.1.

The rest of the proof justifying the interchange of E and $\sum_{i=1}^{\infty}$ requires a bit of measure theory. A student who does not have this background should skip the rest of the proof and proceed to the example below. For those who continue, note

$$E \sum_{i=1}^{\infty} |X_i 1_{[i \leq \alpha]}| = E \sum_{i=1}^{\infty} |X_i| 1_{[i \leq \alpha]}.$$

Since all terms are positive, Fubini or monotone convergence justify the interchange and

$$\sum_{i=1}^{\infty} E|X_i|1_{[i\leq\alpha]} = E|X_1|E\alpha < \infty$$

by assumption. Therefore the function of two variables i and ω

$$X_i(\omega)1_{[i\leq\alpha]}(\omega)$$

is absolutely integrable with respect to the product of P and counting measure. This justifies a Fubini interchange of the iterated integration. ∎

Example. Consider the simple random walk $\{S_n\}$ with $S_0 = 0$ and set

$$N = \inf\{n \geq 1 : S_n = 1\}.$$

Recall $P[X_1 = 1] = p = 1 - P[X_1 = -1]$ so that $EX_1 = p - q$. On $[N < \infty]$ we have $S_N = 1$. If $EN < \infty$ then Wald's identity holds and

$$ES_N = E(X_1)EN = (p - q)EN.$$

If $p = q$, we get a contradiction: If $EN < \infty$ then $P[N < \infty] = 1$; moreover, on the one hand

$$ES_N = E1 = 1 \qquad \text{(since } S_N = 1)$$

and on the other, by Wald

$$ES_N = (\frac{1}{2} - \frac{1}{2})EN = 0.$$

Hence $EN = \infty$. If $EN < \infty$ and $p < q$ then Wald implies $1 = (p-q)EN < 0$, a contradiction. So we get the weak conclusion $EN = \infty$, whereas we know from (1.6) that in fact $P[N = \infty] > 0$. If $p > q$ and $EN < \infty$ we conclude from Wald $EN = (p - q)^{-1}$, in agreement with (1.6.6), but this argument does not prove $EN < \infty$.

1.8.2. SPLITTING AN IID SEQUENCE AT A STOPPING TIME*.

Suppose $\{X_n, n \geq 0\}$ are iid random elements and set

$$\mathcal{F}_n = \sigma(X_0, \ldots, X_n),$$
$$\mathcal{F}'_n = \sigma(X_{n+1}, X_{n+2}, \ldots).$$

---

*This section contains advanced material which may be skipped on first reading by beginning readers. However, the main result Proposition 1.8.2 is easy to understand in the case that α is finite a.s.

$\mathcal{F}_n$ represents the history up to time n and $\mathcal{F}'_n$ is the future after n. For iid random elements, these two σ-fields are independent. The same is true when n is replaced by a stopping time α; however care must be taken to handle the case that $P[\alpha = \infty] > 0$. As before, we must define

$$\mathcal{F}_\infty = \bigvee_{n=0}^{\infty} \mathcal{F}_n = \sigma(\bigcup_{n=0}^{\infty} \mathcal{F}_n).$$

If $\alpha = \infty$ it makes no sense to talk about splitting $\{X_n\}$ into two pieces—the pre- and post-α pieces. Instead we restrict attention to the trace probability space. If $(\Omega, \mathcal{A}, P)$ is the probability space on which $\{X_n\}$ and α are defined, the trace probability space is

$$(\Omega^\#, \mathcal{A}^\#, P^\#) = (\Omega \cap [\alpha < \infty], \mathcal{A} \cap [\alpha < \infty], P\{\cdot | \alpha < \infty\})$$

(assuming $P[\alpha < \infty] > 0$). If $P[\alpha < \infty] = 1$, then there is no essential difference between the original and the trace space.

Proposition 1.8.2. *Let $\{X_n, n \geq 0\}$ be iid and suppose α is a stopping time of the sequence (i.e., with respect to $\{\mathcal{F}_n\}$). In the trace probability space $(\Omega^\#, \mathcal{F}^\#, P^\#)$, the pre- and post-$\alpha$ σ-fields $\mathcal{F}_\alpha, \mathcal{F}'_\alpha$ are independent and*

$$\{X_n, n \geq 0\} \overset{d}{=} \{X_{\alpha+k}, k \geq 1\}$$

in the sense that for $B \in \mathcal{R}^\infty$

(1.8.2.1) $$P^\#[(X_{\alpha+k}, k \geq 1) \in B] = P[(X_n, n \geq 0) \in B].$$

Proof. Suppose $\Lambda \in \mathcal{F}_\alpha$. Then

$$P\{\Lambda \cap [\alpha < \infty] \cap [\{X_{\alpha+k}, k \geq 1\} \in B]\}$$

(1.8.2.2) $$= \sum_{n=0}^{\infty} P\{\Lambda \cap [\alpha = n] \cap [(X_{n+k}, k \geq 1) \in B]\}.$$

From the definition of $\mathcal{F}_\alpha$ we have

$$\Lambda \cap [\alpha = n] \in \mathcal{F}_n.$$

Since $[(X_{n+k}, k \geq 1) \in B] \in \mathcal{F}'_n$ which is independent of $\mathcal{F}_n$ we get (1.8.2.2) equal to

$$\sum_{n=0}^{\infty} P[\Lambda \cap [\alpha = n]] P[\{X_{n+k}, k \geq 1\} \in B]$$

$$= \sum_{n=0}^{\infty} P[\Lambda \cap [\alpha = n]]\, P[\{X_k, k \geq 0\} \in B]$$

$$= P[\Lambda \cap [\alpha < \infty]]\, P[\{X_k, k \geq 0\} \in B].$$

By dividing (1.8.2.2) by $P[\alpha < \infty]$ we conclude

(1.8.2.3) $P^{\#}\left[\Lambda \cap \left[\{X_{\alpha+k}, k \geq 1\} \in B\right]\right] = P^{\#}(\Lambda)P\left[\{X_k, k \geq 0\} \in B\right].$

Let $\Lambda = \Omega$. We conclude that (1.8.2.1) is true. Once we know (1.8.2.1) is true we may rewrite (1.8.2.3) as

(1.8.2.4) $P^{\#}\left[\Lambda \cap \left[\{X_{\alpha+k}, k \geq 1\} \in B\right]\right] = P^{\#}(\Lambda)P^{\#}\left[\{X_{\alpha+k}, k \geq 1\} \in B\right]$

which gives the required independence. ∎

Example. Let $\{X_n, n \geq 1\}$ be iid Bernoulli,

$$P[X_1 = 1] = p = 1 - P[X_1 = -1]$$

and let the associated simple random walk be

$$S_0 = 0, S_n = X_1 + \cdots + X_n, n \geq 1.$$

We derive the quadratic equation for $\Phi(s) = Es^N$ where

$$N = \inf\{n \geq 1 : S_n = 1\}$$

without first deriving a recursion for $\{P[N = k], k \geq 1\}$. We have

$$\Phi(s) = Es^N 1_{[X_1=1]} + Es^N 1_{[X_1=-1]}$$
$$= sp + Es^N 1_{[X_1=-1]}.$$

Define

$$N_1 = \inf\{n \geq 1 : \sum_{j=1}^{n} X_{1+j} = 1\}$$

and on $[N_1 < \infty]$ define

$$N_2 = \inf\{n \geq 1 : \sum_{j=1}^{n} X_{N_1+1+j} = 1\}$$

so that on $[N_1 < \infty, X_1 = -1]$

$$N = 1 + N_1 + N_2.$$

Define $P^{\#} = P(\cdot | N_1 < \infty)$. Now on $[N_1 < \infty]$, N_2 is the same functional of $\{X_{N_1+1+j}, j \geq 1\}$ as N_1 is of $\{X_{1+j}, j \geq 1\}$. So from the previous results, for any k,

$$P[N_1 = k] = P^{\#}[N_2 = k].$$

Thus for $0 < s < 1$

$$Es^N 1_{[X_1=-1]} = Es^{1+N_1+N_2} 1_{[X_1=-1,N_1<\infty]}$$

(since on $[X_1 = -1, N_1 = \infty]$ we have $N = \infty$ and $s^N = 0$). Let $E^\#$ be expectation with respect to the measure $P^\#$. Then

$$Es^N 1_{[X_1=-1]} = sqEs^{N_1+N_2} 1_{[N_1<\infty]}$$
$$= sqE^\# s^{N_1+N_2} P[N_1 < \infty],$$

and since N_1, N_2 are independent with respect to $P^\#$ (by 1.8.2.4), this is

$$= sqE^\# s^{N_1} E^\# s^{N_2} P[N_1 < \infty].$$

Using (1.8.2.1) we get

$$= sqE^\# s^{N_1} Es^{N_1} P[N_1 < \infty]$$
$$= sqEs^{N_1} 1_{[N_1<\infty]} Es^{N_1}$$
$$= sqEs^{N_1} Es^{N_1}$$
$$= sq\Phi^2(s)$$

and we conclude as in Section (1.6) that

$$\Phi(s) = sp + sq\Phi^2(s).$$

EXERCISES

1.1. (a) Let X be the outcome of tossing a fair die. What is the gf of X? Use the gf to find EX.

(b) Toss a die repeatedly. Let μ_n be the number of ways to throw die until the sum of the faces is n. (So $\mu_1 = 1$ (first throw equals 1), $\mu_2 = 2$ (either the first throw equals 2 or the first 2 throws give 1 each), and so on. Find the generating function of $\{\mu_n, n \geq 1\}$.

1.2. Let $\{X_n, n \geq 1\}$ be iid Bernoulli random variables with

$$P[X_1 = 1] = p = 1 - P[X_1 = 0]$$

and let $S_n = \sum_{i=1}^n X_i$ be the number of successes in n trials. Show S_n has a binomial distribution by the following method:

(1) Prove for $n \geq 0, 1 \leq k \leq n+1$

$$P[S_{n+1} = k] = pP[S_n = k - 1] + qP[S_n = k].$$

(2) Solve the recursion using generating functions.

1.3. Let $\{X_n, n \geq 1\}$ be iid non-negative integer valued random variables independent of the non-negative integer valued random variable N and suppose

$$E(X_1) < \infty, \mathrm{Var}(X_1) < \infty, EN < \infty, \mathrm{Var}(N) < \infty.$$

Set $S_n = \sum_{i=1}^{n} X_i$. Use generating functions to check

$$\mathrm{Var}(S_N) = E(N)\mathrm{Var}(X_1) + (EX_1)^2\mathrm{Var}(N).$$

1.4. What are the range and index set for the following stochastic processes:

(a) Let X_i be the quantity of beer ordered by the ith customer at Happy Harry's and let $N(t)$ be the number of customers to arrive by time t. The process is $\{X(t) = \sum_{i=1}^{N(t)} X_i, t \geq 0\}$ where $X(t)$ is the quantity ordered by time t.

(b) Thirty-six points are chosen randomly in Alaska according to some probability distribution. A circle of random radius is drawn about each point yielding a random set S. Let $X(A)$ be the value of the oil in the ground under region $A \cap S$. The process is $\{X(B), B \subset \text{Alaska}\}$.

(c) Sleeping Beauty sleeps in one of three positions:

(1) On her back looking radiant.
(2) Curled up in the fetal position.
(3) In the fetal position, sucking her thumb and looking radiant only to an orthodontist.

Let $X(t)$ be Sleeping Beauty's position at time t. The process is $\{X(t), t \geq 0\}$.

(d) For $n = 0, 1, \ldots$, let X_n be the value in dollars of property damage to West Palm Beach, Florida and Charleston, South Carolina by the nth hurricane to hit the coast of the United States.

1.5. If X is a non-negative integer valued random variable with

$$X \sim \{p_k\}, \quad P(s) = Es^X,$$

express the generating functions if possible, in terms of $P(s)$, of (a) $P[X \leq n]$, (b) $P[X < n]$, (c) $P[X \geq n]$.

1.6. (Karlin and Taylor, 1975, p. 38) Let X and Y be jointly distributed non-negative integer valued random variables. For $|s| \leq 1, |t| \leq 1$, define the joint generating function

$$P_{X,Y}(s,t) = \sum_{i,j=0}^{\infty} s^i t^j P[X = i, Y = j].$$

Prove that X and Y are independent iff

$$P_{X,Y}(s,t) = P_X(s)P_Y(t) \qquad \forall s,t.$$

1.7. A Skip Free Negative Random Walk. Suppose $\{X_n, n \geq 1\}$ is independent, identically distributed. Define $S_0 = X_0 = 1$ and for $n \geq 1$

$$S_n = X_0 + X_1 + \cdots + X_n.$$

For $n \geq 1$ the distribution of X_n is specified by

$$P[X_n = j - 1] = p_j, \quad j = 0, 1, \ldots$$

where

$$\sum_{j=0}^{\infty} p_j = 1, \quad f(s) = \sum_{j=0}^{\infty} p_j s^j, 0 \leq s \leq 1.$$

(The random walk starts at 1; when it moves in the negative direction, it does so only by jumps of -1. The walk cannot jump over states when moving in the negative direction.) Let

$$N = \inf\{n : S_n = 0\}.$$

If $P(s) = Es^N$, show $P(s) = sf(P(s))$. (Note what happens at the first step: Either the random walk goes from 1 to 0 with probability p_0 or from 1 to j with probability p_j.) If $f(s) = p/(1 - qs)$ corresponding to a geometric distribution, find the smallest solution.

1.8. In a branching process

$$P(s) = as^2 + bs + c$$

where $a > 0, b > 0, c > 0, P(1) = 1$. Compute π. Give a condition for sure extinction.

1.9. In a binomial replacement branching model with $P(s) = q + ps$, let $T = \inf\{n : Z_n = 0\}$.

(1) Find $P[T = n]$ for $n \geq 1$.
(2) Find $P[T = n]$ assuming $Z_0 = i > 0$.

1.10. Harry lets his health habits slip during a depressed period and discovers spots growing between his toes according to a branching process with generating function

$$P(s) = .15 + .05s + .03s^2 + .07s^3 + .4s^4 + .25s^5 + .05s^6.$$

Will the spots survive? With what probability?

1.11. A Point Process. Let $N(A)$ be the number of points in region A. Assume that for any n, the set A can be decomposed, $A = \cup_{j=1}^n A_j^{(n)}$, such that $A_1^{(n)}, \ldots, A_n^{(n)}$ are disjoint, $N(A) = \sum_{j=1}^n N(A_j^{(n)})$ and $N(A_1^{(n)}), \ldots, N(A_n^{(n)})$ are independent. Assume

$$P[N(A_j^{(n)}) = 0] = \exp\{-\lambda/n\}, \quad P[N(A_j^{(n)}) \geq 2] \leq \frac{\lambda}{n}\delta(\frac{\lambda}{n})$$

where $\delta(x)$ is a positive function such that $\delta(x) \to 0$ as $x \to 0$. Show $N(A)$ has a Poisson distribution.

1.12. For a branching process $\{Z_n\}$, let $S = 1 + \sum_{n=1}^\infty Z_n$ be the total population ever born. Find a recursion which is satisfied by the generating function of S. Solve this in the case $P(s) = q + ps$ and $P(s) = p/(1 - qs)$. What is $E(S)$?

1.13. Let $[x]$ be the greatest integer $\leq x$. Check by integral comparison or another such method that

$$\lim_{N\to\infty} \sum_{j=N+1}^{[Ne]} j^{-1} = \log e = 1.$$

Let $\{X_j, j \geq 1\}$ be independent random variables with

$$P[X_j = 1] = 1/j = 1 - P[X_j = 0]$$

and set $S_n = \sum_{i=1}^n X_i$, $n \geq 1$.

(1) What is the generating function of

$$S_{[Ne]} - S_N = \sum_{j=N+1}^{[Ne]} X_j?$$

(2) Use the continuity theorem for generating functions to show

$$\lim_{N\to\infty} P[S_{[Ne]} - S_N = k] = p(k, 1) = e^{-1}/k!.$$

(3) Define

$$L(1) = \inf\{j > 1 : X_j = 1\}.$$

Compute the generating function of $\{P[L(1) > n], n \geq 2\}$. What is $EL(1)$?

(4) If $\{Z_n, n \geq 1\}$ is a sequence of iid random variables with a continuous distribution, show that

$$\{1_{[Z_n > \vee_{i=1}^{n-1} Z_i]}, n \geq 1\} \overset{d}{=} \{X_j, j \geq 1\}.$$

1.14. Harry comes from a long line of descendents who do not get along with their parents. Consequently each generation vows to be different from their elders. Assume the offspring distribution for the progenitor has generating function $f(s)$ and that the offspring distribution governing the number of children per individual in the first generation has generating function $g(s)$. The next generation has offspring governed by $f(s)$ and the next has $g(s)$ so that the functions alternate from generation to generation. Determine the extinction probability of this process and the mean number of individuals in the nth (assume n is even) generation.

1.15. (a) Suppose X is a non-negative integer valued random variable. Conduct a compound experiment. Observe X items and mark each of the X items independently with probability s where $0 < s < 1$. What is the probability that all X items are marked?

(b) Suppose X is a non-negative integer valued random variable with mass function $\{p_k\}$. Suppose T is a geometric random variable independent of X with

$$P[T \geq n] = s^n, \quad 0 < s < 1.$$

Compute $P[T \geq X]$.

1.16. Stopping Times. (a) If α is a stopping time with respect to the σ-fields $\{\mathcal{F}_n\}$ then prove $\mathcal{F}_\alpha$ is a σ-field.

(b) If $\alpha_k, k \geq 1$ are stopping times with respect to $\{\mathcal{F}_n\}$, show $\vee_k \alpha_k$ and $\wedge_k \alpha_k$ are stopping times. (Note $\vee$ means "max" and $\wedge$ means "min".) If $\{\alpha_k\}$ is a monotone increasing family of stopping times then $\lim_{k \to \infty} \alpha_k$ is a stopping time.

(c) If $\alpha_1 \leq \alpha_2$ show $\mathcal{F}_{\alpha_1} \subset \mathcal{F}_{\alpha_2}$.

1.17. For a simple random walk $\{S_n\}$ let $u_0 = 1$ and for $n \geq 1$, let

$$u_n = P[S_n = 0].$$

Compute by combinatorics the value of u_n. Find the generating function $U(s) = \sum_{n=0}^{\infty} u_n s^n$ in closed form. To get this in closed form you need the identity

$$\binom{2n}{n} = 4^n (-1)^n \binom{-\frac{1}{2}}{n}.$$

1.18. Happy Harry's Fan Club. Harry's restaurant is located near Orwell University, a famous institution of higher learning. Because of the crucial culinary, social and intellectual role played by Harry and the restaurant in the life of the University, a fan club is started consisting of two types of members: students and faculty. Due to the narrow focus of the club, membership automatically terminates after one year. Student members of the *Happy Harry's Fan Club* are so fanatical they recruit other members when their membership expires. Faculty members never recruit because they are too busy. A student recruiter will recruit two students with probability 1/4, one student and one faculty member with probability 2/3 and 2 faculty with probability 1/12. Assume the club was started by one student. After n years, what is the probability that no faculty will have yet been recruited? What is the probability the club will eventually have no members?

1.19. At 2 AM business is slow and Harry contemplates closing his establishment for the night. He starts flipping an unfair coin out of boredom and decides to close when he gets r consecutive heads. Let T be the number of flips necessary to obtain r consecutive heads. Suppose the probability of a head is p and the probability of a tail is q. Define $p_k = P[T = k]$ so that $p_k = 0$ for $k < r$.

(1) Prove for $k > r$

$$p_k = P[T = k] = p^r q[1 - p_0 - p_1 - \cdots - p_{k-r-1}].$$

(2) Compute the generating function of T and verify $P(1) = 1$.
(3) Compute ET. If you are masochistic, try $\text{Var}(T)$.

The next night at midnight, Harry is bored, so again he starts flipping coins. To vary the routines he looks for a pattern of HT (head then tail). For $n \geq 2$, let

$$f_n = P[\text{ the pattern HT first appears at trial number } n \,].$$

Compute the generating function of $\{f_n\}$ and find the mean and variance.

1.20. In a branching process, suppose $P(s) = q + ps^2, 0 < q, p < 1, q + p = 1$. Let T be the time the population first becomes extinct:

$$T = \inf\{n \geq 1 : Z_n = 0\}.$$

(1) Find the probability of eventual extinction.
(2) Suppose the population starts with one individual. Find $P[T > n]$.

1.21. Let $\{N(t), t \geq 0\}$ be a process with independent increments which means that for any k and times $0 \leq t_1 \leq \cdots \leq t_k$ $N(t_1), N(t_2) - N(t_1), \ldots, N(t_k) - N(t_{k-1})$ are independent random variables. Suppose for each t that $N(t)$ is non-negative integer valued with

$$P_t(s) = Es^{N(t)}.$$

For $\tau < t$ express $Es^{N(t)-N(\tau)}$ in terms of $P_\tau(s), P_t(s)$.

1.22. Harry and the Mutant Creepazoids. The neighborhood where Harry's restaurant is located starts to deteriorate, largely due to the rise of a weird teenage gang called the *Mutant Creepazoids*. By eavesdropping in the bar of the restaurant and similar techniques, Harry learns of the gang's plans for neighborhood domination. Each week, every creepazoid is required to recruit new members. The number recruited per member is random with generating function $P(s)$. (Note that this number includes the recruiter; if the number is zero, this means the recruiter died or left the gang.) However, each week a certain proportion of the recruits and members, namely $q, 0 < q < 1$, drop out since they apparently decide the weird life style is not for them. Let Z_n be the number of gang members in the nth generation.

(1) Assuming the gang started with one member. What is the generating function of the number of gang members after a week; i.e., what is the gf of Z_1?

(2) If we start with one recruit (someone who is not sure he will stay with the group), what is the distribution of the number of recruits rounded up by the initial recruit after one week?

(3) Show the means of the two distributions found above are the same. When is the variance the same? If the variance is the same, are the distributions the same?

(4) Under what conditions is the disturbance to the neighborhood transitory?

(5) Finding that the gang strains his nerves, Harry decides to leave for Florida every other week. He therefore observes the population of the gang to be $\{Z_{2n}, n \geq 0\}$. Is this a branching process? If so, what is the offspring distribution which generates this process?

1.23. For a branching process with offspring distribution

$$p_n = pq^n, \quad n \geq 0, p + q = 1, 0 < p < 1,$$

find the extinction probability and give conditions for sure extinction.

1.24. Branching in Varying Environments. This is the same model as the simple branching process except that individuals in the nth generation reproduce according to a reproduction law $\{p_{nk}, k \geq 0\}$ with generating function $\phi_n(s) = \sum_k p_{nk} s^k$. As before, let Z_n be the number in the nth generation.

(1) Construct a model for this population analogous with the construction of Section 1.4.

(2) Express the generating function

$$f_n(s) = Es^{Z_n}$$

in terms of $\phi_k(s), k \geq 0$ where $\phi_0(s) = s$.
(3) Express $m_n = EZ_n$ in terms of $\mu_i, i \geq 0$ where $\mu_i = \phi'_j(1)$.

1.25. Harry and the Management Software. Eager to give Happy Harry's Restaurant every possible competitive advantage, Harry writes inventory management software that is supposedly geared to restaurants. Harry, sly fox that he is, has designed the software to contain a virus that wipes out all computer memory and results in a restaurant being unable to continue operation. He starts by crossing the street and giving a copy to the trendy sprouts bar. The software is presented with the condition that the recipient must give a copy to two other restaurateurs, thus spreading the joy of technology. The time it takes a recipient to find someone else for the software is random. Upon receipt of the software, the length of time until it wipes out a restaurant's computer memory is also random. Of course, once a restaurant's computer memory is wiped out, the owner would not continue to disburse the software. Thus a restaurateur may distribute the software to 0, 1 or 2 other restaurants.

For $j = 0, 1, 2$, define

$$p_j = P[\text{ a restaurateur distributes the software to } j \text{ other restaurants }].$$

Suppose $p_0 = .2, p_1 = .1, p_2 = .7$. What is the probability that Harry's plans for world domination of the restaurant business will succeed?

1.26.* Suppose X_1, X_2 are independent, $N(0, 1)$ random variables on the space $(\Omega, \mathcal{A}, P)$.
  (a) Prove $X_1 \overset{d}{=} -X_1$; i.e., prove that $P \circ X_1^{-1} = P \circ (-X_1)^{-1}$ on $\mathcal{R}$.
  (b) Prove

$$(X_1, X_1 + X_2) \overset{d}{=} (X_1, X_1 - X_2)$$

in R^2; i.e., prove

$$P \circ (X_1, X_1 + X_2)^{-1} = P \circ (X_1, X_1 - X_2)^{-1}$$

on $\mathcal{R}^2$.

Now suppose $\{X_n, n \geq 1\}$ is an iid sequence of $N(0, 1)$ random variables.

---

* This problem requires some advanced material which should be skipped on the first reading by beginning readers.

(c) Prove

$$(X_1, X_1 + X_2, \dots) \overset{d}{=} (-X_1, -X_1 - X_2, \dots)$$

in R^∞.

(d) If X, Y are random elements of a metric space S, and $g : S \mapsto S'$ is a mapping from S to a second metric space S', show that $X \overset{d}{=} Y$ implies $g(X) \overset{d}{=} g(Y)$.

1.27. If X_r has a negative binomial distribution with parameters p, r (cf. Example 1.3.6, Section 1.3.3), show that if $r \to \infty$ and $rq \to \lambda > 0$, then the negative binomial random variable X_r converges in distribution to a Poisson random variable with parameter λ.

1.28. Consider the simple branching process $\{Z_n\}$ with offspring distribution $\{p_k\}$ and generating function $P(s)$.

(a) When is the total number of offspring $\sum_{n=0}^{\infty} Z_n < \infty$?

(b) When the total number of offspring is finite, give the functional equation satisfied by the generating function $\Psi(s)$ of $\sum_{n=0}^{\infty} Z_n < \infty$.

(c) Zeke initiates a family line which is sure to die out. Lifetime earnings of each individual in Zeke's line of descent (including Zeke) constitute iid random variables which are independent of the branching process and have common distribution function $F(x)$, where F concentrates on $[0, \infty)$. Thus to each individual in the line of descent is associated a non-negative random variable. What is the probability $H(x)$ that no one in Zeke's line earns more than x in his/her lifetime, where of course $x > 0$.

(d) When

$$P(s) = \frac{.5}{1 - .5s}$$

find $\Psi(s)$. If in addition,

$$F(x) = 1 - e^{-x}, \quad x > 0,$$

find $H(x)$.

CHAPTER 2

Markov Chains

IN TRYING to make a realistic stochastic model of any physical situation, one is forced to confront the fact that real life is full of dependencies. For example, purchases next week at the supermarket may depend on satisfaction with purchases made up to now. Similarly, an hourly reading of pollution concentration at a fixed monitoring station will depend on previous readings; tomorrow's stock inventory will depend on the stock level today, as well as on demand. The number of customers awaiting service at a facility depends on the number of waiting customers in previous time periods.

The dilemma is that dependencies make for realistic models but also for unwieldy or impossible probability calculations. The more independence built into a probability model, the more possibility for explicit calculations, but the more questionable is the realism of the model. Imagine the absurdity of a probability model of a nuclear reactor which assumes each component of the complex system fails independently. The independence assumptions would allow for calculations of the probability of a core meltdown, but the model is so unrealistic that no government agency would be so foolish as to base policy on such unreliable numbers—at least not for long.

When constructing a stochastic model, the challenge is to have dependencies which allow for sufficient realism but which can be analytically tamed to permit sufficient mathematical tractability. Markov processes frequently balance these two demands nicely. A Markov process has the property that, conditional on a history up to the present, the probabilistic structure of the future does not depend on the whole history but only on the present. Dependencies are thus manageable since they are conditional on the present state; the future becomes conditionally independent of the past. Markov chains are Markov processes with discrete index set and countable or finite state space.

We start with a construction of a Markov chain process $\{X_n, n \geq 0\}$. The process has a discrete state space denoted by S. Usually we take the state space S to be a subset of integers such as $\{0, 1, \dots\}$ (infinite state space) or $\{0, 1, \dots, m\}$ (finite state space). When considering stationary Markov chains, it is frequently convenient to let the index set be

$\{\ldots, -1, 0, 1, \ldots\}$, but for now the non-negative integers suffice for the index set.

How does a Markov chain evolve? To fix ideas, think of the following scenario. During a decadent period of Harry's life he used to visit a bar every night. The bars were chosen according to a random mechanism. Harry's random choice of a bar was dependent only on the bar he had visited the previous night, not on the choices prior to the previous night. What would be the ingredients necessary for the specification of a model of bar selection? We would need an initial distribution $\{a_k\}$ so that when Harry's decadent period commenced he chose his initial bar to be the kth with probability a_k. We would also need *transition probabilities* p_{ij} which **could** determine the probability of choosing the jth pub if on the prior night the ith was visited.

Section 2.1 begins with a construction of a Markov chain and a discussion of elementary properties. The construction also describes how one would simulate a Markov chain.

2.1. CONSTRUCTION AND FIRST PROPERTIES.

Let us first recall how to simulate a random variable with non-negative integer values $\{0, 1, \ldots\}$. Suppose X is a random variable with

$$P[X = k] = a_k, \quad k \geq 0, \quad \sum_{i=0}^{\infty} a_i = 1.$$

Let U be uniformly distributed on $[0, 1]$. We may simulate X by observing U, and if U falls in the interval $(\sum_{i=0}^{k-1} a_i, \sum_{i=0}^{k} a_i]$ then we pick the value k. (As a convention here and in what follows, set $\sum_{i=0}^{-1} a_i = 0$.) Now if we define

$$Y = \sum_{k=0}^{\infty} k 1_{(\Sigma_{i=0}^{k-1} a_i, \Sigma_{i=0}^{k} a_i]}(U)$$

so that $Y = k$ iff $U \in (\sum_{i=0}^{k-1} a_i, \sum_{i=0}^{k} a_i]$, then Y has the same distribution as X, and we have simulated X.

We now construct a Markov chain. For concreteness we assume the state space S is $\{0, 1, \ldots\}$. Only minor modifications are necessary if the state space is finite, for example $S = \{0, 1, \ldots, m\}$. We need an initial distribution $\{a_k\}$ where $a_k \geq 0, \sum_{k=0}^{\infty} a_k = 1$ to govern the choice of an initial state. We also need a transition matrix to govern transitions from state to state. A transition matrix is a matrix which in the infinite state space case is $P = (p_{ij}, i \geq 0, j \geq 0)$ or, written out,

$$P = \begin{pmatrix} p_{00} & p_{01} & \cdots \\ p_{10} & p_{11} & \cdots \\ \vdots & & \ddots \end{pmatrix}$$

and where the entries satisfy

$$p_{ij} \geq 0, \quad \sum_{j=0}^{\infty} p_{ij} = 1, \quad i = 0, 1, \ldots.$$

(In the case where the state space S is finite and equal to $\{0, 1, \ldots, m\}$, P is $(m+1) \times (m+1)$ dimensional.)

We now construct a Markov chain $\{X_n, n \geq 0\}$. We need a scheme which will choose an initial state k with probability a_k that will generate transitions from i to j with probability p_{ij}. Let $\{U_n, n \geq 0\}$ be iid uniform random variables on $(0, 1)$. Define

$$X_0 = \sum_{k=0}^{\infty} k 1_{(\Sigma_{i=0}^{k-1} a_i, \Sigma_{i=0}^{k} a_i]}(U_0).$$

This is the construction given above, which produces a random variable that takes the value k with probability a_k. The rest of the process is defined inductively. Define the function $f(i, u)$ with domain $S \times [0, 1]$ by

$$f(i, u) = \sum_{k=0}^{\infty} k 1_{(\Sigma_{j=0}^{k-1} p_{ij}, \Sigma_{j=0}^{k} p_{ij}]}(u)$$

so that $f(i, u) = k$ iff $u \in (\sum_{j=0}^{k-1} p_{ij}, \sum_{j=0}^{k} p_{ij}]$. Now for $n \geq 0$ define

$$X_{n+1} = f(X_n, U_{n+1}).$$

Note that if $X_n = i$, we have constructed X_{n+1} so that it equals k with probability p_{ik}. Also observe that X_0 is a function of U_0, X_1 is a function of X_0 and U_1 and hence is a function of U_0 and U_1, and so on so, that in general we have X_{n+1} is a function of $U_0, U_1, \ldots, U_{n+1}$.

Some **elementary properties** of the construction follow.

1. We have

(2.1.1) $$P[X_0 = k] = a_k, \quad k \geq 0,$$

and for any $n \geq 0$

(2.1.2) $P[X_{n+1} = j | X_n = i] = p_{ij}.$

This follows since the conditional probability in (2.1.2) is equal to

$$P[f(X_n, U_{n+1}) = j | X_n = i] = P[f(i, U_{n+1}) = j | X_n = i]$$
$$= P[f(i, U_{n+1}) = j]$$

since U_{n+1} and X_n are independent. By the construction at the beginning of Section 2.1, this probability is p_{ij}.

2. As a generalization of (2.1.2) we show we may condition on a history with no change in the conditional probability provided the history ends in state i. More specifically we have

(2.1.3) $P[X_{n+1} = j | X_0 = i_0, \ldots, X_{n-1} = i_{n-1}, X_n = i] = p_{ij}$

for integers $i_0, i_1, \ldots, i_{n-1}, i, j$ in the state space (provided $P[X_0 = i_0, \ldots, X_{n-1} = i_{n-1}, X_n = i] > 0$).

As with property 1, this conditional probability is

$$P[f(i, U_{n+1}) = j | X_0 = i_0, \ldots, X_n = j]$$

and since $X_0, \ldots, X_n$ are independent of U_{n+1}, the foregoing probability is

$$P[f(i, U_{n+1}) = j] = p_{ij}.$$

Processes satisfying (2.1.2) and (2.1.3) possess the *Markov property* meaning

$$P[X_{n+1} = j | X_0 = i_0, \ldots, X_{n-1} = i_{n-1}, X_n = i]$$
$$= P[f(X_n, U_{n+1}) = j | X_n = i].$$

3. As a generalization of (2.1.3), we show that the probability of the future subsequent to time n given the history up to n, is the same as the probability of the future given only the state at time n; and this conditional probability is independent of n (but dependent on the state). Precisely, we have for any integer m and any states $k_1, \ldots, k_m$

$$P[X_{n+1} = k_1, \ldots, X_{n+m} = k_m | X_0 = i_0, \ldots, X_{n-1} = i_{n-1}, X_n = i]$$
$$= P[X_{n+1} = k_1, \ldots, X_{n+m} = k_m | X_n = i]$$
(2.1.4) $$= P[X_1 = k_1, \ldots, X_m = k_m | X_0 = i].$$

In shorthand notation, denote the event $[X_{n+1} = k_1, \ldots, X_{n+m} = k_m]$ by $[(X_j, j \geq n+1) \in B]$. Note that in the probability of (2.1.4) we can replace X_{n+1} by $f(i, U_{n+1})$, and we can replace X_{n+2} by $f(X_{n+1}, U_{n+2}) = f(f(i, U_{n+1}), U_{n+2})$ and so on. Thus in the probability of (2.1.4) we can replace $(X_j, j \geq n+1)$ by something depending only on $U_j, j \geq n+1$ which is independent of $X_0, \ldots, X_n$. Therefore the conditional probability is

$$P[(f(i, U_{n+1}), f(f(i, U_{n+1}), U_{n+2}), \ldots) \in B].$$

Since this also equals

$$P[(f(i, U_1), f(f(i, U_1), U_2), \ldots) \in B],$$

the result follows.

The three properties above are the essential characteristics of a Markov chain.

Definition. *Any process $\{X_n, n \geq 0\}$ satisfying (2.1.2)—(2.1.3) is called a Markov chain with initial distribution $\{a_k\}$ and transition probability matrix P.*

Sometimes a transition probability matrix is called a *Markov* or a *stochastic* matrix.

The constructed Markov chain has *stationary transition probabilities* since the conditional probability in (2.1.2) is independent of n. Sometimes a Markov chain with stationary transition probabilities is called *homogeneous*.

Warning. Although the constructed process possesses stationary transition probabilities, the process in general is not stationary. For the process $\{X_n\}$ to be stationary, the following condition, describing a translation property of the finite dimensional distributions, must hold: For any non-negative integers m, ν and any states $k_0, \ldots, k_m$ we have

$$P[X_0 = k_0, \ldots, X_m = k_m] = P[X_\nu = k_0, \ldots, X_{\nu+m} = k_m].$$

(Roughly speaking, this says the statistical evolution of the process over an interval is the same as that of the process over a translated interval.) The concept of a Markov chain being a stationary stochastic process and having stationary transition probabilities should not be confused. Conditions for the Markov chain to be stationary are discussed in Section 2.12.

The process constructed above will sometimes be referred to as the *simulated Markov chain*. We will show in Proposition 2.1.1 that any Markov chain $\{X_n^{\#}, n \geq 0\}$ satisfying (2.1.1), (2.1.2) will be indistinguishable from the simulated chain $\{X_n\}$ in the sense that

$$\{X_n, n \geq 0\} \stackrel{d}{=} \{X_n^{\#}, n \geq 0\},$$

that is, the finite dimensional distributions of both processes are the same. Together, the ingredients $\{a_k\}$ and $\mathbf{P}$ in fact determine the distribution of the process as shown next.

Proposition 2.1.1. *Given a Markov chain satisfying (2.1.1)–(2.1.3), the finite dimensional distributions are of the form*

$$(2.1.5) \qquad P[X_0 = i_0, X_1 = i_1, \ldots, X_k = i_k] = a_{i_0} p_{i_0 i_1} p_{i_1 i_2} \cdots p_{i_{k-1} i_k}$$

for $i_0, \ldots, i_k$ integers in the state space and $k \geq 0$ arbitrary. Conversely given a density $\{a_k\}$ and a transition matrix $\mathbf{P}$ and a process $\{X_n\}$ whose finite dimensional distributions are given by (2.1.5), we have that $\{X_n\}$ is a Markov chain satisfying (2.1.1)–(2.1.3).

So the *Markov property*, i.e., (2.1.2)–(2.1.3), can be recognized by the form of the finite dimensional distributions given in (2.1.5).

Proof. Recall the *Chain Rule* of conditional probability. If $A_0, \ldots, A_k$ are events then

$$P\left(\bigcap_{i=1}^{k} A_i\right) = P(A_k | \bigcap_{i=0}^{k-1} A_i) P(A_{k-1} | \bigcap_{i=0}^{k-2} A_i) \ldots P(A_1 | A_0) P(A_0)$$

provided $P(\bigcap_{i=0}^{j} A_i) > 0, j = 0, 1, \ldots, k - 1$.
  Suppose (2.1.1)–(2.1.3) hold and set $A_j = [X_j = i_j]$ so that if

$$(2.1.6) \qquad P[X_0 = i_0, \ldots, X_j = i_j] > 0, j = 0, \ldots, k - 1$$

then

$$P[X_0 = i_0, \ldots, X_k = i_k]$$
$$= \prod_{j=1}^{k} P[X_j = i_j | X_0 = i_0, \ldots, X_{j-1} = i_{j-1}] P[X_0 = i_0]$$

and applying (2.1.3) to the right side we get

$$\prod_{j=1}^{k} P[X_j = i_j | X_{j-1} = i_{j-1}] a_{i_0} = a_{i_0} \prod_{j=1}^{k} p_{i_{j-1} i_j}.$$

What if (2.1.6) fails for some j? Let

$$j^* = \inf\{j \geq 0 : P[X_0 = i_0, \ldots, X_j = i_j] = 0\}.$$

If $j^* = 0$ then $a_{i_0} = 0$ and (2.1.5) holds trivially. If $j^* > 0$ then by what was already proved

$$P[X_0 = i_0, \ldots, X_{j^*-1}] = a_{i_0} p_{i_0 i_1} \cdots p_{i_{j^*-2} i_{j^*-1}} > 0.$$

Consequently,

$$p_{i_{j^*-1} i_{j^*}} = P[X_0 = i_0, \ldots, X_{j^*} = i_{j^*}] / P[X_0 = i_0, \ldots, X_{i^*-1} = i_{j^*-1}] = 0$$

so again (2.1.5) holds.

     Conversely, suppose for all k and choices of $i_0, \ldots, i_k$ that (2.1.5) holds. For

$$a_{i_0} p_{i_0 i_1} \cdots p_{i_{k-2} i_{k-1}} > 0$$

we have

$$P[X_k = i_k | X_0 = i_0, \ldots, X_{k-1} = i_{k-1}]$$
$$= P[X_0 = i_0, \ldots, X_k = i_k] / P[X_0 = i_0, \ldots, X_{k-1} = i_{k-1}]$$
$$= a_{i_0} p_{i_0 i_1} \cdots p_{i_{k-1} i_k} / a_{i_0} p_{i_0 i_1} \cdots p_{i_{k-2} i_{k-1}} = p_{i_{k-1} i_k}$$

showing that the Markov property holds. ∎

2.2. EXAMPLES.

Here are some examples of Markov chains. Some will be used to illustrate concepts discussed later and some show the range of applications of Markov chains.

Example 2.2.1. Independent Trials. Independence is a special case of Markov dependence. If $\{X_n\}$ are iid with

$$P[X_0 = k] = a_k, \quad k = 0, 1, \ldots, m,$$

then

$$P[X_{n+1} = i_{n+1} | X_0 = i_0, \ldots, X_n = i_n] = P[X_{n+1} = i_{n+1}] = a_{i_{n+1}}$$
$$= P[X_{n+1} = i_{n+1} | X_n = i_n]$$

and

$$\mathbf{P} = \begin{pmatrix} a_0 & a_1 & \cdots & a_m \\ \vdots & & \ddots & \\ a_0 & a_1 & \cdots & a_m \end{pmatrix}.$$

Example 2.2.2. The Simple Branching Process. Consider the simple branching process of Section 1.4. Recall $\{Z_{n,j}\}$ are iid with common distribution $\{p_k\}$ and $Z_0 = 1$ and

$$Z_n = Z_{n,1} + \ldots + Z_{n,Z_{n-1}}.$$

So

$$P[Z_n = i_n | Z_0 = i_0, \ldots, Z_{n-1} = i_{n-1}]$$

$$= P[\sum_{j=1}^{i_{n-1}} Z_{n,j} = i_n | Z_0 = i_0, \ldots, Z_{n-1} = i_{n-1}]$$

$$= P[\sum_{j=1}^{i_{n-1}} Z_{n,j} = i_n],$$

giving the Markov property since this depends only on i_{n-1} and i_n. Thus

$$P[Z_n = j | Z_{n-1} = i] = P[\sum_{k=1}^{i} Z_{n,k} = j] = p_j^{*i}$$

where $*i$ denotes i-fold convolution.

Example 2.2.3. Random Walks. Let $\{X_n, n \geq 1\}$ be iid with

$$P[X_n = k] = a_k, \quad -\infty < k < \infty.$$

Define the random walk by

$$S_0 = 0, \quad S_n = \sum_{i=1}^{n} X_i, \quad n \geq 1.$$

Then $\{S_n\}$ is a Markov chain since

$$P[S_{n+1} = i_{n+1} | S_0 = 0, S_1 = i_1, \ldots, S_n = i_n]$$
$$= P[X_{n+1} + i_n = i_{n+1} | S_0 = 0, \ldots, S_n = i_n]$$
$$= P[X_{n+1} = i_{n+1} - i_n] = a_{i_{n+1} - i_n}$$
$$= P[S_{n+1} = i_{n+1} | S_n = i_n]$$

since X_{n+1} is independent of $S_0, \ldots, S_n$.

A common special case is where only increments of ± 1 and 0 are allowed and where 0 and m are absorbing barriers. The transition matrix is then of the form

$$
P = \begin{pmatrix}
1 & 0 & 0 & 0 & \cdots & 0 & 0 \\
q_1 & r_1 & p_1 & 0 & \cdots & 0 & 0 \\
0 & q_2 & r_2 & p_2 & \cdots & 0 & 0 \\
\vdots & \ddots & & \ddots & & & \vdots \\
0 & \cdots & 0 & 0 & q_m & r_m & p_m \\
0 & 0 & 0 & 0 & \cdots & 0 & 1
\end{pmatrix}.
$$

The tri-diagonal structure is indicative of a random walk with steps $\pm 1, 0$. Note

$$P[S_n = 0 | S_{n-1} = 0] = P[S_n = m | S_{n-1} = m] = 1,$$

which models the hypothesized absorbing nature of states 0 and m and

$$
\begin{aligned}
P[X_{n+1} = i + 1 | X_n = i] &= p_i \\
P[X_{n+1} = i - 1 | X_n = i] &= q_i \\
P[X_{n+1} = i | X_n = i] &= r_i
\end{aligned}
$$

for $1 \le i \le m - 1$.

The case where $r_i = 0, p_i = p, q_i = q$ is the *Gamblers Ruin*: Harry starts with initial fortune i and his opponent Zeke has $m - i$. A coin tossing game is played and the state of the system is Harry's fortune. When Harry wins on a toss, his fortune increases by 1; when he loses a toss, his fortune decreases by 1. If the process enters state m, Zeke is ruined; if the process enters state 0, Harry is ruined.

Example 2.2.4. Success Runs. This is marvelous as a source of examples. The state space is $\{0, 1, 2, \dots\}$ and the transition matrix is of the form

$$
P = \begin{pmatrix}
q_0 & p_0 & 0 & 0 & 0 & \cdots \\
q_1 & 0 & p_1 & 0 & 0 & \cdots \\
q_2 & 0 & 0 & p_2 & 0 & \cdots \\
\vdots & & & & \ddots &
\end{pmatrix}.
$$

To see why the name *success run chain* is suitable, concentrate on the case where $p_i = p$ for all $i \ge 0$. During Harry's semi-pro basketball days, his free-throw shooting constituted independent Bernoulli trials with success probability p. Given a success run of n shots, Harry can extend this success run to length $n + 1$ if he makes the next foul shot (with probability p); but if he misses the next foul shot (with probability q) the length of the current success run becomes 0.

Example 2.2.5. The Deterministically Monotone Markov Chain.
This example in the context of discrete time seems hopelessly trivial but
supplies useful counterexamples and turns out to be not so trivial when we
discuss birth processes in continuous time in Chapter 5. The state space is
$\{1, 2, \ldots\}$. $a_1 = P[X_0 = 1] = 1$, and, for $i \geq 1$, we have $p_{i,i+1} = 1$, so that
the process marches deterministically through the integers towards $+\infty$.

A common method for generating Markov chains with state space S,
discussed in Exercise 2.6, is the following: Suppose $\{V_n, n \geq 0\}$ are iid
random elements in some space E. For instance, E could be R or R^d or
R^∞. Given two functions

$$g_i : S \times E \mapsto S, \quad i = 1, 2,$$

define

$$X_0 = g_1(j, V_0),$$

and for $n \geq 1$

$$X_n = g_2(X_{n-1}, V_n).$$

The branching process and random walks follow this paradigm, as do the
following examples.

Example 2.2.6. An Inventory Model. Let $I(t)$ be the inventory level
of an item at time t. Stock levels are checked at fixed times $T_0, T_1, T_2, \ldots$.
A commonly used restocking policy is that there be two critical values of
inventory s and S where $0 \leq s < S$. If at time T_n, the inventory level
$I(T_n) =: X_n$ is less than or equal to s, immediate procurement is done to
bring the stock level up to level S. If the stock level $X_n = I(T_n) \in (s, S]$,
then no replenishment is undertaken. Let D_n be the total demand during
the time interval $[T_{n-1}, T_n)$, $n = 1, 2, \ldots$, assume $\{D_n, n \geq 1\}$ is iid and
independent of X_0, and suppose $X_0 \leq S$. Then

$$(2.2.1) \qquad X_{n+1} = \begin{cases} (X_n - D_{n+1})_+, & \text{if } s < X_n \leq S, \\ (S - D_{n+1})_+, & \text{if } X_n \leq s, \end{cases}$$

where as usual

$$x_+ = \begin{cases} x, & \text{if } x > 0, \\ 0, & \text{if } x \leq 0. \end{cases}$$

This follows the paradigm $X_{n+1} = g(X_n, D_{n+1})$, $n \geq 0$, and hence $\{X_n\}$
is a Markov chain.

For this inventory model, descriptive quantities of interest include:

1. Long run average stock level

$$\lim_{N\to\infty} N^{-1}\sum_{j=0}^{N} X_j.$$

From the law of large numbers for Markov chains (Section 2.12.1) this will turn out to be calculated as

$$\lim_{n\to\infty} \sum_{j=1}^{S} jP[X_n = j].$$

In Sections 2.13 and 2.14 we discuss how to calculate $\lim_{n\to\infty} P[X_n = j]$.

2. Long run cumulative unsatisfied demand. For $n \geq 1$ let U_n be the unsatisfied demand in period $[T_{n-1}, T_n)$, $n \geq 1$ so that

$$U_n = \begin{cases} (D_n - X_{n-1}) \wedge 0, & \text{if } s < X_{n-1} \leq S, \\ (D_n - S) \wedge 0, & \text{if } X_{n-1} \leq s. \end{cases}$$

We are interested in $\sum_{j=1}^{N} U_j$ for large N.

3. Long run fraction of periods when demand is not satisfied. This can be represented as

$$\lim_{N\to\infty} N^{-1}\sum_{j=1}^{N} 1_{[U_j > 0]}.$$

Inventory models are further discussed in Exercises 2.58 and 2.59. A simple example of a calculation of the long run average stock level is given in Example 2.14.2 of Section 2.14.

Example 2.2.7. The Moran Storage Model. Consider a reservoir of capacity c. The levels $\{X_n\}$ of the reservoir are observed at time points $0, 1, \ldots$. During the interval $[n, n+1)$ there is random input A_{n+1} to the reservoir. This input may result in spillage. At the end of the interval $[n, n+1)$, m units of water (if available) are instantaneously removed. These m units include material not stored due to spillage and we assume $m < c$. If the reservoir contains less than m units of water, the total contents are removed. The inputs $\{A_n\}$ are assumed to be iid and independent of the initial level X_0. Thus the contents process $\{X_n\}$ satisfies the recursion

$$(2.2.2) \qquad X_{n+1} = (X_n + A_{n+1} - m)_+ \wedge c,$$

where $a \wedge b$ means the minimum of a and b. Thus the contents process satisfies a recursion of the type

$$X_{n+1} = g(X_n, V_{n+1})$$

and, therefore, is Markov. The state space is $\{0, 1, \ldots, c\}$. Note that the recursion merely says that the new contents level equals the old contents level plus input minus outflow with some adjustments for the boundary conditions at 0 and c.

If

$$P[A_1 = n] = a_n, \quad P[A_1 \le n] = a_{\le n}, \quad P[A_1 \ge n] = a_{\ge n},$$

then we can write the transition matrix of the contents process as

$$P = \begin{pmatrix} a_{\le m} & a_{m+1} & a_{m+2} & \cdots & a_c & \cdots & a_{c+m-1} & a_{\ge c+m} \\ a_{\le m-1} & a_m & a_{m+1} & \cdots & a_{c-1} & \cdots & a_{c+m-1} & a_{\ge c+m-1} \\ \vdots & & & & \ddots & & & \vdots \\ 0 & 0 & & & a_0 & \cdots & a_{m-1} & a_{\ge m} \end{pmatrix}.$$

Note that the column with entries $(a_c, a_{c-1}, \ldots, a_0)$ is column number $c - m + 1$.

Further material on storage models is discussed in Section 2.13 and in Exercises 2.60, 2.61, 2.62.

Example 2.2.8. Discrete Queueing Models. There are two types of models which we will categorize roughly as type M/G/1 type and type G/M/1. This reason for this terminology will not be clear until the next chapter.

Customers arrive at a facility and wait for service on a first-come first-served basis. Assume there is one server. Let $X(t)$ be the number of customers in the system at time t, that is, the number waiting or in service.

For the queueing model of type M/G/1, we assume that service completions occur at times $T_0, T_1, \ldots$ so these times are when departures from the system occur. Set $X_n = X(T_n+)$ where the "+" reminds us that we measure the number in the system just subsequent to a departure. Let A_{n+1} be the number of arrivals during the service period of the customer who departs at time T_{n+1}. Thus $\{X_n\}$ satisfies the recursion

$$X_{n+1} = (X_n - 1)_+ + A_{n+1},$$

since the number in the system at T_{n+1} is the number at T_n plus arrivals minus the customer who departed when his service was completed. If the assumptions on input to the system and service times make $\{A_n\}$ independent and identically distributed and independent of X_0, then $\{X_n\}$ is a Markov chain. If

$$P[A_1 = k] = a_k, \quad k \ge 0,$$

then one readily checks that the transition matrix P of this chain is

$$P = \begin{pmatrix} a_0 & a_1 & a_2 & & \cdots \\ a_0 & a_1 & a_2 & & \cdots \\ 0 & a_0 & a_1 & & \cdots \\ 0 & 0 & a_0 & a_1 & \cdots \\ \vdots & & \ddots & & \end{pmatrix}.$$

Stationary distributions and recurrence criteria for this model are studied in Section 2.15. See also Exercise 2.6.

Now we consider the queueing system of type G/M/1. As before, let $X(t)$ be the number in the system at time t and suppose customers arrive at epochs $\tau_0, \tau_1, \ldots$. Let S_{n+1} be the number of potential service completions in the interval $[\tau_n, \tau_{n+1})$ and let $X_n = X(\tau_n -)$ be the number in the system just prior to the nth arrival. Then

$$X_{n+1} = (X_n - S_{n+1} + 1)_+,$$

since the number in the system prior to the $(n+1)$st arrival is 1 plus the number in the system prior to the nth arrival minus the number who have completed service and left. Assume assumptions on input and service have been made to assure that the variables $\{S_n\}$ are iid and independent of X_0 and assume

$$P[S_1 = j] = a_j.$$

Then one readily calculates the transition matrix P as

$$P = \begin{pmatrix} \sum_{i=1}^{\infty} a_i & a_0 & 0 & & \cdots \\ \sum_{i=2}^{\infty} a_i & a_1 & a_0 & 0 & \cdots \\ \sum_{i=3}^{\infty} a_i & a_2 & a_1 & a_0 & \cdots \\ \vdots & & & \ddots & \end{pmatrix}.$$

2.3. Higher Order Transition Probabilities.

The tractability of Markov chain models is based on the fact that probabilities of interest may be computed by matrix manipulations.

Let $\mathbf{P} = (p_{ij})$ be the transition matrix of a Markov chain $\{X_n, n \geq 0\}$ and suppose the initial probabilities are $P[X_0 = j] = a_j$. Matrix powers of $\mathbf{P}$ are defined as usual by

$$\mathbf{P}^2 = \mathbf{P} \cdot \mathbf{P}$$

with the (i, j)th-entry $p_{ij}^{(2)}$ given by

$$p_{ij}^{(2)} = \sum_k p_{ik} p_{kj}.$$

(Note that when the state space S is infinite, the series above converges since $\sum_k p_{ik} p_{kj} \le \sum_k p_{ik} = 1$.) Similarly $\mathbf{P}^3 = \mathbf{P}^2 \cdot \mathbf{P} = \mathbf{P} \cdot \mathbf{P}^2$ has the (i, j)th-entry

$$p_{ij}^{(3)} = \sum_k p_{ik}^{(2)} p_{kj} = \sum_k p_{ik} p_{kj}^{(2)}$$

and in general $\mathbf{P}^{n+1} = \mathbf{P}^n \cdot \mathbf{P} = \mathbf{P} \cdot \mathbf{P}^n$ has the (i, j)th-entry

$$p_{ij}^{(n+1)} = \sum_k p_{ik}^{(n)} p_{kj} = \sum_k p_{ik} p_{kj}^{(n)}.$$

Finally, define $\mathbf{P}^0$ to be the identity matrix.

Define the n-step transition probabilities for the Markov chain by ($n \ge 1$, i, k in the state space)

$$P[X_n = k | X_0 = i],$$

and it follows immediately from the second equality of (2.1.4) that for any $m \ge 0$

(2.3.1) $$P[X_n = k | X_0 = i] = P[X_{n+m} = k | X_m = i].$$

So the probability that a path started at i ends at j after n steps does not depend on the time at which the path is initiated.

Proposition 2.3.1. *We have for all $n \ge 0$, and i, j in the state space*

(2.3.2) $$p_{ij}^{(n)} = P[X_n = j | X_0 = i].$$

We compute the transition probabilities by taking matrix powers.

Proof. The formula is obviously true for $n = 0, 1$ and as a warm-up, let us check it for $n = 2$. We have

$$P[X_2 = j | X_0 = i] = \sum_k P[X_2 = j, X_1 = k | X_0 = i].$$

Assuming $P[X_0 = i] = a_i > 0$, this is

$$\sum_k P[X_2 = j, X_1 = k, X_0 = i]/a_i$$

$$= \sum_k a_i p_{ik} p_{kj}/a_i = \sum_k p_{ik} p_{kj} = p_{ij}^{(2)}.$$

Now suppose (2.3.2) is true for $n = 0, 1, \ldots, N$ and verify it for $n = N + 1$:

$$P[X_{N+1} = j|X_0 = i]$$

$$= \sum_k P[X_{N+1} = j, X_1 = k, X_0 = i]/a_i$$

$$= \sum_k P[X_{N+1} = j|X_1 = k, X_0 = i]P[X_1 = k, X_0 = i]/a_i.$$

From (2.1.4) this is

$$\sum_k P[X_N = j|X_0 = k]P[X_1 = k|X_0 = i],$$

which, by the induction hypothesis, is

$$\sum_k p_{ik} p_{kj}^{(N)} = p_{ij}^{(N+1)},$$

as required. ∎

The obvious matrix identity

$$\mathbf{P}^{n+m} = \mathbf{P}^n \cdot \mathbf{P}^m$$

for $n, m \geq 0$, when written in component form

$$p_{ij}^{(n+m)} = \sum_k p_{ik}^{(n)} p_{kj}^{(m)}$$

is sometimes called the *Chapman-Kolmogorov equation*. It expresses the fact that a transition from i to j in $n+m$ steps can be achieved by moving from i to an intermediate state k in n steps (with probability $p_{ik}^{(n)}$), and then given that the process is in state k (the Markov property allows indifference to the arrival route), a transition from k to j in m-steps must be achieved (with probability $p_{kj}^{(m)}$). For the computation of the probability of the $n + m$ step transition, the law of total probability requires summing over the intermediate states k.

Corollary 2.3.2. *The unconditional probabilities $P[X_n = j]$ are computed from*

$$(2.3.4) \qquad a_j^{(n)} := P[X_n = j] = \sum_i a_i p_{ij}^{(n)}.$$

Proof. We have

$$P[X_n = j] = \sum_i P[X_n = j | X_0 = i] P[X_0 = i] = \sum_i a_i p_{ij}^{(n)}. \quad \blacksquare$$

We now pause and collect some **conventions**.

1. For the purposes of matrix manipulations, vectors are column vectors or a matrix with a single column. Thus we write

$$\mathbf{a}^{(n)} = (a_0^{(n)}, a_1^{(n)}, \ldots)', \; \mathbf{a} = (a_0, a_1, \ldots)'.$$

and (2.3.4) can be rewritten in equivalent matrix form

$$(2.3.4') \qquad\qquad (\mathbf{a}^{(n)})' = \mathbf{a}' \mathbf{P}^n.$$

Since $(\mathbf{a}^{(n-1)})' = \mathbf{a}' \mathbf{P}^{n-1}$ an alternate form of (2.3.4') is the recursion

$$(\mathbf{a}^{(n)})' = (\mathbf{a}^{(n-1)})' \mathbf{P}.$$

2. Write

$$P_i(\cdot) = P\{\cdot | X_0 = i\}$$

for the probability measure conditioned by $X_0 = i$. Alternatively we may imagine

$$a_i = 1, \quad a_k = 0, k \neq i,$$

or, equivalently, $a_j = \delta_{ij}$.

Calculation of P^n must either be done by a computer or by using eigenvalue expansions. In the 2×2 case elementary means suffice as shown next.

Proposition 2.3.3. *Suppose $0 < a < 1$, $0 < b < 1$ and let*

$$\mathbf{P} = \begin{pmatrix} 1 - a & a \\ b & 1 - b \end{pmatrix}$$

be a transition matrix corresponding to a Markov chain with state space $\{0, 1\}$. Then

$$\mathbf{P}^n = (a + b)^{-1} \left\{ \begin{pmatrix} b & a \\ b & a \end{pmatrix} + (1 - a - b)^n \begin{pmatrix} a & -a \\ -b & b \end{pmatrix} \right\}.$$

Proof. Let the initial probability vectors be (a_0, a_1) and note that if

$$(a_0, a_1) = (1, 0) \text{ then } P[X_{n+1} = 0] = p_{00}^{(n+1)}$$

while if

$$(a_0, a_1) = (0, 1) \text{ then } P[X_{n+1} = 0] = p_{10}^{(n+1)}.$$

Proceed recursively. We have

$$
\begin{aligned}
P[X_{n+1} = 0] &= P[X_{n+1} = 0, X_n = 0] + P[X_{n+1} = 0, X_n = 1] \\
&= P[X_{n+1} = 0 | X_n = 0] P[X_n = 0] \\
&\quad + P[X_{n+1} = 0 | X_n = 1] P[X_n = 1] \\
&= p_{00} P[X_n = 0] + p_{10} P[X_n = 1] \\
&= (1 - a) P[X_n = 0] + b P[X_n = 1] \\
&= (1 - a) P[X_n = 0] + b(1 - P[X_n = 0]),
\end{aligned}
$$

which is where the assumption of only two states is used. We then get

$$P[X_{n+1} = 0] = (1 - a - b) P[X_n = 0] + b.$$

Therefore

$$
\begin{aligned}
P[X_{n+1} = 0] &= b + (1 - a - b)\{b + (1 - a - b) P[X_{n-1} = 0]\} \\
&= b + b(1 - a - b) + (1 - a - b)^2 P[X_{n-1} = 0] \\
&= b + b(1 - a - b) + (1 - a - b)^2 \{b + (1 - a - b) P[X_{n-2} = 0]\}
\end{aligned}
$$

$$\vdots$$

$$= b \sum_{j=0}^{n} (1 - a - b)^j + (1 - a - b)^{n+1}$$

$$P[X_0 = 0]$$

from which, replacing $n + 1$ by n, we get

$$
\begin{aligned}
P[X_n = 0] &= (1 - a - b)^n a_0 + b \left(\frac{1 - (1 - a - b)^n}{1 - (1 - a - b)} \right) \\
&= (1 - a - b)^n a_0 + b \left(\frac{1 - (1 - a - b)^n}{a + b} \right),
\end{aligned}
$$

giving

(2.3.6)

$$P[X_n = 0] = \frac{b}{a+b} + (1-a-b)^n(a_0 - \frac{b}{a+b}).$$

Since $P[X_n = 0] + P[X_n = 1] = 1$

(2.3.7) $$P[X_n = 1] = \frac{a}{a+b} - (1-a-b)^n(a_0 - \frac{b}{a+b}).$$

Setting $(a_0, a_1) = (1, 0)$ yields

$$p_{00}^{(n)} = \frac{b}{a+b} + (1-a-b)^n(1 - \frac{b}{a+b}) = \frac{b}{a+b} + (1-a-b)^n\frac{a}{a+b},$$

and setting $(a_0, a_1) = (0, 1)$ yields

$$p_{10}^{(n)} = \frac{b}{a+b}(1 - (1-a-b)^n).$$

The remaining values in the second column of $\mathbf{P}^n$ follow from (2.3.7). ∎

Note, if $0 < a < 1$, $0 < b < 1$ then $0 < 1-a < 1$ and $|1-a-b| < 1$, so, for the 2×2 case,

$$\mathbf{P}^n \to (a+b)^{-1} \begin{pmatrix} b & a \\ b & a \end{pmatrix}.$$

A limiting matrix exists and has constant columns. This is noteworthy and we will later explore the relation between existence of a limit matrix and stationary distributions for more general chains. Also note that the above calculations show that convergence to a limit is geometrically fast with

$$\sup_{i,j} |p_{ij}^{(n)} - \pi_j| \leq (\text{ const })(1-a-b)^n,$$

where $\pi_j = \lim_{n \to \infty} p_{ij}^{(n)}$.

2.4. DECOMPOSITION OF THE STATE SPACE.

Let $\{X_n, n \geq 0\}$ be a Markov chain with discrete state space S. To understand the evolution of the system it is critical to understand which paths through the state space are possible and to understand the allowable movements of the process. For $B \subset S$ let

(2.4.1) $$\tau_B = \inf\{n \geq 0 : X_n \in B\}$$

be the hitting time of B. Abuse the notation a bit and set $\tau_j = \tau_{\{j\}}$.

To understand which states can be reached from a starting state i, the following is basic.

Definition. *For* $i, j \in S$ *we say* j *is accessible from* i, *written* $i \to j$, *if*

$$P_i[\tau_j < \infty] > 0.$$

In other words, starting from i, *with positive probability the chain hits state* j. *Synonyms:* j *is a consequent of* i, i *leads to* j, j *can be reached from* i.

Because $n = 0$ is allowed in (2.4.1), we get $i \to i$ for all $i \in S$ since $P_i[\tau_i < \infty] = 1$; in fact, $P_i[\tau_i = 0] = P_i[X_0 = i] = 1$.

Here is the most useful criterion for accessiblility. We have $i \to j$ iff

(2.4.2) $\exists\, n \geq 0:\; p_{ij}^{(n)} > 0.$

The sufficiency of (2.4.2) is easy. Note

$$[X_n = j] \subset [\tau_j \leq n] \subset [\tau_j < \infty],$$

so that

$$0 < p_{ij}^{(n)} \leq P_i[\tau_j < \infty].$$

Conversely, if for all $n \geq 0$, $p_{ij}^{(n)} = 0$, then

$$P_i[\tau_j < \infty] = \lim_{n \to \infty} P_i[\tau_j \leq n] = \lim_{n \to \infty} P_i\{\cup_{k=0}^{n}[X_k = j]\}$$

$$\leq \limsup_{n \to \infty} \sum_{k=0}^{n} P_i[X_k = j]$$

$$= \limsup_{n \to \infty} \sum_{k=0}^{n} p_{ij}^{(k)} = 0.$$

Here are some simple examples which illustrate the notion of accessibility:

(1) The deterministically monotone Markov chain: Since for $i \geq 0$, $p_{i,i+1} = 1$, we have $i \to i + 1$, and, in fact, for any $j > i$, we get $i \to j$.

(2) Gamblers ruin on $\{0, 1, \dots, m\}$. We have $m \to m$, $0 \to 0$. No other consequents of $0, m$ exist. 0 is a consequent of every state except m.

(3) Simple branching: $0 \to 0$, and 0 has no other consequents.

The notion of accessibility tells us which states can ultimately be reached from a given state i. The following definition addresses the question: If a path of positive probability exists from one state to a second, is there a return path from the second state to the first?

Definition. *States i and j communicate, written $i \leftrightarrow j$, if $i \to j$ and $j \to i$.*

Communication is an equivalence relation which means

(1) $i \leftrightarrow i$ (the relation is reflexive) since $i \to i$,

(2) $i \leftrightarrow j$ iff $j \leftrightarrow i$ (the relation is symmetric)

(3) If $i \leftrightarrow j$ and $j \leftrightarrow k$ then $i \leftrightarrow k$ (the relation is transitive).

Only the last property needs comment: If $i \leftrightarrow j$ and $j \to k$ we show $i \to k$. If $i \to j$ there is n such that $p_{ij}^{(n)} > 0$; similarly $p_{jk}^{(m)} > 0$ for some m since $k \to j$. So by Chapman-Kolmogorov:

$$p_{ik}^{(m+n)} = \sum_{\nu} p_{i\nu}^{(m)} p_{\nu k}^{(m)} \geq p_{ij}^{(n)} p_{jk}^{(m)} > 0$$

so that $i \to k$.

The state space S may now be decomposed into disjoint exhaustive equivalence classes modulo the relation "$\leftrightarrow$". We pick a state, say 0, and we put 0 and all states communicating with 0 in a class, say C_0. Then we pick a state in $S \backslash C_0$, call it i, and put it and all states communicating with i into another class which we name C_1. Continue on in this manner until all states have been assigned. We have

$$C_i \bigcap C_j = \emptyset, i \neq j, \text{ and } \bigcup_i C_i = S.$$

The sets $C_0, C_1, \ldots$ are called (equivalence) *classes*.

Here are some **examples:**

(1) The Deterministically Monotone Markov chain: $C_i = \{i\}$, $i \geq 0$.

(2) Gambler's ruin on $\{0, 1, 2, 3\}$ with matrix

$$\mathbf{P} = \begin{pmatrix} 1 & 0 & 0 & 0 \\ 1/2 & 0 & 1/2 & 0 \\ 0 & 1/2 & 0 & 1/2 \\ 0 & 0 & 0 & 1 \end{pmatrix}.$$

There are three classes: $\{0\}, \{3\}, \{1, 2\}$.

(3) Consider the Markov chain with matrix

$$P = \begin{pmatrix} 1/2 & 1/2 & 0 & 0 \\ 1/2 & 1/2 & 0 & 0 \\ 0 & 0 & 1/2 & 1/2 \\ 0 & 0 & 1/2 & 1/2 \end{pmatrix}$$

on states $\{1, 2, 3, 4\}$. There are two classes

$$C_1 = \{1, 2\}, \quad C_2 = \{3, 4\}.$$

A Markov chain is *irreducible* if the state space consists of only one class; i.e., for any $i, j \in S$ we have $i \leftrightarrow j$.

None of the Markov chains in the three previous examples is irreducible. The success run chain is irreducible if $0 < p_i < 1$. There is a finite path of positive probability linking any two states. For example, $3 \to 2$ since

$$p_{32}^{(3)} = P_3[X_1 = 0, X_2 = 1, X_3 = 2] = q_3 p_0 p_1 > 0.$$

Now we discuss which sets of states lead to other sets of states. The key concept is *closure*. A set of states $C \subset S$ is *closed* if for any $i \in C$ we have

$$P_i[\tau_{C^c} = \infty] = 1.$$

So if the chain starts in C, it never escapes outside C. If $\{j\}$ is closed, we call state j *absorbing*.

Here are two criteria which are useful:

(1) C is closed iff

(2.4.3)   for all $i \in C, j \in C^c$: $p_{ij} = 0.$

(2) j is absorbing iff

(2.4.4) $p_{jj} = 1.$

Note that (2.4.4) is a special case of (2.4.3) with $C = \{j\}$. Observe that if (2.4.3) holds, then for $i \in C$, we have

$$P_i[\tau_{C^c} = 1] = \sum_{j \in C^c} p_{ij} = 0.$$

Similarly

$$
\begin{aligned}
P_i[\tau_{C^c} \leq 2] &= P_i[\tau_{C^c} = 1] + P_i[\tau_{C^c} = 2] \\
&= 0 + P_i[X_1 \in C, X_2 \in C^c] \\
&= \sum_{j \in C^c} \sum_{k \in C} p_{ik} p_{kj} = 0.
\end{aligned}
$$

Continuing on by induction, we get $P_i[\tau_{C^c} \leq n] = 0$, and letting $n \to \infty$ gives $P_i[\tau_{C^c} < \infty] = 0$, showing C is closed.

Note that it is possible to enter a closed set, but it is impossible to leave. In the deterministically monotone Markov chain, $\{n, n+1, \dots\}$ is closed but $n - 1 \to n$. Similarly in the gamblers ruin on $\{0, 1, 2, 3\}$ we have 0 absorbing but $1 \to 0$. Consider the example on $\{0, 1, 2, 3\}$:

$$P = \begin{pmatrix} 1/2 & 1/2 & 0 & 0 \\ 1/2 & 1/2 & 0 & 0 \\ 0 & 0 & 1/2 & 1/2 \\ 0 & 0 & 1/2 & 1/2 \end{pmatrix}.$$

Here $C_1 = \{0, 1\}$ is closed as is $C_2 = \{2, 3\}$. It is impossible to exit C_1 or C_2, and, in this case, it is also impossible to enter C_1 from C_2 or to enter C_2 from C_1. So if $\{X_n\}$ starts in C_1, it stays there forever. The same holds for C_2. The two pieces of the state space ignore each other.

Note if C is closed then $(p_{ij}, i \in C, j \in C)$ is a stochastic matrix: We have $p_{ij} > 0$, and, for $i \in C$,

$$\sum_{j \in C} p_{ij} = 1,$$

since

$$\sum_{j \in C^c} p_{ij} = 0.$$

We close this section with two remarks:

1. There may be an infinite number of closed sets in the state space and closed sets need not be disjoint. In the deterministically monotone Markov chain, the set $\{n, n + 1, \ldots\}$ is closed for every n.

2. A class of states need not be closed. As remarked for the gamblers ruin on $\{0, 1, 2, 3\}$ with $\{0\}, \{3\}$ absorbing, we have $1 \to 0$ but $p_{00} = 1$. Hence $\{1, 2\}$ is a class but it is not closed.

2.5. THE DISSECTION PRINCIPLE.

In preparation for a study of recurrence we have to think about how to decompose the process into independent, identically distributed blocks. The blocks consist of the pieces of path between consecutive visits to a fixed state, say i. It is frequently useful to think of the process as these iid pieces knitted together.

Let $\{X_n, n \geq 0\}$ be a Markov chain, and suppose $P[X_0 = i] = 1$, so that we assume the process starts from state i. Define $\tau_i(0) = 0$ and

$$\tau_i(1) = \inf\{m \geq 1 : X_m = i\}.$$

On $[\tau_i(1) < \infty]$ define

$$\tau_i(2) = \inf\{m > \tau_i(1) : X_m = i\}.$$

Continuing in this way, on $[\tau_i(1) < \infty, \ldots, \tau_i(n) < \infty]$ define

$$\tau_i(n+1) = \inf\{m > \tau_i(n) : X_m = i\}.$$

The times $\{\tau_i(n)\}$ are the times the Markov chain hits state i. We will prove that successive excursions between visits to state i are iid. Define $\alpha_0 = 0, \alpha_1 = \tau_i(1)$, and on $[\tau_i(1) < \infty]$ define $\alpha_2 = \tau_i(2) - \tau_i(1)$ and so on. The mth excursion between visits to state i is the block

$$(X_{\tau_i(m-1)+1}, \ldots, X_{\tau_i(m)}),$$

assuming $\tau_i(m-1) < \infty$. Define

$$\xi_1 = (\alpha_1, X_1, \ldots, X_{\tau_i(1)}),$$

and on $[\alpha_1 = \tau_i(1) < \infty]$ define

$$\xi_2 = (\alpha_2, X_{\tau_i(1)+1}, \ldots, X_{\tau_i(2)}).$$

Continuing, define

$$\xi_{n+1} = (\alpha_{n+1}, X_{\tau_i(n)+1}, \ldots, X_{\tau_i(n+1)})$$

on $[\tau_i(1) < \infty, \ldots, \tau_i(n) < \infty]$.

In order to give precision to the statement "$\{\xi_n\}$ is independent, identically distributed" we must see in which space the ξ's are random elements. Define $N = \{0, 1, \ldots\}$ and

$$E = \left(\cup_{k=1}^{\infty}(\{k\} \times N^k)\right) \cup (\{\infty\} \times N^{\infty})$$

so E is all finite vectors of integers with the length of the vector appended as well as infinite sequences of integers with ∞ appended. The ξ_i's live in E.

We now prove that the Markov path can be broken into iid pieces.

Proposition 2.5.1. *With respect to*

$$P_i^{\#} := P_i(\cdot | \tau_i(1) < \infty, \ldots, \tau_i(k) < \infty),$$

we have that $\xi_1, \ldots, \xi_k$ are iid random elements of E.

Remark. In the important case that the Markov chain is irreducible and recurrent (cf. Section 2.6), we will see that $P_i = P_i^{\#}$.

Proof. For typographical ease, suppose $k = 2$. Consider

$$P_i[\xi_1 = (k, i_1, \ldots, i_k), \xi_2 = (\ell, j_1, \ldots, j_\ell), \tau_i(1) < \infty, \tau_i(2) < \infty]$$
$$= P_i[\alpha_1 = k, X_1 = i_1, \ldots, X_k = i_k, \alpha_2 = \ell, X_{k+1} = j_1, \ldots, X_{k+\ell} = j_\ell,$$
$$(2.5.1) \quad \alpha_1 < \infty, \alpha_2 < \infty].$$

This probability is 0 unless

$$i_k = i, j_\ell = i$$

and unless

$$i_1 \neq i, \ldots, i_{k-1} \neq i; j_1 \neq i, \ldots, j_{\ell-1} \neq i,$$

and these are the only cases of interest. Therefore, the probability is

$$P_i[X_1 = i_1, \ldots, X_{k-1} = i_{k-1}, X_k = i,$$
$$X_{k+1} = j_1, \ldots, X_{k+\ell-1} = j_{\ell-1}, X_{k+\ell} = i]$$
$$= P_i[X_1 = i_1, \ldots, X_k = i]$$
$$\cdot P_i[X_{k+1} = j_1, \ldots, X_{k+\ell} = i | X_1 = i_1, \ldots, X_k = i]$$
$$= P_i[X_1 = i_1, \ldots, X_k = i] P_i[X_1 = j_1, \ldots, X_\ell = i],$$

the last step following by the Markov property (2.1.4). The last product can be written

$$P_i[X_1 = i_1, \ldots, X_{k-1} = i_{k-1}, X_k = i, \tau_i(1) = k]$$
$$(2.5.2)$$
$$\times P_i[X_1 = j_1, \ldots, X_{\ell-1} = j_{\ell-1}, X_\ell = i, \tau_i(1) = \ell].$$

Sum over $i_1, \ldots, i_k$ and $j_1, \ldots, j_\ell$ in (2.5.1) and (2.5.2) to get

$$(2.5.3) \qquad P_i[\alpha_1 = k, \alpha_2 = \ell] = P_i[\alpha_1 = k] P_i[\alpha_1 = \ell],$$

so we conclude

$$P_i[\alpha_1 < \infty, \alpha_2 < \infty] = P_i[\tau_i(1) < \infty, \tau_i(2) < \infty] = (P_i[\alpha_1 < \infty])^2.$$

Divide this through in (2.5.1) and (2.5.2) to get

$$P_i^{\#}[\xi_1 = (k, i_1, \ldots, i_k), \xi_2 = (\ell, j_1, \ldots, j_\ell)]$$
$$= P_i[\xi_1 = (k, i_1, \ldots, i_k) | \alpha_1 < \infty] P_i[\xi_1 = (\ell, j_1, \ldots, j_\ell) | \alpha_1 < \infty].$$

Since the joint distribution of (ξ_1, ξ_2) factors, this becomes

$$P_i^\#[\xi_1 = (k, i_1, \ldots, i_k)] P_i^\#[\xi_1 = (\ell, j_1, \ldots, j_\ell)],$$

and this identity suffices to give the desired result. ∎

Optional Exercise. Try proving this using the construction in Section 2.1 and Proposition 1.8.2.

There is a fairly obvious extension which is needed when the process does not start in state i: If the Markov chain starts in state $j \neq i$, or, equivalently, if P_i is replaced by P_j, then, with respect to

$$P_j^\# = P_j[\cdot | \tau_i(1) < \infty, \ldots, \tau_i(k) < \infty],$$

we have $\xi_2, \ldots, \xi_k$ independent, identically distributed random elements of E and independent of ξ_1, but it is no longer true that $\xi_1 \overset{d}{=} \xi_2$.

Here is a useful corollary.

Corollary 2.5.2. *With respect to $P_i^\#$,*

$$(\alpha_1, \ldots, \alpha_k) = (\tau_i(1), \tau_i(2) - \tau_i(1), \ldots, \tau_i(k) - \tau_i(k-1))$$

are independent, identically distributed $\{1, 2, \ldots\}$-valued random variables. In particular,

$$(2.5.3) \qquad P_i^\#[\alpha_1 = \ell_1, \ldots, \alpha_k = \ell_k] = \prod_{i=1}^{k} P_i^\#[\alpha_i = \ell_i],$$

and

$$
\begin{aligned}
P_i[\tau_i(1) = k, &(X_{\tau_i(1)+\ell}, 1 \leq \ell \leq p) = (n_1, \ldots, n_p)] \\
&= P_i[\tau_i(1) = k] P_i[X_\ell = n_\ell, \ell = 1, \ldots, p]
\end{aligned}
$$

$(2.5.4)$

for any $k, p, n_1, \ldots, n_p$.

The method of proof of (2.5.4) is the same and is thus omitted.

2.6. TRANSIENCE AND RECURRENCE.

We now discuss several classifications of states which lead to useful decompositions of the state space. As we shall see, the most fundamental classification of a state depends on how often the chain returns to that state.

A state i is called *recurrent* if the chain returns to i with probability 1 in a finite number of steps. Otherwise the state is *transient*. If we recall the hitting time variables $\{\tau_i(n), n \geq 1\}$ introduced previously, we can define these concepts precisely as follows: State i is recurrent if

$$P_i[\tau_i(1) < \infty] = 1.$$

In the contrary case,

$$P_i[\tau_i(1) = \infty] > 0$$

and state i is transient; in this case there is positive probability of never returning to state i. State i is *positive recurrent* if

$$E_i(\tau_i(1)) < \infty.$$

So for a positive recurrent state, not only is the return time finite almost surely, but the expected return time is finite.

For $n \geq 1$ define

$$f_{jk}^{(n)} = P_j[\tau_k(1) = n]$$

to be the distribution of the hitting time of k starting from j. Note that since $\tau_k(1) \geq 1$, we have $f_{jk}^{(0)} = 0$. We have that

$$f_{jk} := \sum_{n=0}^{\infty} f_{jk}^{(n)} = P_j[\tau_k(1) < \infty]$$

is the probability of hitting k in finite time starting from j. In particular, state i is recurrent iff $f_{ii} = 1$, and a recurrent state i is positive recurrent iff

$$m_i := E_i(\tau_i(1)) = \sum_{n=0}^{\infty} n f_{ii}^{(n)} < \infty.$$

It is important to have the best possible criteria for transience and recurrence and to interpret the meaning of these concepts as fully as possible. As a start to a better understanding of these concepts, define the following generating functions for $0 < s < 1$:

$$F_{ij}(s) = \sum_{n=0}^{\infty} f_{ij}^{(n)} s^n$$

$$P_{ij}(s) = \sum_{n=0}^{\infty} p_{ij}^{(n)} s^n,$$

and notice that the last generating function is not the generating function of a probability mass function.

Proposition 2.6.1. *(a) We have for $i \in S$*

$$p_{ii}^{(n)} = \sum_{k=0}^{n} f_{ii}^{(k)} p_{ii}^{(n-k)}, \quad n \geq 1,$$

and for $0 < s < 1$

$$P_{ii}(s) = \frac{1}{1 - F_{ii}(s)}.$$

(b) We have for $i \neq j$

$$p_{ij}^{(n)} = \sum_{k=0}^{n} f_{ij}^{(k)} p_{jj}^{(n-k)}, \quad n \geq 0,$$

and for $0 < s < 1$

$$P_{ij}(s) = F_{ij}(s) P_{jj}(s).$$

In principle, this determines the first passage probabilities $F_{ij}(s)$ from P. However, the relation between the generating functions does not always provide a practical scheme for obtaining the first passage probabilities. A technique which helps to compute the f's will be given at the end of this section.

Proof. (a) Since $[X_n = i] \subset [\tau_i(1) \leq n]$ we have that

$$p_{ii}^{(n)} = P_i[X_n = i] = \sum_{k=1}^{n} P_i[X_n = i, \tau_i(1) = k]$$

$$= \sum_{k=1}^{n} P_i[\tau_i(1) = k, X_{\tau_i(1)+n-k} = i].$$

From Proposition 2.5.1 we can split this probability at $\tau_i(1)$ to get

$$\sum_{k=1}^{n} P_i[\tau_i(1) = k] P_i[X_{n-k} = i] = \sum_{k=1}^{n} f_{ii}^{(k)} p_{ii}^{(n-k)},$$

as desired. To obtain the generating function statement, multiply through by s^n and sum from 1 to ∞:

$$P_{ii}(s) - 1 = \sum_{n=1}^{\infty} p_{ii}^{(n)} s^n.$$

Since $f_{ii}^{(0)} = 0$ this is the same as

$$= \sum_{n=0}^{\infty} \sum_{k=0}^{n} f_{ii}^{(k)} p_{ii}^{(n-k)} s^n$$

$$= \sum_{k=0}^{\infty} (\sum_{n=k}^{\infty} p_{ii}^{(n-k)} s^{n-k}) f_{ii}^{(k)} s^k$$

$$= F_{ii}(s) P_{ii}(s).$$

from which comes the result.

The proof of (b) is similar. ∎

An easy corollary of these relations is next.

Proposition 2.6.2. *We have*

$$i \text{ is recurrent iff } f_{ii} = 1 \text{ iff } \sum_{n=0}^{\infty} p_{ii}^{(n)} = \infty$$

so that

$$i \text{ is transient iff } f_{ii} < 1 \text{ iff } \sum_{n=0}^{\infty} p_{ii}^{(n)} < \infty.$$

Proof. We have i recurrent iff $F_{ii}(1) = 1$ iff

$$\lim_{s \to 1} P_{ii}(s) = \lim_{s \to 1} \frac{1}{1 - F_{ii}(s)} = \infty,$$

and since $P_{ii}(1) = \sum_{n=0}^{\infty} p_{ii}^{(n)}$ we are done. ∎

This result has the following illuminating interpretation. Define for $j \in S$

$$N_j = \sum_{n=1}^{\infty} 1_{[X_n = j]}$$

to be the number of visits by the process to state j after time 0 so that for any i, j we have

$$E_i N_j = \sum_{n=1}^{\infty} E_i 1_{[X_n = j]} = \sum_{n=1}^{\infty} P_i[X_n = j] = \sum_{n=1}^{\infty} p_{ij}^{(n)}.$$

Letting $j = i$, the previous result says that when i is the initial state, i is recurrent iff the expected number of visits by the chain to state i is infinite.

More can be said about the connection between recurrence/transience and the number of visits to a state.

Proposition 2.6.3. *We have for any* $i, j \in S$ *and non-negative integer* k

$$P[N_j = k | X_0 = i] = \begin{cases} 1 - f_{ij}, & \text{if } k = 0 \\ f_{ij} f_{jj}^{k-1}(1 - f_{jj}), & \text{if } k \geq 1. \end{cases}$$

Thus, *if* j *is transient, then for all states* i

$$P_i[N_j < \infty] = 1,$$

and

$$E_i(N_j) = f_{ij}/(1 - f_{jj}) = \sum_{n=1}^{\infty} p_{ij}^{(n)} < \infty,$$

and N_j *is geometrically distributed with respect to* P_j:

$$P_j[N_j = k] = (1 - f_{jj})(f_{jj})^k, \quad k \geq 0.$$

If j *is recurrent, then*

$$P_j[N_j = \infty] = 1,$$

and, for any i,

$$P_i[N_j = \infty] = f_{ij}.$$

Proof. Observe that for states i, j

$$P_i[N_j \geq 1] = P_i[\tau_j(1) < \infty] = f_{ij}.$$

From Proposition 2.5.1 (cf. 2.5.3), for any $k \geq 1$,

$$\begin{aligned} P_i[N_j \geq k] = P_i[\tau_j(k) < \infty] &= P_i[\tau_j(1) < \infty, \ldots, \tau_j(k) < \infty] \\ &= P_i[\tau_j(1) < \infty] P_j[\tau_j(1) < \infty]^{k-1} \\ &= f_{ij}(f_{jj})^{k-1}. \end{aligned}$$

(2.6.1)

In particular, with respect to P_j, we see that for $k \geq 1$

$$P_j[N_j \geq k] = (f_{jj})^k.$$

Suppose j is transient. We have

$$\begin{aligned} P_i[N_j = \infty] &= \lim_{k \to \infty} P_i[N_j \geq k] \\ &= \lim_{k \to \infty} f_{ij}(f_{jj})^{k-1} = 0 \end{aligned}$$

since $f_{jj} < 1$. Also from Lemma 1.1.1

$$E_i(N_j) = \sum_{m=0}^{\infty} P_i[N_j > m]$$

$$= \sum_{m=0}^{\infty} f_{ij}(f_{jj})^m = f_{ij}/(1 - f_{jj}). \quad \blacksquare$$

So if the chain starts from a recurrent state i, then it visits i infinitely often, which can also be written

$$P_i\{[X_n = i] \text{ i.o. }\} = 1,$$

and if i is transient then the chain visits i finitely often.

Although the generating function relationships provide a link between $\{p_{ij}^{(n)}\}$ and $\{f_{ij}^{(n)}\}$, it is primarily of theoretical use. In practice, if the quantities $\{f_{ij}^{(n)}\}$ are needed, one can use a recursive scheme for computation. For $i, j \in S$ we have $f_{ij}^{(1)} = p_{ij}$, while for $n > 1$ we have, by a first jump decomposition,

$$f_{ij}^{(n)} = P_i[X_1 \neq j, \ldots, X_{n-1} \neq j, X_n = j]$$

$$= \sum_{k \neq j, \, k \in S} P_i[X_1 = k, X_2 \neq j, \ldots, X_{n-1} \neq j, X_n = j]$$

$$= \sum_{k \neq j} P_i[X_2 \neq j, \ldots, X_{n-1} \neq j, X_n = j | X_1 = k] P_i[X_1 = k],$$

which from (2.1.4) is

$$= \sum_{k \neq j} P_k[X_1 \neq j, \ldots, X_{n-2} \neq j, X_{n-1} = j] P_i[X_1 = k]$$

$$= \sum_{k \neq j} p_{ik} f_{kj}^{(n-1)}.$$

To summarize,

$$(2.6.2) \qquad f_{ij}^{(n)} = \begin{cases} p_{ij}, & \text{if } n = 1 \\ \sum_{k \neq j} p_{ik} f_{kj}^{(n-1)}, & \text{if } n > 1. \end{cases}$$

This is best expressed as a matrix recursion. Set $^{(j)}\mathbf{P} = (^{(j)}p_{ik})$, where

$$(2.6.3) \qquad {}^{(j)}p_{ik} = \begin{cases} p_{ik}, & \text{if } k \neq j \\ 0, & \text{if } k = j, \end{cases}$$

so that we get $^{(j)}\mathbf{P}$ by setting the jth column of $\mathbf{P}$ equal to 0. For fixed $j \in S$ define the column vector

$$\mathbf{f}^{(n)} = (f_{ij}^{(n)}, i \in S)'.$$

Then (2.6.2) becomes

$$(2.6.3')\qquad \mathbf{f}^{(n)} = \begin{cases} (p_{ij}, i \in S)', & \text{if } n = 1 \\ {}^{(j)}\mathbf{P}f^{(n-1)}, & \text{if } n > 1, \end{cases}$$

which can also be expressed

$$\mathbf{f}^{(n)} = {}^{(j)}\mathbf{P}^{n-1}\mathbf{f}^{(1)}.$$

Example 2.6.1. A famous U.K. study of occupational mobility across generations was conducted after World War II. Three occupational levels were identified:

(1) upper level (executive, managerial, high administrative, professional)

(2) middle level (high grade supervisor, non-manual, skilled manual)

(3) lower level (semi-skilled or unskilled).

Transition probabilities from generation to generation were estimated to be

$$\mathbf{P} = \begin{array}{c} 1 \\ 2 \\ 3 \end{array}\begin{pmatrix} .45 & .48 & .07 \\ .05 & .70 & .25 \\ .01 & .5 & .49 \end{pmatrix}.$$

We are interested in $(f_{i1}^{(n)}, i = 1, 2, 3)'$. We have

$$^{(1)}\mathbf{P} = \begin{pmatrix} 0 & .48 & .07 \\ 0 & .70 & .25 \\ 0 & .5 & .49 \end{pmatrix}$$

and

$$\mathbf{f}^{(1)} = (.45, .05, .01)'.$$

Thus

$$\mathbf{f}^{(2)} = {}^{(1)}Pf^{(1)} = (.0247, .0375, .0299)'$$
$$\mathbf{f}^{(3)} = {}^{(1)}Pf^{(2)} = (.02009, .03372, .0334)',$$

and by powering up $^{(1)}\mathbf{P}$ we find

$$\mathbf{f}^{(9)} = (^{(1)}\mathbf{P})^8\mathbf{f} = (.01519, .02644, .0279)'.$$

Because all entries of P are positive, the chain is irreducible. As we shall see, the state space being finite implies that all states are recurrent and thus, as will be checked later, $f_{i1} = P_i[\tau_1(1) < \infty] = 1$, $i = 1, 2, 3$; i.e., the chain reaches state 1, the highest economic level, in a finite time. Remember the time scale here is generations. By continuing the recursive scheme we find

$$(P_i[\tau_1(1) \le 5], i = 1, 2, 3)' = (\sum_{m=1}^{5} f_{i1}^{(m)}, \ i = 1, 2, 3)'$$

$$= \begin{pmatrix} .45 \\ .05 \\ .01 \end{pmatrix} + \begin{pmatrix} .0247 \\ .0375 \\ .0299 \end{pmatrix} + \begin{pmatrix} .02009 \\ .03372 \\ .0334 \end{pmatrix} + \begin{pmatrix} .0185261 \\ .0319577 \\ .033229 \end{pmatrix} + \begin{pmatrix} .0176657 \\ .0306777 \\ .0322611 \end{pmatrix}$$

$$= \begin{pmatrix} .530985 \\ .183860 \\ .138791 \end{pmatrix}.$$

Note $P_3[\tau_1(1) \le 5]$ is small.

2.7. PERIODICITY.

The concept of *periodicity* is necessary for understanding the motion of a stochastic system. It may well be the case that certain movements of the system can only be completed in paths whose lengths are multiples of, say, a certain number d. As a paradigm, think of the simple random walk where steps are ± 1. Returns to 0 can only occur along paths whose lengths are even since every positive step must be compensated for by a negative step.

We now explain how to classify states as either periodic or aperiodic. Define the *period* of state i to be

$$d(i) := \gcd\{n \ge 1 : p_{ii}^{(n)} > 0\},$$

where gcd means greatest common divisor. (If $\{n \ge 1 : p_{ii}^{(n)} > 0\} = \emptyset$, then set $d(i) = 1$.) If $d(i) = 1$, call i aperiodic and if $d(i) > 1$ call i periodic with period $d(i)$. The definition means that if $p_{ii}^{(n)} > 0$ then n is an integer multiple of $d(i)$, and $d(i)$ is the largest integer with this property. Returns to state i are only possible via paths whose lengths are multiples of $d(i)$.

Example 2.7.1. (1) Unrestricted simple random walk (see Section 1.6) $\{S_n = \sum_{k=1}^{n} X_k, n \ge 0\}$ with state space $\{\ldots, -1, 0, 1, \ldots\}$. The period of 0 is 2 since $p_{00}^{(n)} = 0$ unless n is even.

(2) Random walk with steps $\{X_n\}$ having possible values $\pm 1, 0$ and satisfying

$$P[X_n = 1] = p > 0, \quad P[X_n = 0] = r > 0, \quad P[X_n = -1] = q > 0.$$

Then $d(0) = 1$ since $p_{00} = r > 0$ and therefore $1 \in \{n \geq 1 : p_{00}^{(n)} > 0\}$.

Note that whenever $p_{ii} > 0$, i is aperiodic.

(3) Consider the chain on $\{1, 2, 3\}$ with matrix

$$P = \begin{matrix} 1 \\ 2 \\ 3 \end{matrix} \begin{pmatrix} 0 & 1 & 0 \\ 1/2 & 0 & 1/2 \\ 1 & 0 & 0 \end{pmatrix}.$$

Then $p_{11}^{(1)} = 0$, $p_{11}^{(2)} \geq p_{12}p_{21} > 0$, $p_{11}^{(3)} = p_{12}p_{23}p_{31} > 0$ so

$$\{2, 3\} \subset \{n : p_{11}^{(n)} > 0\},$$

and, since $\gcd\{2, 3\} = 1$, we have $d(1) = 1$, even though $p_{11}^{(1)} = 0$.

Remark. It is possible to leave a periodic state and never return. For example, in gamblers ruin on $\{0, 1, 2, 3\}$, states 1 and 2 have period 2.

2.8. SOLIDARITY PROPERTIES.

A property of states is called a solidarity or class property if whenever i has the property and $i \leftrightarrow j$, then j also has the property. Put another way, if C is an equivalence class of states and $i \in C$ has the property, then every state $j \in C$ has the property.

The good news:

Proposition 2.8.1. *Recurrence, transience and the period of a state are solidarity properties.*

The practical impact is that these properties need to be checked for only one representative of a class, not every element of the class. Thus, for example, if $i \leftrightarrow j$ then $d(i) = d(j)$.

Proof. Suppose $i \leftrightarrow j$ and i is recurrent. Since $i \rightarrow j$ there exists n such that $p_{ij}^{(n)} > 0$ and since $j \rightarrow i$ there exists m such that $p_{ji}^{(m)} > 0$. From the matrix identity

$$P^{m+n+k} = P^m P^k P^n,$$

we get

$$(2.8.1) \qquad p_{jj}^{(n+m+k)} = \sum_{\alpha,\beta \in S} p_{j\alpha}^{(m)} p_{\alpha\beta}^{(k)} p_{\beta j}^{(n)}$$

$$\geq p_{ji}^{(m)} p_{ii}^{(k)} p_{ij}^{(n)} = \left(p_{ji}^{(m)} p_{ij}^{(n)} \right) p_{ii}^{(k)}$$

$$= c p_{ii}^{(k)},$$

where $c > 0$. Since i is recurrent we have from Proposition 2.6.2 that $\sum_k p_{ii}^{(k)} = \infty$, and therefore

$$\sum_{\ell=1}^{\infty} p_{jj}^{(\ell)} \geq \sum_{k=1}^{\infty} p_{jj}^{(m+n+k)} \geq c \sum_{k=1}^{\infty} p_{ii}^{(k)} = \infty,$$

from which j is also recurrent.

This argument is symmetric in i, j, so if $i \leftrightarrow j$ then i is recurrent if and only if j is recurrent. Since transient means not recurrent, we also have that if $i \leftrightarrow j$, then i transient if and only if j is transient.

Suppose $i \leftrightarrow j$ and i has period $d(i)$ and j has period $d(j)$. From (2.8.1) we have for $c > 0$

$$(2.8.2) \qquad p_{jj}^{(n+m+k)} \geq c p_{ii}^{(k)}.$$

Now $p_{ii}^{(0)} = 1$, so from (2.8.2) we get $p_{jj}^{(n+m)} > 0$, which means $n + m = k_1 d(j)$ for some positive integer k_1. For any $k > 0$ such that $p_{ii}^{(k)} > 0$, we have $p_{jj}^{(n+m+k)} \geq c p_{ii}^{(k)} > 0$ so that

$$n + m + k = k_2 d(j)$$

for a positive integer k_2. Now for k such that $p_{ii}^{(k)} > 0$ we have

$$k = n + m + k - (n + m) = k_2 d(j) - k_1 d(j) = (k_2 - k_1) d(j).$$

So $d(j)$ is a divisor of

$$\{ n \geq 1 : p_{ii}^{(n)} > 0 \}$$

Since the greatest common divisor of this set is by definition $d(i)$, we know that $d(j)$ is a divisor of $d(i)$ and hence $d(i) \geq d(j)$. By the symmetry of this argument between i, j we also get that $d(i)$ is a divisor of $d(j)$ so that $d(i) \leq d(j)$. Hence $d(i) = d(j)$. ∎

2.9. EXAMPLES.

Here we consider some examples and obtain criteria for recurrence or transience using the basic definitions. More sophisticated techniques for deciding on recurrence or transience will be considered later.

First a reminder about the significance of the concept of recurrence. Recurrence may be thought of as a stability property for a stochastic system. It describes the strong tendency of the model to return to the center of the state space. Transience may be associated with a tendency toward the extremes of the state space: Queue lengths build up without bound, busy periods may become infinite, branching processes explode, random walks drift to infinities, etc.

Example 2.9.1. Success Run Chain. Recall the transition matrix for this chain is of the form

$$P = \begin{pmatrix} q_0 & p_0 & 0 & 0 & \cdots \\ q_1 & 0 & p_1 & 0 & \cdots \\ q_2 & 0 & 0 & p_2 & \\ \vdots & & \vdots & & \ddots \end{pmatrix}$$

where $0 < p_i < 1, i \geq 0$. When is a state recurrent? This chain is irreducible and therefore $i > 0$ is recurrent iff 0 is recurrent and thus it suffices to determine a criterion for the recurrence of 0. We have $f_{00}^{(1)} = q_0$, and for $n \geq 2$ we get

$$f_{00}^{(n)} = P_0[X_1 = 1, X_2 = 2, \ldots, X_{n-1} = n - 1, X_n = 0]$$
$$= p_0 p_1 \cdots p_{n-2} q_{n-1}.$$

Write

$$u_n = \prod_{i=0}^{n} p_i, \quad n \geq 0,$$

and we obtain from $q_{n-1} = 1 - p_{n-1}$

$$f_{00}^{(n)} = u_{n-2} - u_{n-1}, \quad n \geq 2,$$

from which

$$\sum_{n=1}^{N+1} f_{00}^{(n)} = q_0 + (u_0 - u_1) + (u_1 - u_2) + \cdots + u_{N-1} - u_N$$

$$= q_0 + u_0 - u_N = 1 - u_N.$$

So 0 is recurrent iff $u_N = \prod_{i=0}^{N} p_i \to 0$ as $N \to \infty$. A condition for this can be obtained by the the following lemma from our toolbox.

Lemma 2.9.1. *If* $0 < p_i < 1$, *for* $i \geq 0$ *then*

$$u_N = \prod_{i=0}^{N} p_i \to 0 \text{ iff } \sum_i q_i = \sum_i (1 - p_i) = \infty$$

and

$$\prod_{i=0}^{\infty} p_i > 0 \text{ iff } \sum_i q_i = \sum_i (1 - p_i) < \infty.$$

Proof. Recall that if $a_n \sim b_n$ as $n \to \infty$, then $\sum_n a_n < \infty$ iff $\sum_n b_n < \infty$, since $a_n \sim b_n$ means that $\lim_{n \to \infty} a_n / b_n = 1$. Equivalently, for any $\epsilon > 0$, we have for all large n that $(1 - \epsilon)b_n < a_n < (1 + \epsilon)b_n$.

$$\prod_{i=0}^{\infty} p_i > 0 \text{ iff } \sum_{i=0}^{\infty} -\log(1 - q_i) < \infty \text{ iff } \sum_{i=0}^{\infty} q_i < \infty.$$

The last equivalence follows from

$$-\log(1 - x) \sim x, \quad x \to 0$$

since by L'Hôpital's rule or Taylor expansion

$$\lim_{x \to 0} \frac{-\log(1 - x)}{x} = 1.$$

Note, both $\sum_i -\log(1 - q_i) < \infty$ and $\sum_i q_i < \infty$ imply $q_i \to 0$. ∎

We conclude 0 is recurrent iff $\sum_i (1 - p_i) = \infty$, which says that the p_i's are not too close to 1; the q_i's, measuring pressure toward 0, are substantial.

Example 2.9.2. Simple Random Walk. Recall the setup in Section 1.6. Let $\{X_n\}$ be iid, and

$$S_n = \sum_{i=1}^{n} X_i, \quad P[X_1 = 1] = p = 1 - P[X_1 = -1].$$

If $p > q$ then by the strong law of large numbers

$$P[\lim_{n \to \infty} \frac{S_n}{n} = EX_1] = 1.$$

Since $EX_1 = p - q > 0$, we have

$$P[\lim_{n \to \infty} S_n = \infty] = 1,$$

so there is a last visit to 0 almost surely, and 0 is transient. Similarly, if $p < q$, we find 0 transient. If $p = q$, we have already seen in the discussion of the simple random walk in Section 1.6 that $P_0[\tau_0(1) < \infty] = 1$, hence 0 is recurrent in this case.

Alternatively, note that when $p = q = \frac{1}{2}$, we have $p_{00}^{(2n+1)} = 0$, and

$$p_{00}^{(2n)} = \binom{2n}{n}\left(\frac{1}{2}\right)^n\left(\frac{1}{2}\right)^n = P[n \text{ steps right, } n \text{ steps left }].$$

Pull *Stirling's formula* from your toolbox:

$$n! \sim \sqrt{2\pi}e^{-n}n^{n+1/2}, \quad n \to \infty,$$

from which we get

$$\binom{2n}{n} \sim (\pi n)^{-1/2}4^n, \quad n \to \infty.$$

Since

$$p_{00}^{(2n)} = \binom{2n}{n}\left(\frac{1}{4}\right)^n \sim (\pi n)^{-1/2},$$

we have $\sum_{k=1}^{\infty} p_{00}^{(k)} = \infty$, and therefore 0 is recurrent.

These methods generalize to multidimensional random walks. Let

$$\mathbf{X}_i = (X_i^{(1)}, \dots, X_i^{(d)})$$

be a d-dimensional random walk step and define as before $\mathbf{S}_0 = 0$, $\mathbf{S}_n = \mathbf{X}_1 + \dots + \mathbf{X}_n$. If

$$E\mathbf{X}_1 := (EX_1^{(1)}, \dots, EX_1^{(d)}) \neq \mathbf{0},$$

then again the argument using the strong law of large numbers shows $\mathbf{0}$ is hit finitely often. Suppose the range of $\mathbf{X}_i$ is $\{-1, 1\}^d$ and that $E\mathbf{X}_1 = 0$. Suppose each value $\{-1, 1\}^d$ is equally likely with

$$P[\mathbf{X}_1 = (i_1, \dots, i_d)] = \frac{1}{2^d}$$

for $(i_1, \dots, i_d) \in \{-1, 1\}^d$. This implies the components $X_1^{(1)}, \dots, X_1^{(d)}$ of $\mathbf{X}_1$ are iid and have a symmetric distribution $P[X_1^{(j)} = \pm 1] = 1/2$, $j = 1, \dots, d$. Therefore

$$\mathbf{S}_n = (S_n^{(1)}, \dots, S_n^{(d)}),$$

where $\{S_n^{(1)}, n \geq 0\}, \ldots, \{S_n^{(d)}, n \geq 0\}$ are independent symmetric, simple random walks. Thus

$$p_{00}^{(2n)} = P[\mathbf{S}_{2n} = \mathbf{0}] = P[S_{2n}^{(1)} = 0, \ldots, S_{2n}^{(d)} = 0]$$

$$= (P[S_{2n}^{(1)} = 0])^d = \left(\binom{2n}{n}(\tfrac{1}{4})^n\right)^d$$

$$\sim (\pi n)^{-d/2} \text{ as } n \to \infty,$$

by Stirling's formula. So, since

$$\sum_{n=1}^{\infty} \frac{1}{n^{d/2}} = \infty \text{ for } d = 1, 2$$

$$< \infty \text{ for } d \geq 3,$$

we find

$$\sum_{n=1}^{\infty} p_{00}^{(n)} = \infty \text{ for } d = 1, 2$$

$$< \infty \text{ for } d \geq 3.$$

Therefore for the simple symmetric random walk in R^d we have that $\mathbf{0}$ is recurrent if $d = 1, 2$ and transient if $d \geq 3$.

Example 2.9.3. Simple Branching Process. Assume to avoid a degenerate situation that $p_1 \neq 1$. Since 0 is absorbing, 0 is recurrent. As we now demonstrate, however, $1, 2, \ldots$ are transient. If $p_0 = 0$, then the number of offspring per individual is at least 1, and therefore $\{Z_n\}$ is nondecreasing. If we start in state k, the only possible way to return to state k is if each of the k members of the current generation have exactly one offspring. Thus for $k \geq 1$

$$f_{kk} = P[\text{ eventual return to } k]$$
$$= P[Z_{n+1} = k | Z_n = k]$$
$$= P[Z_{n+1,j} = 1, j = 1, \ldots, k]$$
$$= p_1^k < 1,$$

so that in the case $p_0 = 0$ we have k transient. For the next case consider $p_0 = 1$. Then $p_{k0} = 1$ so $f_{kk} = 0 < 1$ and again state k is transient. Finally consider the case where $0 < p_0 < 1$. Since 0 is absorbing,

$$f_{k,k} \leq P_k[Z_1 \neq 0] = 1 - P_k[Z_1 = 0] = 1 - p_0^k < 1.$$

So for this case too, $k \geq 1$ is transient.

We now consider the instability property remarked on in Section 1.4. Pick N. Since the states $1, 2, \ldots, N$ are all transient, there is a last time $\{Z_n\}$ visits these states. So eventually either $Z_n = 0$ or $Z_n > N$, whence $P[Z_n \to 0$ or $Z_n \to \infty] = 1$. Since

$$P[Z_n \to 0 \text{ or } Z_n \to \infty] = P[Z_n \to 0] + P[Z_n \to \infty]$$

and since

$$P[Z_n \to 0] = P[\text{ extinction }] = \pi,$$

we get

$$P[Z_n \to \infty] = 1 - \pi.$$

Compare this with the results proven by the methods of Section 1.4.

2.10. CANONICAL DECOMPOSITION.

For studying a Markov chain with a large state space, it is helpful to decompose the state space into subclasses and then to study each class separately. We first prove that a recurrent class (a class such that all states are recurrent) is closed. This will allow us to decompose the state space as

$$S = T \cup (\cup_i C_i),$$

where T consists of transient states (T is not necessarily one class) and $C_1, C_2, \ldots$ are closed, recurrent, disjoint classes.

Proposition 2.10.1. *Suppose $j \in S$ is recurrent and for $k \neq j$ we have $j \to k$. Then*

    (1)   *k is recurrent,*
    (2)   *$j \leftrightarrow k$,*
    (3)   *$f_{jk} = f_{kj} = 1$.*

This proposition implies that a recurrent class is closed: If j is recurrent, any consequent is in the same class (from (2)). Repeated use of this argument shows that it is impossible to exit from this class.

Proof. (1) follows from (2) and solidarity. We concentrate on (2), and we need to prove $k \to j$. In order to get a contradiction, suppose j is not a consequent of k, which means

$$P_k[X_n \neq j, \ \forall n \geq 1] = 1.$$

Since $j \to k$, there exists m such that $p_{jk}^{(m)} > 0$. Since j is assumed recurrent, the chain visits state j infinitely often starting from j, and thus

$$0 = P_j[X_l \neq j, \forall l \geq m]$$
$$\geq P[X_l \neq j, \forall l \geq m, X_m = k].$$

Conditioning on the path up to time m and applying (2.1.4), we get

$$= p_{jk}^{(m)} P_k[X_l \neq j, l \geq 1]$$
$$= p_{jk}^{(m)} > 0,$$

which yields the required contradiction.

Now focus on (3). Assume $j \to k$ and thus there exists m such that

$$P_j[X_1 \neq j, \ldots, X_m \neq j, X_m = k] > 0.$$

Then we have from the recurrence of j that

$$0 = 1 - f_{jj} = P_j[\tau_j(1) = \infty]$$
$$\geq P_j[\tau_j(1) = \infty, X_m = k].$$

Again, conditioning on the path up to time m and applying (2.1.4) gives

$$= P_j[X_1 \neq j, \ldots, X_{m-1} \neq j, X_m = k] P_k[\tau_j(1) = \infty]$$
$$= P_j[X_1 \neq j, \ldots, X_{m-1} \neq j, X_m = k](1 - f_{kj}),$$

and therefore we get $1 - f_{kj} = 0$, which is the desired result. Symmetry also gives $f_{jk} = 1$. ∎

The result just proved quickly gives the decomposition of the state space promised in the beginning of the section.

Corollary 2.10.2. *The state space S of a Markov chain may be decomposed as*

$$S = T \cup C_1 \cup C_2 \cup \ldots,$$

where T consists of transient states (but T is not necessarily one class), $C_1, C_2, \ldots$ are closed, disjoint classes of recurrent states, and if $j \in C_\alpha$ then

$$f_{jk} = \begin{cases} 1, & \text{if } k \in C_\alpha \\ 0, & \text{if } k \notin C_\alpha. \end{cases}$$

Furthermore, if we relabel the states so that for $i = 1, 2, \ldots$ states in C_i have consecutive labels, with states in C_1 having the smallest labels, those of C_2 having the next smallest, etc., then the transition matrix P can be rewritten as

$$P = \begin{pmatrix} P_1 & 0 & 0 & 0 & \cdots \\ 0 & P_2 & 0 & 0 & \cdots \\ 0 & 0 & P_3 & 0 & \cdots \\ \vdots & \vdots & & \ddots & \\ Q_1 & Q_2 & Q_3 & & \cdots \end{pmatrix}$$

where $P_1, P_2, \ldots$ are square stochastic matrices with transitions within C_i being governed by P_i. Transitions from states in T are governed by the matrices Q_i.

If S contains an infinite number of states, it is possible for $S = T$ so that there are no closed, recurrent classes. The deterministically monotone Markov chain is a good example. If S is finite, however, not all states can be transient, as shown by the next result.

Proposition 2.10.3. *If S is finite, not all states can be transient.*

Proof. Suppose $S = \{1, 2, \ldots, m\}$, $m < \infty$. If $j \in T$ we know $\sum_{n=1}^{\infty} p_{ij}^{(n)} < \infty$ for any i. Therefore, as $n \to \infty$, $p_{ij}^{(n)} \to 0$. Summing over j gives

$$1 = \sum_{j=1}^{m} p_{ij}^{(n)} \to 0$$

if all states are transient. ∎

For a Markov chain, a basic task is deciding which states are transient and which are recurrent. For a finite Markov chain, the results of this section provide a classification method:

(1)   Decompose S into equivalence classes.
(2)   The closed classes are recurrent.
(3)   The classes which are not closed consist of transient states.

Decomposing S into equivalence classes can be accomplished by the following steps:

(1)   Pick a state i and find all consequents of i and all consequents of the consequents of i, and so on. This will give cl(i), the smallest closed set containing i. Find $k \notin$ cl(i), and determine cl(k). Keep this up until S is exhausted.

(2)   The resulting closed sets may contain more than one equivalence class. Non-closed classes which are subsets of closed sets will contain transient states.

(3)   Writing down directed graphs of the states may help if the number of states is reasonable.

(4) Teaching the computer to do this may be wise if the number of states is large.

Example 2.10.1. Gamblers Ruin on $\{0,1,2,3,4)$. We have

$$P = \begin{matrix} 0 \\ 1 \\ 2 \\ 3 \\ 4 \end{matrix} \begin{pmatrix} 1 & 0 & 0 & 0 & 0 \\ q & 0 & p & 0 & 0 \\ 0 & q & 0 & p & 0 \\ 0 & 0 & q & 0 & p \\ 0 & 0 & 0 & 0 & 1 \end{pmatrix}.$$

In canonical form this is

$$P = \begin{matrix} 0 \\ 4 \\ 1 \\ 2 \\ 3 \end{matrix} \begin{pmatrix} 1 & 0 & 0 & 0 & 0 \\ 0 & 1 & 0 & 0 & 0 \\ q & 0 & 0 & p & 0 \\ 0 & 0 & q & 0 & p \\ 0 & p & 0 & q & 0 \end{pmatrix},$$

so $C_1 = \{0\}, C_2 = \{4\}, T = \{1,2,3\}$.

Example 2.10.2. Let $S = \{1,2,\ldots,5\}$ with

$$P = \begin{matrix} 1 \\ 2 \\ 3 \\ 4 \\ 5 \end{matrix} \begin{pmatrix} 1/2 & 0 & 1/2 & 0 & 0 \\ 0 & 1/4 & 0 & 3/4 & 0 \\ 0 & 0 & 1/3 & 0 & 2/3 \\ 1/4 & 1/2 & 0 & 1/4 & 0 \\ 1/3 & 0 & 1/3 & 0 & 1/3 \end{pmatrix}.$$

The directed graph is given in Figure 2.1.

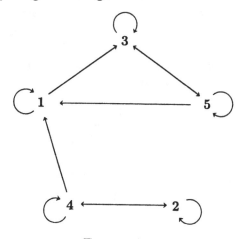

FIGURE 2.1

Note that cl(1) = {1, 3, 5}, and this is a class. Further, {2, 4} is also a class, but it is not closed. Hence $S = T \cup C_1$ where $T = \{2, 4\}$ and $C_1 = \{1, 3, 5\}$. In canonical form the matrix is

$$P = \begin{array}{c} 1 \\ 3 \\ 5 \\ 2 \\ 4 \end{array} \begin{pmatrix} 1/2 & 1/2 & 0 & 0 & 0 \\ 0 & 1/3 & 2/3 & 0 & 0 \\ 1/3 & 1/3 & 1/3 & 0 & 0 \\ 0 & 0 & 0 & 1/4 & 3/4 \\ 1/4 & 0 & 0 & 1/2 & 1/4 \end{pmatrix}.$$

2.11. ABSORPTION PROBABILITIES.

Frequently we are interested in the time until the system goes from some initial state to some terminal critical state. Such a terminal state may represent "breakdown" in a reliability context, bankruptcy in a financial or business context, or simply a state of interest. For instance, in the occupational mobility example at the end of Section 2.6, we are interested in the number of steps necessary to go from the lowest economic class to the highest.

Such problems can often be cast as absorption problems with the following sort of formulation: Let $S = T \cup C_1 \cup C_2 \cup \ldots$ be the canonical decomposition of the state space. T consists of the transient states and the classes C_i are closed and recurrent. Define

$$\tau = \inf\{n \geq 0 : X_n \notin T\}$$

to be the exit time of T. Note that there are cases where $P_i[\tau = \infty] > 0$. For example, in the deterministically monotone Markov chain we have $P_i[\tau = \infty] = 1$ for all $i \in S$, since $T = S$. For now, however, assume that $P_i[\tau < \infty] = 1$ for all i. We shall see that this is the case if the state space is finite. Note that when τ is finite, X_τ is the first state hit outside T.

Partition the transition matrix P as

$$P = \begin{pmatrix} Q & R \\ 0 & P_2 \end{pmatrix},$$

where Q is the restriction of the matrix P to the states corresponding to T; that is,

$$Q = (Q_{ij}, i, j \in T) = (p_{ij}, i, j \in T)$$

and

$$R = (R_{kl}, k \in T, l \in T^c) = (p_{kl}, k \in T, l \in T^c).$$

Now define for $i \in T$ and $k \in T^c$

$$u_{ik} = P_i[X_\tau = k].$$

Recall that we assume τ is finite for all starting states i. Once the chain leaves T, it will hit one of the closed recurrent classes and hence can never return to T. Thus, we can interpret u_{ik} as the probability that the chain leaves T because of absorption at state k in the closed, recurrent class when the initial state is i. Quantities related to absorption are easily computed from the $\{u_{ik}\}$. For example,

$$u_i(C_l) := P_i[X_\tau \in C_l] = \sum_{k \in C_l} u_{ik},$$

the probability that absorption takes place at class C_l, is easily computed by summing the absorption probabilities corresponding to the states in C_l.

We now give some properties of the matrix Q. For $i, j \in T$ and $n \geq 0$ we have

(2.11.1) $$p_{ij}^{(n)} = Q_{ij}^{(n)}.$$

To check this, observe that

(2.11.2) $$\begin{aligned} Q_{ij}^{(n)} &= \sum_{j_1 \in T, \ldots, j_{n-1} \in T} p_{ij_1} p_{j_1 j_2} \cdots p_{j_{n-1} j} \\ &= P_i[X_n = j, \tau > n]. \end{aligned}$$

However, since for $j \in T$

$$[X_n = j] \subset [\tau > n],$$

we get the right side of (2.11.2) is just $p_{ij}^{(n)}$, and thus we have verified (2.11.1). A consequence of (2.11.1) is

(2.11.3) $$\sum_{n=0}^{\infty} Q_{ij}^{(n)} = E_i \sum_{n=0}^{\infty} 1_{[X_n=j]} < \infty,$$

so that $\sum_{n=0}^{\infty} Q_{ij}^{(n)}$ is the expected number of visits to the transient state j starting from transient state i.

Keep in mind the following two examples:

1. Gambler's ruin on $\{0, \ldots, m\}$. Here $T = \{1, \ldots, m-1\}$ and the absorption probabilities of interest are

$$u_{i0} = P_i[\text{ Harry goes broke }], \quad 1 \leq i \leq m-1,$$

and

$$u_{im} = P_i[\text{ Zeke goes broke and Harry wins }], \quad 1 \leq i \leq m - 1.$$

2. Occupational mobility. Recall the state space was $\{1, 2, 3\}$ with 1 representing the highest economic class and 3 the lowest. The matrix is

$$P = \begin{pmatrix} .45 & .48 & .07 \\ .05 & .7 & .25 \\ .01 & .5 & .49 \end{pmatrix}.$$

We are interested in calculating the expected time to go from 3 to 1. At first glance this problem seems to have little to do with absorption. The matrix has all entries positive and hence is irreducible; because the state space is finite the irreducible class is recurrent and there are no transient states. However, by changing P to

$$P' = \begin{pmatrix} 1 & 0 & 0 \\ .05 & .7 & .25 \\ .01 & .5 & .49 \end{pmatrix}$$

and analyzing the expected time to absorption in state 1 starting from 3 for this modified chain, the problem is indeed cast as one of absorption. If we are interested in the expected first passage time to 1, changing the probabilities of how the system leaves 1 does not affect the absorption times. For the Markov chain corresponding to the modified matrix, $T = \{2, 3\}$ and interest centers on

$$w_i = E_i \tau, \quad i = 2, 3$$

where for this problem $\tau = \inf\{n \geq 0 : X_n = 1\}$.

To compute u_{ij} we use *first step analysis* and decompose the event $[X_\tau = j]$ according to what happens at the first transition:

$$[X_\tau = j] = \cup_k [X_\tau = j, X_1 = k].$$

This gives a recursion for the u_{ij}'s. In the finite state space case this recursion can be neatly solved via matrix manipulations. We have for $i \in T, j \in T^c$,

$$u_{ij} = P_i[X_\tau = j] = \sum_{k \in S} P_i[X_\tau = j, X_1 = k]$$

$$= \sum_{k \in T} P_i[X_\tau = j, X_1 = k] + \sum_{k \in T^c} P_i[X_\tau = j, X_1 = k]$$

$$= A + B.$$

To analyze B, observe that if $k \in T^c$ then the events $[X_\tau = j]$ and $[X_1 = k]$ are disjoint unless $j = k$, so we have $B = p_{ij}$. For A we have that $\tau \geq 2$, and by conditioning on X_1 and using the Markov property,

$$
\begin{aligned}
A &= \sum_{k \in T} \sum_{n \geq 2} P_i[\tau = n, X_n = j, X_1 = k] \\
&= \sum_{k \in T} \sum_{n \geq 2} P_i[X_2 \in T, \ldots, X_{n-1} \in T, X_n = j, X_1 = k] \\
&= \sum_{k \in T} \sum_{n \geq 2} P_i[X_2 \in T, \ldots, X_{n-1} \in T, X_n = j | X_1 = k] P_i[X_1 = k] \\
&= \sum_{k \in T} \sum_{n \geq 2} p_{ik} P_k[X_1 \in T, \ldots, X_{n-2} \in T, X_{n-1} = j] \\
&= \sum_{k \in T} \sum_{n \geq 2} p_{ik} P_k[\tau = n - 1, X_\tau = j] \\
&= \sum_{k \in T} p_{ik} P_k[X_\tau = j] = \sum_{k \in T} p_{ik} u_{kj}.
\end{aligned}
$$

Since for $i, k \in T$ we have $p_{ik} = Q_{ik}$. By combining $A + B$, we get that the recursion becomes $(i \in T, j \in T^c)$

$$(2.11.4) \qquad\qquad u_{ij} = \sum_{k \in T} Q_{ik} u_{kj} + p_{ij}.$$

This recursion, of course, merely says that absorption by a recurrent state j can take place in two ways: Either absorption is accomplished in one step (with probability p_{ij}), or, if not in one step, then a transition must be made to an intermediate transient state k (probability Q_{ik}) and then from k the chain must be absorbed by state j (probability u_{kj}).

If we set $U = (u_{ij}, i \in T, j \in T^c)$, then in matrix notation (2.11.4) becomes

$$(2.11.4') \qquad\qquad U = QU + R$$

which is the same as $U - QU = U(I - Q) = R$. If $I - Q$ has an inverse, we get the matrix solution

$$U = (I - Q)^{-1}R.$$

The matrix

$$(I - Q)^{-1}$$

arises frequently in absorption calculations and is known as the *fundamental matrix*. When the state space is finite (or when T is finite) $I - Q$ indeed has an inverse, which can be represented as

$$(I - Q)^{-1} = \sum_{n=0}^{\infty} Q^n$$

so that from (2.11.3) we have that

$$((I - Q)^{-1})_{i,j} = E_i \sum_{n=0}^{\infty} 1_{[X_n = j]}.$$

Example. Amateur Night at Happy Harry's. Friday night is amateur night at Happy Harry's Restaurant where a seemingly infinite stream of performers dreaming of stardom perform in lieu of the usual professional floor show. The quality of the performers falls into five categories with "1" being the best and "5" being unspeakably atrocious, representing for Harry's discriminating clientele an exceedance of the threshold of pain which may cause a riot. The probability a class 5 performer will cause the crowd to riot is .3. After the riot is quelled, performances resume—the show must go on. Since performers tend to bring along friends of similar talent to perform, it is found that the succession of states on Friday night at Happy Harry's can be modelled as a six-state Markov chain, where state 6 represents "riot" and state "i" represents a class "i" performer, $1 \leq i \leq 5$. The transition matrix for this chain is

$$P = \begin{pmatrix} .05 & .15 & .3 & .3 & .2 & 0 \\ .05 & .3 & .3 & .3 & .05 & 0 \\ .05 & .2 & .3 & .35 & .1 & 0 \\ .05 & .2 & .3 & .35 & .1 & 0 \\ .01 & .1 & .1 & .1 & .39 & .3 \\ .2 & .2 & .2 & .2 & .2 & 0 \end{pmatrix}.$$

To play it safe Harry starts the evening off with a class 2 performer. What is the probability that a star is discovered (a class 1 performer) before a riot is encountered? What is the expected number of performers seen before the first riot?

To solve this problem we make states 1 and 6 absorbing and solve for u_{21}. Although it is a bit pedantic, we first write the modified matrix in canonical form so that states 1 and 6 correspond to the last two rows and columns:

$$P' = \begin{array}{c} 2 \\ 3 \\ 4 \\ 5 \\ 1 \\ 6 \end{array} \begin{pmatrix} .3 & .3 & .3 & .05 & .05 & 0 \\ .2 & .3 & .35 & .1 & .05 & 0 \\ .2 & .3 & .35 & .1 & .05 & 0 \\ .1 & .1 & .1 & .39 & .01 & .3 \\ 0 & 0 & 0 & 0 & 1 & 0 \\ 0 & 0 & 0 & 0 & 0 & 1 \end{pmatrix}.$$

Thus we have

$$Q = \begin{pmatrix} .3 & .3 & .3 & .05 \\ .2 & .3 & .35 & .1 \\ .2 & .3 & .35 & .1 \\ .1 & .1 & .1 & .39 \end{pmatrix}, \quad R = \begin{pmatrix} .05 & 0 \\ .05 & 0 \\ .05 & 0 \\ .01 & .3 \end{pmatrix}.$$

With the help of a package like Minitab we find

$$(I - Q)^{-1} = \begin{pmatrix} 3.68 & 3.40 & 3.76 & 1.48 \\ 2.51 & 4.32 & 3.72 & 1.52 \\ 2.51 & 3.31 & 4.72 & 1.52 \\ 1.43 & 1.81 & 2.00 & 2.38 \end{pmatrix},$$

from which

$$U = (I - Q)^{-1}R = \begin{pmatrix} .557 & .443 \\ .542 & .457 \\ .543 & .457 \\ .285 & .714 \end{pmatrix},$$

and the required probability is $u_{21} = .557$. ∎

In order to analyze the expected number of performers seen in this example we first need to develop some new equations. Suppose

$$g : S \mapsto R$$

is a function on the state space, and define for $i \in T$

(2.11.5) $$w_i = E_i\left(\sum_{n=0}^{\tau-1} g(X_n)\right).$$

Think of g as a reward for being in state i; then w_i is the cumulative reward starting from $i \in T$ until absorption in T^c. Some useful examples of g are

(1) If $g \equiv 1$ then

$$w_i = E_i \tau,$$

the expected absorption time.

(2) If $g(i) = \delta_{ij}$ for $i, j \in T$ (that is, $g(i) = 1$ if $i = j$ and $g(i) = 0$ if $i \neq j$), then

$$w_i = E_i \sum_{n=0}^{\infty} 1_{[X_n = j]},$$

the expected number of visits to transient state j, starting from i. We already know this is the (i, j)th entry of the fundamental matrix, if S is finite.

We now derive a recursion for $\{w_i\}$ in a manner similar to the way we obtained the recursion for the absorption probabilities. We have that

$$w_i = E_i \sum_{n=0}^{\tau-1} g(X_n) = g(i) + E_i \sum_{n=1}^{\tau-1} g(X_n) 1_{[\tau > 1]},$$

where the first term on the right results from the reward from being in state $X_0 = i$. The second term on the right is
(2.11.6)
$$E_i \sum_{n=1}^{\infty} g(X_n) 1_{[n < \tau]} = \sum_{j \in T} E_i \left(\sum_{n=1}^{\infty} g(X_n) 1_{[X_1 \in T, \dots, X_n \in T]} \Big| X_1 = j \right) p_{ij}.$$

Now define $f : S \mapsto R^{\infty}$ by

$$f(x_1, x_2, \dots) = \sum_{n=1}^{\infty} g(x_n) 1_{[x_1 \in T, \dots, x_n \in T]},$$

and the right side of (2.11.6) becomes

$$\sum_{j \in T} E_i \left(f(X_1, X_2, \dots) | X_1 = j \right) p_{ij}.$$

From the Markov property (2.1.4) this equals

$$\sum_{j\in T} E_j f(X_0, X_1, \dots) p_{ij}$$

$$= \sum_{j\in T} E_j \sum_{n=1}^{\infty} g(X_{n-1}) 1_{[X_0\in T, \dots, X_{n-1}\in T]} p_{ij}$$

$$= \sum_{j\in T} E_j \sum_{m=0}^{\infty} g(X_m) 1_{[X_0\in T, \dots, X_m\in T]} p_{ij}$$

$$= \sum_{j\in T} E_j \sum_{m=0}^{\infty} g(X_m) 1_{[\tau>m]} p_{ij}$$

$$= \sum_{j\in T} E_j \sum_{m=0}^{\tau-1} g(X_m) p_{ij}$$

$$= \sum_{j\in T} w_j p_{ij}.$$

We conclude that w_i satisfies

$$(2.11.7) \qquad w_i = g(i) + \sum_{j\in T} p_{ij} w_j, \quad i \in T.$$

In the case when S or T is finite, a matrix solution is again possible in terms of the fundamental matrix. Define the column vectors

$$w = (w_i, i \in T)', \quad g = (g(i), i \in T)',$$

and then (2.11.7) becomes

$$w = g + Qw$$

which has the solution

$$w = (I - Q)^{-1} g.$$

Recall that in the case $g \equiv 1$,

$$(E_i(\tau), i \in T)' = (I - Q)^{-1} 1,$$

where $1 = (1, 1, \dots)'$ is a column vector of 1's.

Example continued. To find the expected number of performers starting from state 2 seen before the first riot, make state 6 absorbing. We have

$$Q = \begin{pmatrix} .05 & .15 & .3 & .3 & .2 \\ .05 & .3 & .3 & .3 & .05 \\ .05 & .2 & .3 & .35 & .1 \\ .05 & .2 & .3 & .35 & .1 \\ .01 & .1 & .1 & .1 & .39 \end{pmatrix},$$

and the desired answer is $(I - Q)^{-1}\mathbf{1}$. Using Minitab, we find that the fundamental matrix is

$$(I - Q)^{-1} = \begin{pmatrix} 2.068 & 4.854 & 6.541 & 7.229 & 3.333 \\ 1.152 & 6.389 & 7.046 & 7.788 & 3.333 \\ 1.123 & 5.149 & 7.869 & 7.644 & 3.333 \\ 1.122 & 5.149 & 6.869 & 8.644 & 3.333 \\ 0.591 & 2.815 & 3.678 & 4.066 & 3.333 \end{pmatrix}.$$

Multiplying by the column vector $(1, 1, 1, 1, 1)'$ yields

$$(I - Q)^{-1}\mathbf{1} = (24.026, 25.710, 25.119, 25.119, 14.484)',$$

and the desired expected value is 25.710.

We now discuss the general solution to the famous gamblers ruin problem on the states $\{0, \ldots, m\}$.

Example 2.11.1. Gambler's Ruin. Suppose $p \neq q$ (if $p = q$ modifications are necessary to prevent dividing by 0 in what follows) and set

$$u_i = P_i[X_\tau = 0] = u_{i0}, \quad 1 \leq i \leq m - 1$$

for the probability that Harry loses and Zeke wins. Here $T = \{1, \ldots, m-1\}$. The equations (2.11.4) become

$$u_i = \sum_{k \in T} p_{ik} u_k + p_{i0}, \quad i \in T,$$

so that

$$\begin{aligned} u_1 &= pu_2 + q \\ u_i &= qu_{i-1} + pu_{i+1}, \quad 2 \leq i \leq m - 2. \\ u_{m-1} &= qu_{m-2} \end{aligned}$$

If we set $u_0 = 1, u_m = 0$, these equations can be combined into the system

$$\begin{aligned} &u_0 = 1, u_m = 0 \\ (2.11.8) \qquad &u_i = qu_{i-1} + pu_{i+1}, \quad 1 \leq i \leq m - 1. \end{aligned}$$

This becomes $pu_i + qu_i = qu_{i-1} + pu_{i+1}$, so that

$$p(u_{i+1} - u_i) = q(u_i - u_{i-1}).$$

Let $\rho = q/p \neq 1$, and we have

(2.11.9) $\qquad u_{i+1} - u_i = \rho(u_i - u_{i-1}), \quad 1 \leq i \leq m - 1.$

Iterating back, we see that

(2.11.10) $\qquad u_{i+1} - u_i = \rho^i(u_1 - u_0) = \rho^i(u_1 - 1), \quad 1 \leq i \leq m - 1.$

By inspection, the equation holds also for $i = 0$. Sum over i:

$$\sum_{i=0}^{m-1} (u_{i+1} - u_i) = \sum_{i=0}^{m-1} \rho^i(u_1 - 1).$$

The left side telescopes so

$$u_m - u_0 = 0 - 1 = -1 = \left(\sum_{i=0}^{m-1} \rho^i \right) (u_1 - 1)$$

from which

$$1 - u_1 = 1/\sum_{i=0}^{m-1} \rho^i.$$

From (2.10.10) we get

$$u_{i+1} - u_i = \frac{-\rho^i}{\sum_{j=0}^{m-1} \rho^i}, \quad 1 \leq i \leq m - 1,$$

and summing again yields ($0 \leq j \leq m - 1$)

$$\sum_{i=j}^{m-1} (u_{i+1} - u_i) = u_m - u_j = -u_j$$

$$= \frac{-\sum_{i=j}^{m-1} \rho^i}{\sum_{j=0}^{m-1} \rho^i}$$

from which

(2.11.11) $\qquad u_j = \frac{\rho^j - \rho^m}{1 - \rho^m}, \quad 0 \leq j \leq m.$

We now discuss the case when (2.11.4) has a unique solution and the fundamental matrix exists. When S is finite, we have already remarked that

$$\sum_{n=0}^{\infty} Q^n < \infty,$$

and it is easy to see from this that

$$(I - Q)^{-1} = \sum_{n=0}^{\infty} Q^n.$$

This covers most of the elementary and usual applications. The rest of this section discusses the uniqueness of the solution to (2.11.4) and the existence of the fundamental matrix in more general contexts. This discussion may contain more detail than is necessary for the beginning student; therefore some readers may wish to skip the rest of this section and continue reading at the beginning of Section 2.12.

When S is infinite, (2.11.4) need not have a unique solution, and this case will now be considered in some detail.

Example 2.11.2. Consider the transient success run chain with

$$P = \begin{pmatrix} q_0 & p_0 & 0 & & \cdots \\ q_1 & 0 & p_1 & 0 & \cdots \\ q_2 & 0 & 0 & p_2 \cdots \\ \vdots & & & \ddots \end{pmatrix}$$

and $\prod_{i=0}^{\infty} p_i > 0$, $\sum_i (1 - p_i) < \infty$. (Refer to Lemma 2.9.1.) Make 0 absorbing so the matrix becomes

$$P\prime = \begin{pmatrix} 1 & 0 & 0 & 0 & \cdots \\ q_1 & 0 & p_1 & 0 & \cdots \\ q_2 & 0 & 0 & p_2 & \cdots \\ \vdots & \ddots & & & \end{pmatrix}.$$

Ignoring the initial row and column gives

$$Q = \begin{pmatrix} 0 & p_1 & 0 & 0 & \cdots \\ 0 & 0 & p_2 & 0 & \cdots \\ 0 & 0 & 0 & p_3 & \cdots \\ \vdots & & & \ddots \end{pmatrix}.$$

Thus the system (2.11.4) becomes ($i \geq 1, u_{i0} = u_i$)

(2.11.12) $$u_i = p_i u_{i+1} + q_i.$$

Set $\overline{u_i} = 1 - u_i$ and we get

(2.11.13) $$\overline{u_i} = p_i \overline{u_{i+1}}.$$

One readily checks that for any $0 \leq c \leq 1$, a solution of (2.11.13) is

$$\overline{u_i} = c \prod_{k=i}^{\infty} p_k,$$

and thus a solution of (2.11.12) is

$$u_i = 1 - c \prod_{k=i}^{\infty} p_k$$

for any $0 \leq c \leq 1$. Note that if $c = 1$ we have

$$1 - \prod_{k=i}^{\infty} p_k = 1 - P_i[\tau = \infty] = P_i[\tau < \infty],$$

which seems to be the desired solution of (2.11.12).

The most useful solution of (2.11.4) is what turns out to be the minimal solution:

$$U^{\wedge} = \sum_{n=0}^{\infty} Q^n R,$$

where for $i \in T, j \in T^c$ we have

$$
\begin{aligned}
\left(\sum_{n=0}^{\infty} Q^n R \right)_{i,j} &= \sum_{n=0}^{\infty} \sum_{k \in T} Q_{ik}^{(n)} R_{kj} \\
&= \sum_{n=0}^{\infty} \sum_{k \in T} P_i[X_1 \in T, \ldots, X_n \in T, X_n = k, X_{n+1} = j] \\
&= \sum_{n=0}^{\infty} P_i[X_1 \in T, \ldots, X_n \in T, X_{n+1} = j] \\
&= \sum_{n=0}^{\infty} P_i[\tau = n + 1, X_\tau = j] \\
&= P_i[\tau < \infty, X_\tau = j].
\end{aligned}
$$

This last calculation shows that the infinite series defining $U^{\wedge}$ converges.

To check that $U^\wedge$ is indeed a solution of (2.11.4), simply observe

$$\sum_{n=0}^{\infty} Q^n R = R + \sum_{n=1}^{\infty} Q^n R$$

$$= R + \sum_{n=0}^{\infty} Q^{n+1} R$$

$$= R + \sum_{n=0}^{\infty} QQ^n R$$

$$= R + Q \sum_{n=0}^{\infty} Q^n R.$$

So $U^\wedge$ is a solution of (2.11.4'), and, moreover, we can show it is the minimal solution satisfying $0 \le u_{ij} \le 1$. To see this, suppose U is some other solution satisfying the inequalities $0 \le U \le 1$ interpreted componentwise. Then

$$U = QU + R \ge R$$

so that

$$U = QU + R \ge QR + R.$$

Repeating this procedure

$$U \ge Q(QR + R) + R = Q^2 R + QR + R.$$

In general, after N applications of this procedure, we have

$$U \ge \sum_{n=0}^{N} Q^n R.$$

Letting $N \to \infty$ shows that $U \ge U^\wedge$.

There is a unique solution to (2.11.4') bounded between 0 and 1 iff for any U satisfying (2.11.4') and $0 \le U \le 1$ we have

$$0 \le U - U^\wedge = Q(U - U^\wedge) \text{ implies } U - U^\wedge = 0.$$

To understand the last implication, we examine the system of equations

(2.11.14) $$\sum_{j \in T} Q_{ij} x_j = x_i, \quad 0 \le x_i \le 1, \quad i \in T,$$

or, in matrix form,

(2.11.14') $$Qx = x, \quad 0 \le x \le 1.$$

Observe first that

$$(x_i^\vee, i \in T) := (P_i[\tau = \infty], i \in T)$$

satisfies (2.11.14'), since for $i \in T$ we have

$$\begin{aligned}
P_i[\tau = \infty] &= \sum_{k \in T} P_i[X_1 = k, \tau = \infty] \\
&= \sum_{k \in T} P_i[\tau = \infty | X_1 = k] P_i[X_1 = k] \\
&= \sum_{k \in T} p_{ik} P_k[\tau = \infty].
\end{aligned}$$

Also $x^\vee$ has the analytical expression $(i \in T)$

$$\begin{aligned}
x_i^\vee = P_i[\tau = \infty] &= \lim_{n \to \infty} P_i[\tau > n] \\
&= \lim_{n \to \infty} \sum_{k \in T} P_i[X_n = k, \tau > n]
\end{aligned}$$

(2.11.15)
$$= \lim_{n \to \infty} \sum_{k \in T} Q_{ik}^{(n)},$$

the last line following from (2.11.2). We have shown

$$x^\vee = \lim_{n \to \infty} Q^n \mathbf{1}.$$

In fact, $x^\vee$ is the *maximal* solution of (2.11.14'). This is easily seen as follows: If x is some other solution of (2.11.14') then $x \leq \mathbf{1}$, so that multiplying by Q we get $x = Qx \leq Q\mathbf{1}$; repeating this procedure n times yields

$$x = Q^n x \leq Q^n \mathbf{1}$$

and letting $n \to \infty$ and using (2.11.15) we get

$$x \leq \lim_{n \to \infty} Q^n \mathbf{1} =: x^\vee.$$

We summarize these findings.

Proposition 2.11.1. *There is a unique solution of*

(2.11.4.)
$$u_{ij} = \sum_{k \in T} Q_{ik} u_{kj} + p_{ij}, \quad 0 \leq u_{ij} \leq 1, \; i \in T, j \in T^c,$$

or, equivalently, in matrix notation

(2.11.4′) $U = QU + R, \ 0 \le U \le 1,$

iff the system

(2.11.14) $\displaystyle\sum_{j \in T} Q_{ij} x_j = x_i, \quad 0 \le x_i \le 1, \quad i \in T,$

or, in matrix form,

(2.11.14′) $Qx = x, \quad 0 \le x \le 1,$

has only the solution $x_i = 0, i \in T$. *This last condition is equivalent to*

$$P_i[\tau = \infty] = P_i[\ \{X_n\} \ \textit{stays forever in } T] = 0, \quad i \in T.$$

Proof. We have already checked that if the only solution of (2.11.4′) is $x = 0$ then (2.11.4′) has a unique solution. We only need show that if (2.11.4′) has a unique solution, then the only solution of (2.11.14′) is $x = 0$. If (2.11.14′) has a non-zero solution then $x^{\vee} \ne 0$. Note if $i \in T, k \in T^c$,

$$\sum_{l \in T} Q_{il}(U_{lk}^{\wedge} + x_l^{\vee}) + R_{ik} = \sum_{l \in T} Q_{il} U_{lk}^{\wedge} + R_{ik} + \sum_{l \in T} Q_{il} x_l^{\vee}$$
$$= U_{ik}^{\wedge} + x_i^{\vee}.$$

Also,

$$0 \le U_{il}^{\wedge} + x_i^{\vee} = P_i[\tau < \infty, X_\tau = l] + P_i[\tau = \infty] \le 1$$

by definition of $U_{il}^{\wedge}$ and $x_i^{\vee}$, so $(U^{\wedge}(i,l) + x_i^{\vee}, l \in T^c, i \in T)$ is another solution of (2.11.4). ∎

2.12. INVARIANT MEASURES AND STATIONARY DISTRIBUTIONS.

In this section we begin the study of stationary distributions for Markov chains. Stationary distributions are a crucial characteristic of a Markov chain because, as we will see, they control the long run behavior of the chain in many ways. When a stationary distribution exists and is used as the initial distribution of a Markov chain, the Markov chain becomes a *stationary* stochastic process (cf. Proposition 2.12.1); we thus remind the reader of the definition of stationarity: A stochastic process $\{Y_n, n \ge 0\}$ is stationary (sometimes called strictly stationary) if for any integers $m \geq 0$ and $k > 0$ we have

$$(Y_0, \ldots, Y_m) \stackrel{d}{=} (Y_k, \ldots, Y_{m+k}),$$

that is, the two vectors have the same joint distributions whatever the length m of the vector and whatever the translation k may be.

We now present the basic definition. Let $\pi = \{\pi_j, j \in S\}$ be a probability distribution. It is called a *stationary distribution* for the Markov chain with transition matrix P if

$$\pi' = \pi'P,$$

i.e., $\pi_j = \sum_{k \in S} \pi_k p_{kj}$, $j \in S$.

We now check that if we start the chain with a stationary distribution, we get a stationary process. Denote by P_π the distribution of the Markov chain when the initial distribution is π. Thus

$$P_\pi(\cdot) = \sum_{i \in S} P\{(\cdot)|X_0 = i\}\pi_i.$$

Proposition 2.12.1. *With respect to P_π we have that $\{X_n, n \geq 0\}$ is a stationary stochastic process. It follows that*

$$P_\pi[X_n = i_0, X_{n+1} = i_1, \ldots, X_{n+k} = i_k] = \pi_{i_0} p_{i_0 i_1} \cdots p_{i_{k-1} i_k}$$
(2.12.1)
$$= P_\pi[X_0 = i_0, \ldots, X_k = i_k]$$

for any $n \geq 0$ and $k \geq 0$ and $i_0, \ldots, i_k \in S$. In particular ,

$$P_\pi[X_n = j] = \pi_j$$

for any $n \geq 0, j \in S$.

Proof. Since the initial distribution is π, the left side of (2.12.1) is

$$\sum_{i \in S} \pi_i p_{i i_0}^{(n)} p_{i_0 i_1} \cdots p_{i_{k-1} i_k}.$$

Now $\pi' = \pi'P$ implies (right multiply by P successively) $\pi' = \pi'P^n$, so the foregoing is

$$\pi_{i_0} p_{i_0 i_1} \cdots p_{i_{k-1} i_k},$$

which is the right side of (2.12.1). ∎

Questions about the uniqueness and existence of stationary distributions must be resolved. For now we concentrate on interpretations.

If $\nu = \{\nu_j, j \in S\}'$ is a sequence of non-negative constants (think of ν determining a measure on the subsets of S), we call ν an *invariant measure* if

$$\nu' = \nu'P.$$

If ν is invariant and also a probability distribution, then it is a stationary distribution. There are invariant measures ν, however, such that $\sum_{j \in S} \nu_j = \infty$, so that it is impossible to scale such a measure to get a probability distribution.

When a recurrent state exists, the following is useful in manufacturing invariant measures.

Proposition 2.12.2. *Let $i \in S$ be recurrent, and define for $j \in S$*

$$(2.12.2) \qquad \nu_j = E_i \sum_{0 \leq n \leq \tau_i(1)-1} 1_{[X_n=j]} = \sum_{n=0}^{\infty} P_i[X_n = j, \tau_i(1) > n].$$

Then ν is an invariant measure. If i is positive recurrent so that $E_i \tau_i(1) < \infty$, then

$$(2.12.3) \qquad \pi_j = \frac{\nu_j}{E_i(\tau_i(1))} = \frac{E_i \sum_{0 \leq n \leq \tau_i(1)-1} 1_{[X_n=j]}}{E_i(\tau_i(1))}$$

is a stationary distribution.

Remark. ν_j given in (2.12.2) is the expected number of visits to j between two visits to i and π_j is this expected number normalized by the expected cycle length $E_i(\tau_i(1))$.

Proof. We begin by showing $\nu' = \nu'P$. In the next proposition, we check that $\nu_j < \infty$. We have that $\nu_i = 1$, and, with respect to P_i, we have $X_0 = X_{\tau_i(1)} = i$ (note $\tau_i(1) < \infty$ since i is recurrent) so, for $j \neq i$,

$$\nu_j = E_i \sum_{1 \leq n \leq \tau_i(1)} 1_{[X_n=j]} = E_i \sum_{n=1}^{\infty} 1_{[X_n=j, n \leq \tau_i(1)]}$$

$$= p_{ij} + \sum_{n=2}^{\infty} P_i[X_n = j, \tau_i(1) \geq n]$$

$$= p_{ij} + \sum_{k \in S,\ k \neq i} \sum_{n=2}^{\infty} P_i[X_n = j, \tau_i(1) \geq n, X_{n-1} = k]$$

$$= p_{ij} + \sum_{k \in S,\ k \neq i} \sum_{n=2}^{\infty} P_i[X_n = j | \tau_i(1) \geq n, X_{n-1} = k]$$

$$P_i[\tau_i(1) \geq n, X_{n-1} = k].$$

Since $[\tau_i(1) \geq n, X_{n-1} = k] = [X_1 \neq i, \ldots, X_{n-1} \neq i, X_{n-1} = k]$, we have from $\nu_i = 1$ and the Markov property

$$\nu_j = \nu_i p_{ij} + \sum_{k \in S, \, k \neq i} \sum_{n=2}^{\infty} p_{kj} P_i[\tau_i(1) \geq n, X_{n-1} = k]$$

$$= \nu_i p_{ij} + \sum_{k \in S, \, k \neq i} \sum_{m=1}^{\infty} p_{kj} P_i[\tau_i(1) \geq m+1, X_m = k]$$

$$= \nu_i p_{ij} + \sum_{k \in S, \, k \neq i} p_{kj} E_i \sum_{m=1}^{\infty} 1_{[\tau_i(1) > m, X_m = k]}$$

$$= \nu_i p_{ij} + \sum_{k \neq i} p_{kj} E_i \sum_{m=1}^{\tau_i(1)-1} 1_{[X_m = k]}.$$

Since $k \neq i$, this is

$$= \nu_i p_{ij} + \sum_{k \neq i} p_{kj} E_i \sum_{m=0}^{\tau_i(1)-1} 1_{[X_m = k]}$$

$$= \nu_i p_{ij} + \sum_{k \neq i} \nu_k p_{kj}$$

$$= \sum_{k \in S} \nu_k p_{kj}$$

as desired.

If i is positive recurrent then

$$\sum_{j \in S} \nu_j = \sum_{j \in S} E_i \sum_{0 \leq n \leq \tau_i(1)-1} 1_{[X_n = j]}$$

$$= E_i \sum_{0 \leq n \leq \tau_i(1)-1} \sum_{j \in S} 1_{[X_n = j]}$$

$$= E_i \sum_{0 \leq n \leq \tau_i(1)-1} 1 = E_i \tau_i(1) < \infty,$$

so $\{\nu_j / E_i(\tau_i(1)), i \in S\}$ is a probability distribution. ∎

Now we consider the existence and uniqueness of invariant measures.

Proposition 2.12.3. *If the Markov chain is irreducible and recurrent, then an invariant measure ν exists and satisfies $0 < \nu_j < \infty, \forall j \in S$, ν is unique up to multiplicative constants: If $\nu_i' = \nu_i' P, \ i = 1, 2,$ then*

*there exists $c > 0$ such that $\nu_1 = c\nu_2$. Furthermore, if the Markov chain is
positive recurrent and irreducible, we set $m_j := E(\tau_j(1))$, and there exists
a unique stationary distribution π where*

$$\pi_j = 1/E_j(\tau_j(1)) = 1/m_j.$$

Remarks.

    (1)   The most effective method of computing m_j is to solve $\pi' = \pi'P$. Do not overlook this computational recipe given in Proposition 2.12.3.

    (2)   Without irreducibility, uniqueness of π cannot be expected.

Remark (2) is illustrated by Examples 2.12.1 and 2.12.2.

Example 2.12.1. Gamblers ruin on $\{0, 1, \ldots, m\}$. Set $\pi_\alpha = (\alpha, 0, \ldots, 0, 1 - \alpha)'$ for any $0 \le \alpha \le 1$. Since π_α concentrates on absorbing states, the picture at time 0 is frozen for all time, so not only is the stationary distribution not unique, there are uncountably many stationary distributions.

Example 2.12.2. Consider the Markov chain on $\{1, 2, 3, 4\}$, where

$$P = \begin{pmatrix} P_1 & 0 \\ 0 & P_2 \end{pmatrix}$$

and where

$$P_1 = P_2 = \begin{pmatrix} 1/2 & 1/2 \\ 1/2 & 1/2 \end{pmatrix}.$$

For each P_i we have $(1/2, 1/2)$ is a stationary distribution. For any $\alpha, 0 \le \alpha \le 1$, we have

$$(\frac{\alpha}{2}, \frac{\alpha}{2}, \frac{1-\alpha}{2}, \frac{1-\alpha}{2})$$

is a stationary distribution for P.

Remark. In a positive recurrent, irreducible Markov chain,

$$\pi_j/\pi_i = \pi_j E_i(\tau_i(1)) = E_i \sum_{n=0}^{\tau_i(1)-1} 1_{[X_n=j]}$$

is the expected number of visits to j between two visits to i.

    The proof of Proposition 2.12.3 does not use any advanced tools. However, it is a bit long and somewhat tedious. Beginning students may skip to the beginning of Section 2.12.1. For the more mature student, here is the proof.

Proof. Existence follows from the construction in Proposition 2.12.2. For ν constructed there, $\nu_i = 1$ and for any $j \in S$, there exists m such that $p_{ji}^{(m)} > 0$ (by irreducibility). Since $\nu' = \nu'P$ implies $\nu' = \nu'P^m$, we have

$$\nu_i = 1 = \sum_{k \in S} \nu_k p_{ki}^{(m)} \geq \nu_j p_{ji}^{(m)},$$

so $\nu_j < \infty$. To check $\nu_j > 0$, note that the irreducibility assumption implies $i \to j$, and therefore there exists an integer m such that $p_{ij}^{(m)} > 0$. Since $\nu' = \nu'P$ implies $\nu' = \nu'P^m$ we get

$$\nu_j = \sum_{k \in S} \nu_k p_{kj}^{(m)} \geq \nu_i p_{ij}^{(m)} = 1 p_{ij}^{(m)} > 0.$$

Now let $\mu = \{\mu_j, j \in S\}$ be another positive sequence satisfying $\mu' = \mu'P$. A modification of the foregoing shows (rule out $\mu_j \equiv 0$, $\forall j$)

$$0 < \mu_j < \infty, \quad \forall j \in S.$$

We may divide through by μ_i to get a new sequence also called μ with the property $\mu_i = 1$, $\mu' = \mu'P$. We need to show $\nu = \mu$.

We begin by showing $\mu_j \geq \nu_j$ for all $j \in S$. Note that

$$\mu_i = 1$$
$$\mu_j = \sum_{k \in S} \mu_k p_{kj}, \quad j \neq i.$$

These two statements can be summarized by

(2.12.4) $$\mu_j = \delta_{ij} + \sum_{k \in S} \mu_k {}^{(i)}p_{kj},$$

where ${}^{(i)}P$ is the P-matrix with its ith column set equal to 0. Note also that

$$\left(({}^{(i)}P)^n \right)_{k,j} = P_k[X_n = j, \tau_i(1) > n], \quad k \neq j,$$

and keep in mind that

(2.12.5) $$\nu_j = \sum_{n=0}^{\infty} P_i[X_n = j, \tau_i(1) > n] = \sum_{n=0}^{\infty} \left(({}^{(i)}P)^n \right)_{i,j}.$$

Let $\delta_i = (\delta_{ij}, j \in S)'$, and the translation to vector notation of (2.12.4) is

$$\mu' = \delta_i' + \mu'^{(i)}P$$
$$= \delta_i' + (\delta_i' + \mu'^{(i)}P)^{(i)}P = \delta_i' + \delta_i'^{(i)}P + \mu'(^{(i)}P)^2$$
$$\vdots$$
$$= \sum_{n=0}^{N} \delta_i'(^{(i)}P)^n + \mu'(^{(i)}P)^{N+1}.$$

Neglect the last term, and let $N \to \infty$ to get

$$\mu' \geq \sum_{n=0}^{\infty} \delta_i'(^{(i)}P)^n,$$

so

$$\mu_j \geq \sum_{n=0}^{\infty} \left((^{(i)}P)^n\right)_{i,j} = \nu_j$$

from (2.12.5).

To show that $\mu_j = \nu_j$, let $\Delta_1 = \mu_j - \nu_j \geq 0$, and observe $\Delta' = \Delta'P$. Since $\Delta_i = \mu_i - \nu_i = 1 - 1 = 0$, we have $\Delta_j = 0$ for all j, since otherwise, if there were some j_0 such that $\Delta_{j_0} > 0$, we would find that $\Delta_j > 0$ for all j, which would contradict $\Delta_i = 0$.

If the chain is positive recurrent, the invariant distribution is unique up to the multiplicative constants. There can be only one stationary probability distribution. We see that

$$\pi_i = 1/m_i = 1/E_i\tau_i(1).$$

Since the reference state i is arbitrary, the results follow. ∎

A simple numerical illustration of the results in this section is given in the example of Section 2.14. A nice theoretical application of these results is contained in Exercise 2.57, where the law of large numbers for Markov chains provides just the right tool for proving almost sure consistency of non-parametric maximum likelihood estimators for the transition probabilities.

2.12.1. TIME AVERAGES.

Here we consider the strong law of large numbers for a Markov chain, so it is wise to recall the strong law of large numbers for independent, identically distributed random variables. Suppose $\{Y_j\}$ are iid random variables with $E|Y_1| < \infty$. Then

$$P[\lim_{n\to\infty} \frac{\sum_{i=1}^n Y_i}{n} = E(Y_1)] = 1.$$

Sometimes this is written

$$\frac{\sum_{i=1}^n Y_i}{n} \to E(Y_1) \text{ a.s.,}$$

where *a.s.* stands for *almost surely*, meaning with probability 1.

Now for the application to Markov chains. Suppose f has domain S:

$$f : S \mapsto R,$$

and suppose f is well behaved; say $f \geq 0$ or f is bounded. Think of $f(i)$ as a reward for being in state i. We are interested in

$$\lim_{N\to\infty} \sum_{n=0}^N f(X_n)/N,$$

which is the average long run reward rate or the rate at which the system earns.

Observe that if $f(k) = \delta_{ik}$, then $\sum_{n=0}^N 1_{[X_n=i]}$ is the number of visits to i, and $\lim_{N\to\infty} \sum_{n=0}^N f(X_n)/N$ is the relative frequency that the chain visits i.

Proposition 2.12.4. *Suppose the Markov chain is irreducible and positive recurrent, and let π be the unique stationary distribution. Then*

$$\lim_{N\to\infty} \sum_{n=0}^N f(X_n)/N = \pi(f) := \sum_{j\in S} f(j)\pi_j,$$

almost surely for any initial distribution.

Remark. For later purposes, it is more convenient to have the limit in a different form: From (2.12.3)

$$\sum_{j\in S} f(j)\pi_j = \sum_{j\in S} f(j)E_i \sum_{n=0}^{\tau_i(1)-1} 1_{[X_n=j]}/E_i(\tau_i(1))$$

$$= E_i \sum_{j\in S} \sum_{0\leq n<\tau_i(1)} f(j)1_{[X_n=j]}/E_i(\tau_i(1))$$

$$= E_i \sum_{0\leq n<\tau_i(1)} \left(\sum_{j\in S} f(j)1_{[X_n=j]}\right)/E_i(\tau_i(1))$$

(2.12.1.1)

$$= \frac{E_i \sum_{0\leq n<\tau_i(1)} f(X_n)}{E_i(\tau_i(1))} = E_i \sum_{0<n\leq\tau_i(1)} \frac{f(X_n)}{E_i(\tau_i(1))},$$

since $\sum_{j \in S} f(j) 1_{[X_n=j]} = f(X_n)$.

Remark. If $f(k) = \delta_{ik}$ then the long run frequency of being in state i is $\sum_{j \in S} f(j) \pi_j = \pi_i$.

Proof. The proof follows readily from the dissection principle Proposition 2.5.1 and the strong law of large numbers for iid random variables. For simplicity, suppose $f \geq 0$; the case where f is bounded is not much harder. Define

$$B(N) = \sup\{k \geq 0 : \tau_i(k) \leq N\},$$

then $B(N)$ is the number of blocks one can squeeze into $[0, N]$.

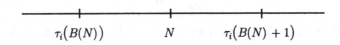

$$\tau_i(B(N)) \qquad N \qquad \tau_i(B(N)+1)$$

FIGURE 2.2. TIME LINE.

Since the variables

$$\eta_k := \sum_{n=\tau_i(k)+1}^{\tau_i(k+1)} f(X_n), \ \ k \geq 1$$

are iid, we have

(2.12.1.2)
$$\sum_{k=1}^{m} \sum_{n=\tau_i(k)+1}^{\tau_i(k+1)} f(X_n)/m \to E_i \sum_{n=1}^{\tau_i(1)} f(X_n)$$

almost surely as $m \to \infty$. Now write

(2.12.1.3)
$$\sum_{n=0}^{\tau_i(B(N))} f(X_n) \leq \sum_{n=0}^{N} f(X_n) \leq \sum_{n=0}^{\tau_i(B(N)+1)} f(X_n).$$

The left side of (2.12.1.3) is

$$\sum_{0 \leq n \leq \tau_i(1)} f(X_n) + \sum_{k=1}^{B(N-1)} \eta_k = J + \sum_{k=1}^{B(N-1)} \eta_k,$$

and the right side is

$$J + \sum_{k=1}^{B(N)} \eta_k,$$

where J is a finite random variable, and hence $J/N \to 0$, as $N \to \infty$. Now we have

$$\frac{\sum_{k=1}^{B(N)} \eta_k}{N} = \frac{\sum_{k=1}^{B(N)} \eta_k}{B(N)} \cdot \frac{B(N)}{N},$$

and, provided you can be convinced that $B(N)/N \to 1/E_i(\tau_i(1))$, we get from (2.12.1.3) and (2.12.1.2) and the strong law of large numbers the following:

$$\lim_{N \to \infty} N^{-1} \sum_{n=0}^{N} f(X_n) = \lim_{N \to \infty} N^{-1} \sum_{k=1}^{B(N)} \eta_k$$

$$= \lim_{N \to \infty} N^{-1} \sum_{k=1}^{B(N-1)} \eta_k = E_i\eta_1 / E_i\tau_i(1)$$

$$= \sum_{j \in S} f(j)\pi_j.$$

It remains to show

(2.12.1.4) $B(N)/N \to 1/E_i(\tau_i(1)).$

Now, from the time line,

(2.12.1.5) $\tau_i(B(N)) \leq N < \tau_i(B(N) + 1).$

Recall $\tau_i(n) = \sum_{k=0}^{n} \alpha_k$ where $\alpha_0, \ldots, \alpha_n$ are independent and $\alpha_1, \ldots, \alpha_n$ are iid with $E_i\alpha_1 = E_i\tau_i(1)$. So $\tau_i(n)/n \to E_i\tau_i(1)$ as $n \to \infty$. In (2.12.1.4) divide by $B(N)$ and let $N \to \infty$ (which implies $B(N) \to \infty$) and

$$\frac{\tau_i(B(N))}{B(N)} \leq \frac{N}{B(N)} \leq \frac{\tau_i(B(N) + 1)}{B(N) + 1} \cdot \frac{B(N) + 1}{B(N)}.$$

Both extremes of these inequalities converge to $E_i(\tau_i(1))$ and hence (2.12.1.4) follows. ∎

Corollary 2.12.5. *If f is bounded,*

$$\lim_{N \to \infty} N^{-1} \sum_{n=1}^{N} E_i f(X_n) = \pi(f), \quad i \in S.$$

In particular, for $f(k) = \delta_{kj}$ we have $E_i f(X_n) = P_i[X_n = j] = p_{ij}^{(n)}$ so

$$\lim_{N\to\infty} N^{-1} \sum_{n=1}^N p_{ij}^{(n)} = \pi_j, \quad i \in S$$

or, in matrix notation,

$$\lim_{N\to\infty} N^{-1} \sum_{n=1}^N P^n = \Pi,$$

where $\Pi_{ij} = \pi_j$ is a matrix with constant columns.

Proof. If $|f(i)| \le M$ then

$$|\sum_{n=1}^N f(X_n)|/N \le M$$

and the result follows from dominated convergence. ∎

2.13. Limit Distributions.

We saw in Section 2.3 that for $a \in (0,1), b \in (0,1)$

$$\lim_{n\to\infty} \begin{pmatrix} 1-a & a \\ b & 1-b \end{pmatrix}^n \to \begin{pmatrix} \frac{b}{a+b} & \frac{a}{a+b} \\ \frac{b}{a+b} & \frac{a}{a+b} \end{pmatrix}.$$

We wonder when such behavior is true in general; i.e., when is it true that

(2.13.1) $$\lim_{n\to\infty} p_{ij}^{(n)} = \pi_j, \quad i,j \in S; \pi_j \ge 0, \sum_{j\in S} \pi_j = 1?$$

If the π_j's were easy to obtain by means other than evaluating limits, they could serve as approximations to the hard-to-obtain entries of P^n and allow rapid qualitative conclusions. A sequence π satisfying (2.13.1) will be called a limit distribution. How easy is it to find limit distributions? The following helps.

Proposition 2.13.1. *A limit distribution is a stationary distribution.*

Proof. We have

$$\pi_j = \lim_{n\to\infty} p_{ij}^{(n+1)} = \lim_{n\to\infty} \sum_{k\in S} p_{ik}^{(n)} p_{kj}.$$

It is tempting to interchange limit and sum, but dominated convergence does not apply when S is infinite since $\{p_{kj}, k \in S\}$ is not a probability distribution in k. If $S = \{0, 1, 2, \dots\}$, we proceed in an elementary way as follows:

$$\pi_j \geq \lim_{n \to \infty} \sum_{k=0}^{M} p_{ik}^{(n)} p_{kj}$$

$$= \sum_{k=0}^{M} \lim_{n \to \infty} p_{ik}^{(n)} p_{kj}$$

$$= \sum_{k=0}^{M} \pi_k p_{kj}.$$

Since for all $M, j \in S$,

$$\pi_j \geq \sum_{k=0}^{M} \pi_k p_{kj},$$

we let $M \to \infty$ to conclude

$$(2.13.2) \qquad \pi_j \geq \sum_{k=0}^{\infty} \pi_k p_{kj}, \forall j \in S.$$

If for some j_0 we had the strict inequality

$$(2.13.3) \qquad \pi_{j_0} > \sum_{k=0}^{\infty} \pi_k p_{kj_0},$$

then summing (2.13.2) over $j \in S$ yields

$$\sum_{j \in S} \pi_j > \sum_{j \in S} \sum_{k \in S} \pi_k p_{kj}$$

$$= \sum_{k} \pi_k \sum_{j} p_{kj} = \sum_{k} \pi_k 1$$

$$= 1,$$

a contradiction. Hence (2.13.3) can happen for no j_0. ∎

If the limits exist in (2.13.1), we know how to calculate them: Solve $\pi' = \pi' P$. The question is when the limits exist.

Support for the existence of limits is provided by Corollary 2.12.5, which tells us

$$\lim_{N \to \infty} \sum_{n=1}^{N} P^n / N = \Pi$$

when the chain is irreducible and positive recurrent. This Cesaro average limit is weaker than the desired $P^n \to \Pi$. Since if $j \in T$ we have $\sum_n p_{ij}^{(n)} < \infty$, which implies $p_{ij}^{(n)} \to 0$, and since $p_{ij}^{(n)} = 0$ if $i \in C_\alpha$, $j \in C_\beta, \alpha \neq \beta$ where C_α, C_β are closed, recurrent classes, it is sensible to focus on irreducible recurrent chains.

It is too much to expect that $P^n \to \Pi$ will always be true. In a chain with period 2, for instance, $p_{jj}^{(2n+1)} = 0$ for all n and for all $j \in S$. We focus therefore on aperiodic, recurrent, irreducible chains.

Theorem 2.13.2. *Suppose the Markov chain is irreducible and aperiodic and that a stationary distribution π exists:*

$$\pi' = \pi'P, \quad \sum_{j \in S} \pi_j = 1, \pi_j \geq 0.$$

Then

    (1)   *The Markov chain is positive recurrent*
    (2) π *is a limit distribution:*

$$\lim_{n \to \infty} p_{ij}^{(n)} = \pi_j, \quad \forall i, j \in S.$$

    (3)   *For all $j \in S$, $\pi_j > 0$.*
    (4)   *The stationary distribution is unique.*

Remark. One of the most useful things about this result is that it provides a practical test for an irreducible chain to be positive recurrent: If a stationary distribution exists, then the chain is positive recurrent. We will give examples of how to apply this in Section 2.15.

Proof. If the chain were transient, then for all $i, j \in S$

$$p_{ij}^{(n)} \to 0, \, n \to \infty,$$

and so for all j

$$\pi_j = \sum_{i \in S} \pi_i p_{ij}^{(n)} \to 0$$

by dominated convergence. Thus $\pi_j = 0$ for all $j \in S$, which contradicts the fact that $\sum_j \pi_j = 1$. Therefore, the chain must be recurrent.

Since the chain is irreducible and recurrent, an invariant measure, unique up to multiplicative constants, must exist (Proposition 2.12.3). From Proposition 2.12.2 for some $c > 0$

$$\nu_j = E_i \sum_{0 \leq n < \tau_i(1)} 1_{[X_n = j]} = c\pi_j,$$

and therefore

$$\infty > \sum_{j \in S} \nu_j = \sum_{j \in S} E_i \sum_{0 \le n < \tau_i(1)} 1_{[X_n = j]}$$

$$= E_i \sum_{0 \le n < \tau_i(1)} \sum_{j \in S} 1_{[X_n = j]}$$

$$= E_i \sum_{0 \le n < \tau_i(1)} 1$$

$$= E_i \tau_i(1),$$

from which we obtain positive recurrence. (Note that this argument did not use aperiodicity.) So (1) follows, and (3) and (4) are covered by Proposition 2.12.2.

For (2) we need the following lemma. Beginning students should note the statement of the lemma and skip the proof.

Lemma 2.13.3. *Let the chain be irreducible and aperiodic. Then for* $i, j \in S$, *there exists* $n_0 = n_0(i, j)$ *such that*

$$\forall n \ge n_0 : \quad p_{ij}^{(n)} > 0.$$

Proof of the Lemma. Let

$$\Lambda = \{n : p_{jj}^{(n)} > 0\}.$$

Some important properties of Λ are the following:

(1) gcd $\Lambda = 1$ since the chain is aperiodic.
(2) $m, n \in \Lambda$ implies $m + n \in \Lambda$ since

$$p_{ij}^{(m+n)} = \sum_{k \in S} p_{jk}^{(m)} p_{kj}^{(n)} \ge p_{jj}^{(m)} p_{jj}^{(n)} > 0.$$

Now, (1) and (2) imply Λ contains all sufficiently large integers (a cheery number theoretic fact; cf. Billingsley, 1986, p. 569, for example), $n \ge n_1$, say.

Given $i, j \in S$, there exists r such that $p_{ij}^{(r)} > 0$. Then for $n \ge r + n_1$

$$p_{ij}^{(n)} = \sum_k p_{ik}^{(r)} p_{kj}^{(n-r)} \ge p_{ij}^{(r)} p_{jj}^{(n-r)} > 0$$

by choice of r and due to $n - r \ge n_1$. ∎

Proof of (2) by the "coupling method". Let $\{X_n\}$ be the original Markov chain. Let $\{Y_n\}$ be independent of $\{X_n\}$ and have transition matrix P and initial vector π so that $\{Y_n\}$ is stationary and $P[Y_n = j] = \pi_j$. Define $\xi_n = (X_n, Y_n)$ so that $\{\xi_n\}$ is a Markov chain on $S \times S$ with the transition matrix

$$P[\xi_{n+1} = (k, \ell)|\xi_n = (i, j)] = p_{ik}p_{j\ell}$$

(see Exercise 2.2) and

$$P[\xi_n = (k, \ell)|\xi_0 = (i, j)] = p_{ik}^{(n)}p_{j\ell}^{(n)}.$$

From the previous lemma, given any (k, ℓ) and $(i, j) \in S \times S$ we have for all sufficiently large n that $p_{ik}^{(n)}p_{j\ell}^{(n)} > 0$. Thus $\{\xi_n\}$ is irreducible.

A stationary probability distribution exists for $\{\xi_n\}$, namely $\pi_{(k,\ell)} := \pi_k\pi_\ell$. To check this, note

$$\sum_{(i,j)\in S\times S} \pi_{i,j}P[\xi_{n+1} = (k, \ell)|\xi_n = (i, j)] = \sum_{(i,j)\in S\times S} \pi_i\pi_j p_{ik}p_{j\ell}$$

$$= \sum_i \pi_i p_{ik} \sum_j \pi_j p_{j\ell} = \pi_k\pi_\ell = \pi_{(k,\ell)}$$

as required. We have an irreducible chain with a stationary distribution, hence (ξ_n) is positive recurrent.

Pick a state i_0 and let

$$\tau_{(i_0,i_0)} = \inf\{n \geq 0 : \xi_n = (i_0, i_0)\}$$

be the hitting time of (i_0, i_0). Since (ξ_n) is recurrent,

$$P[\xi_n = (i_0, i_0) \text{ i.o }] = 1,$$

and thus $P[\tau_{i_0,i_0} < \infty] = 1$. (Recall that i.o stands for infinitely often.)

The idea is this: Imagine two frogs, Sam and Suzie, hopping from rock to rock. Sam hops according to Markov chain $\{X_n\}$ and Suzie follows chain $\{Y_n\}$, but there is a wrinkle. If they both land at rock i_0 together (at time τ_{i_0,i_0}), then Sam jumps on Suzie's back and follows $\{Y_n\}$ from time $\tau_{i_0 i_0}$ onward (the coupling time). Since $X_{\tau_{i_0,i_0}} = Y_{\tau_{i_0,i_0}}$, Sam's total evolution should be equal in distribution to what it would have been if he followed $\{X_n\}$, but since after $\tau_{i_0 i_0}$ Sam rides Suzie on a stationary sequence of states, his state probabilities at time $n > \tau_{i_0 i_0}$ should be governed by π. To make this precise, let P be the probability measure conditional on $P[\xi_0 = (k, \ell)] = \delta_{ki}\pi_\ell$ (i.e., $\{X_n\}$ starts at i and $\{Y_n\}$ starts according to π). Write $\tau = \tau_{i_0,i_0}$ for the coupling time. Observe

$$P[X_n = j, \tau \leq n] = \sum_{m=0}^n P[X_n = j, \tau = m] = \sum_k \sum_{m=0}^n P[\xi_n = (j, k), \tau = m].$$

Applying dissection to the Markov chain $\{\xi_n\}$ at the state (i_0, i_0), we have the last expression equal to

$$\sum_k \sum_{m=0}^n P[\tau = m] P_{i_0 i_0}[\xi_{n-m} = (j,k)] = \sum_k \sum_{m=0}^n P[\tau = m] p_{i_0 j}^{(n-m)} p_{i_0 k}^{(n-m)}$$

$$= \sum_{m=0}^n P[\tau = m] p_{i_0 j}^{(n-m)}.$$

Similarly,

$$P[Y_n = j, \tau \le n] = \sum_k \sum_{m=0}^n P[\xi_n = (k,j), \tau = m],$$

which by dissection of $\{\xi_n\}$ is

$$\sum_k \sum_{m=0}^n P[\tau = m] P_{i_0 i_0}[\xi_{n-m} = k, j] = \sum_k \sum_{m=0}^n P[\tau = m] p_{i_0 k}^{(n-m)} p_{i_0 j}^{(n-m)}$$

$$= \sum_{m=0}^n P[\tau = m] p_{i_0 j}^{(n-m)}.$$

We conclude

$$P[X_n = j, \tau \le n] = P[Y_n = j, \tau \le n],$$

so the state probabilities after coupling are identical. Therefore,

$$\begin{aligned}
|p_{ij}^{(n)} - \pi_j| &= |P[X_n = j] - P[Y_n = j]| \\
&\le |P[X_n = j, \tau \le n] - P[Y_n = j, \tau \le n] \\
&\quad + P[X_n = j, \tau > n] - P|Y_n = j, \tau > n]| \\
&= |P[X_n = j, \tau > n] - P[Y_n = j, \tau > n]| \\
&= |E(1_{[X_n=j]} 1_{[\tau>n]} - 1_{[Y_n=j]} 1_{[\tau>n]})| \\
&\le E|1_{[X_n=j]} - 1_{[Y_n=j]}| 1_{[\tau>n]} \le E1_{[\tau>n]} \\
&= P[\tau > n] \to 0, \text{ as } n \to \infty
\end{aligned}$$

since $P[\tau < \infty] = 1$. ∎

The connection between stationary distributions, limit distributions and positive recurrence is summarized next.

Corollary 2.13.4. *Assume the Markov chain is irreducible and aperiodic. A (the) stationary distribution exists iff the chain is positive recurrent iff a limit distribution exists. If the chain is irreducible and periodic, existence of a stationary distribution is equivalent to positive recurrence.*

Proof. We merely put the pieces together. If the chain is aperiodic, we have positive recurrence equivalent to the existence of a stationary distribution (Proposition 2.12.3 and Theorem 2.13.2), which implies that a limit distribution exists (Theorem 2.13.2), which implies that a stationary distribution exists (Proposition 2.13.1). If the chain is periodic, combine Proposition 2.12.2 and the argument at the beginning of the proof of Proposition 2.13.2. ∎

The phrase *positive recurrent, aperiodic and irreducible* is sometimes subsumed under the name *ergodic*.

The story continues in Section 2.13.1. This may be skipped by beginning students. Those skipping will miss a proof of the useful fact that *a finite state, irreducible, aperiodic Markov chain is always positive recurrent and the stationary distribution always exists*.

We close this section with an example.

Example 2.13.1. The Infinite Capacity Storage Model. Consider the Moran storage model of Section 2.2, but assume infinite storage capacity, namely that the parameter $c = \infty$, and also suppose $m = 1$, so that there is unit release. In this case, the contents process $\{X_n\}$ satisfies the recursion

$$(2.13.4) \qquad X_{n+1} = (X_n + A_{n+1} - 1)_+, \quad n \geq 0.$$

Recall that $\{A_n\}$ are the input variables and are assumed independent, identically distributed and A_{n+1} is independent of X_n for every n. For the distribution of A_1 we have

$$P[A_1 = k] = a_k, \quad k \geq 0$$

and A_1 has generating function

$$A(s) = Es^{A_1} = \sum_{k=0}^{\infty} a_k s^k, \quad 0 \leq s \leq 1.$$

We seek the stationary distribution $\{\pi_k, k \geq 0\}$ of the contents Markov chain $\{X_n\}$ when it exists. If the stationary distribution exists, we know from Theorem 2.13.2 that it is also a limit distribution. Letting $n \to \infty$ in (2.13.4), we get both sides converging in distribution to limit random variables. If X_n converges in distribution to X_∞, say, then X_∞ should have the stationary distribution and satisfy the equation

$$(2.13.5) \qquad X_\infty \overset{d}{=} (X_\infty + A_\infty - 1)_+,$$

where A_∞ has the same distribution as A_1 and is independent of X_∞. Recall that the notation $\xi \overset{d}{=} \eta$ means that ξ has the same distribution as η.

To solve, let

$$\Pi(s) = \sum_{k=0}^{\infty} \pi_k s^k = E s^{X_\infty},$$

and take generating functions on both sides of (2.13.5). This yields

$$\Pi(s) = s^0 P[(X_\infty + A_\infty - 1)_+ = 0]$$
$$+ \sum_{n=1}^{\infty} s^n P[X_\infty + A_\infty - 1 = n]$$
$$= P[X_\infty + A_\infty \le 1] + \sum_{n=1}^{\infty} P[X_\infty + A_\infty = n + 1] s^n$$
$$= P[X_\infty + A_\infty \le 1] + \sum_{j=2}^{\infty} P[X_\infty + A_\infty = j] s^{j-1}$$
$$= P[X_\infty + A_\infty = 0] + P[X_\infty + A_\infty = 1]$$
$$+ \sum_{j=2}^{\infty} P[X_\infty + A_\infty = j] s^j s^{-1}$$
$$= P[X_\infty + A_\infty = 0] + P[X_\infty + A_\infty = 1]$$
$$+ s^{-1} \left(E s^{X_\infty + A_\infty} - P[X_\infty + A_\infty = 0] - P[X_\infty + A_\infty = 1] s \right)$$
$$= P[X_\infty + A_\infty = 0] + s^{-1} \Pi(s) A(s) - s^{-1} P[X_\infty + A_\infty = 0]$$
$$= P[X_\infty + A_\infty = 0](1 - s^{-1}) + s^{-1} \Pi(s) A(s).$$

Solving for $\Pi(s)$ yields

$$\Pi(s) = \frac{P[X_\infty + A_\infty = 0](1 - s^{-1})}{1 - s^{-1} A(s)}$$
$$= \frac{P[X_\infty + A_\infty = 0](s - 1)}{s - A(s)}$$
$$= \frac{P[X_\infty + A_\infty = 0]}{\frac{A(s) - s}{1 - s}}$$
$$= \frac{P[X_\infty + A_\infty = 0]}{1 - \left(\frac{1 - A(s)}{1 - s} \right)}.$$

In order to have a stationary distribution, we must have $\Pi(1) = 1$, so letting $s \uparrow 1$ in the foregoing relation yields

$$1 = \Pi(1) = \frac{P[X_\infty + A_\infty = 0]}{1 - EA_1}.$$

Obviously, for this to be possible we need $EA_1 < 1$, and in this case we see that

$$P[X_\infty + A_\infty = 0] = 1 - EA_1.$$

Using this fact, we arrive at the generating function of the stationary distribution:

$$\Pi(s) = \frac{(1 - EA_1)(1 - s)}{A(s) - s},$$

so, for instance, the long run percentage of time that the reservoir is empty is

$$\pi_0 = \Pi(0) = \frac{1 - EA_1}{a_0}.$$

2.13.1. MORE ON NULL RECURRENCE AND TRANSIENCE*.

As a complement to the discussion in section 2.13, we discuss what happens to limits when the chain is null recurrent or transient.

Proposition 2.13.5. *If the Markov chain is irreducible and aperiodic and either null recurrent or transient, then*

$$\lim_{n \to \infty} p_{ij}^{(n)} = 0, \quad \text{for all } i, j \in S.$$

We know that in the transient case $\sum_{n=0}^{\infty} p_{ij}^{(n)} < \infty$, so of course $\lim_{n \to \infty} p_{ij}^{(n)} = 0$. The new information in this result, therefore, is what happens when the chain is null recurrent.

Proof. Assume the chain is null recurrent. Then a unique invariant measure $\nu = \{\nu_j, j \in S\}'$ exists with the property $\sum_{j \in S} \nu_j = \infty$ since if the sum were finite, the stationary distribution would exist and the chain would be positive recurrent.

Suppose the assertion of the proposition were not true. Then $P^n \to 0$ would be false and there would exist i, j such $p_{ij}^{(n')} \to \delta > 0$ along some subsequence $\{n'\}$. Now use a compactness argument to manufacture subsequential limits for P^n which are not identically zero. Do it like this: Think of the collection of numbers $\{p_{ij}, i, j \in S\}$ as an element in the sequence space $[0,1]^{S \times S}$. This space, being a product of the compact sets

---

* This section may be skipped on first reading by beginning readers.

$[0, 1]$, is also compact. Therefore $\{P^n, n \geq 1\}$ is a sequence in this compact set and hence must have subsequential limits. One of these subsequential limits must be non-zero; just go to the limit along a subsequence of the identified subsequence $\{n'\}$. At least for i, j the limit is non-zero.

Suppose $P^{n''} \to L$, where L is a limit matrix which does not have all its entries zero. We must convince ourselves that L has constant columns; i.e., that L_{kl} is independent of k. Go back to the coupling argument, and suppose $\{X_n\}$ starts in state u and $\{Y_n\}$ starts in state v. As in the original coupling argument, the chain $\{\xi_n\}$ is irreducible. If it were transient, then the assertion of the proposition that we are trying to prove would be obvious, since then

$$0 = \lim_{n \to \infty} P_{uv}[\xi_n = (s, t)] = \lim_{n \to \infty} p_{us}^{(n)} p_{vt}^{(n)}.$$

If the coupled chain is transient for one starting state (u, v), then it is transient for every starting state, and we would have

$$0 = \lim_{n \to \infty} (p_{kl}^{(n)})^2,$$

which gives the desired conclusion. Accordingly, assume the coupled chain is recurrent. The conclusion of the coupling argument is then (τ is again the coupling time)

$$\lim_{n \to \infty} |p_{uk}^{(n)} - p_{vk}^{(n)}| = 0, \forall k \in S,$$

so the subsequential limit matrix L has constant columns. Now mimic the proof of Proposition 2.13.1 to conclude that the rows of L form a stationary probability distribution for P. This is a contradiction, however, to the fact that the invariant distribution ν for P is unique up to multiplicative constants and $\sum_{l \in S} \nu_l = \infty$. The contradiction arose because we assumed there were states i, j such that the limit of $p_{ij}^{(n)}$ was not zero. ∎

We may now conclude that in a finite state space Markov chain which is irreducible and aperiodic we cannot have all states null recurrent. Since we now know in a null recurrent chain that

$$\lim_{n \to \infty} p_{ij}^{(n)} = 0$$

for all $i, j \in S$, the proof of this fact is exactly as the proof of the fact that not all states can be transient (Proposition 2.9.4). We are therefore led to the conclusion that *a finite state, irreducible, aperiodic Markov chain is always positive recurrent and the stationary distribution always exists.*

We know in the positive recurrent case that a stationary distribution exists (Proposition 2.12.3). When the chain is null recurrent, a stationary distribution does not exist, but an invariant measure ν, unique up to multiplicative constants, exists which satisfies $\sum_{j \in S} \nu_j = \infty$. If the chain is transient, then an invariant measure may or may not exist; if it exists, it is not a finite measure, and it need not be unique.

Example 2.13.2. Let us now consider the unrestricted random walk on $S = \{ \ldots, -1, 0, 1, \ldots \}$ with $(p > 0, p + q = 1)$

$$p_{i-1,i} = p, \quad p_{i+1,i} = q.$$

The transition matrix is doubly stochastic, which means that not only is it true that row sums equal 1 but also that the column sums equal 1: $\sum_{i \in S} p_{ij} = 1$ for all $j \in S$. From the doubly stochastic property it is easy to check that $\mathbf{1} = (\ldots, 1, 1, 1, \ldots)$ is an invariant measure. (See also Exercise 2.23. Since this invariant measure is not summable, we have independent corroboration of the fact that this chain is never positive recurrent.) However, if we try to solve the system $\mu' = \mu' P$ in the case $p \neq q$, i.e., in the transient case, we get

$$\mu_i = \sum_{k \in S} \mu_k p_{ki} = \mu_{i-1} p_{i-1,i} + \mu_{i+1} p_{i+1,i} = \mu_{i-1} p + \mu_{i+1} q.$$

Using our experience solving this type of system for the absorption probabilities of the gamblers ruin problem, we might guess that a solution is

$$\mu_i = (\frac{p}{q})^i, \quad i \in S.$$

Indeed, it is an easy verification to check that this is an invariant measure. Note that this μ is not a multiplicative variant of $\mathbf{1}$, so when the random walk is transient we find invariant measures exist but they are not unique.

Example 2.13.3. In this example of a transient Markov chain, we find that an invariant measure does not exist. Take the transient success run chain on the state space $S = \{0, 1, \ldots \}$ with $p_{i,i+1} = p_i$ and $p_{i0} = q_i$. Recall that transience means that $\prod_i p_i > 0$ or, equivalently, $\sum_i q_i < \infty$. The equation $\mu' = \mu' P$ yields

$$(2.13.1.1) \qquad \mu_0 = \sum_{k=0}^{\infty} \mu_k p_{k0} = \sum_{k=0}^{\infty} \mu_k q_k$$

and, for $i \geq 1$,

$$(2.13.1.2) \qquad \mu_i = \sum_{k=0}^{\infty} \mu_k p_{ki} = \mu_{i-1} p_{k-1}.$$

This last system is easily solved to yield, for $i \geq 1$,

$$(2.13.1.3) \qquad \mu_i = \mu_0 \prod_{j=0}^{i-1} p_j.$$

If an invariant measure exists, we may always set $\mu_0 = 1$. Doing so gives, from (2.13.1.1), that

$$1 = \sum_{k=0}^{\infty} \mu_k q_k = q_0 + \sum_{k=1}^{\infty} (\prod_{j=0}^{k-1} p_j) q_k$$

$$= q_0 + \sum_{k=1}^{\infty} (\prod_{j=0}^{k-1} p_j - \prod_{j=0}^{k} p_j)$$

$$= q_0 + \lim_{N \to \infty} \sum_{k=1}^{N} (\prod_{j=0}^{k-1} p_j - \prod_{j=0}^{k} p_j)$$

$$= q_0 + \lim_{N \to \infty} (p_0 + \prod_{j=0}^{N} p_j)$$

$$= q_0 + p_0 + \lim_{N \to \infty} \prod_{j=0}^{N} p_j$$

$$= 1 + \lim_{N \to \infty} \prod_{j=0}^{N} p_j.$$

We get a contradiction in the transient case where $\prod_{j=0}^{\infty} p_j > 0$, so no invariant measure exists.

In the recurrent case the invariant distribution is

$$\mu_0 = 1, \quad \mu_i = \prod_{j=0}^{i-1} p_j, \quad i \geq 1.$$

2.14. COMPUTATION OF THE STATIONARY DISTRIBUTION.

For a finite state Markov chain, solutions of the equation $\pi' = \pi' P$ can be found by hand, although this is tedious with even a moderate number of states. Out of deference to tradition, the outline of this procedure, applicable when the Markov chain has a small number m of states, is as follows:

(1) The vector equation $\pi' = \pi' P$ yields m equations in the m unknowns $\pi_1, \ldots, \pi_m$, but another equation, namely $\sum_i \pi_i = 1$, is

also available. So survey the m equations, $\pi_i = \sum_k \pi_k p_{ki}, 1 \leq i \leq m$, and delete one that looks ugly. Try to keep equations with lots of zero coefficients. Add to this batch of $m-1$ equations, the mth equation $\sum_k \pi_k = 1$.

(2) Replace π_i by x_i, and solve the resulting system of equations by giving x_1, say, an arbitrary but convenient value. Solve for $x_2, \ldots, x_m$ in terms of x_1.

(3) Set $\pi_i = x_i / (\sum_k x_k)$.

This method will be illustrated after discussion of a more machine oriented approach which is applicable when the solution by hand is overly tedious.

Proposition 2.14.1. *Let P be an $m \times m$ irreducible stochastic matrix and suppose ONE is the $m \times m$ matrix all of whose entries are 1. Then, if π is the stationary distribution, we have*

$$(2.14.1) \qquad \pi' = (1, \ldots, 1)(I - P + ONE)^{-1}.$$

Proof. To check this, suppose temporarily that we know that $I - P + ONE$ has an inverse. Since π' satisfies $\pi'(I - P) = 0'$, we have

$$\pi'(I - P + ONE) = 0' + \pi'(ONE) = (1, \ldots, 1).$$

Solving for π' yields

$$\pi' = (1, \ldots, 1)(I - P + ONE)^{-1},$$

as desired.

We now verify that $I - P + ONE$ has an inverse. We can do this by showing that if $(I - P + ONE)x = 0$ then $x = 0$. But if π' is a stationary vector it satisfies $\pi'(I - P) = 0'$, so if $(I - P + ONE)x = 0$, we get by left multiplying by π' that

$$\pi'(I - P + ONE)x = 0' + \pi'(ONE)x = 0.$$

Thus we conclude $\pi'(ONE)x = 0$. But

$$\pi'(ONE) = (1, \ldots, 1)$$

so $(1, \ldots, 1)x = 0$, which implies $(ONE)x = 0$. We conclude $(I - P)x = 0$, which is the same as $Px = x$. This implies for any n that $x = P^n x$, and therefore that $N^{-1} \sum_{n=1}^N P^n x = x$. From Corollary 2.12.5 we have

$$N^{-1} \sum_{n=1}^N P^n \to \Pi,$$

where Π is the matrix with constant columns: $\Pi_{i,j} = \pi_j$. Thus we have as $N \to \infty$ that

$$\mathbf{x} = N^{-1} \sum_{n=1}^{N} P^n \mathbf{x} \to \Pi \mathbf{x},$$

so $\mathbf{x} = \Pi \mathbf{x}$, from which in coordinate form,

$$x_i = \sum_{\alpha=1}^{m} \pi_\alpha x_\alpha,$$

and the right side is independent of i. This means that for some constant c,

$$\mathbf{x}' = c(1, \dots, 1)'.$$

Since we also have

$$0 = (1, \dots, 1)\mathbf{x} = c(1, \dots,)(1, \dots, 1)' = cm,$$

we get $c = 0$ and consequently $\mathbf{x} = 0$. Therefore, since

$$(I - P + \text{ONE})\mathbf{x} = 0$$

implies $\mathbf{x} = 0$, we conclude that $(I - P + \text{ONE})$ is invertible. ∎

Example. Harry, the Semipro.   Our hero, Happy Harry, used to play semipro basketball where he was a defensive specialist. His scoring productivity per game fluctuated between three states: 1 (scored 0 or 1 points), 2 (scored between 2 and 5 points), 3 (scored more than 5 points). Inevitably, if Harry scored a lot of points in one game, his jealous teammates refused to pass him the ball in the next game, so his productivity in the next game was nil. The team statistician, Mrs. Doc, upon observing the transitions between states, concluded these transitions could be modelled by a Markov chain with transition matrix

$$P = \begin{pmatrix} 0 & \frac{1}{3} & \frac{2}{3} \\ \frac{1}{3} & 0 & \frac{2}{3} \\ 1 & 0 & 0 \end{pmatrix}.$$

(1)   What is the long run proportion of games that our hero had high scoring games?

(2)   The salary structure in the semipro leagues includes incentives for scoring. Harry was paid \$40/game for a high scoring performance, \$30/game when he scored between 2 and 5 points and only \$20/game when he scored nil. What was the long run earning rate of our hero?

Solution by hand: The system $x' = x'P$ yields

$$\frac{1}{3}x_2 + x_3 = x_1$$

$$\frac{1}{3}x_1 = x_2$$

$$\frac{2}{3}x_1 + \frac{2}{3}x_2 = x_3.$$

The second equation says

$$x_2 = x_1/3,$$

and the third says

$$x_3 = \frac{2}{3}(x_1 + x_2) = \frac{2}{3}(x_1 + \frac{x_1}{3}) = \frac{2}{3} \cdot \frac{4}{3} \cdot x_1 = \frac{8}{9}x_1.$$

Now set

$$\pi_1 = \frac{x_1}{x_1 + x_2 + x_3} = \frac{x_1}{x_1 + \frac{x_1}{3} + \frac{8}{9}x_1} = \frac{9}{9 + 3 + 8} = \frac{9}{20} = .45$$

$$\pi_2 = \frac{x_2}{x_1 + x_2 + x_3} = \frac{x_1/3}{x_1 + x_1/3 + \frac{8}{9}x_1} = \frac{3}{9 + 3 + 8} = \frac{3}{20} = .15$$

$$\pi_3 = 1 - (\pi_1 + \pi_2) = .4,$$

the answer to (1) is $\pi_3 = .40$. For (2), recall from Proposition 2.12.4 that if $f : S \mapsto R$ is bounded then

$$\lim_{n \to \infty} \sum_{n=0}^{N} f(X_n)/N = \pi(f) = \sum_{i=1}^{3} \pi_i f(i).$$

In our case $f(1) = 20, f(2) = 30, f(3) = 40$ and $\pi(f) = (20)\frac{9}{20} + (30)\frac{3}{20} + (40)\frac{8}{20} = 9 + \frac{9}{2} + 16 = 29.5$.

Solution by machine: We have, with help from Minitab ,

$$(I - P + \text{ONE}) = \begin{pmatrix} 2 & .66667 & .33333 \\ .66667 & 2 & .33333 \\ 0 & 1 & 2 \end{pmatrix}$$

so that

$$(I - P + \text{ONE})^{-1} = \begin{pmatrix} .55 & -.15 & -.066667 \\ -.2 & .6 & .066667 \\ .1 & -.3 & .53333 \end{pmatrix}$$

and
$$(\pi_1, \pi_2, \pi_3) = (1, 1, 1)(I - P + \text{ONE})^{-1} = (.45, .15, .4).$$

The long run earning rate is

$$\pi(f) = (.45, .15, .4) \begin{pmatrix} 20 \\ 30 \\ 40 \end{pmatrix} = 29.5. \qquad \blacksquare$$

By the way, this chain is aperiodic since $p_{11}^{(2)} > 0$ and $p_{11}^{(3)} > 0$ and it is irreducible.

Example 2.14.2. An Inventory Model. Recall Example 2.2.6 of Section 2.2. We let $s = 0$ and $S = 2$. Suppose now that we have a simple distribution of demands in any period, namely,

$$P[D_1 = 0] = .5, \quad P[D_1 = 1] = .4, \quad P[D_1 = 2] = .1.$$

From the recursions defining the inventory level Markov chain, we can quickly check that the state space is $\{0, 1, 2\}$ with transition matrix

$$P = \begin{pmatrix} 0.1 & 0.4 & 0.5 \\ 0.5 & 0.5 & 0.0 \\ 0.1 & 0.4 & 0.5 \end{pmatrix}.$$

We seek the stationary distribution and must therefore compute $(I - P + \text{ONE})^{-1}$. Using Minitab, we obtain

$$I - P + \text{ONE} = \begin{pmatrix} 1.9 & 0.6 & 0.5 \\ 0.5 & 1.5 & 1.0 \\ 0.9 & 0.6 & 1.5 \end{pmatrix}.$$

Inverting, we get

$$(I - P + \text{ONE})^{-1} = \begin{pmatrix} 0.611111 & -0.222222 & -0.055556 \\ 0.055556 & 0.888889 & -0.611111 \\ -0.388889 & 0.222222 & 0.944444 \end{pmatrix}.$$

Thus, the stationary distribution is

$$(\pi_0, \pi_1, \pi_2) = (1, 1, 1)(1 - P + \text{ONE})^{-1}$$
$$= (0.277778, 0.444444, 0.277778).$$

The mean of this distribution is $2.(.277778) + .444444 = 1$. And, by Proposition 2.12.4, with $f(i) = i$, $i = 0, 1, 2$, this is also the long run average inventory level: $\lim_{N \to \infty} N^{-1} \sum_{n=0}^{N} X_n$.

2.15. CLASSIFICATION TECHNIQUES.

Every finite state, irreducible Markov chain is positive recurrent. However, when the state space is infinite, it can be challenging to classify the model as positive recurrent, null recurrent or transient. Furthermore, for a model depending on certain parameters (e.g., input rate, output rate, and so forth) it is of qualitative interest to obtain these classifications as a function of model parameters.

To test for positive recurrence, the most straightforward approach is to test for the existence of a stationary distribution since we know for an irreducible aperiodic chain that positive recurrence and existence of a stationary distribution are equivalent. This is the method used to obtain a criterion for positive recurrence in the queueing example.

Queueing Example. Recall the queueing example from Exercise 2.6 or from Section 2.2. The matrix is

$$P = \begin{pmatrix} a_0 & a_1 & a_2 & \cdots \\ a_0 & a_1 & a_2 & \cdots \\ 0 & a_0 & a_1 & a_2 & \cdots \\ 0 & 0 & a_0 & a_1 & \cdots \\ \vdots & & & \ddots \end{pmatrix},$$

and the state space is $\{0, 1, 2, ...\}$. Recall further that $a_i \geq 0, \sum_{i=0}^{\infty} a_i = 1$. The system $\pi' = \pi' P$ yields

$$\pi_0 = \pi_0 a_0 + \pi_1 a_0$$
$$\pi_1 = \pi_0 a_1 + \pi_1 a_1$$
$$\pi_2 = \pi_0 a_2 + \pi_1 a_2 + \pi_2 a_1 + \pi_3 a_0$$
$$\pi_3 = \pi_0 a_3 + \pi_1 a_3 + \pi_2 a_2 + \pi_3 a_1 + \pi_4 a_0$$
$$\vdots$$

Since the ith column of P is $(a_i, a_i, a_{i-1}, ..., a_0, 0, ...)'$, we find for $i \geq 0$

$$(2.15.1) \qquad \pi_i = \pi_0 a_i + \sum_{j=1}^{i+1} \pi_j a_{i+1-j}.$$

Set $\Pi(s) = \sum_{i=0}^{\infty} \pi_i s^i$. We attempt to solve (2.15.1) by generator function methods. Multiply (2.15.1) by s^i and sum to get

$$(2.15.2) \qquad \Pi(s) = \sum_{i=0}^{\infty} \pi_i s^i = \pi_0 \sum_{i=0}^{\infty} a_i s^i + \sum_{i=0}^{\infty} \sum_{j=1}^{i+1} \pi_j a_{i+1-j} s^i.$$

We need to reverse the order of summation in (2.15.2). Note $1 \leq j \leq i+1$ implies $i \geq j - 1$ and $j \geq 1$, so, setting $A(s) = \sum_{i=0}^{\infty} a_i s^i$, the right side of (2.15.2) is

$$\pi_0 A(s) + \sum_{j=1}^{\infty} \pi_j s^{j-1} \sum_{i=j-1}^{\infty} a_{i-j+1} s^{i-j+1} = \pi_0 A(s) + s^{-1} \left(\sum_{j=1}^{\infty} \pi_j s^j \right) A(s)$$

$$= \pi_0 A(s) + s^{-1} \left(\Pi(s) - \pi_0 \right) A(s).$$

Therefore

$$\Pi(s) = \pi_0 A(s)(1 - s^{-1}) + s^{-1} \Pi(s) A(s),$$

from which

$$\Pi(s) = \pi_0 A(s)(1 - s^{-1})/(1 - s^{-1} A(s))$$

$$= \frac{\pi_0 A(s)}{\frac{1 - s^{-1} A(s)}{1 - s^{-1}}}$$

$$= \frac{\pi_0 A(s)}{\frac{1 - s^{-1} + s^{-1} - s^{-1} A(s)}{1 - s^{-1}}},$$

thus we conclude

(2.15.3)
$$\Pi(s) = \frac{\pi_0 A(s)}{1 - \frac{1 - A(s)}{1 - s}}.$$

This solves for $\Pi(s)$ as a function of π_0 and $A(s)$. Now the question is, when is it possible to specify π_0 so $\Pi(1) = \sum_{0}^{\infty} \pi_k = 1$? In such cases a stationary distribution exists.

In (2.15.3) let $s \uparrow 1$ on the left side to get

$$\Pi(1) = \sum_{k=0}^{\infty} \pi_k.$$

Let

$$\lim_{s \uparrow 1} \frac{1 - A(s)}{1 - s} = A'(1) = \rho = \sum_{k=0}^{\infty} k a_k$$

be the mean number of arrivals per service interval. Since we assume $\{a_k\}$ is a probability distribution, we have $A(1) = 1$. If we take the limit on the right side of (2.15.3), we get

$$\frac{\pi_0 A(1)}{1 - \lim_{s \uparrow 1} \frac{1 - A(s)}{1 - s}} = \frac{\pi_0}{1 - \rho},$$

and we see that it is possible to choose π_0 so that $\Pi(1) = \sum_{0}^{\infty} \pi_k = 1$ iff $0 < \rho < 1$ and in this case $\pi_0 = 1 - \rho$.

We conclude that this queueing model is positive recurrent iff $\rho < 1$, which says the number of arrivals does not overwhelm the service facility.

Now consider the following criterion for transience or recurrence.

Proposition 2.15.1. *Consider an irreducible Markov chain with state space S. Pick a reference state, say 0, and set $Q = (p_{ij}, i, j \in S \setminus \{0\})$. Then the Markov chain is transient iff the system*

$$(2.15.4) \qquad Qx = x, \quad 0 \le x_i \le 1, \ i \in S \setminus \{0\},$$

has a solution not identically 0. The Markov chain is recurrent iff the only solution of (2.15.4) is 0.

Proof. Pretend 0 is an absorbing state with $T = S \setminus \{0\}$ and $\tau = \tau_0 = \inf\{n \ge 0 : X_n \in T^c\} = \inf\{n \ge 0 : X_n = 0\}$. We need the following cheery fact about the original chain:

$$(2.15.5) \qquad 0 \text{ is recurrent iff } \forall \, i \neq 0, f_{i0} = 1.$$

(Certainly if 0 is recurrent, then we know from Proposition 2.9.1 that $f_{i0} = 1$, $\forall \, i \in S$. Conversely, suppose $f_{i0} = 1$, $\forall \, i \neq 0$. Then

$$(2.15.6) \qquad f_{00} = p_{00} + \sum_{j \neq 0} p_{0j} f_{j0},$$

since either we go to 0 in one step or to an intermediate state j from which we ultimately pass to 0 (cf. Section 2.10.1 for the mathematics of this argument). The right side then becomes

$$p_{00} + \sum_{j \neq 0} p_{0j} \cdot 1 = \sum_{j \in S} p_{0j} = 1 = f_{00},$$

implying recurrence.)

Observe that for $i \neq 0$

$$1 - f_{i0} = P_i[\tau = \infty] =: x_i^{\vee}$$

in the notation introduced following (2.11.14'), and also recall from Section (2.11) that $x^{\vee} = \{x_i^{\vee}, i \in T\}'$ satisfies (2.15.4) and is, in fact, the maximal solution satisfying (2.14.4) and $0 \le x \le 1$.

The following holds.

Lemma 2.15.2. $x^{\vee}$ *is the maximal solution of (2.15.4) and either $x^{\vee} = 0$ or $\sup_{i \in T} x_i^{\vee} = 1$.*

Proof. We already know that $x^{\vee}$ is a solution of (2.15.4), and that it is maximal. If $x^{\vee} \neq 0$, then $\sup_{i \neq 0} x_i^{\vee} = c > 0$ and, in matrix form, $x^{\vee} \le c\mathbf{1}$. $x^{\vee} = Q^n x^{\vee} \le cQ^n \mathbf{1}$ from which, for $i \in T$,

$$x_i^{\vee} \le cP_i[\tau > n] \to cx_i^{\vee}.$$

But $x^\vee \neq \mathbf{0}$ means that for some $i \in T$ we have $x_i^\vee \neq 0$, and, since $x_i^\vee \leq cx_i^\vee$, we divide by the nonzero $x_i^\vee$ to get $c \geq 1$. Since $c = \sup_{i \in T} x_i^\vee \leq 1$, we have $c = 1$. ∎

Now we are in a position to prove the proposition speedily. If the original chain is transient, then (2.15.5) informs us that there exists $i \in T$ with $f_{i0} < 1$. This means $x_i^\vee = 1 - f_{i0} > 0$, and there exists a nonzero solution of (2.15.4). Conversely, if a non-zero solution of (2.15.4) exists, then for some $i \in T$, $x_i^\vee > 0$, from which $f_{i0} < 1$. Another application of (2.15.5) yields the conclusion that the chain is transient. This completes the proof of Proposition 2.15.1.

Consider again the queueing example. The matrix, assumed irreducible, is

$$P = \begin{pmatrix} a_0 & a_1 & a_2 & & \cdots \\ a_0 & a_1 & a_2 & & \cdots \\ 0 & a_0 & a_1 & a_2 & \cdots \\ 0 & 0 & a_0 & a_1 & \cdots \\ \vdots & & & \ddots & \end{pmatrix},$$

where $a_i \geq 0$, $\sum_{i=0}^{\infty} a_i = 1$, $\sum_{k=0}^{\infty} ka_k = \rho$.

Proposition 2.15.3. *The queueing example is*

$$\text{transient iff } \rho > 1$$
$$\text{null recurrent iff } \rho = 1$$
$$\text{positive recurrent iff } \rho < 1.$$

Proof. We only need to focus on the statement about transience, since the criterion for positive recurrence is already established.

If $\rho > 1$ we show $x = Qx$, $\mathbf{0} \leq x \leq \mathbf{1}$ has a non-zero solution. With $T = \{1, 2, 3, \ldots\}$ the system $x = Qx$ is

(2.15.7)
$$x_1 = \sum_{i=1}^{\infty} a_i x_i$$

$$x_2 = a_0 x_1 + a_1 x_2 + \ldots$$
$$x_3 = a_0 x_2 + a_1 x_3 + \ldots$$

$$\vdots$$

(2.15.8)
$$x_n = \sum_{i=0}^{\infty} a_i x_{i+n-1}, \quad n \geq 2.$$

Try a solution of the form $x_i = 1 - s^i$, $0 < s < 1$. (The more obvious choice $x_i = s^i$ works fine for (2.15.8) but not for (2.15.7).) From (2.15.8) we get

$$1 - s^n = \sum_{i=0}^{\infty} a_i(1 - s^{i+n-1})$$

$$= 1 - \left(\sum_{i=0}^{\infty} a_i s^i\right) s^{n-1}.$$

Setting $A(s) = \sum_{i=0}^{\infty} a_i s^i$ yields

$$s^n = A(s)s^{n-1}$$

or

$$s = A(s).$$

This is also the equation that results from (2.15.7) with $x_i = 1 - s^i$, $i \geq 1$. But since we are experienced in the art of branching processes, we know that if $\rho > 1$ the equation $s = A(s)$ has a solution in $0 \leq s < 1$. Therefore $\rho > 1$ implies a nonzero solution to $x = Qx$ exists, and hence transience ensues.

It is not clear how to get the fact that the chain being transient implies that 2.15.4 has a non-zero solution. The following simple approach suffices. If $\{X_n\}$ is transient, then for each $j \in \{0, 1, 2 \ldots\}$ there is a last visit. Hence there is a last visit to $\{0, 1, 2, \ldots, M\}$ for any M. This implies that there exists $n_0 = n_0(M, \omega)$ such that for $n \geq n_0$ we have $X_n(\omega) > M$. Then $X_n(\omega) \to \infty$ as $n \to \infty$. Since

$$X_{n+1} = (X_n - 1)^+ + A_{n+1},$$

where A_{n+1} is the number of arrivals in a service period and

$$P[A_{n+1} = k] = a_k, \quad EA_{n+1} = \rho,$$

we have for large n ($n \geq n_0$)

$$X_{n+1}(\omega) = (X_n(\omega) - 1) + A_{n+1}(\omega)$$

and for $N \geq n_0$

$$\sum_{n=n_0}^{N} (X_{n+1}(\omega) - X_n(\omega)) = -(N - n_0) + \sum_{n=n_0}^{N} A_{n+1}(\omega).$$

Simplifying,

$$X_{N+1}(\omega) - X_{n_0}(\omega) = -(N - n_0) + \sum_{n=n_0+1}^{N+1} A_n(\omega),$$

and

$$X_{N+1}(\omega) - \sum_{n=1}^{N+1} (A_n(\omega) - 1) = X_{n_0}(\omega) + n_0 - \sum_{n=1}^{n_0} A_n(\omega).$$

The right side is constant, so, since $X_{N+1} \to \infty$, we have

$$\sum_{n=1}^{N+1} (A_n - 1) \to \infty.$$

Sums of iid random variables with finite mean $\rho - 1$ converge to $+\infty$ iff $\rho - 1 > 0$, i.e., $\rho > 1$. (You have not really seen a proof of this here but supporting evidence comes from the simple random walk; when the mean is 0 the walk is recurrent. Since it keeps coming back to 0 it cannot converge to $+\infty$.)

EXERCISES

2.1. Consider a Markov chain on states $\{0, 1, 2\}$ with transition matrix

$$P = \begin{pmatrix} .3 & .3 & .4 \\ .2 & .7 & .1 \\ .2 & .3 & .5 \end{pmatrix}.$$

Compute $P[X_{16} = 2 | X_0 = 0]$ and $P[X_{12} = 2, X_{16} = 2 | X_0 = 0]$. Try not to do this by hand.

2.2. Let $\{X_n\}$ and $\{Y_n\}$ be two independent Markov chains, each with the same discrete state space S and same transition probabilities. Define the process $\{Z_n\} = \{(X_n, Y_n)\}$ with state space $S \times S$. Show $\{Z_n\}$ is a Markov chain and give the transition probability matrix.

2.3. Show for a Markov chain that, for any $n \geq 1$ and subsets $A_0, \ldots, A_{n-1}$ of the state space,

$$P[X_{n+1} = j | X_0 \in A_0, \ldots, X_{n-1} \in A_{n-1}, X_n = i] = p_{ij}.$$

Verify by giving an example that the following statement is incorrect: For subsets $A_0, \ldots, A_n$ where A_n is not a singleton, we have

$$P[X_{n+1} = j | X_0 \in A_0, \ldots, X_n \in A_n] = P[X_{n+1} = j | X_n \in A_n].$$

2.4. Suppose $p_{ii} > 0$, and let η_i be the exit time from state i:

$$\eta_i = \inf\{n \geq 1 : X_n \neq i\}.$$

Show that η_i has a geometric distribution with respect to P_i.

2.5. If $\{X_n, n \geq 0\}$ is a Markov chain, then by example show that $\{f(X_n), n \geq 0\}$ need not be a Markov chain. (Hint: If f is $1 - 1$ then $\{f(X_n), n \geq 0\}$ is a Markov chain.)

2.6. Suppose E is some space, say a metric space, and that $\{V_n, n \geq 0\}$ are iid random elements in E. For instance, E could be the sequence space R^∞. Imagine that we have functions $g_i, i = 1, 2$ such that

$$g_i : S \times E \to S.$$

Define

$$X_0 = g_1(j, V_0) \quad \text{for some } j \in S$$
$$X_1 = g_2(X_0, V_1),$$

$$\vdots \quad \vdots$$

$$X_{n+1} = g_2(X_n, V_{n+1}),$$

and so on.

(a) Show that $\{X_n\}$ is a Markov chain. (cf. the proof that the "simulated chain" is Markov).

(b) Apply this to the simple branching process.

(c) Apply this to the following single server queueing model: Customers arrive and wait until being served in a first come, first served basis. Between times $n - 1$ and n, the number of arrivals is a random variable A_n with distribution

$$P[A_n = k] = a_k, \quad a_k \geq 0, \quad \sum_{k=0}^{\infty} a_k = 1.$$

Assume the random variables $\{A_n\}$ are iid, and suppose the length of each service is one unit. Let X_n be the number of customers in the system at the start of the nth service period. Write a recursion linking X_{n+1} and

X_n. Apply (1) to conclude that $\{X_n\}$ is a Markov chain. What is the transition matrix?

2.7. For a subset $S_0 \subset S$ define the *closure* of S_0, written $cl(S_0)$, to be the smallest closed set containing S_0.

(a) Prove $cl(\{j\}) = \{k \in S : j \to k\}$.

(b) For the deterministically monotone Markov chain, what is $cl(\{j\})$?

(c) If j is recurrent, show $cl(\{j\})$ is the equivalence class of j.

(d) In the gambler's ruin chain on $\{0, 1, 2, 3\}$, what is the closure of $\{1, 2\}$?

2.8. Consider a Markov chain on $\{1, 2, 3\}$ with transition matrix

$$P = \begin{pmatrix} 1 & 0 & 0 \\ \frac{1}{2} & \frac{1}{6} & \frac{1}{3} \\ \frac{1}{3} & \frac{3}{5} & \frac{1}{15} \end{pmatrix}.$$

Find $f_{13}^{(n)}$ for $n = 1, 2, 3, \ldots$.

2.9. Consider a Markov chain on the states $\{1, \ldots, 9\}$ with transition matrix

$$P = \begin{pmatrix} 0 & .5 & 0 & 0 & .5 & 0 & 0 & 0 & 0 \\ 0 & 0 & 1 & 0 & 0 & 0 & 0 & 0 & 0 \\ 0 & 0 & 0 & 1 & 0 & 0 & 0 & 0 & 0 \\ 1 & 0 & 0 & 0 & 0 & 0 & 0 & 0 & 0 \\ 0 & 0 & 0 & 0 & 0 & 1 & 0 & 0 & 0 \\ 0 & 0 & 0 & 0 & 0 & 0 & 1 & 0 & 0 \\ 0 & 0 & 0 & 0 & 0 & 0 & 0 & 1 & 0 \\ 0 & 0 & 0 & 0 & 0 & 0 & 0 & 0 & 1 \\ 1 & 0 & 0 & 0 & 0 & 0 & 0 & 0 & 0 \end{pmatrix}.$$

Is this chain irreducible? Find the period of state 1.

2.10. If $\{X_n, n \geq 0\}$ is an irreducible Markov chain with period $d \geq 1$, show that $\{X_{nd}, n \geq 0\}$ is aperiodic. Is is irreducible?

2.11. Given is a Markov chain on $S = \{0, 1, 2, 3, 4, 5\}$. In the following two cases give the classes, and determine which states are transient and which are recurrent. In each case compute $f_{40}^{(5)}$.

(a)

$$P = \begin{pmatrix} \frac{1}{3} & 0 & \frac{2}{3} & 0 & 0 & 0 \\ 0 & \frac{1}{4} & 0 & \frac{3}{4} & 0 & 0 \\ \frac{2}{3} & 0 & \frac{1}{3} & 0 & 0 & 0 \\ 0 & \frac{1}{5} & 0 & \frac{4}{5} & 0 & 0 \\ \frac{1}{4} & \frac{1}{4} & 0 & 0 & \frac{1}{4} & \frac{1}{4} \\ \frac{1}{6} & \frac{1}{6} & \frac{1}{6} & \frac{1}{6} & \frac{1}{6} & \frac{1}{6} \end{pmatrix}.$$

(b)

$$P = \begin{pmatrix} 1 & 0 & 0 & 0 & 0 & 0 \\ 0 & \frac{3}{4} & \frac{1}{4} & 0 & 0 & 0 \\ 0 & \frac{1}{8} & \frac{7}{8} & 0 & 0 & 0 \\ \frac{1}{4} & \frac{1}{4} & 0 & \frac{1}{8} & \frac{3}{8} & 0 \\ \frac{1}{3} & 0 & \frac{1}{6} & \frac{1}{4} & \frac{1}{4} & 0 \\ 0 & 0 & 0 & 0 & 0 & 1 \end{pmatrix}.$$

2.12. Consider a Markov chain on $S = \{1,\dots,9\}$ with transition matrix (x signifies a positive entry)

$$P = \begin{matrix} 1 \\ 2 \\ 3 \\ 4 \\ 5 \\ 6 \\ 7 \\ 8 \\ 9 \end{matrix} \begin{pmatrix} 0 & 0 & 0 & x & 0 & 0 & 0 & 0 & x \\ 0 & x & x & 0 & x & 0 & 0 & 0 & x \\ 0 & 0 & 0 & 0 & 0 & 0 & 0 & x & 0 \\ x & 0 & 0 & 0 & 0 & 0 & 0 & 0 & 0 \\ 0 & 0 & 0 & 0 & x & 0 & 0 & 0 & 0 \\ 0 & x & 0 & 0 & 0 & 0 & 0 & 0 & 0 \\ 0 & x & 0 & 0 & 0 & x & x & 0 & 0 \\ 0 & 0 & x & 0 & 0 & 0 & 0 & 0 & 0 \\ 0 & 0 & 0 & x & 0 & 0 & 0 & 0 & x \end{pmatrix}.$$

Put the matrix in canonical form. Classify the states. Give the closed recurrent classes. Which states are transient?

2.13. The Tenure System. A typical assistant professor is hired at one of six levels or "states" which we designate 1, 2, 3, 4, 5, 6. State 7 corresponds to tenure, and state 8 corresponds to "leaving the university." An assistant professor in state i ($i \leq 6$) may move to state $i+1$ or to state 8. An assistant professor in state 6 may move to state 7 or 8.

A study has been done and data collected. On the basis of the data the tenure system is modelled as a Markov chain with transition matrix

$$P = \begin{matrix} 1 \\ 2 \\ 3 \\ 4 \\ 5 \\ 6 \\ 7 \\ 8 \end{matrix} \begin{pmatrix} 0 & .5 & 0 & 0 & 0 & 0 & 0 & .5 \\ 0 & 0 & .6 & 0 & 0 & 0 & 0 & .4 \\ 0 & 0 & 0 & .5 & 0 & 0 & 0 & .5 \\ 0 & 0 & 0 & 0 & .5 & 0 & 0 & .5 \\ 0 & 0 & 0 & 0 & 0 & .8 & 0 & .2 \\ 0 & 0 & 0 & 0 & 0 & 0 & .01 & .99 \\ 0 & 0 & 0 & 0 & 0 & 0 & 1 & 0 \\ 0 & 0 & 0 & 0 & 0 & 0 & 0 & 1 \end{pmatrix}$$

and initial probability vector $(.9, 0, 0, .1, 0, 0, 0, 0)'$.
   (a) What are the closed sets? What are the equivalence classes?
   (b) What is the probability an assistant professor receives tenure?

(c) What is $f_{18}^{(3)}$?

2.14. In the occupational mobility example given at the end of Section 2.6, compute the expected number of steps necessary to go to state 1 for the first time, starting from 3. What is the long run percentage of generations that a family spends in state 3?

2.15. The Media Police have identified six states associated with television watching: 0 (never watch TV), 1 (watch only PBS), 2 (watch TV fairly frequently), 3 (addict), 4 (undergoing behavior modification), 5 (brain dead). Transitions from state to state can be modelled as a Markov chain with the following transition matrix:

$$P = \begin{pmatrix} 1 & 0 & 0 & 0 & 0 & 0 \\ .5 & 0 & .5 & 0 & 0 & 0 \\ .1 & 0 & .5 & .3 & 0 & .1 \\ 0 & 0 & 0 & .7 & .1 & .2 \\ \frac{1}{3} & 0 & 0 & \frac{1}{3} & \frac{1}{3} & 0 \\ 0 & 0 & 0 & 0 & 0 & 1 \end{pmatrix}.$$

(a) Which states are transient and which are recurrent?

(b) Starting from state 1, what is the probability that state 5 is entered before state 0; i.e., what is the probability that a PBS viewer will wind up brain dead?

2.16. Zeke Prevents Bankruptcy. Without benefit of dirty tricks, Harry's restaurant business fluctuates in successive years between three states: 0 (bankruptcy), 1 (verge of bankruptcy) and 2 (solvency). The transition matrix giving the probabilities of evolving from state to state is

$$P = \begin{pmatrix} 1 & 0 & 0 \\ .5 & .25 & .25 \\ .5 & .25 & .25 \end{pmatrix}.$$

(a) What is the expected number of years until Happy Harry's restaurant goes bankrupt assuming that he starts from the state of solvency?

(b) Harry's rich uncle Zeke decides it is bad for the family name if his nephew Harry is allowed to go bankrupt. Thus when state 0 is entered, Zeke infuses Harry's business with cash returning him to solvency with probability 1. Thus the transition matrix for this new Markov chain is

$$P' = \begin{pmatrix} 0 & 0 & 1 \\ .5 & .25 & .25 \\ .5 & .25 & .25 \end{pmatrix}.$$

Is this new Markov chain irreducible? Is it aperiodic? What is the expected number of years between cash infusions from Zeke?

2.17. Students Cope with Depression. A typical graduate student exhibits four states of mind. States 2 and 3 correspond to depression states. State 1 is a suicidal state and state 4 is a state which means the student has decided to seek professional psychiatric help. Changes in state of mind can be modelled as a Markov chain with transition matrix

$$P = \begin{pmatrix} 1 & 0 & 0 & 0 \\ .5 & 0 & .25 & .25 \\ .25 & .5 & 0 & .25 \\ 0 & 0 & 0 & 1 \end{pmatrix}.$$

Compute the probability the student will eventually commit suicide starting from state i $(i = 2, 3)$ and find the expected number of changes of state of mind starting from state i $(i = 2, 3)$ necessary for this result or seeking professional help.

2.18. Learning Experiments with Rats. A rat is put into compartment 4 of the maze. (See Figure 2.3.) He moves through the compartments at random; i.e., if there are k ways to leave a compartment, he chooses each of these with probability $1/k$. What is the probability the rat finds the food in compartment 3 before feeling the electric shock in compartment 7?

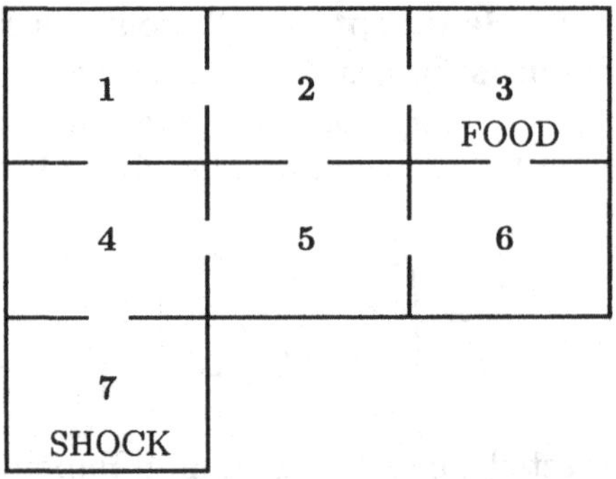

FIGURE 2.3. THE MAZE.

2.19. Harry and the F–Word. Harry has a keen appreciation for precise and elegant use of the English language and cannot abide linguistic crudity. Above all else, he abhors the use of the dreaded f–word. One day, a *Mutant Creepazoid* wanders in for a cup of coffee. The creepazoid's brain has been all but destroyed by too much television and other toxic substances and this has resulted in limited expressive abilities. In fact the conversation of the creepazoid consists primarily of the following phrases:

(1) Ya know.
(2) It was weird.
(3) Like, I dunno, man.
(4) Bummer.
(5) It was awesome.
(6) F–word.

The appearance of these phrases in the mutant creepazoid's speech follows a Markov chain (if ever there was a case for lack of long term memory this is it!) with the following transition matrix:

$$P = \begin{matrix} 1 \\ 2 \\ 3 \\ 4 \\ 5 \\ 6 \end{matrix} \begin{pmatrix} .6 & .1 & .1 & .1 & .1 & 0 \\ .3 & 0 & .3 & .2 & .1 & .1 \\ .3 & .5 & 0 & .1 & .1 & 0 \\ .1 & .2 & .2 & 0 & .1 & .4 \\ .2 & .2 & 0 & .1 & 0 & .5 \\ .2 & .2 & .2 & .2 & .2 & 0 \end{pmatrix}.$$

Suppose the first phrase out of this fellow's mouth is "Like, I dunno, man."

(a) Give the long term frequency that each phrase appears in the creepazoid's conversation.

(b) What is the expected number of transitions until the dreaded f–word is spoken?

(c) Harry gets a migraine at the fifth mention of the f–word. What is the expected number of transitions until the onset of the migraine?

(d) Harry bets Zeke $10 that the phrase "It was weird" would appear before the phrase "It was awesome." What is Harry's expected earnings from this bet? Give details when you answer.

2.20. Harry and the Living Theatre. Harry and a lady friend attend a performance by an acting company called *The Living Theatre* in Bailey Hall, the local theatre. Bailey Hall has six entrances on the main floor, numbered $1, \ldots, 6$, and two entrances in the balcony numbered 7 and 8. This particular performance consists of a seemingly endless stream of actors entering the hall sequentially from random entrances. Upon entering the hall, each actor emits screams of pain that would waken the dead and falls down on the floor writhing in agony until overcome by theatrical death.

The successive entry points of the actors can be modelled as a Markov chain with eight states and transition matrix

$$
P = \begin{array}{c} 1 \\ 2 \\ 3 \\ 4 \\ 5 \\ 6 \\ 7 \\ 8 \end{array}
\begin{pmatrix}
0 & .3 & .4 & .2 & 0 & .1 & 0 & 0 \\
0 & 0 & 0 & .3 & 0 & .3 & .2 & .2 \\
.1 & 0 & 0 & 0 & .3 & .3 & .2 & .1 \\
.4 & .3 & 0 & 0 & 0 & .2 & .1 & 0 \\
.5 & .3 & .2 & 0 & 0 & 0 & 0 & 0 \\
.4 & .2 & .2 & .1 & 0 & 0 & .1 & 0 \\
0 & .4 & 0 & .4 & .2 & 0 & 0 & 0 \\
.3 & 0 & .4 & 0 & .3 & 0 & 0 & 0
\end{pmatrix}.
$$

The performance begins by the first actor entering through door 8. Harry and his friend sit close to door 4.

(a) Is the chain irreducible? Is the chain periodic or aperiodic? Why?

(b) What is the long run percentage of actors who enter via the balcony?

(c) What is the long run percentage of actors who enter by means of door 4 closest to Harry and friend?

(d) What is the expected waiting time between two entrances via door 4?

(e) Suppose, in addition, that Harry's date is slightly psychotic and is unable to cope with chilling screams close to where she is sitting, so that when an actor enters via door 4 and starts screaming and dying she bursts into tears. Compute the expected number of entrances until Harry's date bursts into tears.

(f) Now suppose there is only probability p $(0 < p < 1)$ that Harry's date will burst into tears each time an actor enters via door 4 and starts screaming and dying. Under these conditions, compute the expected number of entrances until she starts crying hysterically.

2.21. Suppose an irreducible Markov chain with a not necessarily finite state space has a transition matrix with the property that $P^2 = P$.

(1) Prove the chain is aperiodic.

(2) Prove $p_{ij} = p_{jj}$ for all i, j in the state space. Find a stationary distribution in terms of P.

2.22. Consider a Markov chain $\{Z_n\}$ with state space $\{0, 1, 2, \ldots,\}$ and transition matrix P. Given two generating functions $A(s)$ and $B(s)$ with $1 > \alpha = A'(1), \infty > \beta = B'(1)$, related to the transition matrix by

$$
\sum_{j=0}^{\infty} p_{ij} s^j = B(s) A(s)^i.
$$

(a) Show that these transition probabilities correspond to a branching process with immigration, namely a process constructed as follows: Let

$\{Z_{nj}, n \geq 1, j \geq 1\}$ be independent, identically distributed non-negative integer valued random variables with a common generating function $A(s)$. Let $\{I_n, n \geq 1\}$ be iid, independent of $\{Z_{nj}\}$, with a common generating function $B(s)$. Define for $n \geq 0$

$$Z_{n+1} = I_{n+1} + \sum_{j=1}^{Z_n} Z_{nj},$$

so that the random variables $\{I_n\}$ count the number of immigrants per generation.

(b) Show that $\{Z_n\}$ has a stationary distribution $\{\Pi_k, k \geq 0\}$ whose generating function $\Pi(s)$ satisfies

$$\Pi(s) = B(s)\Pi(A(s)),$$

and solve to get an expression for Π in terms of iterates of A and B.

(c) This chain is *reversible* (cf. Section 5.9), that is,

$$\pi_i p_{ij} = \pi_j p_{ji},$$

if and only if

$$B(s)\Pi(\theta A(s)) = B(\theta)\Pi(sA(\theta)).$$

(d) Show

$$E(Z_{n+1}|Z_n) = \beta + \alpha Z_n.$$

(e) Compute

$$E(s^{\sum_{j=0}^n Z_j} 1_{[Z_n=j]}|Z_0 = i).$$

(f) If $A(s) = q + ps$ with $p + q = 1$, $0 \leq p \leq 1$ and $B(s) = e^{\lambda(s-1)}$, compute and identify $\Pi(s)$.

(g) Give an example where $\Pi(s)$ corresponds to a geometric distribution (Pakes, 1973; Pyke Tin and Phatarfod, 1976).

2.23. Suppose a Markov chain has m states and is doubly stochastic; i.e., $\sum_{i \in S} p_{ij} = 1, \forall j \in S$. Show that the vector $(1/m, \ldots, 1/m)$ is a stationary distribution.

2.24. If

$$\lim_{N \to \infty} \sum_{n=0}^N P^n / N \to \Pi,$$

where Π has constant columns and row sums that add to 1, is a row of Π a stationary distribution?

2.25. Consider a random walk on $S = \{0, 1, \ldots\}$ with a reflecting barrier at 0. This is a Markov chain with transition matrix

$$P = \begin{pmatrix} 0 & 1 & 0 & 0 & \cdots \\ q & 0 & p & 0 & \cdots \\ 0 & q & 0 & p & \cdots \\ \vdots & & \ddots & & \ddots \end{pmatrix}.$$

Classify when this chain is positive recurrent, null recurrent, transient.

2.26. In Example 2.11.1, what is the long run percentage of transitions which result in a riot? Which yield a star? On the average, how many riots must be endured between two class "1" performances?

2.27. If $\{X_n\}$ is Markov with stationary distribution π, show that $\{(X_n, X_{n+1}), n \geq 0\}$ is Markov. Give its stationary distribution.

2.28. Consider a positive recurrent, irreducible Markov chain $\{X_n\}$ with stationary distribution $\pi' = (\pi_0, \pi_1, \ldots)$. Suppose each time there is a transition from state i to state j there is a reward of $g(i,j)$ which is received. (Assume the function g is nice; say it is bounded or non-negative.)

(a) What is the long term reward rate

$$\lim_{n \to \infty} \frac{1}{n} \sum_{m=0}^{n} E_i g(X_m, X_{m+1})?$$

Why does this limit exist? Does it depend on the initial state i?

(b) In problem 2.19, what is the long term frequency that the phrase "It was weird" is followed by "Ya know"?

2.29. Harry Visits the Dentist. Like a good boy, Harry visits the dentist every six months. Because of a sweet tooth and fetish for chocolate, the condition of his teeth varies according to a Markov chain on the states $\{0, 1, 2, 3\}$ where 0 means no work is required, 1 means a cleaning is required, 2 means a filling is required and 3 means root canal work is needed. Charges for each visit to the dentist depend on the work done. State 0 has a charge of $20, state 1 has a charge of $30, state 2 has a charge of $50 and state 3 has the disastrous charge of $300. Transitions from state to state are governed by the matrix ·

$$\begin{pmatrix} .6 & .2 & .1 & .1 \\ .4 & .4 & .1 & .1 \\ .3 & .3 & .2 & .2 \\ .4 & .5 & .1 & 0 \end{pmatrix}.$$

What is the percentage of visits that are disastrous? What is Harry's long run cost rate for maintaining his teeth?

2.30. Classify the states for the Markov chain with matrix

$$P = \begin{pmatrix} .5 & 0 & .5 & 0 \\ 0 & .2 & 0 & .8 \\ .25 & 0 & .75 & 0 \\ 0 & .6 & 0 & .4 \end{pmatrix}.$$

Compute a stationary distribution for this chain. Is it unique? Compute $m_i = E_i\left(\tau_i(1)\right)$ for each $i \in S$. Assume $S = \{1, 2, 3, 4\}$.

2.31. An airline reservation system has two computers, only one of which is in operation at any given time. A computer may break down on any given day with probability p. There is a single repair facility which takes two days to restore a computer to normal. The facilities are such that only one computer at a time can be dealt with. Form a Markov chain by taking as states the pairs (x, y) where x is the number of machines in operating condition at the end of a day and y is 1 if a day's labor has been expended on a machine not yet repaired and 0 otherwise. What is the transition matrix? (This should be 4×4.) Find the stationary distribution. What percentage of time are there no machines operable? What percentage of time is exactly 1 machine operable? (Karlin and Taylor, 1975.)

2.32. Consider a finite Markov chain $\{X_n\}$ on the state space $S = \{0, 1, \ldots, N\}$ with transition matrix P consisting of three classes $\{0\}$, $\{1, 2, \ldots, N-1\}$ and $\{N\}$ where 0 and N are absorbing states, both accessible from any $k \in \{1, 2, \ldots, N-1\}$. Pick a reference state, say 1, and define an auxiliary Markov chain $\{Y_n\}$, called the return chain, by altering the first and last row of P so that

$$p_{01} = p_{n1} = 1,$$

and leave the other rows unchanged. The return process $\{Y_n\}$ is irreducible. Prove that the expected time w_1 until absorption starting from 1 for the $\{X_n\}$ process equals $1/(\pi_0 + \pi_N) - 1$ where $\pi_0 + \pi_N$ is the stationary probability of being in state 0 or N for the $\{Y_n\}$ process. (Karlin and Taylor, 1975.)

2.33. If A is an $m \times m$ square matrix (m finite) check that if $A^n \to 0$ as $n \to \infty$ then $I - A$ has an inverse and

$$(I - A)^{-1} = \sum_{j=0}^{\infty} A^j.$$

(Kemeny and Snell, 1976, page 22.)

2.34. For a finite irreducible and aperiodic Markov chain, suppose $P^n \to \Pi$ where $\Pi_{ij} = \pi_j$.
  (a) Show $Z := (I - (P - \Pi))^{-1}$ exists and

$$Z = I + \sum_{n=1}^{\infty} (P^n - \Pi).$$

Call Z the fundamental matrix for the irreducible chain.
  (b) Show
   (1) $PZ = ZP$
   (2) $\pi' Z = \pi'$
   (3) $I - Z = \Pi - PZ$.
  (c) Show for any initial vector **a**:

$$\{E_\mathbf{a} \sum_{n=0}^{N-1} 1_{[X_n=j]}, \, j \in S\}' - N\pi \to \mathbf{a}(Z - \Pi) = \mathbf{a}Z - \pi$$

as $N \to \infty$. (Kemeny and Snell, 1976.)

2.35. A process moves on the integers $S = \{1, 2, \ldots, 8\}$. Starting from 1, and on each successive step, it moves to an integer greater than its present position, moving with equal probabilities to each of the remaining larger integers. State 8 is absorbing. Find the expected number of steps to reach state 8. Replace 8 by the arbitrary integer N and redo the exercise.

2.36. Harry and the Comely Young Lady. Harry has his eye on a comely young lady whom he originally spotted at a high cholesterol cooking course. During the course, however, Harry never had the courage to introduce himself as a famous restaurateur. The young lady eats lunch regularly on Optima Street, and Harry, quite smitten with her, observes that she visits the four restaurants on the street labelled 0, 1, 2, 3 according to the Markov chain with transition matrix

$$P = \begin{matrix} 0 \\ 1 \\ 2 \\ 3 \end{matrix} \begin{pmatrix} .1 & .2 & .7 & .0 \\ .2 & .3 & .3 & .2 \\ .3 & .1 & .1 & .5 \\ .4 & .3 & .2 & .1 \end{pmatrix}.$$

Harry's restaurant is labelled 0, and the hated sprouts bar across the street is labelled 3. Assume initially she is equally likely to visit any of the four restaurants.

  (a) What is the probability she visits Harry's before the hated sprouts bar?

(b) What is the expected waiting time until she visits Harry's for the first time?

(c) What is the expected waiting time between two visits to Harry's?

(d) If each time she visits Harry's, there is only probability .3 that he will have the courage to talk to her and ask her for a date, what is the probability that eventually he asks her for a date, and what is the expected waiting time until this occurs?

2.37. Suppose $\{X_n\}$ is Markov with stationary distribution $\{\pi_j, j \in S\}$. Let $\tau = \inf\{n \geq 1 : X_n = X_0\}$ be the time of the first return to the initial state. Evaluate $E_\pi \tau$.

2.38. Let the state space S be infinite, and set $\tau = \inf\{n \geq 0 : X_n \notin T\}$. Suppose $P_i[\tau < \infty] = 1$ for all $i \in S$. Check that $w_i = E_i \tau$ satisfies the equation

$$w_i = 1 + \sum_{j \in T} p_{ij} w_j, \quad i \in T.$$

Does this system have a unique solution? If not, characterize the solution which gives $\{E_i \tau, i \in T\}$. If you believe a unique solution does not exist, demonstrate this by example.

2.39. More Social Mobility. Glass and Hall (1949) distinguish seven states in their social mobility study:

(1) professional, high administrative;
(2) managerial
(3) inspectional, supervisory, non-manual (high);
(4) non-manual low grade;
(5) skilled manual;
(6) semi-skilled manual;
(7) unskilled manual.

From their data the following transition matrix emerges:

$$P = \begin{pmatrix} .386 & .147 & .202 & .062 & .140 & .047 & .016 \\ .107 & .267 & .227 & .120 & .207 & .052 & .020 \\ .035 & .101 & .188 & .191 & .357 & .067 & .061 \\ .021 & .039 & .112 & .212 & .431 & .124 & .061 \\ .009 & .024 & .075 & .123 & .473 & .171 & .125 \\ .000 & .013 & .041 & .088 & .391 & .312 & .155 \\ .000 & .008 & .036 & .083 & .364 & .235 & .274 \end{pmatrix}.$$

Compute the mean first passage time from state i to 1, $2 \leq i \leq 7$.

2.40. Show for a Markov chain that $(X_0, \ldots, X_{n-1})$ and $(X_{n+1}, \ldots, X_{n+N})$ are conditionally independent given X_n for any N. That is, show

$$P[X_0 = j_0, \ldots, X_{n-1} = j_{n-1}, X_{n+1} = j_{n+1}, \ldots, X_{n+N} = j_{n+N} | X_n = i]$$
$$= P[X_0 = j_0, \ldots, X_{n-1} = j_{n-1} | X_n = i]$$
$$\cdot P[X_{n+1} = j_{n+1}, \ldots, X_{n+N} = j_{n+N} | X_n = i].$$

Show

$$P[X_n = j | X_{n+m} = i_{n+m}, \ldots, X_{n+1} = i_{n+1}, X_{n-1} = i_{n-1}, \ldots, X_0 = i_0]$$
$$= P[X_n = j | X_{n+1} = i_{n+1}, X_{n-1} = i_{n-1}],$$

and hence

$$E(X_n = j | X_{n+m} = i_{n+m}, \ldots, X_{n+1} = i_{n+1}, X_{n-1} = i_{n-1}, \ldots, X_0 = i_0)$$
$$= E(X_n = j | X_{n+1} = i_{n+1}, X_{n-1} = i_{n-1}).$$

2.41. In a sequence of Bernoulli trials with outcomes S or F, at index n the state 1 is observed if the trials indexed $n-1$ and n resulted in SS. Similarly, states 2, 3, 4 stand for the patterns SF, FS, FF. Find the transition matrix P and all of its powers. (Note $P^2 = P^3 = \ldots$.) (Feller, 1968.)

2.42. Suppose $\{Z_n, n \geq 1\}$ are iid representing outcomes of successive throws of a die. Define

$$X_n = \max\{Z_1, \ldots, Z_n\}.$$

Show $\{X_n, n \geq 1\}$ is a Markov chain and give its transition matrix P. Calculate from structure of $\{X_n\}$ the higher powers of P (Feller, 1968.)

2.43. In a finite Markov chain, j is transient iff there exists some state k such that $j \to k$ but j is not a consequent of k. Give an example to show this is false if the Markov chain has an infinite number of states.

2.44. Give an example to show that for a Markov chain to be irreducible, it is sufficient *but not necessary* that for some $n \geq 1$

$$p_{ij}^{(n)} > 0, \quad \text{for all } i, j \in S.$$

(Hint: It suffices to consider a two-state Markov chain.)

2.45. In a Markov chain with $S = \{0, 1, \ldots\}$, suppose 0 is absorbing. For $j > 0$ suppose that $p_{jj} = p$, $p_{j,j-1} = q$ where $p + q = 1$. Find $f_{j0}^{(n)}$, the probability of absorption at the nth step from j and find the mean time until absorption. Calculate the generating function of $\{f_{j0}^{(n)}, n \geq 1\}$.

2.46. For a Markov chain recall $f_{ij} = \sum_n f_{ij}^{(n)}$. Prove

(a) $\sup_{n \geq 1} p_{ij}^{(n)} \leq f_{ij} \leq \sum_{n=1}^{\infty} p_{ij}^{(n)}$.

(b) Consequently show

(i) $i \to j$ iff $f_{ij} > 0$.

(ii) $i \leftrightarrow j$ iff $f_{ij} f_{ji} > 0$.

2.47. Given an irreducible Markov chain with matrix $P = \{p_{ij}\}$ and state space $S = \{0, 1, \dots\}$. Modify P to obtain P' by making 0 absorbing:

$$P' = \begin{pmatrix} 1 & 0 & 0 & \cdots \\ p_{10} & p_{11} & p_{12} & \cdots \\ p_{20} & p_{21} & p_{22} & \cdots \\ \vdots & & & \ddots \end{pmatrix}.$$

Verify that in the new chain, the states $\{1, 2, \dots\}$ are transient.

2.48. Let $\{\pi_j, j \in S\}$ be the stationary distribution of a not necessarily irreducible Markov chain. Show that if $\pi_i > 0$ and $i \to j$, then $\pi_j > 0$.

2.49. Let $\{Y_n, n \geq 1\}$ be iid each taking values in $S = \{0, 1, 2, 3, 4\}$ with common distribution

$$P[Y_1 = i] = p_i, \quad i = 0, 1, 2, 3, 4.$$

Let $X_0 = 0$, and define $X_{n+1} = X_n + Y_{n+1}$ (modulo 5). For example, if $(Y_1, Y_2, \dots, Y_6) = (2, 4, 4, 1, 3, 4)$, then $(X_0, X_1, \dots, X_6) = (0, 2, 1, 0, 1, 4, 3)$. Then $\{X_n\}$ is a Markov chain with state space S. Find P and show it is doubly stochastic. Give the stationary distribution.

2.50. Let a Markov chain contain m states. Prove if $k \to j$ then j can be reached from k with positive probability in m steps or less.

2.51. Teach a computer how to classify states of a finite state Markov chain. (Hint: One method utilizes 2.50. In the transition matrix P replace all positive entries by 1. Compute $P + P^2 + \cdots + P^m$ where m is the number of states.)

2.52. Consider a three-state Markov chain with $S = \{1, 2, 3\}$ and transition matrix

$$P = \begin{pmatrix} .25 & .75 & 0 \\ .75 & .25 & 0 \\ 0 & 0 & 1 \end{pmatrix}.$$

(a) What is a stationary distribution?

(b) Give
$$m_i = E_i T_i(1), \quad i = 1, 2, 3.$$

(c) What is $\lim_{n \to \infty} P^n$?
(Hint: This can be done in your head.)

2.53. Harry and the Hollywood Mogul. One of Uncle Zeke's business acquaintances is the Hollywood mogul Sam Darling. Harry, disgusted over the success of yuppie oriented shows like *L.A. Law*, contacts Sam Darling about the possibility of doing a real down home show called *Optima Street Restaurateur*. Harry's idea is that the show will combine cooking tips and urban adventure. Initial story lines look promising and negotiations get serious. Harry's niece alertly notices that the negotiations seem to evolve as if they followed a discretely indexed Markov chain. The time scale is in hours. From Harry's point of view, the states are

(1) Royalties and other financial arrangements are totally inadequate.
(2) Rewards are adequate but artistic control is inadequate.
(3) Rewards and artistic control are adequate, but not enough relatives and neighbors will be employed in the show.
(4) Rewards, artistic control, employment of relatives and neighbors are adequate, but Harry has second thoughts about giving up the life of the respected restaurateur for the fast lane of show business.

The evolution of the negotiations seems to follow the transition matrix

$$P = \begin{matrix} 1 \\ 2 \\ 3 \\ 4 \end{matrix} \begin{pmatrix} .4 & .4 & .2 & 0 \\ .3 & .4 & .3 & 0 \\ .2 & .1 & .4 & .3 \\ .1 & .4 & .2 & .3 \end{pmatrix}.$$

Assume negotiations start in state 1.

(a) What is the long run percentage of time that Harry is in the ambivalent state 4?

(b) Suppose Harry decides to agree to terms on the sixth time the negotiations reach state 3. What is the expected time after the start of negotiations that Harry will agree to terms?

(c) Assume that negotiations have been in progress for 36 arduous hours, and Harry finds himself in state 4. He is sick of Sam Darling and his abrasive manner, and unsure that he wants to dedicate his life to show business. After 36 hours, Harry decides to give up the idea of show business if state 1 is entered before state 3. If state 3 is entered before state 1, he will agree to terms and proceed with the show. Find the probability that Harry gives up show business.

2.54. Consider a recurrent discrete time Markov chain $\{X_n, n \geq 0\}$ with transition matrix P and state space S_0. Let $S \subset S_0, S \neq S_0$ be a subset of states and define

$$\nu_0 = \inf\{n \geq 0 : X_n \in S\},$$
$$\nu_1 = \inf\{n > \nu_1 : X_n \in S,$$

$$\vdots \quad \vdots$$

Define

$$\{Y_j, j \geq 0\} = \{X_{\nu_j}, j \geq 0\}.$$

Call $\{Y_j, j \geq 0\}$ the reduced process, and think of it as being the original process observed only when in the states of S. Check that $\{Y_n\}$ is a Markov chain with state space S.

Partition $S_0 = S \cup S^c$, and accordingly, partition the matrix P as

$$P = \begin{matrix} S \\ S^c \end{matrix} \begin{pmatrix} T & U \\ V & Q \end{pmatrix}.$$

Let

$$d_{ij} = P[Y_1 = j | Y_0 = i], \quad i \in S, j \in S.$$

(a) Compute $D = (d_{ij}, i, j \in S)$, and express this matrix in terms of T, U, V and the fundamental matrix corresponding to Q. Assume only in (a) that the state space is finite.

(b) If the process $\{X_n\}$ has stationary distribution $\pi' = (\pi_j, j \in S_0)$, what is a stationary distribution for $\{Y_n\}$? Express it in terms of $\pi_j, j \in S_0\}$.

(c) Under the assumptions of (b), what is

$$\lim_{N \to \infty} \nu_N/N?$$

(Hint: Use (b).)

2.55.* Strong Markov Property. Suppose $\{X_n\}$ is a Markov chain with state space S, and let T be a stopping time so that the event $[T = n]$ is determined by $X_0, \ldots, X_n$. Suppose for all $i \in S$ that $P_i[T < \infty] = 1$.

(a) Prove $\{X_T, X_{T+1}, \ldots\}$ is a Markov chain with initial distribution $a_j = P_i[X_T = j]$ if $\{X_n\}$ starts from state i. What are the transition probabilities?

(b) Verify

$$P[X_{T+1} = k_1, \ldots, X_{T+m} = k_m | X_T = k_0] =$$
$$P[X_1 = k_1, \ldots, X_m = k_m | X_0 = k_0].$$

---

*This problem relies on more advanced material.

(c) Suppose the index set is $\{\ldots, -1, 0, 1, \ldots\}$. Prove

$$P[X_{T+1} = k_1, \ldots, X_{T+m} = k_m | X_T = k_0, X_{T-1} = k_{-1}, \ldots, X_{T-j} = k_{-j}]$$
$$= P[X_1 = k_1, \ldots, X_m = k_m | X_0 = k_0] = .$$

2.56. Let $\{X_n\}$ be a Markov chain with state space S, and let T be a stopping time so that the event $[T = n]$ is determined by $X_0, \ldots, X_n$. Suppose for all $i \in S$ that $P_i[T < \infty] = 1$. Suppose further that the state space is $S^\infty = \{(s_0, s_1, \ldots) : s_i \in S, i = 0, 1, \ldots\}$. Define $T_1 = T$, and for $n \geq 2$ define

$$T_n(s_0, s_1, \ldots) = T(s_{T_{n-1}+1}, s_{T_{n-1}+2}, \ldots),$$

so that T_n is T applied to the segment of $(s_0, s_1, \ldots)$ beyond index T_{n-1}. Show that $\{X_{T_n}, n \geq 1\}$ is a Markov chain. Explain the relevance for problem 2.54.

2.57. Non-Parametric Maximum Likelihood Estimation of Transition Probabilities. How did researchers in, for example, the occupational mobility study discussed in Section 2.6 estimate the transition probabilities? Here is the non-parametric maximum likelihood method: Imagine we have an m-state Markov chain $\{X_n\}$ with the number of states m known and with the transition matrix unknown. We observe the chain during the times $0, 1, \ldots, n$, and we wish to estimate P. If $i_0, \ldots, i_n$ is the succession of states observed, then the likelihood function $l_n = l_n(P)$ is

$$l_n(P) = P[X_0 = i_0, \ldots, X_n = i_n]$$
$$= P[X_0 = i_0] p_{i_0 i_1} \cdots p_{i_{n-1} i_n}.$$

The log likelihood function $L_n(P) = \log l_n(P)$ is

$$L_n(P) = \log(a_{i_0} p_{i_0 i_1} \cdots p_{i_{n-1} i_n})$$
$$= \log\left(a_{i_0} \prod_{(i,j) \in S \times S} p_{ij}^{n_{ij}}\right)$$
$$= \sum_{(i,j) \in S \times S} n_{ij} \log p_{ij},$$

where n_{ij} is the number of times the path makes a transition from i to j:

$$n_{ij} = \text{card}\{0 \leq k \leq n - 1 : i_k = i_{k+1} = j\}.$$

Define random variables which count the number of times the Markov chain makes a transition from i to j:

$$N_{ij} = \sum_{k=0}^{n-1} 1_{[X_k=i, X_{k+1}=j]},$$

so that $\{N_{ij}, (i,j) \in S \times S\}$ are sufficient statistics for P. Show that the maximum likelihood estimator of P is given by

$$\hat{p}_{ij} = \frac{N_{ij}}{N_i},$$

and the maximum likelihood estimate of p_{ij} is n_{ij}/n_i. Here

$$N_i = \sum_{k \in S} N_{ik}, \quad n_i = \sum_{k \in S} n_{ik}.$$

(Hint: Treat a_{i_0} as a constant as it is a nuisance parameter. Use Lagrange multipliers to maximize the likelihood.) Using problem 2.27 and the material on time averages in Section 2.12.1, discuss almost sure consistency of these estimators.

2.58. An Inventory Model. Review the inventory model discussed in Example 2.2.6. Suppose we assume a simple distribution of demands in any period, namely

$$P[D_1 = 0] = .5, \quad P[D_1 = 1] = .4, \quad P[D_1 = 2] = .1.$$

Let $s = 0$ and $S = 2$. Find the long run cumulative unsatisfied demand and the long run fraction of periods when demand is not satisfied. (Hint: You may wish to consider the Markov chain defined by removing the $+$ signs in (2.2.1). In this case, the state space is $\{-1, 0, 1, 2\}$.)

2.59. More Inventory Theory. For Example 2.2.6, suppose that $T_n = n$, and time is measured in days. Let the critical values be set at $s = 3$ and $S = 8$, and suppose $\{D_n\}$ are iid Poisson random variables with mean $\lambda = 4$ per day.

(a) Show the state space of the inventory Markov chain $\{X_n\}$ is $S = \{0, \ldots, 8\}$, and give the transition matrix.

(b) If $X_0 = 8$, what is the probability that there is a shortage before the end of the first day?

(c) If $X_0 = 8$, find the probability that no replenishment will be necessary at 1 and 2.

(d) Starting with $X_0 = 4$, what is the distribution of time until the first replenishment? (Cinlar, 1975.)

2.60. Storage Models with Content Dependent Release Rules.
Consider the Moran model discussed in Example 2.2.7, but with the variation that the release from the reservoir depends on the content. Given a function

$$r : \{0, \ldots, c\} \mapsto \{0, \ldots, c\},$$

called a release rule, we define

$$X_{n+1} = (X_n + A_{n+1} - r(X_n))_+ \wedge c.$$

Show this process is Markov and give its transition matrix.

2.61. Storage Models with Markov Chain Input. Review the classical Moran storage model in Example 7 of Section 2.2. Suppose the inputs $\{A_n\}$ form a finite state ergodic Markov chain with transition matrix Λ and stationary distribution π, and define the contents process as we did in (2.2.2). Verify that the bivariate process $\{(X_n, A_{n+1})\}$ is a Markov chain and show

$$\begin{aligned}
P[X_{n+1} = r, A_{n+2} = r' | X_n = s, A_{n+1} = s'] \\
= \Lambda_{r's'} P[X_{n+1} = r | X_n = s, A_{n+1} = s'] \\
= \Lambda_{r's'} \delta_{r, (s+s'-m)_+ \wedge c}
\end{aligned}$$

(Lloyd, 1963.)

2.62. Consider the Moran storage model

$$X_{n+1} = (X_n + A_{n+1} - m)_+ \wedge c.$$

The reservoir has capacity c, and the iid inputs have distribution

$$P[A_1 = m - 1] = q, \quad P[A_1 = m] = r, \quad P[A_1 = m + 1] = p,$$

where p, q, r are all non-negative and $p + q + 1 = 1$. Prove that the transition matrix is that of a random walk with reflecting barriers at 0 and c and that the stationary distribution is of the form

$$\pi_j = (\text{const})(\frac{p}{q})^j, \quad j = 0, \ldots, c.$$

2.63. Weather and Agricultural Decisions. Harry grows grapes on the outskirts of the city with the intention of starting the Optima Street Winery. Summer weather in his area follows a daily pattern which can be modelled as a four-state Markov chain with states 0 (sunny, clear), 1 (cool, muggy), 2 (gray and dreary), 3 (raining). The transition matrix for this chain is

$$P = \begin{pmatrix} .4 & .2 & .1 & .3 \\ .4 & .3 & .2 & .1 \\ .6 & .1 & .1 & .2 \\ .2 & .4 & .3 & .1 \end{pmatrix}.$$

It is now gray and dreary. The grapes are not quite ready for picking. Another sunny spell would bring them to perfection, but rain would ruin them. Harry must decide whether to pick somewhat immature grapes or risk waiting for a sunny spell.

To help him decide, compute the probability that a sunny day will occur before rain. What is the long run proportion of days which are gray and dreary? (Wolf, 1989.)

2.64. A company desires to operate s identical machines. These machines are subject to failure according to a given probability law. To replace these failed machines, the company orders new machines at the beginning of each week to make up the total s. It takes one week for each new order to be delivered. Let X_n be the number of machines in working order at the beginning of the nth week and let Y_n denote the number of machines that fail during the nth week. Establish the recursive formula

$$X_{n+1} = s - Y_n.$$

Under what conditions is $\{X_n\}$ a Markov chain? Suppose that the failure law is uniform, that is

$$P[Y_n = j | X_n = i] = \frac{1}{i+1}, \quad j = 0, \ldots, i.$$

Find the transition matrix of the chain, its stationary distribution, and the expected number of machines in operation in the steady state.

2.65. A Markov chain $\{X_n, n \geq 0\}$ has state space $\{0, 1, \ldots\}$ and transition probabilities

$$P[X_{n+1} = i + 1 | X_n = i] = p, \quad i \geq 0,$$
$$P[X_{n+1} = i - 2 | X_n = i] = 1 - p, i \geq 2,$$
$$P[X_{n+1} = 0 | X_n = i] = 1 - p, i = 0, 1,$$

where $0 < p < 1$. Establish a necessary and sufficient condition in terms of p for positive recurrence of $\{X_n\}$, and find the stationary distribution when it exists. Show that the process is null recurrent when $p = 2/3$.

2.66. Let $\{Z_n, -\infty < n < \infty\}$ be a sequence of iid random variables with $P[Z_1 = 0] = P[Z_1 = 1] = 1/2$. Define the stochastic process $\{X_n\}$ with state space $\{0, \ldots, 6\}$ by

$$X_n = Z_{n-1} + 2Z_n + 3Z_{n+1}, \quad -\infty < n < \infty.$$

(a) Determine
$$P[X_0 = 1, X_1 = 3, X_2 = 2]$$

and
$$P[X_1 = 3, X_2 = 2].$$

(b) Is $\{X_n\}$ Markov? Why or why not?

2.67. The position of a particle at time n is a Markov chain with state space $\{\Delta, 0, 1, \ldots\}$. The state Δ corresponds to annihilation of the particle, i.e., $X_n = \Delta$ iff the particle has been annihilated at or before time n. The transition probabilities of $\{X_n\}$ are given as follows: For $\neq i$,

$$P[X_{n+1} = j | X_n = i] = \begin{cases} 2/5, & \text{if } j - i = \pm 1, \\ 1/5, & \text{if } j = \Delta, \end{cases}$$

and for $i = 0$,
$$P[X_{n+1} = 0 | X_n = 0] = 1,$$

and for $i = \Delta$,
$$P[X_{n+1} = \Delta | X_n = \Delta] = 1.$$

Let
$$T = \inf\{n : X_n = \Delta\},$$

remembering the convention that $\inf \emptyset = \infty$. For $i = 0, 1, \ldots,$ let

$$p_i = P[X_n = \Delta \text{ for some } n \,| X_0 = i]$$

and
$$g_i(s) = E(s^T | X_0 = i), \quad 0 \le s \le 1.$$

(a) Find the probabilities $p_i, i \ge 0$.
(b) Determine $g_i(s), i \ge 0$.
(c) Does $\{X_n\}$ have a stationary distribution? If so, determine all such distributions.
(d) If $X_0 = i$, express $E_i(T | T < \infty)$ in terms of $g_i(s)$ and hence evaluate the conditional expectation.

2.68. Let P_1 and P_2 be two distinct two-state transition matrices with states $\{1, 2\}$. Set $X_0 = 1$. Toss a coin. If the coin comes up heads (with probability p) generate $X_1, X_2, \ldots$ using transition matrix P_1. If the coin comes up tails (with probability $1 - p$) then generate $X_1, X_2, \ldots$ using transition matrix P_2.

(a) Is $\{X_n\}$ a Markov chain?

(b) Assuming P_1 and P_2 are positive recurrent and aperiodic, evaluate

$$\lim_{n \to \infty} P[X_n = j], \quad j = 1, 2.$$

Now let P_1, P_2 be as above, and set $X_0 = 1$. Determine $X_1, X_2, \ldots$ by tossing the coin (independently) at each trial to determine which transition matrix to use.

(c) Under this new scheme, is $\{X_n\}$ a Markov chain?

(d) Show that $\lim_{n \to \infty} P[X_n = j]$ is not necessarily the same as that in part (b).

2.69. A boy and girl move into a two-bar town on the same day. Each night the boy visits one or the other of the two bars, starting in bar 1, according to a Markov chain with transition matrix

$$\begin{pmatrix} .7 & .3 \\ .3 & .7 \end{pmatrix}.$$

Likewise, the girl visits one or the other of the two bars according to a Markov chain with transition matrix

$$\begin{pmatrix} .4 & .6 \\ .6 & .4 \end{pmatrix}$$

but starting in bar 2. Assume that the two Markov chains are independent. Naturally, the game ends when boy meets girl, i.e., when they go to the same bar.

(a) Argue that the progress of the game can be described by a three-state Markov chain where one state is absorbing representing the end of the game, and the other two-states give the different bar identities of the boy and girl. Exhibit the transition matrix for this chain.

(b) Let N denote the number of the night on which boy meets girl. What is the distribution of N?

(c) Find the probability that boy visits bar 1 and girl visits bar 2 on the nth night.

2.70. (a) Let $\{X_n, n \geq 0\}$ be a random walk on the integers in the closed interval $[-b, a]$, $a > 0, b > 0$, with absorption at a and $-b$. Suppose that, for $-b < i < a$,

$$P[X_{n+1} = i | X_n = i] = 1 - p - q \geq 0,$$
$$P[X_{n+1} = i + 1 | X_n = i] = p > 0,$$
$$P[X_{n+1} = i - 1 | X_n = i] = q > 0,$$

and suppose $p \neq q$. Determine the probability of absorption at a, given $X_0 = i$.

(b) A particle describes a random walk on $k > 2$ points arranged on the circumference of a circle with the step distribution as above, but with $p + q = 1$, so it has probability p of moving to the point on its right, and probability $q = 1 - p$ of moving to the point on the left. Suppose that the initial position of the particle is A. Show that the probability of visiting all other $k - 1$ points on the circle before returning to A for the first time is

$$\frac{(p^{k-1} + q^{k-1})(p - q)}{(p^{k-1} - q^{k-1})}.$$

2.71. Three cards are placed in a row in one of six possible orders, and their order is changed successively as follows: A random choice of either the left-hand card or the right hand card is made, (the left-hand card being chosen with probability p, the right hand card being chosen with probability $q = 1 - p$), and it is then placed between the other two. This process is repeated indefinitely, the successive choices being independent. Let $X_n = k$ if, after n choices, the cards are in order k.

(a) Explain why $\{X_n\}$ is a Markov chain, and write down its transition matrix.

(b) Show it is irreducible and find its period and its stationary distribution.

(c) What is the mean number of choices required first to restore the initial order of the cards?

2.72. An electric light that has survived for n seconds fails during the $(n + 1)$st second with probability q $(0 < q < 1)$. Let $X_n = 1$ if the light is functioning at time n seconds; otherwise let $X_n = 0$.

(a) Let T be the time to failure (in seconds) of the light; i.e.,

$$T = \inf\{n : X_n = 0\}.$$

Determine ET.

A building contains m lights of the type described, which behave independently. At time 0 they are all functioning. Let Y_n denote the number of lights functioning at time n.

(b) Specify the transition matrix of $\{Y_n\}$.

(c) Find the generating function

$$Es^{Y_n} = \phi_n(s)$$

of Y_n. Use it to find $P[Y_n = 0]$ and EY_n. (Hint: Show $\phi_n(s) = \phi_{n-1}(q + ps)$.)

2.73. A Markov chain has state space $\{0, 1, 2, \ldots\}$ and transition probabilities

$$p_{i,i+1} = \frac{\lambda}{i + \nu + 1}, \quad p_{i0} = 1 - p_{i,i+1},$$

where $\lambda > 0$ and $\nu \geq 0$ are constants. State any other necessary restrictions on the values of λ and ν. Show that the chain is irreducible, aperiodic and positive recurrent. Find explicit forms for the stationary distribution in the cases $\nu = 0$ and $\nu = 1$.

2.74. Let Y be a nonnegative integer valued random variable and let S_1 be the sum of i independent copies of Y. Let $\{h_j\}$ be a probability mass function concentrating on the nonnegative integers. Define a Markov chain with state space $\{0, 1, 2, \ldots\}$ with transition matrix

$$p_{ij} = P[S_1 = j], \quad i \geq 1, \ j \geq 0,$$
$$p_{oj} = h_j.$$

Assume that this chain is irreducible and aperiodic. Show that it has a limiting distribution if $m = EY < 1$ and $\sum_j j h_j < \infty$.
 Let

$$f(s) = Es^Y, \quad h(s) = \sum_{j=0}^{\infty} h_j s^j, \quad 0 \leq s \leq 1,$$

and let $\Pi(s)$ be the generating function of the stationary distribution. Show that $\Pi(\cdot)$ solves the functional equation

$$\Pi(s) = \Pi(f(s)) - \Pi(0)(1 - h(s)).$$

2.75. Choice Theory–Markov Chain Models of Brand Switching.
Suppose for a given product, a consumer has a choice between m competing products, labelled $1, 2, \ldots, m$. For example, the m products might be competing brands of laundry detergent. Suppose successive purchases by a consumer are modeled as an ergodic Markov chain $\{X_n\}$ with transition matrix $P = (p_{ij}, 1 \leq i, j \leq m)$.
 The diagonal entries of P, namely p_{jj}, $1 \leq j \leq m$, are measures of consumer loyalty to a brand since p_{jj} measures the tendency of a consumer to repurchase brand j when his last purchase was brand j.
 Companies are interested in market share. The market share of brand j is the percentage of purchases made which are of brand j. Assuming a homogeneous population, the market share of brand j can be evaluated as

$$\lim_{N \to \infty} N^{-1} \sum_{n=0}^{N} 1_{[X_n = j]}$$

which we know by Proposition 2.12.4 is just π_j, the stationary distribution for the chain.

The producer and the advertising agency are interested in the dependence of market share π_j for brand j on consumer loyalty p_{jj} to brand j.

Consider a simple case. Suppose $p_{jj} = 1/2$, $j = 1, \ldots, m$ and for $i \neq j$ we have $p_{ij} = (2(m-1))^{-1}$. Check that this matrix is doubly stochastic so the stationary distribution is uniform.

Now assume, for $j \neq i$,

$$p_{ij} = \frac{1 - p_{ii}}{m - 1},$$

so that a consumer not choosing his old brand i chooses uniformly among the others. With this simple specification of choice probabilities can we see the dependence of market share on consumer loyalty? Verify

$$\pi_j = \frac{(1 - p_{jj})^{-1}}{\sum_{l=1}^{m}(1 - p_{ll})^{-1}},$$

so that $\pi_j \uparrow 1$ as $p_{jj} \uparrow 1$. (Wolf, 1989.)

2.76. Two Genetics Models. Let $\{X_n, n \geq 0\}$ be a Markov chain on $\{0, 1, \ldots, M\}$ with transition matrix $\{p_{ij}, 0 \leq i, j \leq M\}$. For the following two models compute

$$u_{iM} = P_i[\text{ absorbtion at M }],$$

and show that u_{iM} is the same for both models:

(a) Wright model:

$$p_{ij} = \binom{M}{j} \left(\frac{i}{M}\right)^j \left(1 - \frac{i}{M}\right)^{M-j},$$

for $i, j = 0, \ldots, M$.

(b) Moran model: Define

$$p_i = \frac{i(M - i)}{M^2}, \quad i = 0, \ldots, M,$$

and set

$$p_{ij} = p_i, \text{ if } |j - i| = 1, 0 < i < M,$$
$$p_{ii} = 1 - 2p_i, \text{ if } 0 < i < M,$$
$$p_{00} = p_{MM} = 1,$$
$$p_{ij} = 0, \text{ otherwise.}$$

2.77. For a finite state Markov chain $\{X_n\}$ compute

$$\text{Cov}(X_n, X_{n+k})$$

in terms of the transition probabilities and the quantities $\{P[X_n = j]\}$. What if the chain is stationary? If the chain is stationary, what is the behavior of

$$\text{Cov}(X_n, X_{n+k})$$

as $k \to \infty$?

CHAPTER 3

Renewal Theory

RENEWAL PROCESSES model occurrences of events happening at random times where the times between the events can be approximated by independent, identically distributed random variables. With such a simple description, one wonders how flexible and powerful a tool renewal processes can be. Despite the simple description, renewal theory is one of the most basic of the building blocks in applied probability. Often a complex stochastic model has one or more embedded renewal processes, and this fact lies at the heart of the analysis of such processes and is basic to the idea of *regeneration*, which allows a process to be decomposed into independent, identically distributed blocks of random lengths. The dissection principle for Markov chains is an example of how to decompose a process into iid blocks.

In a one-semester survey course, try to cover most of Sections 3.1 through 3.8. Sections 3.9 to 3.12.3 are more advanced and should not be read by beginners.

3.1. BASICS.

Suppose $\{Y_n, n \geq 0\}$ is a sequence of independent random variables which take on only non-negative values. Furthermore, suppose the sequence $\{Y_n, n \geq 1\}$ is identically distributed with common distribution $F(x)$. We always assume

$$F(0-) = 0, \quad F(0) < 1$$

or, equivalently, for each $n \geq 1$,

$$P[Y_n < 0] = 0, \quad P[Y_n = 0] < 1.$$

For $n \geq 0$, define

$$S_n = Y_0 + \cdots + Y_n.$$

The sequence $\{S_n, n \geq 0\}$ is called a *renewal sequence*. The quantities S_n are usually thought of as times of occurrence of some phenomenon and are called renewal times or epochs. The process is called *delayed* if

$P[Y_0 > 0] > 0$; otherwise it is *pure*, and $S_0 = 0 = Y_0$ so that for a pure renewal process time zero is considered a renewal epoch. F is called the interarrival distribution. If F is proper ($F(\infty) = 1$) then the process is called proper. If F is defective ($F(\infty) < 1$), then $\{S_n\}$ is terminating or transient, and there will be a final renewal.

To fix ideas, we consider several simple examples.

Examples. (i) Replacements: Items being tested are observed sequentially. For instance, a light bulb is observed until burnout, and then a magical janitor instantaneously replaces the expired bulb with a fresh one. The process repeats. Failure times are $\{S_n, n \geq 0\}$. If the initial bulb is fresh, a pure renewal process is an appropriate model; otherwise a delayed renewal process is appropriate.

(ii) Imagine the replacements in (i) are proceeding happily. I arrive to visit my friend the janitor. Unless I arrive at exactly a replacement epoch (if F is continuous this event has probability zero), the process I see is delayed. The delay is the time from my arrival until the next replacement.

(iii) Returns to a state: Let $\{X_n, n \geq 0\}$ be a Markov chain with finite state space. Fix a state i. As in Section 2.5 on the dissection principle, define the successive return times to state i by

$$\tau_0(i) = \inf\{n : X_n = i\},$$

and, for $n \geq 0$,

$$\tau_{n+1}(i) = \inf\{m \geq \tau_n(i) : X_m = i\}.$$

Then $\{\tau_n(i), n \geq 0\}$ is a renewal sequence. If the initial state is i, the process is pure; otherwise it is delayed, and the delay distribution is the first passage distribution to state i from the initial state. In Chapter 5 we will observe similar phenomena with continuous time Markov chains.

In addition to Markov chains, many other processes possess regenerative states where the process probabilistically restarts itself in that state. Times of entrance to such states constitute a renewal process. Buzz words such as "probabilistically restarts" should take on precise meaning as we develop the theory. For instance, we will see that under proper assumptions, beginnings of empty periods in dam models or beginnings of busy periods in queueing models constitute renewal epochs. Within complex processes one must be able to identify embedded renewal sequences.

Often there is a choice in how one identifies renewal processes, as the next example shows.

(iv) On-off process: Operative periods of a machine alternate with down periods during which repairs are made. Suppose all periods are independent, operative periods are independent, identically distributed (iid), and so are down periods. Two distinct renewal processes can be identified:

   (1) Times when the machine becomes inoperative. These are denoted by x's.

   (2) Times when service is completed and the machine becomes operative. These are denoted by o's.

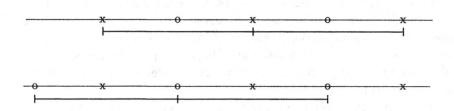

FIGURE 3.1. TWO EMBEDDED RENEWAL PROCESSES

In renewal theory it is often convenient to consider the associated point process which counts renewals. Of interest is the counting function

$$N(t) = \sum_{n=0}^{\infty} 1_{[0,t]}(S_n)$$

giving the number of renewals in $[0, t]$. The expectation

$$U(t) = EN(t)$$

is called the renewal function and plays a key role in asymptotic analysis.

3.2. ANALYTIC INTERLUDE.

This section collects some analytic facts that are needed for comfortable manipulation of quantities related to renewal processes. Some of this material parallels the discussions in Chapter 1. For instance, the discussion on convolution and Laplace transforms is similar to corresponding sections of Chapter 1 which were limited to positive integer valued random variables.

3.2.1. INTEGRATION.

In what follows we will see integrals with respect to a monotone function $U(x)$ on $[0, \infty)$ of the form

$$\int_0^\infty g(x)dU(x)$$

or

$$\int_{[0,\infty)} g(x)U(dx).$$

If you know what Lebesgue-Stieltjes integration is, interpret these integrals in this manner and skip to Section 3.2.2.

If you do not know what a Lebesgue-Stieltjes integral is but know what a Riemann-Stieltjes integral is, interpret these integrals in this manner: Note that if $U(x)$ has a jump at 0 then this must be accounted for in the integration; the integral includes 0, and some authors would write the integral sign as $\int_{0-}^{\infty}$. Now proceed to Section 3.2.2.

If you know neither about Lebesgue-Stieltjes nor about Riemann-Stieltjes integrals, the following is sufficient for the unstarred sections of this book. We distinguish three cases of interest:

(1) U is absolutely continuous (AC). This means there exists a density u satisfying $u(x) \geq 0$, such that $\int_0^T u(x)dx < \infty$, $\forall T > 0$ and for $b > a \geq 0$

$$U(b) - U(a) = \int_a^b u(s)ds.$$

Then interpret

$$\int_0^{\infty} g(x)U(dx) = \int_0^{\infty} g(x)dU(x) = \int_0^{\infty} g(x)u(x)dx.$$

(2) U is discrete. This means there are atoms $\{a_i\}$ and weights $\{w_i\}$, $w_i < \infty$, and $\lim_{h\downarrow 0} U(a_i + h) - U(a_i - h) =: U(\{a_i\}) = w_i$ meaning the measure U places mass w_i at location a_i. So the distribution function $U(x)$ is constant except at points a_i, where it jumps up by an amount w_i. Thus $U(x)$ satisfies

$$U(x) = \sum_{i:0\leq a_i\leq x} w_i.$$

We interpret the integrals by

$$\int g(x)U(dx) = \int g(x)dU(x) := \sum_i g(a_i)w_i.$$

This was the set-up of Chapter 1.

(3) U is a mixture of the form

$$U(x) = \alpha U_{AC}(x) + \beta U_d(x),$$

where U_{AC} is AC with density $u_{AC}(x)$, and U_d is discrete with atoms $\{a_i\}$ and weights $\{w_i\}$ and $\alpha > 0, \beta > 0$. In this case

$$\int g(x)U(dx) = \int g(x)dU(x)$$

$$= \alpha \int g(x)U_{AC}(dx) + \beta \int g(x)U_d(dx)$$

$$= \alpha \int g(x)u_{AC}(x)dx + \beta \sum_i g(a_i)w_i.$$

As an example, consider a case encountered frequently in renewal theory. $U(x)$ is of the form

$$U(x) = 1 + \int_0^x u(s)ds.$$

Here

$$U_{AC}(x) = \int_0^x u(s)ds$$

and

$$U_d(x) = \begin{cases} 1, & \text{if } x > 0 \\ 0, & \text{if } x < 0, \end{cases}$$

so U_d has an atom at 0 with weight 1 and $\alpha = \beta = 1$. Then

$$\int g(x)U(dx) = g(0)U(\{0\}) + \int_0^x g(s)u(s)ds$$

$$= g(0) + \int_0^x g(s)u(s)ds.$$

Take this example to heart since it will recur repeatedly.

For more advanced work, it is convenient to write U both for the distribution function $U(x)$ and for the measure determined by this distribution function and the relation $U(a, b] = U(b) - U(a)$ when $0 \le a \le b$.

3.2.2. CONVOLUTION.

Suppose all functions are defined on $R_+ = [0, \infty)$ and all distributions concentrate on R_+. Call a function g *locally bounded* if g is bounded on finite intervals. For a locally bounded g and a distribution F define the convolution of F and g as the function

$$F * g(t) := \int_0^t g(t - x)F(dx) \quad \text{for } t \ge 0,$$

where for the definition of the integral we rely on Section 3.2.1 and imagine the integration includes the endpoints. (Remember that if there are atoms, one cannot be sloppy about endpoints.) Some properties of convolution now follow.

1. $F * g \geq 0$.

2. $F * g$ is locally bounded; in fact

$$\sup_{0 \leq s \leq t} |F * g(s)| \leq \left(\sup_{0 \leq s \leq t} |g(s)| \right) F(t).$$

To verify this, define $\|g\| =: \sup_{0 \leq s \leq t} |g(s)|$, which is finite for every t by the assumption of local boundedness. Then for any $s \leq t$

$$|F * g(s)| = \left| \int_0^s g(s - x) F(dx) \right|$$
$$\leq \int_0^s |g(s - x)| F(dx)|$$
$$\leq \|g\| \int_0^s F(dx)| = \|g\| F(t) < \infty.$$

3. If g is bounded and continuous, then $F * g$ is continuous since $F * g(t) = Eg(t - Y_1)$ where Y_1 has distribution F. Thus if $t_n \to t$, we get almost surely that $g(t_n - Y_1) \to g(t - Y_1)$, by continuity, and, since g is bounded, dominated convergence yields $Eg(t_n - Y_1) = F * g(t_n) \to Eg(t - Y_n) = F * g(t)$.

4. The convolution operation can be repeated: $F * (F * g)$. As an aid to writing consider the following: Define

$$F^{0*}(x) = 1_{[0,\infty)}(x)$$
$$F^{1*}(x) = F(x)$$

and for $n \geq 1$

$$F^{(n+1)*}(x) = F^{n*} * F(x).$$

Then F^{0*} acts as an identity:

$$F^{0*} * g = g,$$

and an associative property holds:

$$F * (F * g) = (F * F) * g = F^{2*} * g.$$

(Note that $F * F = F^{2*}$ is a distribution—see (5) below.) A similar result holds for higher convolution powers.

5. Convolution of two distributions corresponds to sums of independent random variables (rv's). Let X_1 and X_2 be independent, with X_i having distribution $F_i, i = 1, 2$. Then $X_1 + X_2$ has distribution $F_1 * F_2$ since for $t \geq 0$

$$P[X_1 + X_2 \leq t] = P[(X_1, X_2) \in \{(x, y) \in R_+^2 : x + y \leq t\}]$$

$$= \iint_{\{(x,y) \in R_+^2 : x+y \leq t\}} F_1(dx) F_2(dy),$$

and we may write the double integral as an iterated integral:

$$= \int_0^t \left[\int_{y=0}^{t-x} F_2(dy) \right] F_1(dx) = \int_0^t F_2(t-x) F_1(dx).$$

6. The proof in (5) shows the commutative property: $F_1 * F_2 = F_2 * F_1$.

7. By induction we may show that if $X_1, \ldots, X_n$ are iid with common distribution F, then $X_1 + \cdots + X_n$ has distribution F^{n*}.

8. If F_i is absolutely continuous (AC) with density f_i, $i = 1, 2$ then $F_1 * F_2$ is AC with density (for $t > 0$)

$$f_1 * f_2(t) := \int_0^t f_1(t-y) f_2(y) dy = \int_0^t f_2(t-y) f_1(y) dy.$$

In fact, if F is AC, then for any distribution G, $F * G$ is AC.

To check these last assertions, observe that

$$F_1 * F_2(t) = \iint_{\{(x,y):x+y \leq t\}} f_1(x) f_2(y) dx dy$$

$$= \int_{y=0}^t \left(\int_{x=0}^{t-y} f_1(x) dx \right) f_2(y) dy$$

by writing the double integral as an iterated integral. Changing variables, this is

$$= \int_{y=0}^t \left(\int_{u=y}^t f_1(u-y) du \right) f_2(y) dy,$$

and reversing the order of integration yields

$$= \int_{u=0}^{t} \left(\int_{y=0}^{u} f_2(y) f_1(u-y) dy \right) du$$

$$= \int_{u=0}^{t} f_1 * f_2(y) dy.$$

For the second assertion replace f_1 by f and $f_2(y)dy$ by $G(dy)$, and we find in a similar way

$$F * G(t) = \int_{0}^{t} \left(\int_{y=0}^{u} f(u-y) G(dy) \right) du$$

$$= \int_{0}^{t} G * f(u) du,$$

showing $F * G$ is AC with density $G * f$.

3.2.3. LAPLACE TRANSFORMS.

Suppose X is a non-negative random variable with distribution F. The Laplace transform of X or F is the function defined on R_+ by

$$\hat{F}(\lambda) := E e^{-\lambda X} = \int_{0}^{\infty} e^{-\lambda x} F(dx), \quad \lambda \geq 0.$$

Note that, since $e^{-\lambda X} \leq 1$, we have $\hat{F}(\lambda) < \infty$ for all $\lambda > 0$.

Some useful properties of the transform follow:

1. Distinct distributions have distinct transforms. For the mathematically adept, this can be proven using the Stone-Weierstrass theorem; a proof using elementary tools is developed in Feller, 1971.

2. Suppose X_1, X_2 are independent and that X_i has distribution F_i, $i = 1, 2$. Then

$$(\widehat{F_1 * F_2})(\lambda) = \hat{F}_1(\lambda) \hat{F}_2(\lambda),$$

since

$$(\widehat{F_1 * F_2})(\lambda) = E e^{-\lambda(X_1 + X_2)}$$

$$= E e^{-\lambda X_1} E e^{-\lambda X_2},$$

by independence,

$$= \hat{F}_1(\lambda) \hat{F}_2(\lambda).$$

The transform therefore converts the fairly complex operation of convolution into a simpler operation of multiplication. This phenomenon was already encountered in Chapter 1 where we saw that the generating function of the convolution of two discrete densities was the product of the individual generating functions.

Similarly, for any $n \geq 0$, if F is a distribution,

$$(\widehat{F^{n*}})(\lambda) = (\hat{F}(\lambda))^n.$$

We now give some examples of how to compute Laplace transforms.

Example 3.2.1. If X is uniform on $(0, 1)$, then

$$Ee^{-\lambda X} = \int_0^1 e^{-\lambda x} dx = (1 - e^{-\lambda})/\lambda.$$

Example 3.2.2. If X has an exponential density with parameter α so that $F(dx) = \alpha e^{-\alpha x} 1_{(0,\infty)}(x) dx$, then

$$\hat{F}(\lambda) = \alpha/(\alpha + \lambda).$$

Example 3.2.3. Start by noting that

$$\int_0^\infty e^{-\lambda x} \frac{\alpha(\alpha x)^n e^{-\alpha x}}{n!} dx = \alpha^{n+1} \int_0^\infty \frac{e^{-(\lambda+\alpha)x} x^n}{n!} dx$$

$$= (\frac{\alpha}{\alpha + \lambda})^{n+1} \int_0^\infty \frac{e^{-y} y^n}{n!} dy$$

$$= (\frac{\alpha}{\alpha + \lambda})^{n+1}.$$

If $X_1, \ldots, X_{n+1}$ are iid, each with exponential density $\alpha e^{-\alpha x} 1_{(0,\infty)}(x)$, then $\sum_{i=1}^{n+1} X_i$ has the Laplace transform

$$Ee^{-\lambda \sum_{i=1}^{n+1} X_i} = (\alpha/(\alpha + \lambda))^{n+1}$$

$$= \int_0^\infty e^{-\lambda x} \frac{\alpha(\alpha x)^n e^{-\alpha x}}{n!} dx.$$

Therefore, because a transform uniquely determines the distribution, we obtain the significant fact that $\sum_{i=1}^{n+1} X_i$ has the gamma density $\alpha(\alpha x)^n e^{-\alpha x}/n!$ on R_+.

More properties of the transform now follow:

3. For $\lambda > 0$, $\hat{F}(\lambda)$ has derivatives of all orders, and any order derivative may be obtained by differentiating appropriately under the integral sign. For any $n \geq 1$,

$$(-1)^n \frac{d^n}{d\lambda^n} \hat{F}(\lambda) = \int_0^\infty e^{-\lambda x} x^n F(dx), \quad \lambda > 0.$$

(More advanced readers can easily check this using the dominated convergence theorem.) By monotone convergence,

$$\lim_{\lambda \downarrow 0}(-1)^n \frac{d^n}{d\lambda^n}\hat{F}(\lambda) = \int_0^\infty x^n F(dx) \leq \infty.$$

In particular, if the random variable X has F as its distribution then $EX = -\hat{F}'(0)$ and $EX^2 = \hat{F}''(0)$, and so on.

4. Convenience formulas:

(3.2.3.1)
$$\int_0^\infty e^{-\lambda x}F(x)dx = \lambda^{-1}\hat{F}(\lambda), \quad \int_0^\infty e^{-\lambda x}(1 - F(x))dx = (1 - \hat{F}(\lambda))/\lambda.$$

The second formula follows directly from the first, which is obtained merely by a reversal of the order of integration:

$$\int_0^\infty e^{-\lambda x}F(x)dx = \int_{x=0}^\infty e^{-\lambda x}\left(\int_{u=0}^x F(du)\right)dx$$

$$= \int_{u=0}^\infty \left(\int_{x=u}^\infty e^{-\lambda x}dx\right)F(du)$$

$$= \int_{u=0}^\infty e^{-\lambda u}\lambda^{-1}F(du)$$

$$= \lambda^{-1}\hat{F}(\lambda).$$

We now extend these notions to arbitrary distributions and measures U on R_+. Suppose $U(x)$ is non-decreasing on $[0, \infty)$, but perhaps $U(\infty) := \lim_{x\uparrow\infty} U(x) > 1$. If there exists $a \geq 0$ such that

$$\int_0^\infty e^{-\lambda x}U(dx) < \infty$$

for $\lambda > a$, then

$$\hat{U}(\lambda) := \int_0^\infty e^{-\lambda x}U(dx), \quad \lambda > a$$

is called the Laplace transform of U. If no such a exists, we say the Laplace transform is undefined.

The previous properties of Laplace transforms all have obvious extensions. For example, the Laplace transform of U, if it exists, uniquely determines U.

The reason for dealing with transforms of such functions U is that, as we will see in the next section, we frequently deal with the renewal function

$U(t) = EN(t)$, and it is convenient to be able to take the transform of this function.

Example 3.2.4. $U(dx) = e^{ax}dx$. Then

$$\int_0^\infty e^{-\lambda x}e^{ax}dx = \begin{cases} (\lambda - a)^{-1}, & \text{if } \lambda > a \\ \infty, & \text{if } \lambda \le a. \end{cases}$$

Example 3.2.5. $U(dx) = e^{x^2}dx$. Then

$$\int_0^\infty e^{-\lambda x}e^{x^2}dx = \infty \text{ for all } \lambda > 0,$$

and the Laplace transform does not exist.

Example 3.2.6. $U(dx) = (1 - F(x))dx$ where F is proper on R_+. Then

$$\hat{U}(\lambda) = (1 - \hat{F}(\lambda))/\lambda$$

from the convenience formulas (3.2.3.1).

Example 3.2.7. $U(dx) = x^n dx$. Then

$$\int_0^\infty e^{-\lambda x}x^n dx = \lambda^{-(n+1)}\int_0^\infty e^{-s}s^n ds$$
$$= \Gamma(n+1)\lambda^{-(n+1)}$$
$$= \lambda^{-(n+1)}n!$$

Next we discuss an inversion problem which will help us examine examples later.

Inversion Example 3.2.8. Let $X_1, \ldots, X_n$ be iid rv's uniformly distributed on $(0,1)$. What is the density of $\sum_{i=1}^n X_i$? We have

$$Ee^{-\lambda \sum_{i=1}^n X_i} = (Ee^{-\lambda X_1})^n = (\lambda^{-1}(1 - e^{-\lambda}))^n$$

(from Example 3.2.1)

$$= \sum_{k=0}^n \binom{n}{k}(-1)^k e^{-\lambda k}\lambda^{-n}.$$

Now $e^{-\lambda k} = \int_{x=0}^\infty e^{-\lambda x}\epsilon_k(dx)$, where ϵ_k is the measure putting an atom of size 1 at $\{k\}$. Furthermore,

$$\lambda^{-n} = \int_0^\infty e^{-\lambda x}\frac{x^{n-1}}{(n-1)!}dx =: \int_0^\infty e^{-\lambda x}g(x)dx,$$

so $e^{-\lambda k}\lambda^{-n}$ is the transform of $\epsilon_k * g(x)$, which is

$$= \int_0^x g(x-y)\epsilon_k(dy) = \begin{cases} 0, & \text{if } x < k \\ g(x-k) = \frac{(x-k)^{n-1}}{(n-1)!}, & \text{if } x \geq k \end{cases}$$

$$=: (x-k)_+^{n-1}/(n-1)!,$$

and the required density of $\sum_{i=1}^n X_i$ is

$$\sum_{k=0}^n \binom{n}{k}(-1)^k \frac{(x-k)_+^{n-1}}{(n-1)!}.$$

3.3. COUNTING RENEWALS.

In this section we focus on properties of the random variables $N(t)$, where $N(t)$ is the counting function of the number of renewals in $[0,t]$:

$$N(t) = \sum_{n=0}^\infty 1_{[0,t]}(S_n).$$

The quantity $EN(t)$, called the *renewal function*, is a basic quantity in renewal theory; as we will see in Section 3.5, the renewal function is a convolution factor of solutions of *renewal equations*. Renewal equations are a basic tool for solving for probabilities of interest in renewal theory.

Note that if $S_0 = 0$, then $t = 0$ is counted as a renewal epoch. In this case the renewal function is

$$U(t) := E \sum_{n=0}^\infty 1_{[0,t]}(S_n) = \sum_{n=0}^\infty P[S_n \leq t] = \sum_{n=0}^\infty F^{n*}(t).$$

On the other hand, in the delayed case, if $S_0 = Y_0$ has distribution G, then the renewal function is

$$V(t) = \sum_{n=0}^\infty P[S_n \leq t] = \sum_{n=1}^\infty G * F^{(n-1)*}(t) = G * U(t).$$

The following useful relations between $\{S_n\}$ and $\{N(t)\}$ allow limit properties of $\{N(t)\}$ to be determined from the behavior of $\{S_n\}$:

(3.3.1) $$[N(t) \leq n] = [S_n > t], \quad n \geq 0,$$

(3.3.2) $$S_{N(t)-1} \leq t < S_{N(t)} \text{ on } [N(t) \geq 1],$$

(3.3.3) $$[N(t) = n] = [S_{n-1} \leq t < S_n], \quad n \geq 1.$$

Note that the event $[N(t) \leq n]$ depends only on $S_0, \ldots, S_n$. (Advanced students should observe that for fixed t, $N(t)$ is a stopping time with respect to $\{\mathcal{F}_n = \sigma(Y_0, \ldots, Y_n), n \geq 0\}$.) Here are some moment properties of $N(t)$.

Theorem 3.3.1. *For any $t \geq 0$,*

(1) $\sum_{n=0}^{\infty} \gamma^n F^{n*}(t) < \infty$ *for* $\gamma < 1/F(0)$.

(2) *The moment generating function of $N(t)$ exists , so all moments of $N(t)$ are finite. In particular $U(t) < \infty$.*

Proof. (1) Fix $\gamma < 1/F(0)$. Observe

$$\lim_{\lambda \to \infty} \hat{F}(\lambda) = \lim_{\lambda \to \infty} \left(F(0) + \int_{(0,\infty)} e^{-\lambda x} F(dx) \right)$$
$$= F(0).$$

It is possible to pick a large λ such that $\hat{F}(\lambda)\gamma < 1$, and we have

$$\sum_{n=0}^{\infty} \gamma^n F^{n*}(t) = \sum_{n=0}^{\infty} \gamma^n P[S_n \leq t]$$

(where $S_n = Y_1 + ... + Y_n, Y_0 = 0$),

$$= \sum_{n=0}^{\infty} \gamma^n P[e^{-\lambda S_n} \geq e^{-\lambda t}].$$

By Markov's inequality (Billingsley, 1986, page 74, for example), which is almost the same as Chebychev's inequality, we have the upper bound

$$\sum_{0}^{\infty} \gamma^n E e^{-\lambda S_n} e^{\lambda t} = e^{\lambda t} \sum_{n=0}^{\infty} (\gamma \hat{F}(\lambda))^n < \infty,$$

since $\gamma \hat{F}(\lambda) < 1$.

(2) A positive rv Z has a moment generating function iff the distribution of Z has a tail which is exponentially bounded: For some $K > 0, c > 0$ and all $x > 0$, we have

(3.3.4) $$P[Z > x] \leq K e^{-cx}.$$

We check this by noting that if the moment generating function exists in $(0, \theta_0)$, then for $\theta < \theta_0$, we have

$$P[Z > x] = P[e^{\theta Z} > e^{\theta x}] \leq E(e^{\theta Z}) e^{-\theta x},$$

from Markov's inequality. Conversely, suppose that the distribution of Z has an exponentially bounded tail as in (3.3.4). Then, for $\theta < c$,

$$E e^{\theta Z} < \infty \text{ iff } E(e^{\theta Z} - 1) < \infty.$$

However,

$$E(e^{\theta Z} - 1) = E \int_0^\infty \theta e^{\theta u} 1_{[u<Z]} du$$
$$= \int_0^\infty \theta e^{\theta u} P[Z > u] du$$

(reversing the expectation and integral),

$$\leq \theta K \int_0^\infty e^{\theta u} e^{-cu} du < \infty.$$

To prove (2) then, it suffices to show that $P[N(t) > n]$ is exponentially bounded. Choose $1 < \gamma < 1/F(0)$, and from (1) we have

$$\gamma^n F^{n*}(t) \to 0, \text{ as } n \to \infty.$$

There exists n_0 such that for $n \geq n_0$

$$F^{n*}(t) \leq \gamma^{-n} = e^{-(\log \gamma)n}.$$

Using (3.3.1) with $Y_0 = 0$, we have, for $n \geq n_0$,

$$P[N(t) > n] = P[S_n \leq t] = F^{n*}(t) \leq e^{-(\log \gamma)n}.$$

With suitable choice of K, this can be extended for all n to

$$P[N(t) > n] \leq K e^{-cn}. \blacksquare$$

The previous result assures us that $U(t)$ is always finite. Unfortunately, it is not easy to compute U explicitly. The following two examples are cases where an explicit calculation can be accomplished.

Example 3.3.1. Suppose F is the exponential distribution with density

$$F(dx) = \alpha e^{-\alpha x} dx, \quad x \geq 0.$$

From Example 3.2.3 we have, for $n \geq 1$,

$$F^{n*}(dx) = \alpha(\alpha x)^{n-1} \frac{e^{-\alpha x}}{(n-1)!}.$$

Thus

$$\sum_{n=1}^{\infty} F^{n*}(x) = \sum_{n=1}^{\infty} \int_0^x \alpha(\alpha s)^{n-1} \frac{e^{-\alpha s}}{(n-1)!} ds$$

$$= \int_0^x \sum_{n=1}^{\infty} \alpha(\alpha s)^{n-1} \frac{e^{-\alpha s}}{(n-1)!} ds$$

$$= \int_0^x \alpha ds = \alpha x.$$

Remembering the term F^{0*}, we obtain

$$U(x) = \sum_{n=0}^{\infty} F^{n*}(x) = 1 + \alpha x.$$

Example 3.3.2. Now let F be the gamma density

$$F(dx) = xe^{-x}dx, \quad x > 0.$$

We seek the density of $\sum_{n=1}^{\infty} F^{n*}$. Observe that the Laplace transform of $\sum_{n=1}^{\infty} F^{n*}$ can be written as

$$\left(\widehat{\sum_{n=1}^{\infty} F^{n*}} \right) (\lambda) = \sum_{n=1}^{\infty} \left(\widehat{F^{n*}} \right) (\lambda) = \sum_{n=1}^{\infty} (\hat{F}(\lambda))^n$$

$$= \sum_{n=1}^{\infty} ((1+\lambda)^{-2})^n$$

from Example 3.2.3,

$$= \frac{(1+\lambda)^{-2}}{1 - (1+\lambda)^{-2}} = \frac{1}{1 + 2\lambda + \lambda^2 - 1}$$

$$= \frac{1}{\lambda^2 + 2\lambda} = \frac{1}{\lambda(\lambda + 2)}.$$

By a partial fraction expansion, this is

$$= \frac{1}{2\lambda} - \frac{1}{2(\lambda + 2)}$$

$$= \int_{x=0}^{\infty} e^{-\lambda x} \frac{1}{2} dx - \int_{x=0}^{\infty} e^{-\lambda x} \frac{1}{2} e^{-2x} dx$$

$$= \int_{x=0}^{\infty} e^{-\lambda x} (\frac{1}{2} - \frac{1}{2} e^{-2x}) dx$$

$$= \int_{x=0}^{\infty} e^{-\lambda x} \left(\sum_{n=1}^{\infty} F^{n*} \right) (dx).$$

Therefore, the required density is $\frac{1}{2} - \frac{1}{2}e^{-2x}$, since the transform uniquely determines the measure. Thus we have

$$U(x) = \sum_{n=0}^{\infty} F^{n*}(x)$$

$$= 1 + \int_0^x \frac{1}{2} - \frac{1}{2}e^{-2s} ds$$

$$= \frac{3}{4} + \frac{x}{2} + \frac{1}{4}e^{-2x}.$$

We now look at some limit theorems for $N(t)$ as $t \to \infty$ which can be readily obtained from standard limit theorems for $\{S_n\}$.

Theorem 3.3.2. *Suppose* $\mu = EY_1 = \int_0^\infty xF(dx) < \infty$ *is the mean interarrival time.*

    (1) *If* $P[Y_0 < \infty] = 1$, *then almost surely*

$$N(t)/t \to \mu^{-1}, \ as \ t \to \infty.$$

    (2) *If* $\sigma^2 = \mathrm{Var}(Y_1) < \infty$ *then* $N(t)$ *is* $AN(\mu t^{-1}, t\sigma^2\mu^{-3})$; *i.e.,*

$$\lim_{t\to\infty} P[(N(t) - t\mu^{-1})/(t\sigma^2\mu^{-3})^{1/2} \le x] = N(0, 1, x),$$

    where $N(0, 1, x)$ *is the standard normal distribution function.*

Proof. From the strong law of large numbers,

$$n^{-1}S_n = n^{-1}Y_0 + n^{-1}\sum_{i=1}^{n} Y_i \to 0 + \mu = \mu,$$

as $n \to \infty$. Observe that $N(t) \uparrow \infty$ a.s. (Advanced students can check this by observing that, since $\{N(t)\}$ is monotone, it is enough to show $N(t) \uparrow \infty$ in probability. But this is immediate, since

$$P[N(t) > n] = P[S_n \le t] \quad \text{(from 3.3.1)}$$
$$= G * F^{n*}(t) \to 1 \text{ as } t \to \infty.)$$

We therefore conclude that, with probability 1,

$$\lim_{t\to\infty} S_{N(t)}/N(t) = \mu.$$

Recalling (3.3.2) we have

$$S_{N(t)-1} \le t < S_{N(t)},$$

from which

$$\left(\frac{S_{N(t)-1}}{N(t)-1}\right)\left(\frac{N(t)-1}{N(t)}\right) \le \frac{t}{N(t)} \le \frac{S_{N(t)}}{N(t)}.$$

As $t \to \infty$ the two extremes converge to μ almost surely, and the result follows.

For the proof of (2), let $[x]$ be the greatest integer less than or equal to x. We use (3.3.1) and the central limit theorem for partial sums:

$$\lim_{n\to\infty} P[(S_n - n\mu)/\sigma\sqrt{n} \le x] = N(0,1,x)$$

uniformly in $x \in R$. ($N(0,1,x)$ is the standard normal distribution function with mean 0 and variance 1.)

We have

$$P[N(t) - t\mu^{-1})/(\sigma^2 t\mu^{-3})^{1/2} \le x] = P[N(t) \le [x(\sigma^2 t\mu^{-3})^{1/2} + t\mu^{-1}]].$$

Setting $h(t) = [x(\sigma^2 t\mu^{-3})^{1/2} + t\mu^{-1}]$ and applying (3.3.1), the above is

$$P[S_{h(t)} > t] = P[(S_{h(t)} - \mu h(t))/\sigma h^{1/2}(t) > (t - \mu h(t))/\sigma h^{1/2}(t)].$$

It suffices to show $h(t) \to \infty$ and $z(t) := (t - \mu h(t))/\sigma h^{1/2}(t) \to -x$, since in that case the uniform convergence in the central limit theorem will imply

$$P[(S_{h(t)} - \mu h(t))/\sigma h^{1/2}(t) > z(t)] \to 1 - N(0,1,-x) = N(0,1,x).$$

The first check is easy since $h(t) \sim t\mu^{-1} \to \infty$. For the second we have

$$h(t) = x(\sigma^2 t\mu^{-3})^{1/2} + t\mu^{-1} + \epsilon(t),$$

where $|\epsilon(t)| \le 1$, and hence

$$z(t) = \frac{t - \mu x(\sigma^2 t\mu^{-3})^{1/2} - t - \mu\epsilon(t)}{\sigma h^{1/2}(t)}$$

$$\sim \frac{-\mu x(\sigma^2 t\mu^{-3})^{1/2}}{\sigma(t\mu^{-1})^{1/2}} = -x. \blacksquare$$

We now consider what has come to be called the elementary renewal theorem.

Theorem 3.3.3. *Let* $\mu = EY_1 \leq \infty$. *Then*

$$\lim_{t \to \infty} t^{-1}V(t) = \lim_{t \to \infty} t^{-1}U(t) = \mu^{-1}$$

provided $Y_0 < \infty$ *a.s.*

There are three common ways to proceed with a proof. One is by means of the Tauberian theorem for Laplace transforms (cf. Feller, Vol II, 1971, Chapter XIII). A second proof uses the Blackwell Renewal Theorem, which we consider later. Here is an elementary approach based on truncation.

Proof. First of all, Fatou's lemma and Theorem 3.3.2 (1) imply that

$$\mu^{-1} = E \liminf_{t \to \infty} t^{-1}N(t) \leq \liminf_{t \to \infty} t^{-1}EN(t)$$
$$= \liminf_{t \to \infty} t^{-1}V(t).$$

For an inequality in the reverse direction, set $Y_0^* = 0, Y_i^* = Y_i \wedge b, S_0^* = 0, S_n^* = Y_1^* + \ldots + Y_n^*, N^*(t) = \sum_{n=0}^{\infty} 1_{[0,t]}(S_n^*), V^*(t) = EN^*(t)$. Then $S_n \geq S_n^*$, and $N^*(t) \geq N(t)$.

We now observe that $ES_{N^*(t)}^* = EY_1^* EN^*(t) = EY_1^* V^*(t)$, which is a special case of the Wald Identity since $N^*(t)$ is a stopping time with respect to $\{\sigma(Y_0^*, \ldots, Y_n^*), n \geq 0\}$. Therefore

$$\limsup_{t \to \infty} t^{-1}V(t) \leq \limsup_{t \to \infty} t^{-1}V^*(t)$$
$$= \limsup_{t \to \infty} t^{-1}ES_{N^*(t)}^*/EY_1^*$$
$$= \limsup_{t \to \infty} t^{-1}\left(ES_{N^*(t)-1}^* + Y_{N^*(t)}^*\right)/EY_1^*$$
$$\leq \limsup_{t \to \infty} t^{-1}(t+b)/EY_1^* = 1/EY_1^*$$
$$= 1/E(Y_1 \wedge b).$$

Let $b \uparrow \infty$ so that, by monotone convergence, $E(Y_1 \wedge b) \uparrow EY_1 = \mu$, and we get

$$\limsup_{t \to \infty} t^{-1}V(t) \leq 1/\mu,$$

as desired. ∎

3.4. RENEWAL REWARD PROCESSES.

A slight elaboration of the strong law of large numbers and Theorem 3.3.2 yields basic facts about renewal reward processes. Suppose we have a renewal sequence $\{S_n, n \geq 0\}$ and suppose that at each renewal epoch S_n we are given a possibly random reward R_n. The phrase *reward* should be interpreted in the broad sense, as this could be a negative quantity associated with costs. We assume that the sequence of random variables $\{R_n, n \geq 1\}$ is iid although not necessarily independent of $\{S_n, n \geq 0\}$. (In fact, if the inter-renewal times are $\{Y_n, n \geq 0\}$, one possibility would be for $R_n = cY_n$ so that the reward was proportional to the length of the interarrival interval.) The *renewal reward* process is defined to be

$$R(t) = \sum_{i=0}^{N(t)-1} R_i = \sum_{i=0}^{\infty} R_i 1_{[S_i \leq t]}, \quad t \geq 0,$$

so that $R(t)$ is the total reward accumulated by time t. (If $S_0 > 0$, then for $t \in [0, S_0)$ we interpret $R(t) = 0$.)

To fix ideas, imagine an insurance company receiving claims according to a renewal process $\{S_n\}$. The sizes of the claims are random and represented by the random variables $\{R_n\}$. The total claim against the company in $[0, t]$ is $R(t)$.

We think of $R(t)/t$ as the reward per unit time, and, if the limit

(3.4.1) $$\lim_{t \to \infty} R(t)/t =: r$$

exists almost surely, it is natural to think of r as the long run reward rate. Under mild conditions this limit does indeed exist.

Proposition 3.4.1. If $E|R_j| < \infty$ and $EY_j = \mu \in (0, \infty)$ for $j \geq 1$, then the limit in (3.4.1) exists almost surely, and

(3.4.1) $$\lim_{t \to \infty} \frac{R(t)}{t} =: r = \frac{ER_1}{\mu}.$$

Proof. We have

$$\lim_{t \to \infty} \frac{R(t)}{t} = \lim_{t \to \infty} \frac{\sum_{i=0}^{N(t)-1} R_i}{t}$$

$$= \lim_{t \to \infty} \frac{\sum_{i=0}^{N(t)-1} R_i}{N(t)-1} \frac{N(t)-1}{t}.$$

Since $N(t) - 1 \to \infty$, we have, from the strong law of large numbers and Theorem 3.3.2 (1), that this limit is equal to

$$= ER_1\mu^{-1}. \blacksquare$$

Example. Harry Budgets for Maintenance. In the old days, before Harry knew any Operations Research, he used to budget c_1 per day for maintenance at the restaurant. He now decides to budget less, $c_2 < c_1$, but now risks the possibility of being fined by OSHA (Occupational Safety and Health Administration) for health hazards or safety violations on the premises. OSHA inspects Harry's premises according to a renewal process with mean inter-renewal time $E(Y_1) = 45$ days. At each inspection, there is probability $p > 0$ that a violation will be found resulting in a fine. The events $\{[$ detection occurs at the nth inspection $], n \geq 0\}$ are independent of each other and the times of inspection. The fines vary in severity according to the violation and may be assumed to be independent and identically distributed with a uniform distribution on the interval $(0, 2c_3)$.

(1) What is the expected time until the first violation is detected?
(2) What is Harry's long run cost rate under the new policy?
(3) What is the range of values of c_2 which guarantee that the new policy saves Harry money over the long run?

Let $\{S_i, i \geq 0\}$ be times when violations are detected, and let $\{Y_i\}$ be the times between inspections. Then

$$S_1 = \sum_{i=1}^{N} Y_i,$$

where N is the number of inspections necessary to uncover a violation. N is geometrically distributed: $P[N = n] = q^{n-1}p, n \geq 1$, so $E(N) = 1/p$, and $E(S_1) = E(N)E(Y_1) = 45/p$.

Let F_i be the fine on the ith violation; $E(F_i) = c_3$. Let $N(t)$ count the number of violations up to time t so that the times which $N(t)$ counts are the $\{S_i\}$. Harry's maintenance costs up to time t are

$$C(t) = c_2 t + \sum_{i=1}^{N(t)-1} F_i;$$

the long run average cost rate is therefore

$$C(t)/t = c_2 + t^{-1} \sum_{i=1}^{N(t)-1} F_i$$
$$\to c_2 + E(F_1)/E(S_1) = c_2 + c_3 p/45.$$

We want $c_2 + c_3 p/45 \leq c_1$, that is $c_2 \leq c_1 - c_3 p/45$ $\blacksquare$.

Under reasonable conditions, Proposition 3.4.1 also holds in the expected value sense:

$$ER(t)/t \to ER_1\mu^{-1}, \text{ as } t \to \infty.$$

This would be the case if, for instance, $N(t)$ was independent of $\{R_n, n \geq 0\}$ (cf. Chapter 1) or if, for each $n \geq 0$, we have (R_n, Y_n) independent of $\{Y_j, j \neq n\}$ and $\{R_n, n \geq 1\}$ identically distributed with $E|R_n| < \infty$. Under these assumptions we can write

$$
\begin{aligned}
ER(t) &= ER(t)1_{[N(t)=0]} + ER(t)1_{[N(t)>0]} \\
&= 0 + ER(t)1_{[N(t)>0]} \\
&= E \sum_{i=0}^{N(t)-1} R_i 1_{[N(t)>0]} \\
&= E \sum_{i=0}^{N(t)} R_i 1_{[N(t)>0]} - ER_{N(t)} 1_{[N(t)>0]} \\
&= A + B,
\end{aligned}
$$

and we must show,

$$\lim_{t\to\infty} A/t = ER_1/\mu, \quad \lim_{t\to\infty} B/t = 0.$$

As in the proof of Wald's Identity (Section 1.8.1), we have

$$
\begin{aligned}
A &= E \sum_{i=0}^{N(t)} R_i 1_{[N(t)>0]} = ER_0 1_{[N(t)>0]} + E \sum_{i=1}^{\infty} R_i 1_{[i \leq N(t)]} \\
&= ER_0 1_{[N(t)>0]} + E \sum_{i=1}^{\infty} R_i 1_{[i-1 < N(t)]},
\end{aligned}
$$

which, by applying (3.3.1), is

$$
\begin{aligned}
&= ER_0 1_{[N(t)>0]} + E \sum_{i=1}^{\infty} R_i 1_{[S_{i-1} \leq t]} \\
&= ER_0 1_{[N(t)>0]} + ER_1 E \sum_{i=1}^{\infty} 1_{[S_{i-1} \leq t]} \\
&= ER_0 1_{[N(t)>0]} + ER_1 V(t) \sim tER_1/\mu.
\end{aligned}
$$

We have from (3.3.3) that

$$B = ER_{N(t)}1_{[N(t)>0]} = \sum_{n=1}^{\infty} ER_n 1_{[N(t)=n]}$$

$$= \sum_{n=1}^{\infty} ER_n 1_{[S_{n-1}\leq t<S_n]},$$

and, because (Y_n, R_n) is assumed independent of S_{n-1}, we get by conditioning on S_{n-1}

$$= \sum_{n=1}^{\infty} \int_0^t ER_n 1_{[x\leq t<x+Y_n]}P[S_{n-1} \in dx]$$

$$= \int_0^t ER_1 1_{[t<x+Y_1]}V(dx).$$

Set

$$z(u) = E|R_1|1_{[u<Y_1]}.$$

Then $z(u) \to 0$ as $u \to \infty$ as a consequence of $E|R_1| < \infty$, $z(u)$ is nonincreasing, and $z(u) \leq E|R_1|$. We conclude

$$|B| \leq V * z(t),$$

and it is now quick work to see that $B/t \to 0$ as $t \to \infty$. Given $\epsilon > 0$, there exists $M > 0$ such that $z(u) < \epsilon$ if $u > M$. Write for any large $t \geq M$

$$V * z(t) = \int_0^t z(t-x)V(dx)$$

$$= \int_0^{t-M} z(t-x)V(dx) + \int_{t-M}^t z(t-x)V(dx)$$

$$\leq \epsilon V(t-M) + z(0)(V(t) - V(t-M)).$$

From the elementary renewal theorem we get that

$$\limsup_{t\to\infty} V * z(t)/t \leq \frac{\epsilon}{\mu} + z(0)(\frac{1}{\mu} - \frac{1}{\mu}) = \frac{\epsilon}{\mu}.$$

Since $\epsilon > 0$ is arbitrary, the desired conclusion follows.

A variant on the renewal reward process which is handled similarly allows rewards or penalties to accumulate continuously in time instead of being delivered in discrete bursts at renewal times. Such processes are sometimes called *cumulative processes*. Suppose for simplicity that the renewal process is pure, and imagine that in the nth renewal interval $[S_n, S_{n+1})$ the reward rate is random and equal to I_n. The reward rate process is

$$I(t) = I_n \text{ if } t \in [S_n, S_{n+1})$$

$$= \sum_{n=0}^{\infty} I_n 1_{[S_n, S_{n+1})}(t),$$

and the accumulated reward up to time t is

$$C(t) = \int_0^t I(u)du.$$

This is easily analyzed since

$$C(t) = \sum_{n=1}^{N(t)-1} \int_{S_{n-1}}^{S_n} I(u)du + \int_{S_{N(t)-1}}^{t} I(u)du$$

(when $N(t) = 1$, we interpret $\sum_{n=1}^{0} = 0$)

$$= \sum_{n=1}^{N(t)-1} \int_{S_{n-1}}^{S_n} I_{n-1}du + \int_{S_{N(t)-1}}^{t} I_{N(t)-1}du$$

$$= \sum_{n=1}^{N(t)-1} Y_n I_{n-1} + (t - S_{N(t)-1})I_{N(t)-1}.$$

If the last term can be neglected asymptotically then using the argument of Proposition 3.4.1 shows

$$C(t) \sim t\mu^{-1} EY_1 I_0$$

as $t \to \infty$, provided $E|Y_1 I_0| < \infty$.

Other assumptions of a similar nature on the rate are possible.

Example. Harry and the Public Health Service. It is well known to the Public Health Service that Harry's Restaurant gives only casual attention to matters of cleanliness; consequently, the Public Health Service periodically closes the restaurant and mandates a cleanup. Independent periods of operation and closure form an alternating renewal process. The open periods are iid with means μ_o, and the closed periods are iid with means μ_c. Lengths of open periods and closed periods are independent of each other. What is the long run proportion of time that Happy Harry's Restaurant is open to its adoring public?

Imagine that at time 0 the restaurant is open and that $\{S_n\}$ are the times when the restaurant reopens after being closed so that

$$Y_{n+1} = S_{n+1} - S_n = OP_{n+1} + CL_{n+1}.$$

Denote by $\{OP_n\}$ the iid lengths of open periods and by $\{CL_n\}$ the iid lengths of closed periods. $N(t)$ counts the renewals $\{S_n\}$. Let $I(t)$ be 1 if the restaurant is open at time t and 0 otherwise; then

$$C(t) = \int_0^t I(u)\,du,$$

and

$$\sum_{n=0}^{N(t)-1} OP_n \le C(t) \le \sum_{n=0}^{N(t)} OP_n,$$

and therefore

$$\lim_{t\to\infty} \frac{C(t)}{t} = \frac{E(OP_1)}{E(S_1)} = \frac{\mu_o}{\mu_o + \mu_c},$$

giving the long run proportion of time the restaurant is open. ∎

3.5. THE RENEWAL EQUATION.

The renewal equation is the convolution equation of the form

$$Z = z + F * Z,$$

or, in long hand,

(3.5.1) $$Z(t) = z(t) + \int_0^t Z(t-y)F(dy).$$

All functions are defined on $[0, \infty)$, and it is convenient to set $z(t) = Z(t) = F(t) = 0$ for $t < 0$. Z is an unknown function, z is assumed to be a known function and F is a distribution on $[0, \infty)$ with $F(\infty) = \lim_{x\uparrow\infty} F(x) < \infty$. If $F(\infty) = 1$ then we call the equation *proper*.

In branching processes such equations are common and appear in the form

(3.5.2) $$Z = z + mF * Z,$$

where F is proper. The branching process literature calls this a renewal equation with a parameter (cf. Athreya and Ney (1972)).

We now consider some examples where the renewal equation arises naturally. Generally such equations arise when one considers a probability statement about a function of the path up to time t. The term z arises from conditioning on the first renewal being later than t, and the convolution term of the renewal equation arises by conditioning on the time of the first renewal and moving the origin of time up to this first renewal epoch.

Example 3.5.1. Consider the renewal function $U(t)$. We have

$$U(t) = \sum_{n=0}^{\infty} F^{n*}(t) = F^{0*}(t) + \sum_{n=1}^{\infty} F^{n*}(t)$$

$$= F^{0*}(t) + F * \sum_{n=1}^{\infty} F^{(n-1)*}(t)$$

$$= F^{0*}(t) + F * U(t),$$

and we have the renewal equation with $Z = U, z = F^{0*}$.

Example 3.5.2. Age Dependent Branching Process. Individuals of a population live for a random amount of time determined by a probability distribution G. At the time of death an individual reproduces k offspring with probability p_k, $k \geq 0$. We assume $\sum_{i=0}^{\infty} p_i = 1$, and reproduction is independent of life length. Furthermore, different individuals behave independently of one another. For concreteness, suppose the population is initiated at time 0 by a progenitor of age zero. (For a discussion of family trees and the probability space on which such a process lives, see Jagers, 1975.)

Suppose $X(t)$ is the population size at time t, and set $Z(t) = EX(t)$. Let the lifetime of the progenitor be the random variable L_1, and suppose N is the number of offspring procreated by the progenitor. Then $P[N = k] = p_k$, $k \geq 0$. Set $m = EN = \sum_{k=0}^{\infty} kp_k$. Finally, let $\{\{X_i(t), t \geq 0\}, i \geq 1\}$ be iid copies of X independent of (X, N, L_1). Then

$$Z(t) = EX(t) = EX(t)1_{[L_1 \leq t]} + EX(t)1_{[L_1 > t]}.$$

On the set $L_1 > t$, $X(t) = 1$. On the set $L_1 \leq t$, we have

$$X(t) = \sum_{i=1}^{N} X_i(t - L_1)1_{[L_1 \leq t]}.$$

Therefore,

$$Z(t) = P[L_1 > t] + E\sum_{i=1}^{N} X_i(t - L_1)1_{[L_1 \leq t]}.$$

For the second term (interpreting $\sum_1^0 = 0$),

$$E\sum_{i=1}^{N} X_i(t - L_1)1_{[L_1 \leq t]} = \sum_{k=0}^{\infty} E\sum_{i=1}^{k} X_i(t - L_1)1_{[N=k, L_1 \leq t]}$$

$$= \sum_{k=0}^{\infty} P[N = k]E\sum_{i=1}^{k} X_i(t - L_1)1_{[L_1 \leq t]}$$

$$= \sum_{k=0}^{\infty} kP[N = k]EX_1(t - L_1)1_{[L_1 \leq t]}$$

$$= m\int_0^t E(X_1(t - L_1)|L_1 = y)P[L_1 \in dy]$$

$$= m\int_0^t Z(t - y)P[L_1 \in dy].$$

Summarizing,

$$Z(t) = P[L_1 > t] + m\int_0^t Z(t - y)P[L_1 \in dy],$$

which is (3.5.2).

Example 3.5.3. Forward and Backward Recurrence Times. Consider a renewal sequence $\{S_n, n \geq 0\}$ with $S_n - S_{n-1} = Y_n$, and set

$$B(t) = S_{N(t)} - t.$$

For $t \geq S_0$, set

$$A(t) = t - S_{N(t)-1},$$

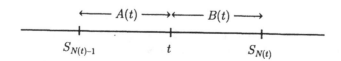

FIGURE 3.2. FORWARD AND BACKWARD RECURRENCE TIMES

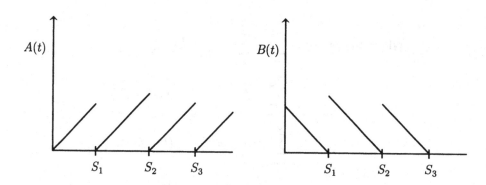

FIGURE 3.3. PATHS OF $A(t), B(t)$.

The *forward recurrence time* $B(t)$, also called the *excess life* or *residual life*, is the time until the next renewal. The *backward recurrence time* $A(t)$ is the time since the last renewal and is also called the *age* process. Paths of B are line segments between jumps, which decrease at an angle of -45 degrees. For A the line segments increase and have slope 45 degrees. See Figure 3.3.

To compute the distributions of $A(t)$ and $B(t)$, we write a renewal equation as a first step. Later, after some theory has been developed, these equations are solved.

For A we assume the process is pure. Let $x > 0$ and fixed:

$$P[A(t) \le x] = P[A(t) \le x, Y_1 \le t] + P[A(t) \le x, Y_1 > t].$$

Now, $A(t) = t$ on $[Y_1 > t]$, so

$$P[A(t) \le x, Y_1 > t] = (1 - F(t))1_{[0,x]}(t).$$

On $[Y_1 \le t]$, note that A "starts from scratch" at Y_1, so

$$P[A(t) \le x, Y_1 \le t] = \int_0^t P[A(t - y) \le x]F(dy).$$

The phrase "starts from scratch" is time-honored but vague. The following elementary argument is precise. We have

$$P[A(t) \le x, Y_1 \le t] = P[t - S_{N(t)-1} \le x, N(t) \ge 2]$$

$$= \sum_{n=2}^{\infty} P[t - S_{n-1} \le x, S_{n-1} \le t < S_n].$$

Condition on Y_1, and the above is

$$= \sum_{n=2}^{\infty} \int_0^t P[t - (y + \sum_{i=2}^{n-1} Y_i) \leq x, y + \sum_{i=2}^{n-1} Y_i \leq t \leq y + \sum_{i=2}^{n} Y_i] F(dy)$$

$$= \sum_{n=2}^{\infty} \int_0^t P[t - y - S_{n-2} \leq x, S_{n-2} \leq t - y \leq S_{n-1}] F(dy)$$

$$= \sum_{n=2}^{\infty} \int_0^t P[t - y - S_{N(t-y)-1} \leq x, N(t-y) = n - 1] F(dy)$$

$$= \sum_{n=1}^{\infty} \int_0^t P[A(t-y) \leq x, N(t-y) = n] F(dy)$$

$$= \int_0^t P[A(t-y) \leq x] F(dy).$$

Thus, in summary,

$$P[A(t) \leq x] = (1 - F(t))1_{[0,x]}(t) + \int_0^t P[A(t-y) \leq x] F(dy),$$

which is of the form (3.5.1) with $z(t) = (1 - F(t))1_{[0,x]}(t)$ and $Z(t) = P[A(t) \leq x]$.

For B we have for $x > 0$ (still supposing $Y_0 = 0$),

$$P[B(t) > x] = P[B(t) > x, Y_1 \leq t] + P[B(t) > x, Y_1 > t].$$

In order for $B(t) > x$ on $[Y_1 > t]$, we must have $Y_1 > t + x$. On $Y_1 \leq t$, B starts over at Y_1, so

$$P[B(t) > x] = \int_0^t P[B(t-y) > x] F(dy) + 1 - F(t + x).$$

This is made precise as with the equation involving $A(t)$. As an exercise, try writing the equation for B when the process is delayed.

A challenging example of risk processes is discussed in Section 3.5.1.

We now consider the solution of the renewal equation (3.5.1).

$$(3.5.1) \qquad\qquad Z(t) = z(t) + \int_0^t Z(t-y) F(dy).$$

Set $m = F(\infty) < \infty$ and $U(t) = \sum_{n=0}^{\infty} F^{n*}(t)$. In the majority of cases, F is a probability distribution and $m = 1$, but the added generality is useful. Assume $F(0) < 1$. Then $U(t) < \infty$ for all $t > 0$, because we can write

$$U(t) = \sum_{n=0}^{\infty} m^n (m^{-1} F)^{n*}(t).$$

Since $m(m^{-1}F)(0) = F(0) < 1$, the finiteness of U follows from Theorem 3.3.1.

Theorem 3.5.1. *Suppose $z(t) = 0$ for $t < 0$, and z is locally bounded.*
Suppose $F(0) < 1$.

(1) *A locally bounded solution of the renewal equation (3.5.1) is*
$U * z(t) = \int_0^t z(t - u)U(du)$.

(2) *There is no other locally bounded solution vanishing on $(-\infty, 0)$.*

Proof. (1) We first check that $U * z$ is locally bounded: For any $T > 0$

$$\sup_{0 \le t \le T} U * z(t) = \sup_{0 \le t \le T} \int_0^t z(t - u)U(du)$$

$$\le \left(\sup_{0 \le t \le T} z(s) \right) U(T) < \infty.$$

To check $U * z$ is a solution, note

$$F * (U * z) = (F * U) * z$$

(by associativity)

$$= (U - F^{0*}) * z$$

(from the renewal equation for U given in Example 3.5.1)

$$= U * z - z.$$

Thus
$$U * z = z + F * (U * z),$$

and $U * z$ is a solution of (3.5.1).

(2) Let Z_1, Z_2 be two locally bounded solutions of (3.5.1) vanishing on
$(-\infty, 0)$ so that

(3.5.3) $Z_i = z + F * Z_i, \quad i = 1, 2.$

Set $H = Z_1 - Z_2$, and H is locally bounded also. Further, if we difference
the two equations in (3.5.5) then

$$H = Z_1 - Z_2 = F * (Z_1 - Z_2) = F * H.$$

Iterating, we get, for any $n \ge 1$,

$$H = F^{n*} * H.$$

Therefore, for any $T > 0$, and by recalling property (2) of Section 3.2.2

$$\sup_{0 \le t \le T} |H(t)| = \sup_{0 \le t \le T} \left| \int_0^t (Z_1(t-y) - Z_2(t-y)) F^{n*}(dy) \right|$$
$$\le \sup_{0 \le t \le T} H(t) F^{n*}(T) \to 0$$

as $n \to \infty$, since H is locally bounded and $U(T) < \infty$ implies $F^{n*}(T) \to 0$. So $H \equiv 0$ and $Z_1 = Z_2$. ∎

Example 3.5.4. If $F(x) = 1 - e^{-\alpha x}$, $x > 0$, then recall from Section 3.3 that

$$U(t) = 1 + \alpha t.$$

Therefore

$$U * z(t) = z(t) + \alpha \int_0^t z(t-y) dy$$
$$= z(t) + \alpha \int_0^t z(u) du.$$

Example 3.5.5. If $F(dx) = xe^{-x} 1_{(0,\infty)}(x) dx$, then, from the inversion example in Section 3.3,

$$\sum_{n=1}^{\infty} F^{n*}(dx) = \frac{1}{2} - \frac{1}{2} e^{-2x}, \quad x > 0,$$

and, in this case,

$$U * z(t) = z(t) + \int_0^t z(t-y) \{ \frac{1}{2} - \frac{1}{2} e^{-2y} \} dy.$$

Example 3.5.2 (continued). If G is the life length distribution and $\mu(t) = EX(t)$, then

$$\mu(t) = \left(\sum_{n=0}^{\infty} m^n G^{n*} \right) * (1 - G)(t).$$

(Make the identification $F = mG, U = \sum_{n=0}^{\infty} m^n G^{n*}$.)

Example 3.5.3 (continued). Solving the renewal equation for the distribution of $A(t)$ gives

$$P[A(t) \le x] = U * ((1 - F) \cdot 1_{[0,x]})(t).$$

For the tail of the distribution of $B(t)$, we get

$$P[B(t) > x] = U * (1 - F(\cdot + x))(t),$$

i.e.,

$$(3.5.4) \qquad P[B(t) > x] = \int_0^t (1 - F(x + t - y))U(dy).$$

A special case is $F(dx) = \alpha e^{-\alpha x} 1_{(0,\infty)}(x)dx$. Then

$$\int_0^t (1 - F(x + t - y))U(dy) = 1 - F(x + t) + \alpha \int_0^t (1 - F(x + t - y))dy$$

$$= e^{-\alpha(x+t)} + \alpha \int_0^t e^{-\alpha(x+t-y)}dy$$

$$= e^{-\alpha(x+t)} + e^{-\alpha x} \int_0^t \alpha e^{-\alpha s}ds$$

$$= e^{-\alpha(x+t)} + e^{-\alpha x}(1 - e^{-\alpha t}) = e^{-\alpha x}$$

so $P[B(t) > x] = e^{-\alpha x}$, and $B(t)$ has the exponential density for each t.

An interpretation of $U(dx)$ or $V(dx)$: Renewal equations are obtained by conditioning on an *initial* renewal or jump in the process. Frequently one can bypass the derivation of the renewal equation and go straight to a deduction of the solution by conditioning on the *last* jump before time t. The following heuristic helps one to do this. Fix $x > 0$. Then

$$P[\text{There is some renewal in } (x, x + dx]] = P\{\bigcup_{n=0}^{\infty} [S_n \in (x, x + dx]]\}.$$

Because $(x, x + dx]$ is such a small interval, at most one renewal could possibly be squeezed into it; hence the events $\{[S_n \in (x, x + dx]], n \geq 0\}$ are disjoint (heh, heh), and therefore,

$$P[\text{There is some renewal in } (x, x + dx]] = \sum_{n=0}^{\infty} P[S_n \in (x, x + dx]] = V(dx).$$

We may now interpret (3.5.4) as follows:

$$P[B(t) > x]$$

$$= \int_{u \leq t} P[\text{Last renewal before } t \text{ in } (u, u + du]; \text{ no renewal in } (t, t + x]]$$

$$= \int_{u \leq t} P[\text{Some renewal at } (u, u + du]; \text{ no renewal in } (u, t + x]]$$

$$= \int_{u \leq t} U(du)(1 - F(t + x - u)).$$

If you are nervous about this heuristic derivation, the derivation of (3.5.4) based on the last renewal before t is more rigorously given as follows (assuming $Y_0 = 0$):

$$P[B(t) > x] = P[S_{N(t)} - t > x]$$
$$= \sum_{n=1}^{\infty} P[S_n - t > x; N(t) = n]$$
$$= \sum_{n=1}^{\infty} P[S_n - t > x, S_{n-1} \le t < S_n],$$

and conditioning on S_{n-1} gives

$$= \sum_{n=1}^{\infty} \int_0^t P[y - t + Y_n > x, t < y + Y_n] F^{(n-1)*}(dy)$$
$$= \int_0^t P[Y_n > t + x - y] \sum_{n=1}^{\infty} F^{(n-1)*}(dy)$$
$$= \int_0^t (1 - F(t + x - y)) U(dy).$$

3.5.1. RISK PROCESSES*.

This is one of the time-honored examples in renewal theory and is more challenging than the previous examples. It is an excellent example of an interesting problem whose solution is obtained by writing and solving a renewal equation.

An insurance company receives claims according to a *Poisson process with rate* α (i.e., a renewal process with interarrival density $\alpha e^{-\alpha x} 1_{(0,\infty)}(x)$). Sizes of claims are given by non-negative iid rv's $X_1, X_2, \ldots$, and $\{X_n\}$ is independent of the Poisson process. Let $f(t)$ be the amount of capital the company has at time t, i.e., the risk reserve. We assume that, provided $f(t) \ge 0$, the risk reserve increases at rate c at time t due to the inflow of premiums as well as to profits from investments. We are interested in the ruin probability for the company, and we consider

$$R(x) = P[f(t) > 0 \text{ for all } t > 0 | f(0) = x] = P[\text{ no ruin } | f(0+) = x].$$

---

* This section contains challenging material which may be skipped on first reading by beginning readers.

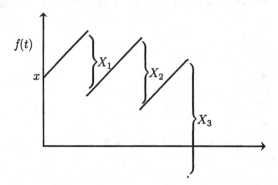

FIGURE 3.4. The Risk Process

After claim X_3, ruin occurs in the above picture.
Let $S_1, S_2, S_3, \ldots$ be the times of claims. Then

$$f(t) = x + ct \qquad\qquad 0 < t < S_1,$$
$$= x + cS_1 - X_1 \qquad\quad t = S_1.$$

If $x + cS_1 - X_1 < 0$, ruin occurs. Otherwise,

$$f(t) = x + cS_1 - X_1 + c(t - S_1)$$
$$= x + ct - X_1 \qquad\qquad S_1 \le t < S_2$$
$$= x + cS_2 - X_1 - X_2 \qquad t = S_2,$$

After the first jump, the process f continues as if it were starting from
time 0 with a new initial state $x + cS_1 - X_1$. Therefore conditioning on the
time and place of the first jump gives

$$R(x) = P[f(t) \ge 0 \text{ for all } t > 0, x + cS_1 - X_1 \ge 0 | f(0) = x]$$

$$= \int_{\{(s,y):x+cs-y\ge 0\}} P[f(t) > 0,\ \forall\, t | f(0) = x+cs-y]\, P[(S_1, X_1) \in (ds, dy)].$$

Using the independence of X_1 and S_1, this becomes

$$= \int_{s=0}^{\infty} \int_{y=0}^{x+cs} R(x + cs - y)\alpha e^{-\alpha s}\, ds P[X_1 \in dy].$$

We manipulate this to get it into the form of a renewal equation. First let
$s' = x + cs$ so $ds' = cds$, and

$$R(x) = \int_{s'=x}^{\infty} \int_{y=0}^{s'} R(s' - y)\alpha e^{-\alpha(s'-x)c^{-1}} c^{-1}\, ds' P[X_1 \in dy].$$

Dropping the prime

$$e^{-axc^{-1}}R(x) = \alpha c^{-1}\int_x^\infty e^{-\alpha c^{-1}s}\left(\int_0^s R(s-y)P[X_1 \in dy]\right)ds.$$

The right side is the integral of a bounded function and hence is absolutely continuous on $(0,\infty)$. This means that the left side is also absolutely continuous, and thus R is continuous. If R is continuous , the integrand on the right side is continuous. So the right side, being the integral of a continuous function, is differentiable. (Review the argument of (8), Section 3.2.2.) Differentiating gives us

$$e^{-axc^{-1}}R'(x)-R(x)e^{-axc^{-1}}\alpha c^{-1} = -\alpha c^{-1}e^{-\alpha c^{-1}x}\int_0^x R(x-y)P[X_1 \in dy];$$

i.e.,

$$(3.5.1.1)\qquad R'(x) = \alpha c^{-1}R(x) - \alpha c^{-1}\int_0^x R(x - y)P[X_1 \in dy].$$

At this point one has a choice of whether to derive an equation for R or R'. We try R'. Noting that

$$R(x) = \int_0^x R'(u)du + R(0),$$

we have

$$\int_0^x R(x - y)P[X_1 \in dy]$$

$$= \int_{y=0}^x\int_{u=0}^{x-y} R'(u)duP[X_1 \in dy] + R(0)P[X_1 \leq x]$$

$$= \int_{y=0}^x\int_{u=y}^x R'(x - u)duP[X_1 \in dy] + R(0)P[X_1 \leq x]$$

$$= \int_{u=0}^x\left(\int_{y=0}^u P[X_1 \in dy]\right) R'(x - y)du + R(0)P[X_1 \leq x]$$

$$= \int_0^x R'(x - u)P[X_1 \leq u]du + R(0)P[X_1 \leq x],$$

so

$$R'(x) = \alpha c^{-1}\left(\{R(x) - R(0)P[X_1 \leq x]\} - \int_0^x R'(x - u)P[X_1 \leq u]du\right)$$

or

$$(3.5.1.2) \quad R'(x) = \alpha c^{-1} R(0) P[X_1 > x] + \int_0^x R'(x - u) \alpha c^{-1} P[X_1 > u] du.$$

Set $F(du) = \alpha c^{-1} P[X_1 > u]$ so that $F(R_+) = \alpha c^{-1} EX_1$, and we have $Z = R'$, $z(x) = \alpha c^{-1} R(0) P[X_1 > x]$ (which still involves an unknown $R(0)$). This gives a renewal equation.

Renewal equations can be solved using Laplace transforms. Set

$$\phi(\lambda) = Ee^{-\lambda X_1}, \theta(\lambda) = \int_0^\infty e^{-\lambda x} R'(x) dx.$$

Then using the convenience formulas of (3.2.3.1) gives the transform version of (3.5.7) as

$$\theta(\lambda) = \alpha c^{-1} R(0)(1 - \phi(\lambda))\lambda^{-1} + \alpha c^{-1}(1 - \phi(\lambda))\lambda^{-1}\theta(\lambda).$$

Solving for θ we get

$$(3.5.1.3) \qquad\qquad \theta(\lambda) = \frac{\alpha c^{-1} R(0)\lambda^{-1}(1 - \phi(\lambda))}{1 - \alpha c^{-1}\lambda^{-1}(1 - \phi(\lambda))}.$$

Let

$$\mu = EX_1 = \int_0^\infty P[X_1 > x] dx = \lim_{\lambda \downarrow 0} \lambda^{-1}(1 - \phi(\lambda)).$$

We seek a solution of (3.5.1.2) which is non-negative, integrable and not identically zero. From (3.5.1.3) we get

$$\lim_{\lambda \downarrow 0} \theta(\lambda) = \int_0^\infty R'(x) dx = \frac{\alpha \mu c^{-1} R(0)}{1 - \alpha \mu c^{-1}},$$

and therefore we see that a necessary condition for such a solution to exist is that

$$(3.5.1.4) \qquad\qquad\qquad \alpha\mu < c.$$

This is a balance or mean drift condition asserting that the upward rate c dominates the downward rate $\alpha\mu$. (If $\alpha\mu c^{-1} \geq 1$, then for a positive finite solution we would need $R(0) = 0$ from which $\theta(\lambda) = 0$ and thus $R' \equiv 0$.)

When (3.5.1.4) holds we find a solution. First we check that

$$\lim_{x \to \infty} \uparrow R(x) = R(\infty) = 1.$$

Let $N(0, t]$ count the Poisson process of claims. Then

$$R(x) = P[x + ct > \sum_{i=1}^{N(0,t]} X_i, \forall\, t]$$

$$= P[x + cS_n > \sum_{i=1}^{n} X_i, \forall\, n]$$

$$= P[x > \sum_{i=1}^{n} X_i - cS_n, \forall\, n]$$

$$= P[x > \bigvee_{n=1}^{\infty} (\sum_{i=1}^{n} X_i - cS_n)].$$

If we can show $\bigvee_{n=1}^{\infty} (\sum_{i=1}^{n} X_i - cS_n) < \infty$, then it will follow that $\lim_{x\to\infty} R(x) = 1$. However,

$$E(X_i - c(S_i - S_{i-1})) = \mu - c\alpha^{-1} < 0,$$

by (3.5.1.4), so, by the strong law of large numbers,

$$\sum_{i=1}^{n} (X_i - c(S_i - S_{i-1})) = \sum_{i=1}^{n} X_i - cS_n \to -\infty \text{ a.s.,}\quad n \to \infty,$$

and hence

$$\bigvee_{n=1}^{\infty} (\sum_{i=1}^{n} X_i - cS_n) < \infty.$$

Now that we know $R(\infty) = 1$, we may determine $R(0)$. Let $\lambda \downarrow 0$ in (3.5.1.3) to get

$$\lim_{\lambda \to 0} \theta(\lambda) = \int_0^{\infty} R'(x)dx = R(\infty) - R(0) = 1 - R(0)$$

$$= \frac{\alpha\mu c^{-1} R(0)}{1 - \alpha\mu c^{-1}}.$$

Solving for $R(0)$, we find

$$R(0) = 1 - \alpha\mu c^{-1},$$

so R is of the form

$$R(x) = R(0) + \int_0^{x} R'(u)du, \quad x \geq 0,$$

and

$$\theta(\lambda) = R(0)\frac{\alpha c^{-1}(1 - \phi(\lambda))\lambda^{-1}}{1 - \alpha c^{-1}(1 - \phi(\lambda))\lambda^{-1}} = R(0)\sum_{n=1}^{\infty}(\alpha c^{-1}(1 - \phi(\lambda))\lambda^{-1})^n.$$

Denote the density

$$\alpha c^{-1} P[X_1 > x] =: g(x),$$

so

$$\int_0^{\infty} e^{-\lambda x} g(x)dx = \alpha c^{-1}(1 - \phi(\lambda))\lambda^{-1}$$

and

$$R'(x) = R(0)\sum_{n=1}^{\infty} g^{n*}(x)$$

(recall $g^{2*}(x) = \int_0^x g(x - y)g(y)dy$, etc.).

For example, if X_1 is exponential, with mean μ, then

$$\phi(\lambda) = \frac{\mu^{-1}}{\lambda + \mu^{-1}}; \quad \frac{1 - \phi(\lambda)}{\lambda} = \frac{1}{\lambda + \mu^{-1}},$$

and

$$\begin{aligned}
\theta(\lambda) &= R(0)\frac{\alpha c^{-1}\left(\frac{1-\phi(\lambda)}{\lambda}\right)}{1 - \alpha c^{-1}\left(\frac{1-\phi(\lambda)}{\lambda}\right)} = R(0)\frac{\alpha c^{-1}\left(\frac{1}{\lambda+\mu^{-1}}\right)}{1 - \alpha c^{-1}\left(\frac{1}{\lambda+\mu^{-1}}\right)} \\
&= R(0)\frac{\alpha c^{-1}}{\lambda + \mu^{-1} - \alpha c^{-1}} = \frac{R(0)\alpha c^{-1}}{\mu^{-1} - \alpha c^{-1}}\left(\frac{\mu^{-1} - \alpha c^{-1}}{\lambda + \mu^{-1} - \alpha c^{-1}}\right) \\
&= \left(\frac{R(0)\alpha c^{-1}}{\mu^{-1} - \alpha c^{-1}}\right)\int_0^{\infty} e^{-\lambda x}(\mu^{-1} - \alpha c^{-1})e^{-(\mu^{-1}-\alpha c^{-1})x}dx \\
&= \int_0^{\infty} e^{-\lambda x} R'(x)dx.
\end{aligned}$$

In this exponential case, we conclude

$$\begin{aligned}
R'(x) &= \frac{R(0)\alpha c^{-1}}{\mu^{-1} - \alpha c^{-1}}(\mu^{-1} - \alpha c^{-1})e^{-(\mu^{-1}-\alpha c^{-1})x} \\
&= (1 - \frac{\alpha\mu}{c})\frac{\alpha}{c}e^{-\mu^{-1}(1-\frac{\alpha\mu}{c})x} \\
&= (\frac{\alpha\mu}{c})(\mu^{-1}(1 - \frac{\alpha\mu}{c}))e^{-\mu^{-1}(1-\frac{\alpha\mu}{c})x}
\end{aligned}$$

from which

$$\begin{aligned}
R(x) &= R(0) + \int_0^x R'(u)du \\
&= (1 - \frac{\alpha\mu}{c}) + \frac{\alpha\mu}{c}\left(1 - e^{-\mu^{-1}(1-\frac{\alpha\mu}{c})x}\right).
\end{aligned}$$

3.6. THE POISSON PROCESS AS A RENEWAL PROCESS.

We pause here to collect some facts about the Poisson process. One convenient way to define a homogeneous Poisson process on $[0, \infty)$ with rate α is to say it is a pure renewal process with interarrival distribution

$$F(x) = 1 - e^{-\alpha x}, \quad x > 0,$$

where we do not count a renewal at time 0. Thus the Poisson process is the counting function $\{N(0, t], t > 0\}$ where $N(0, t] = N(t) - 1 = N(t) - N(0)$ and $\{N(t), t \geq 0\}$ is the pure renewal process corresponding to exponentially distributed interarrival times.

Recall from Example 3.2.2 that

$$(3.6.1) \qquad F^{n*}(dx) = \alpha(\alpha x)^{n-1} \frac{e^{-\alpha x}}{(n-1)!} dx, \quad n \geq 1, x > 0,$$

and therefore

$$(3.6.2) \qquad F^{n*}(x) = 1 - e^{-\alpha x} \sum_{k=0}^{n-1} (\alpha x)^k / k!,$$

since differentiating in (3.6.2) yields $\frac{d}{dx} F^{n*}(x)$, which agrees with the right side of (3.6.1). Therefore, for a pure process

$$
\begin{aligned}
P[N(t) = n+1] &= P[S_n \leq t < S_{n+1}] \\
&= F^{n*}(t) - F^{(n+1)*}(t) \\
&= 1 - e^{-\alpha x} \sum_{k=0}^{n-1} (\alpha x)^k / k! - (1 - e^{-\alpha x} \sum_{k=0}^{n} (\alpha x)^k / k!) \\
&= e^{-\alpha x} (\alpha x)^n / n! = P[N(0, t] = n].
\end{aligned}
$$

The forgetfulness property of the exponential distribution is reflected in the fact that

$$P[B(t) \leq x] = 1 - e^{-\alpha x},$$

and one checks easily

$$P[A(t) \leq x] = \begin{cases} 1 - e^{-\alpha x}, & t \geq x \\ 1, & t < x. \end{cases}$$

Finally, we have

$$U(t) = EN(t) = 1 + EN(0, t] = 1 + \sum_{n=0}^{\infty} n e^{-\alpha t} (\alpha t)^n / n! = 1 + \alpha t.$$

Keep these facts at your fingertips, as they will be used frequently.

3.7. AN INFORMAL DISCUSSION OF RENEWAL LIMIT THEOREMS AND REGENERATIVE PROCESSES.

Here we discuss informally two very important (equivalent) renewal limit theorems and emphasize their significance by applying them to some examples. We also discuss the important class of processes called *regenerative processes*.

Before this discussion, it is useful to distinguish between renewal theory in discrete time (where interarrival random variables take values only in a set of the form $\{0, a, 2a, ...\}$) and the contrary case. Say F is *arithmetic* if for some $a > 0$ we have $F(\{0, a, 2a, ...\}) = 1$. If there is no such value a, then call F non-arithmetic. In the discrete case where F is arithmetic there always exists a largest value of a, say a_0, such that F concentrates on $\{na_0, n \geq 0\}$. This is called the *span* of F. In Chapter 1 we considered non-negative integer valued random variables, and there typically we had $a_0 = 1$.

This chapter focuses mostly on the non-arithmetic case. The soon to be discussed Blackwell theorem and the key renewal theorem are correct for both cases, if, in the arithmetic case, we assume that all variables t, x, etc., however, are multiples of the span a_0. Integrals are then to be interpreted in the arithmetic case as follows: dx is the uniform measure assigning mass a_0 to each point $0, a_0, 2a_0, \ldots$. This is convenient because, with this convention, for instance

$$\mu = \int_0^\infty xF(dx) = \sum_{n=0}^\infty na_0 F\{na_0\} = \sum_{j=0}^\infty a_0(1 - F(ja_0))$$

$$= \int_0^\infty (1 - F(x))dx.$$

As we will discuss in Section 3.8, the arithmetic case can be treated separately via Markov chain limit theorems. Although it is possible to treat both the arithmetic and non-arithmetic cases together, and, for the most part, there is no confusion, in several respects the novice is better off considering the two cases separately.

The two main asymptotic results (which are in fact equivalent) are Blackwell's theorem and what is known as the key renewal theorem originally formulated by Walter Smith. Blackwell's theorem states that if the delay distribution is proper, then

$$V(t, t + a] \to a/\mu, \quad \text{as } t \to \infty.$$

For the key renewal theorem, consider the solution $Z = U * z$ of the renewal equation

$$Z = z + F * Z,$$

here F is a proper distribution. If z satisfies a mild hypothesis called
rect Riemann integrability (defined below in Section 3.10.1; abbreviated
s dRi), then

$$\lim_{t\to\infty} Z(t) = \lim_{t\to\infty} U * z(t) = \mu^{-1} \int_0^\infty z(s)ds.$$

he dRi condition is satisfied, for instance, for functions z which are de-
easing and integrable.

This suggests the *renewal method*: If it is desired to calculate $\lim_{t\to\infty}$
(t), we first write a renewal equation for Z, solve it and then apply the
ey renewal theorem. This plan will be followed for the calculation of the
symptotic distributions of $A(t)$ and $B(t)$ below.

Blackwell's theorem is more refined than the elementary renewal the-
rem. In fact, the Blackwell theorem bears the same relation to the elementary renewal theorem as convergence of a sequence bears to Cesaro
onvergence of the sequence. To see that the Blackwell theorem implies
he elementary renewal theorem, observe first that Blackwell's theorem
nplies

$$V(n-1, n] \to 1/\mu,$$

nd hence the Cesaro averages of $\{V(n-1, n]\}$ converge:

$$\frac{1}{n} \sum_{k=1}^n V(k-1, k] = \frac{1}{n} V(0, n] \to 1/\mu.$$

herefore,

$$\limsup_{t\to\infty} t^{-1}(V(t) - V(0)) \leq \limsup_{t\to\infty} \frac{V([t]+1) - V(0)}{[t]}$$

$$= \limsup_{t\to\infty} \left(\frac{V([t]+1) - V(0)}{[t]+1} \right) \left(\frac{[t]+1}{[t]} \right) = \frac{1}{\mu},$$

ith a reverse inequality obtained similarly.

Detailed discussion of the logical relationships between Blackwell's the-
rem and the key renewal theorem will be given later in Section 3.10.2. For
ow we hint that Blackwell's theorem makes the key renewal theorem plau-
ble because

$$\int_0^t z(t-x)U(dx) = \int_0^t z(y)d(-U(t-y))$$

$$= \int_0^t z(y)U(t-y, t-y+dy],$$

nd, applying Blackwell's theorem, this is

$$\approx \int_0^t z(y)dy/\mu \to \mu^{-1} \int_0^\infty z(y)dy.$$

Example 3.5.3 (continued). Recall for the forward recurrence time we have

$$P[B(t) > x] = U * (1 - F(\cdot + x))(t),$$

where $z(s) = 1 - F(s + x)$. Thus for each $x > 0$, applying the key renewal theorem gives

$$\lim_{t\to\infty} P[B(t) > x] = \mu^{-1} \int_0^\infty z(u)du$$

$$= \mu^{-1} \int_0^\infty (1 - F(u + x))du$$

$$= \mu^{-1} \int_x^\infty (1 - F(s))ds =: 1 - F_0(x).$$

Similarly, for the age process we have

$$P[A(t) \leq x] = U * ((1 - F) \cdot 1_{[0,x]})(t),$$

so that $z(s) = (1 - F(s))1_{[0,x]}(s)$. Letting $t \to \infty$, and applying the key renewal theorem, we get

$$\lim_{t\to\infty} P[A(t) \leq x] = \mu^{-1} \int_0^\infty z(u)du$$

$$= \mu^{-1} \int_0^\infty ((1 - F(u))1_{[0,x]}(u)du$$

$$= \mu^{-1} \int_0^x (1 - F(u))du = F_0(x)$$

so the limit distributions for $A(t)$ and $B(t)$ coincide. Note that if $F(dx) = \alpha e^{-\alpha x}, x > 0$, then

$$1 - F_0(x) = \mu^{-1} \int_x^\infty ((1 - F(u))du = \alpha \int_x^\infty e^{-\alpha u}du = e^{-\alpha x} = 1 - F(x).$$

Therefore, for a Poisson process the limit distribution for $B(t)$ is the same as the distribution of $B(t)$ which is one manifestation of the stationarity of the Poisson process.

Note that the transform of F_0 yields

$$\hat{F}_0(\lambda) = \int_0^\infty e^{-\lambda x}dF_0(x)$$

$$= \mu^{-1} \int_0^\infty e^{-\lambda x}((1 - F(x))ds$$

$$= \frac{1 - \hat{F}(\lambda)}{\lambda\mu}$$

from convenience formulas (3.2.3.1). Suppose we start a delayed renewal process with delay distribution $G = F_0$. What is the corresponding renewal function $V(t)$? We have $V = G * U = G * \sum_{n=0}^{\infty} F^{n*}$ and therefore

$$\hat{V}(\lambda) = \hat{F}_0(\lambda)\hat{U}(\lambda) = \frac{\hat{F}_0(\lambda)}{1 - \hat{F}(\lambda)} = \frac{1}{\lambda\mu}$$

which we recognize from Example 3.2.7 with $n = 0$ as the transform of $V(t) = t/\mu$. Conversely, if $V(t) = t/\mu$, then

$$\hat{V}(\lambda) = 1/(\lambda\mu) = \hat{G}(\lambda)/((1 - \hat{F}(\lambda)),$$

from which

$$\hat{G}(\lambda) = \frac{1 - \hat{F}(\lambda)}{\lambda\mu} = \hat{F}_0(\lambda),$$

and thus $G = F_0$. We conclude that *for a delayed renewal process, the renewal function $V(t)$ is linear, $V(t) = t/\mu$, if and only if the delay distribution G is F_0.*

3.7.1. AN INFORMAL DISCUSSION OF REGENERATIVE PROCESSES.

A broad class of processes exhibit limits for their state probabilities because of the key renewal theorem. Consider a stochastic process $\{X(t), t \in T\}$ where the index set T is either $[0, \infty)$ or $\{0, 1, \ldots\}$. Suppose there is a renewal process $\{S_n\}$ of times where the process regenerates (probabilistically restarts itself) so the future after one of these times, say S_n, looks probabilistically exactly as it did back at time 0. For example consider a positive recurrent Markov chain with state space $\{0, 1, \ldots\}$ started from 0. Renewal epochs are the return times to 0. The evolution of the chain after each return time is that of a Markov chain started from state 0.

To put a bit more precision into this (though still being informal), we say a process $\{X(t), t \in T\}$ is *regenerative* if there are random times $\{S_n\} \subset T$ such that

1. $\{S_n\}$ is a renewal process.

2.   The process after any S_n has the same distribution as the whole process; for any n and $k, 0 < t_1 < \cdots < t_k, t_i \in T, 1 \leq i \leq k$, we have

$$(X(S_n + t_i), 1 \leq i \leq k) \stackrel{d}{=} (X(t_i), 1 \leq i \leq k).$$

3.   For any S_n, the process past S_n is independent of $\{S_0, S_1, \ldots, S_n\}$; i.e., $\{X(t + S_n), t \in T\}$ is independent of $\{S_0, \ldots, S_n\}$.

Consider the pieces of path $(X(t), t \in T \cap [S_n, S_{n+1}))$, as the nth cycle of the process.

Some examples will clarify what is intended.

Example 3.7.1. Markov Chains. Suppose the chain is positive recurrent with state space $\{0, 1, \ldots\}$ and started from 0. Then $S_n = \tau_0(n)$.

Example 3.7.2. Smith's Random Tours. Let $(\{Y_j(t), t \geq 0\}, \xi_j)$, $j \geq 1$, be iid with $\xi_j \geq 0$. Think of $\{Y_j(t), 0 \leq t \leq \xi_j\}$ as a piece of the jth path of a random duration ξ_j. Define $S_0 = 0, S_n = \sum_{i=1}^{n} \xi_i$ and

$$X(t) = \sum_{j=1}^{\infty} Y_j(t - S_{j-1}) 1_{[S_{j-1}, S_j)}(t), \quad t \geq 0.$$

To get $X(t)$, we take the snippets of path $\{Y_j(t), 0 \leq t \leq \xi_j\}$ (or random tours of random length as Smith called them) and stitch them together. This construction gives a method for simulating a regenerative process.

Example 3.7.3. M/G/1 Queue. Consider a queueing model with Poisson arrivals of rate a: the arrival process of customers is modelled as a Poisson process with interarrival distributions which are exponential with parameter a. Customers are served on a first come first served basis, and service times are iid non-negative random variables with distribution G. An arriving customer at time 0 initiates a *busy period* which continues until the initial customer and all his descendents are served; i.e., until the server working continuously looks up, sees an empty queue and can take a break. At the end of the initial busy period, the queue is empty. The next customer arrives after a random exponential (parameter a) length of time which is the forward recurrence time of the Poisson process. This forward recurrence time is an *idle* period, at the end of which, another customer arrives, and the work load builds anew. Let $V(t)$ be the load on the server at time t; if the input is shut off, $V(t)$ is the amount of work facing the server until the queue is empty.

This describes a decomposition of the work load process $V(t)$ into cycles consisting of a busy period and an idle period. Renewal times are the beginnings of busy periods.

There is no specification of the state space in the definition of a regenerative process. For Smith's theorem, given next, we suppose for simplicity that the state space is $\{0, 1, \ldots\}$. More general state spaces are considered in Section 3.12.

Smith's Theorem 3.7.1. *Let $\{X(t), t \in T\}$ be regenerative with renewal times $\{S_n = \sum_{j=1}^{n} Y_j, n \geq 0\}$ which constitute a pure renewal process. Define*

$$q_j(t) = P[X(t) = j, S_1 > t],$$

and suppose $\mu = EY_1 = E(\text{ cycle length }) < \infty$. If the distribution of the cycle length, Y_1, satisfies a mild regularity condition (satisfied for instance if the distribution is absolutely continuous or concentrates on the integers) then the following limits exist:

$$\lim_{t\to\infty, t\in T} P[X(t) = j] = p_j = \frac{1}{EY_1} \int_0^\infty q_j(t)dt$$

$$= \mu^{-1} E \int_0^{Y_1} 1_{\{j\}}(X(t))dt$$

$$= \frac{E(\text{ occupation time in } j \text{ per cycle })}{E(\text{ cycle length })}.$$

Proof. We use the renewal method: We write a renewal equation for

$$P_j(t) =: P[X(t) = j]$$

and solve and then use the key renewal theorem. We have

$$\begin{aligned} P_j(t) &= P[X(t) = j] \\ &= P[X(t) = j, S_1 > t] + P[X(t) = j, S_1 \le t] \\ &= q_j(t) + P[X(t) = j, S_1 \le t]. \end{aligned}$$

In the second term, condition on S_1, move the time origin up to S_1 and use the regenerative property to conclude that the second term is

$$\int_0^t P_j(t - s)F(ds) = F * P_j(t).$$

We therefore obtain the renewal equation

$$P_j(t) = q_j(t) + F * P_j(t),$$

which has the solution

$$P_j(t) = U * q_j(t).$$

Applying the key renewal theorem gives

$$\lim_{t\to\infty, t\in T} P_j(t) = \mu^{-1} \int_0^\infty q_j(s)ds.$$

(Note that regularity conditions on F are necessary to ensure the fact that $q_j(t)$ is dRi, which is needed to apply the key renewal theorem.)

It remains to analyze the limit and express it in a different form. We note

$$\int_0^\infty q_j(s)ds = \int_0^\infty P[X(s) = j, S_1 > s]ds$$

$$= \int_0^\infty E1_{[X(s)=j,s<S_1]}ds$$

$$= E\left(\int_0^\infty 1_{[X(s)=j,s<S_1]}ds\right)$$

$$= E\left(\int_0^{S_1} 1_{[X(s)=j]}ds\right)$$

$$= E(\text{ occupation time in } j \text{ per cycle }),$$

which completes the derivation. ∎

Frequently this result serves primarily as an existence result but does not help calculate the limits explicitly. Sometimes, however, symmetry or other special structure provides a way to calculate the limits explicitly.

Example 3.7.3 (continued).   The M/G/1 Queue. Let $Q(t)$ be the number in the system at time t. This is regenerative with renewal times $\{S_n\}$ which are the beginnings of busy periods, so we know $\lim_{t\to\infty} P[Q(t) = j]$ exists. For $j = 0$ we have

$$p_0 = \frac{E(\text{ occupation time in } 0 \text{ per cycle })}{E(\text{ cycle length })}.$$

We know that the cycle consists of a busy period and an idle period. The idle period is exponential with parameter a, so $E(\text{ cycle length }) = E(\text{BP}) + a^{-1}$. The occupation time in state 0 is just the idle time with expectation $1/a$, so

$$p_0 = \frac{1/a}{E(\text{BP}) + 1/a}.$$

This still leaves us the task of computing $E(\text{BP})$. The computation of p_j for $j \geq 1$ is harder.

Example.   Harry Visits Haifa. While visiting Haifa, Harry discovers that people who wish to travel quickly from the port area up the mountain to The Carmel frequently take a taxi known as a *sherut*. The system operates as follows: *Sherut-eem* are lined up in a row at a taxi stand. The capacity of each car is K people. Potential passengers arrive according to a renewal process and enter the taxi at the head of the row. As soon as K

people are in the car, it departs in a cloud of diesel emissions. The next car moves up, accepts passengers until it is full, then departs for The Carmel, and so on. Let $X(t)$ be the number of people in the car at the head of the line at time t, and set $P_j(t) = P[X(t) = j]$. Assume the range of $X(t)$ is $\{0, 1, \ldots, K - 1\}$.

Care must be taken to identify the correct renewal processes. Let $\{T_n, n \geq 0\}$ be an ordinary renewal process of epochs at which customers arrive. If we assume $X(0) = 0$, then the renewal epochs at which $X(t)$ regenerates are $\{S_n\}$ where

$$S_0 = 0, S_1 = T_K, S_2 = T_{2K}, \ldots,$$

since these are the times when cars become full and depart. See Figure 3.5 for a visual aide.

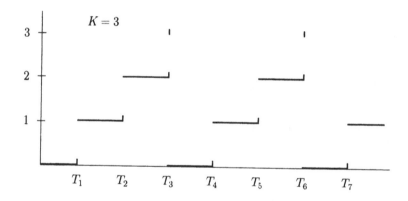

FIGURE 3.5. THE *Sherut* PROCESS

For $j = 0, \ldots, K - 1$ we have

$$\lim_{t \to \infty} P_j(t) = \frac{E(\text{occupation time in } j \text{ per cycle})}{E(\text{cycle length})}$$

$$= \frac{ET_1}{KET_1} = \frac{1}{K}.$$

This is one example where we can compute $P_j(t)$ fairly explicitly. We have, for $j = 0, 1, \ldots, K - 1$,

$$q_j(t) = P[X(t) = j, S_1 > t] = P[M(0, t] = j],$$

where M is the counting function for the renewal process $\{T_n\}$. Therefore

$$P_j(t) = P[M(0, t] = j] * U(t),$$

where U is the renewal function of the process $\{S_n\}$. If we restrict ourselves to the case of Poisson arrivals with rate a, then

$$P[M(0,t) = j] = \frac{e^{-at}(at)^j}{j!}$$

and

$$\hat{U}(\lambda) = \sum_{n=0}^{\infty} \left((\frac{a}{a+\lambda})^K \right)^n$$

$$= \frac{1}{1 - (\frac{a}{a+\lambda})^K}.$$

Therefore, for the Laplace transform $\hat{P}_j(\lambda)$ of $P_j(t)$,

$$\hat{P}_j(\lambda) = \int_0^{\infty} e^{-\lambda x} P_j(x) dx$$

$$= \int_0^{\infty} e^{-\lambda x} U * q_j(x) dx = \hat{U}(\lambda) \int_0^{\infty} e^{-\lambda x} q_j(x) dx.$$

Now

$$\int_0^{\infty} e^{-\lambda x} q_j(x) ds = \int_0^{\infty} e^{-\lambda x} \frac{e^{-ax}(ax)^j}{j!} dx$$

$$= a^{-1} \int_0^{\infty} e^{-\lambda x} \frac{a(ax)^j e^{-ax}}{j!} dx.$$

From Example 3.2.3 this is

$$= a^{-1}(\frac{a}{a+\lambda})^{j-1},$$

from which

$$\hat{P}_j(\lambda) = \frac{a^{-1}(\frac{a}{a+\lambda})^{j-1}}{1 - (\frac{a}{a+\lambda})^K}. \quad \blacksquare$$

3.8. DISCRETE RENEWAL THEORY.

Suppose the interarrival distribution of the renewal process is arithmetic with span 1. The sequence $\{Y_n, n \geq 1\}$ consists of independent, identically distributed, non-negative, integer valued random variables, and we set

$$P[Y_1 = k] = f_k, \quad k \geq 0,$$

and assume $P[Y_1 < \infty] = \sum_{k=0}^{\infty} f_k = 1$. Define for $n \geq 0$

$$u_n = U(\{n\}) = E \sum_{k=0}^{\infty} 1_{\{n\}}(S_k)$$

$$= E(\text{ number of } S_k \text{ equal to } n).$$

If we assume (which we will) that $f_0 = 0$, then $P[Y_1 \geq 1] = 1$, and, consequently, an alternate interpretation of u_n is

$$u_n = P\{\bigcup_{k=0}^{\infty} [S_k = n]\} = P[S_k = n \text{ for some } k \geq 0].$$

Note that $U(n) = U[0, n] = \sum_{i=0}^{n} u_i$ and $u_0 = 1$.

We may relate the ordinary renewal process to a Markov chain as follows. Let

$$n_0 = \sup\{n : P[Y_1 = n] > 0\}.$$

Define

$$p_n = \begin{cases} P[Y_1 > n | Y_1 > n - 1], & \text{if } 1 \leq n < n_0 \\ 0, & \text{if } n \geq n_0 , \end{cases}$$

and set $q_n = 1 - p_n$. Now consider a success run Markov chain $\{A_n\}$ with transition matrix

$$P = \begin{matrix} 0 \\ 1 \\ 2 \\ \vdots \end{matrix} \begin{pmatrix} q_1 & p_1 & 0 & 0 & \cdots \\ q_2 & 0 & p_2 & 0 & \cdots \\ q_3 & 0 & 0 & p_3 & \cdots \\ \vdots & & & \ddots & \end{pmatrix}.$$

The point of this construction is that, for $n \leq n_0$,

$$f_{00}^{(n)} = P_0[A_1 = 1, A_2 = 2, \ldots, A_{n-1} = n - 1, A_n = 0]$$

$$= p_1 p_2 \cdots p_{n-1} q_n$$

$$= \frac{P[Y_1 > 1]}{P[Y_1 > 0]} \frac{P[Y_1 > 2]}{P[Y_1 > 1]} \cdots \frac{P[Y_1 > n - 1]}{P[Y_1 > n - 2]} \left(1 - \frac{P[Y_1 > n]}{P[Y_1 > n - 1]}\right)$$

$$= \frac{P[Y_1 = n]}{P[Y_1 > 0]} = P[Y_1 = n]$$

$$= f_n,$$

and, for $n > n_0$,

$$f_{00}^{(n)} = f_n = 0.$$

Thus, the first return probabilities to state 0 for the Markov chain are given by $\{f_n\}$. If we start this Markov chain in state 0, then, by dissection, the return times to 0 have the same distribution as the original renewal process $\{S_n\}$:

$$\{S_n, n \geq 0\} \stackrel{d}{=} \{\tau_0(n), n \geq 0\}.$$

(In fact, $\{A_n, n \geq 0\}$ is the age process of the ordinary renewal sequence $\{S_n\}$, but it is not necessary to verify this here.)

We may now apply our knowledge of Markov chains to the renewal sequence. We have

$$p_{00}^{(n)} = P_0[A_n = 0] = P[\text{ renewal at time } n] = u_n.$$

From Proposition 2.6.1 we get

$$p_{00}^{(n)} = \sum_{j=0}^{n} f_{00}^{(j)} p_{00}^{(n-j)},$$

which in renewal theory language becomes

$$u_n = \sum_{j=0}^{n} f_j u_{n-j}.$$

This shows that the sequences $\{u_n\}$ and $\{f_n\}$ determine one another, and techniques comparable to what is used in Proposition 2.6.1 yield

$$U(s) = \frac{1}{1 - F(s)}, \quad 0 \leq s < 1,$$

where $U(s) = \sum_{n=0}^{\infty} u_n s^n$ and $F(s) = F_{00}(s) = \sum_{n=1}^{\infty} f_n s^n$. Furthermore, since the distribution of Y_1 is arithmetic with span 1, 0 is an aperiodic state of the Markov chain, and if $EY_1 = \mu < \infty$, then from Theorem 2.13.2 we get

(3.8.1) $$u_n = p_{00}^{(n)} \to 1/\mu, \quad n \to \infty,$$

and this is the discrete version of Blackwell's theorem (cf. Section 3.10). Of course it now follows that if $h > 0$ is an integer

$$U(n, n+h] = u_n + \cdots + u_{n+h} \to h/\mu$$

as $n \to \infty$, and this is the exact analogue of Blackwell's theorem.

For the key renewal theorem we have the following: Let $z(k), k \geq 0$, be a non-negative function defined on the integers satisfying $\sum_{k=0}^{\infty} z(k) < \infty$ so that in particular $z(k) \to 0$ as $k \to \infty$. Then the asymptotic behavior of the convolution $U * z$ is

$$\lim_{n \to \infty} U * z(n) = \mu^{-1} \sum_{k=0}^{\infty} z(k).$$

This is readily seen from (3.8.1) as follows. Given ϵ there exists k_0 such that if $k > k_0$, then

$$\mu^{-1} - \epsilon \leq u_k \leq \mu^{-1} + \epsilon,$$

since $u_k \to 1/\mu$ as $k \to \infty$. So for large n we have

$$U * z(n) = \sum_{k=0}^{n} z(n-k)u_k = \sum_{k=0}^{k_0} z(n-k)u_k + \sum_{k=k_0+1}^{n} z(n-k)u_k$$

$$\leq \sum_{k=0}^{k_0} z(n-k)u_k + (\mu^{-1} + \epsilon) \sum_{k=k_0+1}^{n} z(n-k)$$

$$= \sum_{k=0}^{k_0} z(n-k)u_k + (\mu^{-1} + \epsilon) \sum_{m=0}^{n-k_0-1} z(m),$$

and since $z(n) \to 0$ as $n \to \infty$ we get

$$\limsup_{n \to \infty} U * z(n) \leq (\mu^{-1} + \epsilon) \sum_{m=0}^{\infty} z(m).$$

We may let $\epsilon \to 0$ since the left side does not depend on ϵ. In a similar way, we find

$$\liminf_{n \to \infty} U * z(n) \geq (\mu^{-1} - \epsilon) \sum_{m=0}^{\infty} z(m).$$

So we see that in the discrete case both Blackwell's theorem and the key renewal theorem follow from the Markov chain limit theorem.

Suppose now that $\{S_n, n \geq 0\}$ is a delayed renewal process with

$$P[Y_0 = n] = g_n, \ n \geq 1; \quad P[Y_1 = n] = f_n, \ n \geq 1,$$

with $\sum_{n=1}^{\infty} f_n = 1$. If, as usual ,

$$V(\{n\}) =: v_n = P[S_k = n, \text{ for some } k\,]$$

then for $n \geq 1$

$$v_n = \sum_{k=1}^{n} g_k u_{n-k},$$

so that $v_n = U * g(n)$, and the key renewal theorem yields

$$\lim_{n \to \infty} v_n = \sum_{k=1}^{\infty} g_k / EY_1 = \mu^{-1} \sum_{k=1}^{\infty} g_k.$$

If $\{Y_n, n \geq 1\}$ are interarrival times with arithmetic distribution of span a_0, then by adjusting time scales we get back to the case of span 1. Set $Y_n^{\#} = Y_n / a_0$ so that the distribution of $Y_n^{\#}$ is arithmetic with span 1. Then for a pure renewal process

$$U(\{na_0\}) =: u_{na_0} = P\left\{ \bigcup_{k=0}^{\infty} [S_k = na_0] \right\}$$

$$= P\left\{ \bigcup_{k=0}^{\infty} [S_k / a_0 = n] \right\}$$

$$= P\left\{ \bigcup_{k=0}^{\infty} [S_k^{\#} = n] \right\}$$

$$= u_n^{\#} \to \frac{1}{EY_1^{\#}} = \frac{1}{EY_1 / a_0} = \frac{a_0}{EY_1}.$$

3.9. STATIONARY RENEWAL PROCESSES*.

The homogeneous Poisson process has the interesting property of a constant renewal rate meaning that the renewal function is linear. If the Poisson process is built from interarrival intervals which are iid exponentially distributed with parameter a, then

$$EN(t) - 1 = at.$$

Recall from the end of Section 3.7, however, that there is a much broader class of processes with this property. We checked in Section 3.7 that when

---

* This section contains advanced material which may be skipped on first reading by beginning readers.

$\mu = EY_1 < \infty$, the renewal function $V(t)$ of a delayed renewal process is linear, $V(t) = t/\mu$, $t > 0$, if and only if the delay distribution G is

$$F_0(x) = \mu^{-1} \int_0^x (1 - F(s))ds.$$

With this in mind, define a stationary renewal process as a delayed renewal process with $\{Y_n, n \geq 1\}$ iid with common distribution F with $\mu = \int_0^\infty x dF(x) < \infty$ and delay distribution F_0 for the distribution of Y_0. So far, we know a delayed renewal process has constant renewal rate but much more is true: We explore in the next theorem additional facts about stationary renewal processes which provide further justification for the adjective *stationary*.

Theorem 3.9.1. *Assume $\mu < \infty$ and*

$$F_0(x) = \mu^{-1} \int_0^x (1 - F(u))du.$$

(a) *A stationary renewal process gives rise to a strictly stationary point process: For any $h > 0$, integer $k > 0$, time points $0 \leq t_1 < \cdots < t_k$ and $s_i > 0$, $i = 1, \ldots, k$, we have equality in distribution of the k-dimensional random vectors*

$$\{N(t_i, t_i + s_i], i = 1, \ldots, k\} \stackrel{d}{=} \{N(t_i + h, t_i + s_i + h], i = 1, \ldots, k\}.$$

(b) *In a stationary renewal process $V(t, t + b] = \mu^{-1}b$ independent of t for any $b > 0$.*

(c) *In a stationary renewal process $\{B(t), t \geq 0\}$ is a Markov process with stationary transition probabilities and F_0 is a stationary probability measure for this Markov process. Thus if a renewal process is stationary (so that $P[Y_0 \leq x] = F_0(x) = P[B(0) \leq x])$ then B is a strictly stationary process, meaning for any $k, h > 0, 0 \leq t_1 < \cdots < t_k$ we have*

$$P[B(t_i) \leq x_i, 1 = 1, \ldots, k] = P[B(t_i + h) \leq x_i, 1 = 1, \ldots, k].$$

(d) *In a stationary renewal process F_0 is a stationary distribution for the Markov process $\{A(t), t \geq 0\}$.*

(e) *If F_0 is the distribution of the age of the item present at time zero then $V(t, t + b] = \mu^{-1}b$ independent of t for any $b > 0$.*

Any of these properties characterizes a stationary renewal process and F_0.

A discussion of Theorem 3.9.1 occupies the rest of this section. As a preliminary, we prove a result which is very similar to Proposition 1.8.2.

Lemma 3.9.2. *Suppose* $\{S_n, n \geq 0\}$ *is a stationary renewal sequence. Then for any* $h > 0$

(3.9.1) $\{S_{N(h)+n} - h, n \geq 0\} \stackrel{d}{=} \{S_n, n \geq 0\}$

in R^∞ and
(3.9.2)

$\{Y_0, Y_1, \dots\} \stackrel{d}{=} \{S_{N(h)} - h, Y_{N(h)+k}, k \geq 1\} = \{B(h), Y_{N(h)+k}, k \geq 1\}$

in R^∞.

Proof. It suffices to show (3.9.2) since, by applying the map

$$g(y_0, y_1, \dots) = (y_0, y_0 + y_1, y_0 + y_1 + y_2, \dots)$$

(from $R^\infty \mapsto R^\infty$), (3.9.1) follows from (3.9.2).

For real $x_0, \dots, x_k$ we have

$$P[S_{N(h)} - h \leq x_0, Y_{N(h)+i} \leq x_i, i \leq k]$$

$$= \sum_{n=0}^{\infty} P[S_{N(h)} - h \leq x_0, N(h) = n]$$

$$= \sum_{n=0}^{\infty} P[S_n - h \leq x_0, Y_{n+i} \leq x_i, i \leq k, N(h) = n].$$

Since $[N(h) = n] = [S_{n-1} \leq h < S_n]$, we have $[N(h) = n] \in \sigma(S_0, \dots, S_n) = \sigma(Y_0, \dots, Y_n)$, so

$$[S_n - h \leq x_0] \cap [N(h) = n] \text{ and } \cap_{i=1}^{k} [Y_{n+i} \leq x_i]$$

are independent. Hence the above is

$$\sum_{n=0}^{\infty} P[S_n - h \leq x_0, N(h) = n] P[\cap_{i=1}^{k} [Y_{n+i} \leq x_i]]$$

$$= \prod_{i=1}^{k} F(x_i) \sum_{n=0}^{\infty} P[S_n - h \leq x_0, N(h) = n]$$

$$= \prod_{i=1}^{k} F(x_i) P[S_{N(h)} - h \leq x_0]$$

$$= P[B(h) \leq x_0] \prod_{i=1}^{k} F(x_i)$$

$$= P[Y_0 \leq x_0, Y_i \leq x_i, i \leq k]. \blacksquare$$

Before proceeding with the proof of Theorem 3.9.1, we consider again the forward recurrence time $B(t)$.

Proposition 3.9.3. *For a stationary renewal process, the forward recurrence time process* $\{B(t), t \geq 0\}$ *is a homogeneous Markov process, meaning that, for* $t > 0, s > 0, x > 0$, *it satisfies*

$$(3.9.3) \qquad P[B(t+s) \leq x | B(u), u \leq t] = P[B(t+s) \leq x | B(t)]$$
$$(3.9.4) \qquad\qquad\qquad\qquad\qquad\qquad = P[B(s) \leq x | B(0)].$$

Remark. A comparable result can be stated and proven for the age process $\{A(t)\}$. However, since $A(t) = t - S_{N(t)-1}$ is only defined on $[S_0, \infty)$, a provision must be made for extending the definition of $A(t)$ to $[0, S_0)$. See Asmussen (1987), for example.

Proof. We only consider the statement about B. We proceed to show that we can write $B(t+s)$ as a function of $B(t)$ and quantities independent of $B(t)$. See Exercise 2.6 for a discussion of this method.

Define the forward recurrence functional on sequences of non-negative elements. This is a mapping $g_t : [0, \infty)^\infty \mapsto [0, \infty)$ defined by

$$g_t(x_0, x_1, \ldots) = \inf\{\sum_{i=0}^{n} x_i - t : \sum_{i=0}^{n} x_i > t\}.$$

Now there are two cases to consider, namely, $B(t) \geq s$ or $B(t) < s$. If $B(t) \geq s$, then
$$B(t+s) = B(t) - s.$$

On the other hand, if $B(t) < s$, then there is a renewal at epoch $t + B(t)$, and we may imagine the process starting from this time. Thus in this case

$$B(t+s) = g_{t+s-t-B(t)}(Y_{N(t)+k}, k \geq 1) = g_{s-B(t)}(Y_{N(t)+k}, k \geq 1).$$

Summarizing the two cases together, we have

$$B(t+s) = (B(t) - s)1_{[B(t) \geq s]} + (g_{s-B(t)})1_{[B(t) < s]}.$$

The result now follows from Lemma 3.9.2 and Exercise 2.6. ∎

We now turn to the proof of Theorem 3.9.1.

Proof of Theorem 3.9.1. We discussed (b) adequately at the beginning of this section. Consider (c). Recall for a pure process that

$$P[B(t) > x] = \int_0^t (1 - F(t + x - s))U(ds).$$

Therefore, for a delayed process with delay distribution G,

$$P[B(t) > x] = P[B(t) > x, N(t) = 0] + P[B(t) > x, N(t) > 0]$$

$$= 1 - G(t + x) + \int_0^t P[B(t - u) \geq x | B(0) = 0]G(du)$$

$$= 1 - G(t + x) + G * U * 1 - F(\cdot + x)(t)$$

$$= 1 - G(t + x) + V * 1 - F(\cdot + x)(t).$$

If $G = F_0$, then $V(dy) = \mu^{-1}dy$, and, for all $t > 0$,

$$P[B(t) > x] = 1 - F_0(t + x) + \mu^{-1} \int_0^t (1 - F(t + x - y))dy.$$

Making the change of variable $u = t + x - y$, this is

$$= \mu^{-1} \int_{t+x}^\infty (1 - F(y))dy + \mu^{-1} \int_x^{t+x} (1 - F(u))dy$$

$$= \mu^{-1} \int_x^\infty (1 - F(y))dy = 1 - F_0(x)$$

$$= P[B(0) > x],$$

so $B(t) \stackrel{d}{=} B(0)$ for any $t > 0$. To check that B is strictly stationary (a general fact about Markov processes started by a stationary distribution), note that, for any k and $x_1, \ldots, x_k$ and $h > 0$, we have

$$P[B(t_i + h) \leq x_i, i \leq k] = \int_{R_+} P[B(t_i + h \leq x_i, i \leq k | B_h = x]F_0(dx).$$

Since B is a homogeneous Markov process, this is

$$= \int_{R_+} P[B(t_i) \leq x_i, i \leq k | B(0) = x]F_0(dx)$$

$$= P[B(t_i) \leq x_i, i \leq k]$$

showing B to be strictly stationary.

We now prove (a). For $y \in R^\infty$ define $f : R^\infty \mapsto R^k$ by

$$f(y) = \left(\sum_{n=0}^\infty 1_{(t_i, t_i + s_i]} (\sum_{j=0}^n y_j), i = 1, \ldots, k \right).$$

Then from Lemma 3.9.2

$$f(Y_0, Y_1, \dots) \stackrel{d}{=} f(S_{N(h)} - h, Y_{N(h)+k}, k \geq 1).$$

The left side is

$$(N(t_i, t_i + s_i], i \leq k).$$

The right side is

$$(\sum_{n=0}^{\infty} 1_{(t_i, t_i + s_i]}(S_{N(h)+n} - h), i \leq k)$$

$$= (\sum_{n=0}^{\infty} 1_{(t_i+h, t_i+s_i+h]}(S_{N(h)+n}), i \leq k)$$

$$= (\sum_{n=N(h)}^{\infty} 1_{(t_i+h, t_i+s_i+h]}(S_n), i \leq k),$$

and, since $S_0, S_1, \dots, S_{N(h)-1}$ are all in $[0, h)$ and hence less than $t_i + h$, the above sum is

$$\left(\sum_{n=0}^{\infty} 1_{(t_i+h, t_i+s_i+h]}(S_n), i \leq k \right) = (N(t_i + h, t_i + s_i + h], i \leq k).$$

Conversely, if N is stationary, then $N(t, t+b] \stackrel{d}{=} N(0, b]$, so $EN(t, t+b] = V(t, t+b]$ is independent of t; therefore, stationarity follows from (b).

The converse of (c) is most easily verified by using the (not yet proven) key renewal theorem to prove

$$\lim_{t \to \infty} P[B(t) \leq x] = F_0(x).$$

If $B(t) \stackrel{d}{=} B(0)$, then $P[B(0) \leq x] = F_0(x)$ as desired.

We understand (e) as follows: Suppose an item whose life length L has distribution F is placed in service at time $-A < 0$, where A has distribution F_0. Then

$$P[Y_0 > x | A = a] = P[L > x + a | L > a]$$

$$= \frac{1 - F(x + a)}{1 - F(a)},$$

and therefore

$$P[Y_0 > x] = \int_0^\infty P[Y_0 > x | A = a] F_0(da)$$

$$= \int_0^\infty \frac{1 - F(a + x)}{1 - F(a)} (1 - F(a)) \mu^{-1} da$$

$$= \mu^{-1} \int_0^\infty (1 - F(a + x)) da$$

$$= 1 - F_0(x).$$

So the process is stationary. ∎

3.10. THE BLACKWELL AND KEY RENEWAL THEOREMS*

The two main asymptotic results in renewal theory are Blackwell's theorem and what is known as the key renewal theorem originally formulated by Walter Smith. Blackwell's theorem states that if the delay distribution is proper, then $V(t, t + a] \to a/\mu$ as $t \to \infty$. For the key renewal theorem, consider the solution $Z = U * z$ of the renewal equation

$$Z = z + F * Z.$$

If the function z is directly Riemann integrable (defined below; abbreviated as dRi) then

$$\lim_{t \to \infty} Z(t) = \lim_{t \to \infty} U * z(t) = \mu^{-1} \int_0^\infty z(s) ds.$$

Recall from Section 3.7 that this suggests the *renewal method*: If in a renewal theory argument it is desired to calculate $\lim_{t \to \infty} Z(t)$, for some function $Z(t)$ we first write a renewal equation for Z, solve it and then apply the key renewal theorem. Recall the calculation of the asymptotic distributions of $A(t)$ and $B(t)$ in Section 3.7. Remember also that in Section 3.7 we checked that Blackwell's theorem was stronger than the elementary renewal theorem.

Reminder: The conventions regarding the arithmetic case described at the beginning of Section 3.7 are in force and are needed for interpreting the two theorems in the arithmetic case. Alternatively, for the arithmetic case, rely on Section 3.8.

---

*This section contains advanced material which may be skipped on first reading by beginning readers.

3.10.1. DIRECT RIEMANN INTEGRABILITY.

Additional references for this material are the following excellent books: Feller, 1971, pp. 361–362; Cinlar, 1975, p. 294–6; Jagers, 1975, p. 108ff.

The key renewal theorem was originally discussed by W. Smith under a variety of hypotheses. It seems that Feller recognized that the equivalence of the key renewal theorem and Blackwell's theorem was achieved under the condition he termed direct Riemann integrability. Continue as usual to suppose that the function z satisfies $z \geq 0$ and $z(t) = 0$, $t < 0$. (We suppose that $z \geq 0$ because this is what is needed in applications. A theory could also be developed allowing negative values for z, but we do not pursue this.)

For what follows, heavy use is made of the usual notational convention: We set as usual

$$\bigvee_{t \in A} f(t) = \sup_{t \in A} f(t), \quad \bigwedge_{t \in A} f(t) = \inf_{t \in A} f(t)$$

for the supremum and infimum of a real valued function f defined on a domain A.

We begin by reviewing the classical definition of Riemann (R) integration on $[0, \infty)$. First we define what it means for z to be R-integrable on $[0, a]$. Define for $k \geq 1$

$$\underline{m}_k(h) = \bigwedge_{(k-1)h \leq t < kh} z(t)$$

$$\overline{m}_k(h) = \bigvee_{(k-1)h \leq t < kh} z(t)$$

$$\overline{\sigma}(h) = \sum_{k:kh \leq a} h\overline{m}_k(h)$$

$$\underline{\sigma}(h) = \sum_{k:kh \leq a} h\underline{m}_k(h).$$

Observe $\underline{\sigma}(h)$ is nondecreasing as $h \downarrow 0$, and $\overline{\sigma}(h)$ is non-increasing as $h \downarrow 0$. We define z to be R-integrable on $[0, a]$ if

$$\overline{\sigma}(h) - \underline{\sigma}(h) \to 0$$

as $h \to 0$, and, in this case, we set

$$\int_0^a z(s)ds := \lim_{h \downarrow 0} \overline{\sigma}(h).$$

An important and useful classical fact is that z is R-integrable on $[0, a]$ iff z is bounded and a.e. continuous.

We then define R-integrability of z on $[0, \infty)$ to mean z is R-integrable on $[0, a]$ for all $a > 0$ and

$$\lim_{a \to \infty} \int_0^a z(s)ds \text{ exists.}$$

In this case

$$\int_0^\infty z(s)ds := \lim_{a \to \infty} \int_0^a z(s)ds.$$

In contrast, we have the notion of direct Riemann integration: Define $\overline{m}_k(h)$, $\underline{m}_k(h)$ and $\overline{\sigma}(h), \underline{\sigma}(h)$ as above except now for $\overline{\sigma}$ and $\underline{\sigma}$ we have

$$\underline{\sigma}(h) = \sum_{k=1}^\infty h\underline{m}_k(h), \text{ and } \overline{\sigma}(h) = \sum_{k=1}^\infty h\overline{m}_k(h).$$

Then z is dRi if $\overline{\sigma}(h) < \infty$ for all h and

$$\lim_{h \to 0} \overline{\sigma}(h) - \underline{\sigma}(h) = 0.$$

Note that if $\overline{\sigma}(h) = \infty$ there is no hope of finding $h' < h$ such that $\overline{\sigma}(h') < \infty$.

Example 3.10.1. Since $z \geq 0$, $\int_0^\infty z(s)ds$ is the area between the graph of z and the horizontal axis. Consider a function z consisting of triangles. The nth triangle is centered at n and has base of length μ_n and height h_n. See Figure 3.6. Suppose the μ_n are such that the triangles do not overlap. We may allow $\mu_n \to 0$, $h_n \uparrow \infty$ in such a way that $\sum h_n \mu_n < \infty$. Then

$$\int_0^\infty z(s)ds = \sum_{n=1}^\infty \frac{1}{2}h_n\mu_n < \infty$$

and z is R-integrable. (Recall for a function to be Riemann integrable on $[0, \infty)$, it does not have to be bounded.) However z is not dRi since

$$\overline{\sigma}(1) = \sum_{n=1}^\infty \bigvee_{(n-l) \leq s < n} z(s) = \sum_{n=1}^\infty h_n = \infty.$$

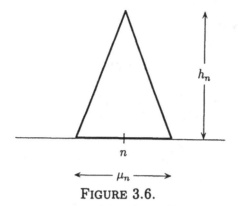

FIGURE 3.6.

By doctoring this example a bit, Feller (1971) shows that the key renewal theorem can fail if the function z oscillates too much in a neighborhood of infinity.

Example 3.10.2. Let F concentrate on $\{1 - \alpha, \alpha\}$ where $0 < \alpha < 1$ and α is irrational. (We pick α irrational to get F non-arithmetic.) Then the renewal function U concentrates on

$$\{k - n\alpha : k, n \text{ integers such that } k - n\alpha \geq 0\}.$$

(To check this, think of adding iid random variables taking values in $\{1 - \alpha, \alpha\}$.) Define z by triangles centered at $k - n\alpha$ of height 1 with bases small enough that the triangles do not overlap and the sum of the areas of the triangles is finite. See Figure 3.7.

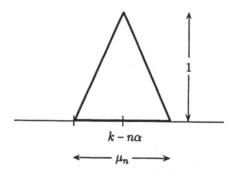

FIGURE 3.7.

The key renewal theorem predicts

$$\lim_{t \to \infty} U * z(t) = \mu^{-1} \int_0^\infty z(s)ds < \infty$$

but for r an integer

$$
\begin{aligned}
Z(r) &= \int_0^r z(r - y)U(dy) \\
&= \sum_{k,n:k-n\alpha\leq r} z(r - (k - n\alpha))U\{k - n\alpha\} \\
&= \sum_{k,n:k-n\alpha\leq r} 1U(\{k - n\alpha\}) \\
&= U(r) \to \infty
\end{aligned}
$$

as $r \to \infty$. Thus the key renewal theorem fails.

Moral: If z gets large in neighborhoods of infinity at embarrassing times, the key renewal theorem may fail. The concept of dRi prevents this.

The following remarks and criteria flesh out the definition of direct Riemann integrability.

Remark 3.10.1. *If z has compact support then Riemann integrability is the same as direct Riemann integrability.*

Remark 3.10.2. *If z is dRi then z is R-integrable on $[0, \infty)$ and*

$$
\lim_{h\downarrow 0} \overline{\sigma}(h) = \int_0^\infty z(s)ds,
$$

where $\int_0^\infty z(s)ds$ is the Riemann integral.

Proof. The method is to approximate $[0, \infty)$ by $[0, a)$ and then approximate the integral by a sum: First observe

$$
\begin{aligned}
0 &= \lim_{h\downarrow 0} \downarrow \overline{\sigma}(h) - \underline{\sigma}(h) \\
&= \lim_{h\downarrow 0} \sum_{k=1}^\infty h\overline{m}_k(h) - \sum_{k=1}^\infty h\underline{m}_k(h) \\
&= \lim_{h\downarrow 0} \sum_{k=1}^\infty h(\overline{m}_k(h) - \underline{m}_k(h)) \\
&\geq \lim_{h\downarrow 0} h \sum_{k:kh\leq a} (\overline{m}_k(h) - \underline{m}_k(h)),
\end{aligned}
$$

which shows z is R-integrable on $[0, a]$ for any a.

Next, for given ϵ there exists $a = a(\epsilon)$ such that $\sum_{n>a} \overline{m}_n(1) < \epsilon$. For $h < 1$,

$$(3.10.1.1) \qquad \overline{\sigma}(h) - \sum_{k:kh\leq a} \overline{m}_k(h)h = \sum_{k:kh>a} \overline{m}_k(h)h \leq \sum_{k>a} \overline{m}_k(1) < \epsilon.$$

Since z is dRi,

$$\lim_{h \downarrow 0} \downarrow \overline{\sigma}(h) =: \sigma_0$$

exists and therefore there exists h_0 such that if $h \leq h_0$

(3.10.1.2) $|\sigma_0 - \overline{\sigma}(h)| < \epsilon.$

Also there is h_1 such that if $h \leq h_1$

(3.10.1.3) $|\sum_{k:kh \leq a} \overline{m}_k(h)h - \int_0^a z(s)ds| < \epsilon.$

The last step follows since z is R-integrable on $[0, a]$. Combining (3.10.1)–(3.10.3), we get for $h \leq h_0 \wedge h_1 \wedge 1$

$$|\sigma_0 - \int_0^a z(s)ds| \leq |\sigma_0 - \overline{\sigma}(h)| + |\overline{\sigma}(h) - \sum_{k:kh \leq a} \overline{m}_k(h)h|$$

$$+ |\sum_{h:kh \leq a} \overline{m}_k(h)h - \int_0^a z(s)ds| < 3\epsilon.$$

This shows

$$\lim_{a \to \infty} \int_0^a z(s)ds = \sigma_0,$$

so we conclude z is R-integrable on $[0, \infty)$ and $\sigma_0 = \int_0^\infty z(s)ds$. ∎

The easiest and most popular criterion for z to be dRi is next.

Remark 3.10.3. *If $z \geq 0$ is non-increasing, then z is dRi iff z is R-integrable.*

For non-increasing functions, the concepts of Riemann integrability and direct Riemann integrability are the same. This result also provides a partial converse to Remark 2.10.1.2.

Proof. Because of Remark 3.10.2, it suffices to prove the result in one direction. Suppose z is non-increasing and Riemann integrable. Because of R-integrability, we have

$$\infty > \int_0^\infty z(s)ds = \sum_{n=1}^\infty \int_{(n-1)h}^{nh} z(s)ds,$$

and, because of monotonicity, this is bounded below by

$$\geq \sum_1^\infty z(nh)h,$$

so z R-integrable implies $\underline{\sigma}(h) < \infty$.

Next, observe that for any N, since z is non-increasing

$$\sum_{n=1}^N \overline{m}_n(h)h - \sum_{n=1}^n \underline{m}_n(h)h \leq h \sum_1^N (z((n-1)h) - z(nh))$$

$$= h(z(0) - z(Nh)).$$

As $N \to \infty$, this expression has a limit:

$$\to h(z(0) - z(\infty)).$$

Thus $\overline{\sigma}(h) < \infty$ iff $\underline{\sigma}(h) < \infty$ and

$$\overline{\sigma}(h) - \underline{\sigma}(h) \leq h(z(0) - z(\infty)).$$

Since we know $\underline{\sigma}(h) < \infty$, we conclude $\overline{\sigma}(h) < \infty$, from which as $h \to 0$

$$\overline{\sigma}(h) - \underline{\sigma}(h) \to 0.$$

Thus z is dRi. ∎

Remark 3.10.4. *If*

   a) z is R-integrable on $[0, a]$ for all $a > 0$

and

   b) $\overline{\sigma}(1) < \infty$,

then z is dRi.

Proof. Since $\overline{\sigma}(1) < \infty$, for $h < 1$, we have that $\underline{\sigma}(h) \leq \overline{\sigma}(h) \leq \overline{\sigma}(1) < \infty$ so the infinite sums to follow converge. Given ϵ, there is N_0 such that $\sum_{n > N_0} \overline{m}_n(1) < \epsilon$. Write

$$\overline{\sigma}(h) - \underline{\sigma}(h) = h \sum_{n:nh\leq N_0} (\overline{m}_n(h) - \underline{m}_n(h)) + h \sum_{n:nh > N_0} (\overline{m}_n(h) - \underline{m}_n(h)).$$

As $h \downarrow 0$ we have

$$|h \sum_{n:nh > N_0} (\overline{m}_n(h) - \underline{m}_n(h))| \leq 2h \sum_{n:nh > N_0} \overline{m}_n(h) \leq 2\epsilon,$$

and, since z is R-integrable on $[0, N_0]$,

$$|h \sum_{n:nh\leq N_0} (\overline{m}_n(h) - \underline{m}_n(h))| \to |\int_0^{N_0} z(s)ds - \int_0^{N_0} z(s)ds| = 0.$$

Therefore $\overline{\sigma}(h) - \underline{\sigma}(h) \to 0$, and z is dRi. ∎

Here is an easy Corollary.

Remark 3.10.5. *If z is R-integrable on $[0, \infty)$ and $z \leq g$ where g is dRi, then z is dRi.*

Proof. This follows from Remark 3.10.4 since (a) is satisfied and

$$\sum_{n=1}^{\infty} \overline{m}_n(1) \leq \sum_{n=1}^{\infty} \bigvee_{n-1 \leq s < n} g(s) < \infty,$$

since g is dRi. ∎

3.10.2. Equivalent Forms of the Renewal Theorem*.

In this section we discuss various equivalent forms for the main renewal theorems. In the next section we contemplate the mysteries of the proof of one of the equivalent forms, namely Blackwell's theorem.

Theorem 3.10.1. *Suppose that F is proper and that $F(0) < 1$. Define as usual*

$$\mu = \int_0^\infty x F(dx) \leq \infty, \quad F_0(x) = \mu^{-1} \int_0^x (1 - F(y)) dy$$

(where if $\mu = \infty$ then $F_0 \equiv 0$). Recall that F is the interarrival distribution, U is the renewal function of a pure renewal process and V is the renewal function of a delayed renewal process. The following are equivalent:

(i) The Blackwell renewal theorem: If G is any proper delay distribution, then

$$\lim_{t \to \infty} V(t, t+b] = \mu^{-1} b$$

for $b > 0$.

(ii) The key renewal theorem: Suppose $z(t)$ is dRi. Then

$$\lim_{t \to \infty} U * z(t) = \mu^{-1} \int_0^\infty z(s) ds.$$

(iii) For any proper delay distribution G

$$\lim_{t \to \infty} P[B(t) \leq x] = F_0(x)$$

for all $x > 0$.

(iv) For any proper delay distribution G

$$\lim_{t \to \infty} P[A(t) \leq x] = F_0(x)$$

---

* This section contains advanced material which may be skipped on first reading by beginning readers.

for all $x > 0$.

Before the proof, we remind the reader about the convention discussed in Section 3.7 governing arithmetic distributions which tells how to interpret the results in the arithmetic case. Also consider the following. Since for a stationary renewal process $V(t, t+b] = \mu^{-1}b$ and $P[B(t) \le x] = F_0(x)$, it is tempting in the case of $\mu < \infty$ to interpret these results by stating that the process becomes *asymptotically stationary*. The phrase *asymptotically stationary* can be understood in a variety of ways. Here is one: The renewal sequence past t should converge in distribution to a stationary renewal sequence as $t \to \infty$. By the renewal sequence past t we mean renewal epochs $\{S_n - t : S_n > t\}$ measured with t as a new origin. Another way to write this formally is

$$(S_{N(t)} - t, S_{N(t)+k} - t, k \ge 1).$$

By differencing successive terms of this sequence we get

$$(B(t), Y_{N(t)+k}, k \ge 1).$$

Examining Proposition 1.8.2. and Lemma 3.9.2 we see that

$$(B(t), Y_{N(t)+k}, k \ge 1) \overset{d}{=} (Y_0^*, Y_k, k \ge 1),$$

where $(Y_0^*, Y_k, k \ge 1)$ are independent, $(Y_k, k \ge 1)$ is iid with common distribution $F(x)$ and $P[Y_0^* \le x] = P[B(t) \le x]$. From (iii) above we get

$$(Y_0^*, Y_k, k \ge 1) \Rightarrow (\tilde{Y}_0, Y_k, k \ge 1)$$

in R^∞ where $P[\tilde{Y}_0 \le x] = F_0(x)$ and $\tilde{Y}_0$ is independent of $(Y_k, k \ge 1)$. (Here "$\Rightarrow$" stands for convergence in distribution in R^∞ which means all finite dimensional distributions converge.) Therefore, if we apply to the previous convergence the map

$$(x_1, x_2, \dots) \mapsto (x_1, x_1 + x_2, x_1 + x_2 + x_3, \dots),$$

as $t \to \infty$, we obtain

$$(S_{N(t)} - t, S_{N(t)+k} - t, k \ge 1) \Rightarrow (\tilde{S}_n, n \ge 0),$$

where $(\tilde{S}_n)$ is a stationary process. This convergence can be interpreted to mean the renewal sequence past t becomes *asymptotically stationary* as $t \to \infty$.

Another interpretation of asymptotic stationarity might be that, for any k and $0 \le t_1 < \cdots < t_k$,

$$(B(t + t_i), i \le k) \Rightarrow (\tilde{B}(t + t_i), \, i \le k),$$

where $\tilde{B}$ is the stationary version of B. One cannot quite prove this using renewal theory (Breiman, 1968, pages 134–5).

Proofs. To begin with, it is easy to see the equivalence of (iii) and (iv). Write

$$P[B(t) \le x] = P[N(t, t + x] \ge 1] = P[A(t + x) < x],$$

and the result follows. We now consider the other implications.

(ii) $\rightarrow$(iv): Recall in the pure case

$$P[A(t) \le x] = U * ((1 - F(\cdot))1_{[0,x]}(\cdot))(t) =: Z(t)$$
$$= U * z(t)$$

where $z(t) = (1 - F(t))1_{[0,x]}(t)$. Since z has compact support, z is dRi and therefore the key renewal theorem asserts

$$\lim_{t \to \infty} Z(t) = \mu^{-1} \int_0^\infty z(s)ds$$
$$= \mu^{-1} \int_0^\infty (1 - F(s))1_{[0,x]}(s)ds$$
$$= \mu^{-1} \int_0^x (1 - F(s))ds = F_0(x).$$

If the process is delayed with delay distribution G, then

$$P[A(t) \le x] = P[A(t) \le x, S_0 > t] + \int_0^t Z(t - y)G(dy).$$

Note $P[A(t) \le x, S_0 > t] \le P[S_0 > t] \to 0$ as $t \to \infty$. Also define $f_t(y) = Z(t - y)1_{[0,t]}(y)$. Since $Z(t)$ is a probability, we have $f_t(y) \le 1$ for all $t > 0$, $y > 0$, and since $Z(t) \to F_0(x)$ as $t \to \infty$ implies $\lim_{t \to \infty} f_t(y) = F_0(x)$ for all $y > 0$, it follows from dominated convergence that

$$\int_0^t Z(t - y)G(dy) = \int_0^\infty f_t(y)G(dy)$$
$$\to \int_0^\infty F_0(x)G(dy) = F_0(x).$$

Remark. Ponder why it is better to prove (ii)$\rightarrow$(iv) rather than (ii)$\rightarrow$(iii).

(iii)→(i): Check that

$$V(t, t+b] = G_t * U(b),$$

where $G_t(x) := P[B(t) \leq x] \to F_0(x)$ as $t \to \infty$ for each $x > 0$. Write

$$G_t * U(b) = \int_0^b G_t(b-s)U(ds).$$

On $[0, b]$, U is a finite measure. Since $G_t(b-s) \leq 1$ and for each s we have $G_t(b-s) \to F_0(b-s)$ as $b \to \infty$, it follows by dominated convergence that

$$V(t, t+b] \to \int_0^b F_0(b-s)U(ds)$$
$$= F_0 * U(b) = \mu^{-1}b,$$

from Section 3.7 or Theorem 3.9.1(c).

(i)→(ii): We begin with a lemma needed for the proof.

Lemma 3.10.2. *If $F(b) < 1$ then*

$$U(t-b, t] \leq (1 - F(b))^{-1}$$

for all $t \geq b$, and, therefore, for $t \geq b$,

$$\sup_{t \geq 0} U(t, t+b] \leq (1 - F(b))^{-1} = c(b) < \infty.$$

Proof of the lemma. We have the renewal equation

$$U = F^{0*} + F * U,$$

so that

$$U - F * U = F^{0*}.$$

Therefore

$$1 = \int_0^t (1 - F(t-s))U(ds)$$
$$\geq \int_{t-b}^t (1 - F(t-s))U(ds)$$
$$\geq (1 - F(b))U(t-b, t]. \quad \blacksquare$$

Proof. We are now prepared to prove that Blackwell's theorem implies the key renewal theorem. We check the key renewal theorem for successively more complex z, assuming the validity of the Blackwell theorem.

Step 1: Suppose $z(t) = 1_{[(n-1)h,nh)}(t)$. Then

$$z(t-s) = 1 \text{ iff } (n-1)h \le t-s < nh$$
$$\text{iff } t - nh < s \le t - (n-1)h,$$

so

$$U * z(t) = \int_0^t z(t-s)U(ds) = U(t-nh, t-(n-1)h],$$

and, as $t \to \infty$, Blackwell's theorem yields

$$U * z(t) \to \mu^{-1}h = \mu^{-1} \int_0^\infty z(s)ds.$$

Step 2: Suppose $z(t) = \sum_{n=1}^\infty c_n 1_{[(n-1)h,nh)}(t)$ where $c_n \ge 0$, $\sum c_n < \infty$ and h is chosen so that $F(h) < 1$. Then

$$U * z(t) = \sum_{n=1}^\infty c_n(U(t-nh, t-(n-1)h]).$$

For each n we have

$$U(t-nh, t-(n-1)h] \to \mu^{-1}h$$

as $t \to \infty$, and

$$\sup_{t,n} U(t-nh, t-(n-1)h] \le c(h) < \infty$$

by Lemma 3.10.2. Therefore, by dominated convergence

$$U * z(t) \to \sum c_n\mu^{-1}h = \mu^{-1} \int_0^\infty z(s)ds.$$

Step 3: Let z be dRi, and define

$$\overline{z}(t) = \sum_1^\infty \overline{m}_n(h)1_{[(n-1)h,nh)}(t)$$

$$\underline{z}(t) = \sum_1^\infty \underline{m}_n(h)1_{[(n-1)h,nh]}(t),$$

where

$$\overline{m}_n(h) = \bigvee_{(n-1)h \le s < nh} z(s), \quad \underline{m}_n(h) = \bigwedge_{(n-1)h \le s < nh} z(s).$$

Then $\overline{z}$, $\underline{z}$ are functions of the type considered in Step 2 since, by the definition of direct Riemann integrability,

$$\infty > \sum_1^\infty \overline{m}_n(h) \ge \sum_1^\infty \underline{m}_n(h).$$

Therefore, from Step 2,

$$\lim_{t \to \infty} U * \overline{z}(t) = \mu^{-1} \sum_1^\infty \overline{m}_n(h)h =: \overline{\sigma}(h)/\mu$$

and

$$\lim_{t \to \infty} U * \underline{z}(t) = \mu^{-1} \sum_1^\infty \underline{m}_n(h)h =: \underline{\sigma}(h)/\mu.$$

Finally, since $\underline{z} \le z \le \overline{z}$, we have

$$
\begin{aligned}
\mu^{-1}\underline{\sigma}(h) = \liminf_{t \to \infty} U * \underline{z}(t) &\le \liminf_{t \to \infty} U * z(t) \\
&\le \limsup_{t \to \infty} U * z(t) \\
&\le \limsup_{t \to \infty} U * \overline{z}(t) = \mu^{-1}\overline{\sigma}(h).
\end{aligned}
$$

This holds for all h. Let $h \downarrow 0$. From the definition of direct Riemann integration $\overline{\sigma}(h) - \underline{\sigma}(h) \to 0$ and $\overline{\sigma}(h) \to \int_0^\infty z(s)ds$. The result follows. ∎

Before closing this section, we give one more example of the renewal method: If one desires the asymptotic properties of $Z(t)$, write a renewal equation that Z satisfies and solve it. If $Z = U * z$ and z is dRi then $Z(t) \to \mu^{-1} \int_0^\infty z(s)ds$.

Example 3.10.3. Second Order Properties of U. If $\mu < \infty$ we know $t^{-1}U(t) \to \mu^{-1}$. What is the behavior of

$$Z(t) = U(t) - t/\mu?$$

Recall $t/\mu = F_0 * U(t)$ so

$$Z(t) = U(t) - F_0 * U(t) = (1 - F_0) * U(t).$$

If $1 - F_0$ is dRi then the key renewal theorem applies. But $1 - F_0$ is non-increasing, so, from Remark 3.10.3, $1 - F_0$ is dRi iff $1 - F_0$ is Riemann integrable. Observe that, with several applications of Fubini's Theorem justifying changes in the order of integration,

$$
\begin{aligned}
\int_0^\infty (1 - F_0(t))dt &= \mu^{-1} \int_{t=0}^\infty \int_{s=t}^\infty (1 - F(s))ds \, dt \\
&= \mu^{-1} \int_{s=0}^\infty \left(\int_{t=0}^s dt \right) (1 - F(s)) \, ds \\
&= \mu^{-1} \int_0^\infty s(1 - F(s))ds \\
&= \mu^{-1} \int_{s=0}^\infty \left(\int_{u=s}^\infty F(du) \right) s \, ds \\
&= \mu^{-1} \int_{u=0}^\infty \left(\int_{s=0}^u s \, ds \right) F(du) \\
&= \mu^{-1} \int_0^\infty \frac{1}{2} u^2 F(du).
\end{aligned}
$$

Thus $1 - F_0$ is dRi iff $\int_0^\infty u^2 F(du) < \infty$ in which case

$$
U(t) - \mu^{-1}t \to (2\mu)^{-1} \int_0^\infty u^2 F(du).
$$

3.10.3. PROOF OF THE RENEWAL THEOREM*.

Having shown the equivalences of various forms of the renewal theorems in Section 3.10.2 we are free to prove whichever of the equivalent statements is most appealing. We follow Lindvall's (1975) coupling procedure (see also Asmussen (1987)) and prove in the non-arithmetic case that for a pure renewal process

$$
(3.10.3.1) \qquad \lim_{t \to \infty} P[B(t) \le x] = F_0(x).
$$

(Note that if we can prove this in the pure case, then a simple dominated convergence argument gives the result in the delayed case.) Recall that if B were stationary, then, for every t, equality would hold in (3.10.3.1). This suggests the following coupling proof.

Let $\{S_n, n \ge 0\}$ be a pure process with $\mu < \infty$. Let $\{\tilde{S}_n, n \ge 0\}$ be a stationary renewal sequence independent of $\{S_n\}$. All quantities related to $\{\tilde{S}_n\}$ will have tildas attached. Recall

$$
F_0(x) = P[\tilde{S}_0 \le x] = P[\tilde{B}(t) \le x]
$$

---

* This section contains advanced material which may be skipped on first reading by beginning readers.

for all t.

Run $\{S_n\}$ and $\{\tilde{S}_n\}$ side by side. Construct a pasted together process $\{S_n^*\}$ by observing $\{S_n\}$ until the first epoch where a renewal from $\{S_n\}$ is within ϵ distance of a renewal in $\{\tilde{S}_n\}$. Then switch over to observing $\{\tilde{S}_n\}$. The pasted together process $\{S_n^*\}$ has the property

$$\{S_n^*\} \overset{d}{=} \{S_n\}$$

in R^∞, but after the switch the process is stationary and the forward recurrence times should have distribution F_0. Relating the forward recurrence times of $\{S_n\}$ to those of $\{S_n^*\}$ and letting $\epsilon \to 0$ should yield the result. See Figure 3.8.

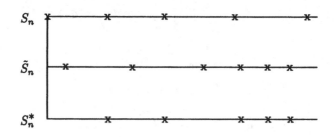

FIGURE 3.8.

The above description, though appealing, glosses over several problems. Why is $\{S_n\} \overset{d}{=} \{S_n^*\}$? and why does there exist an epoch of $\{S_n\}$ close to that of $\{\tilde{S}_n\}$ which can be used for the switch? We consider the proof in a series of steps.

Step 1: We consider why an epoch of $\{S_n\}$ is eventually close to an epoch of $\{\tilde{S}_n\}$. Given $\delta > 0$, consider $\tilde{B}(S_i)$, which is the waiting time until the next renewal in $\{\tilde{S}_n\}$ measured from S_i. We must show

$$(3.10.3.2) \qquad P[\tilde{B}(S_i) < \delta \text{ i.o. }] = 1,$$

for then there will be lots of suitable switching times. See Figure 3.9.

Let $A_i = \cup_{j \geq i}[\tilde{B}(S_j) < \delta]$. Then

$$A_\infty = \lim_{i \to \infty} \downarrow A_i = \cap_{i=1}^\infty A_i = [\tilde{B}(S_j) < \delta \text{ i.o. }].$$

We first show that

$$(3.10.3.3) \qquad PA_i \text{ is independent of } i.$$

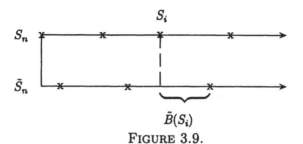

FIGURE 3.9.

For this we check

(3.10.3.4) $$(\tilde{B}(S_j), j \geq 1) \overset{d}{=} (\tilde{B}(S_j), j \geq 0).$$

Observe that for any $k > 0$, and fixed i, Borel sets $\Lambda_0, \ldots, \Lambda_k$

$$P[\tilde{B}(S_i) \in \Lambda_0, \tilde{B}(S_{i+1}) \in \Lambda_1, \ldots, \tilde{B}(S_{i+k}) \in \Lambda_k]$$
$$= \int P[\tilde{B}(s_{i+j}) \in \Lambda_j, 0 \leq j \leq k] P[S_{i+j} \in ds_{i+j}, 0 \leq j \leq k].$$

Because $\tilde{B}$ is strictly stationary, this is

$$\int P[\tilde{B}(s_{i+j} - s_i) \in \Lambda_j, 0 \leq j \leq k] P[S_{i+j} \in ds_{i+j}, 0 \leq j \leq k]$$
$$= P[\tilde{B}(0) \in \Lambda_0, \tilde{B}(S_{i+1} - S_i) \in \Lambda_1, \ldots, \tilde{B}(S_{i+n} - S_i) \in \Lambda_k],$$

and, since $\{S_n\}$ is independent of $\{\tilde{S}_n\}$, this is

$$P[\tilde{B}(0) \in \Lambda_0, \tilde{B}(S_1) \in \Lambda_1, \ldots, \tilde{B}(S_k) \in A_k].$$

This proves (3.10.3.4) and thus $PA_i = PA_0$, as asserted.

We therefore conclude

(3.10.3.5) $$\lim_{t \to \infty} PA_i = PA_\infty = PA_0.$$

For what follows we need two lemmas. The first is from Feller, 1971, page 147.

For a measure G on R, say x is a point of increase of G if $G(x-\epsilon, x+\epsilon) > 0$ for all $\epsilon > 0$. The assumption that F is not arithmetic means the points of increase of F are not a subset of

$$L(h) := \{0, h, 2h, \ldots\}$$

for any $h > 0$. We write $pi(G)$ for the set consisting of points of increase of G. Let $\Sigma = \cup_{n=0}^{\infty} pi(F^{n*})$ so that $\Sigma \subset pi(U)$.

Lemma 3.10.3. *Suppose F is non-arithmetic. Then for each δ there exists T such that if $t \geq T$*

$$U(t, t + \delta) > 0.$$

Equivalently, for a pure renewal process,

$$P[N(t, t + \delta) > 0] > 0$$

for all $t \geq T$.

Proof. First observe that if F_1, F_2 are distributions on R then $a \in pi(F_1), b \in pi(F_2)$ implies $a + b \in pi(F_1 * F_2)$. If ξ_i is a rv with distribution F_i, $i = 1, 2$, then

$$P[\xi_1 + \xi_2 \in (a + b - \epsilon, a + b + \epsilon)]$$
$$\geq P[\xi_1 \in (a - \frac{\epsilon}{2}, a + \frac{\epsilon}{2}), \xi_2 \in (b - \frac{\epsilon}{2}, b + \frac{\epsilon}{2})]$$
$$= F_1(a - \frac{\epsilon}{2}, a + \frac{\epsilon}{2}), F_2(b - \frac{\epsilon}{2}, b + \frac{\epsilon}{2}) > 0.$$

From this it follows that if

(3.10.3.6) $\qquad a, b \in \Sigma, \ a < b, \ h = b - a > 0$ then $na + mh \in \Sigma$

for $0 \leq m \leq n$. To check this, observe that if $a, b \in \Sigma$ then for some integers r, s

$$a \in pi(F^{r*}), \quad b \in pi(F^{s*})$$

and therefore

$$na + mh = (n - m)a + mb \in pi(F^{((n-m)r+ms)*}) \subset \Sigma.$$

Now suppose $t \in [na, na + nh]$. From (3.10.3.6) it follows t must be within distance h of some point in Σ. But the intervals $[na, na+nh]$ overlap for n large; i.e., $(n + 1)a < na + nh$ if $ah^{-1} < n$ so any $t \geq T = a^2h$ is within distance h of a point in Σ. So if $\delta > h$

$$U(t, t + \delta) > 0$$

for $t \geq T$. It now remains to show h can be chosen arbitrarily small. We proceed by contradiction. We wish to show

$$\inf\{|b - a|, a, b \in \Sigma\} = 0.$$

Suppose otherwise and that $\inf\{|b - a|, a, b \in \Sigma\} = h_0 > 0$. There exist, then, $a, b \in \Sigma$, $a < b$ and $h_0 \leq b - a = h \leq 2h_0$. It follows that

$$(3.10.3.7) \qquad [na, na + nh] \cap \Sigma = \{na + mh, m = 0, 1, \ldots, n\},$$

since if there is a point p in $[na, na + nh] \cap \Sigma$ not of the form given by the right side of (3.10.3.7), it would be at distance less than $h/2 \leq h_0$ from some point $na + k_0 h \in \Sigma$, which contradicts the definition of h_0.

Suppose $n \geq a/h$. Then on the one hand $(n+1)a \in \Sigma$ and on the other $(n+1)a \in [na, na + nh]$. Thus $(n+1)a$ belongs to the left side of (3.10.3.7) and hence to the right side.

Therefore, $(n + 1)a = na + mh$ for some $m \in \{0, 1, \ldots, n\}$, so $a = mh$ and $a \in L(h)$. We conclude from (3.10.3.7) that

$$(3.10.3.8) \qquad [na, na + nh] \bigcap \Sigma \subset L(h).$$

To finish, take $c \in pi(F)$. Since the itervals $[na, na + nh]$ overlap for large n, there exist k and n such that

$$c + ka \in [na, na + nh].$$

Since $c \in pi(F)$ and $a \in \Sigma$, we have $c + ka \in \Sigma$, and, further, from (3.10.3.8), we have $c + ka \in L(h)$. Since $a \in L(h)$, we conclude that $c \in L(h)$. Hence $pi(F) \subset L(h)$ which contradicts the fact that F is non-arithmetic.

To prove $U(t, t + \delta) > 0$ for $t \geq T$ iff $P[N(t, t + \delta) > 0) > 0$ for $t \geq T$ we observe $U(t, t + \delta) > 0$ iff there is some r such that $F^{r*}(t, t + \delta) > 0$. Since

$$\begin{aligned} P[N(t, t + \delta) > 0] &= P\{\cup_0^\infty [S_n \in (t, t + \delta)]\} \\ &\geq P[S_n \in (t, t + \delta)] = F^{r*}(t, t + \delta), \end{aligned}$$

the result follows. ■

The second needed preliminary is the Hewitt-Savage zero-one law. Call a set $A \subset R^\infty$ invariant if

$$(x_1, x_2, \ldots, x_i, \ldots, x_j, \ldots) \in A \text{ iff } (x_1, x_2, \ldots, x_j, \ldots, x_i, \ldots) \in A$$

for any i, j. This is equivalent to supposing A is invariant under permutations involving finitely many coordinates:

$$(x_1, x_2, \ldots, x_k, \ldots) \in A \text{ iff } (x_{\pi(1)}, x_{\pi(2)}, \ldots, x_{\pi(k)}, \ldots) \in A$$

for any k and any permutation π of $\{1, 2, \ldots, k\}$.

Lemma 3.10.4. *(Hewitt-Savage)* *If* $\xi = \{\xi_n, n \geq 1\}$ *is an iid sequence, then for any invariant set* A

$$P[\xi \in A] = 0 \text{ or } 1.$$

See Feller, 1971, page 124, or Billingsley, 1986, page 304, for a proof. For us, it is important to realize that the $\{\xi_n\}$ do not have to be R-valued but can be iid random elements of more general spaces. For instance, ξ_n could have range R^2; that is, $\{\xi_n\}$ could be an iid sequence of random vectors.

We now continue our hunt for coupling times. The next step is to show

$$(3.10.3.9) \qquad P[A_0|\tilde{Y}_0 = t] > 0, \quad \text{for all } t \geq 0.$$

(The reason for conditioning on $\tilde{Y}_0$ becomes obvious later, but suffice it to say for now that we wish to apply Hewitt-Savage to the iid sequence $(Y_1, \tilde{Y}_1), (Y_2, \tilde{Y}_2), \ldots$. Conditioning on $\tilde{Y}_0$ eliminates the offending $\tilde{Y}_0$ which has distribution F_0 and not the common distribution F.) To check (3.10.3.9), fix a value of t and write

$$P[A_0|\tilde{Y}_0 = t] = P\left\{\cup_{j=0}^{\infty}[\tilde{B}(S_j) < \delta]|\tilde{Y}_0 = t\right\}$$

$$= P\left\{\cup_{j=0}^{\infty}[t + \sum_{i=1}^{n} \tilde{Y}_i \in (S_j, S_j + \delta) \text{ for some } n \geq 1]\right\}.$$

For any j, this is

$$\geq P[t + \sum_{i=1}^{n} \tilde{Y}_i \in (S_j, S_j + \delta) \text{ for some } n \geq 1]$$

$$= \int_0^{\infty} P\left\{\cup_{n=1}^{\infty}[t + \sum_{i=1}^{n} \tilde{Y}_i \in (s, s + \delta)]\right\} P[S_j \in ds].$$

With T as in Lemma (3.10.3.1), this is

$$\geq \int_{T+t}^{\infty} P\left\{\cup_{n=1}^{\infty}[\sum_{i=1}^{n} \tilde{Y}_i \in (s - t, s - t + \delta)]\right\} P[S_j \in ds].$$

The integrand equals $P[N(s - t, s - t + \delta) > 0]$ which on $\{s : s > T + t\}$ is strictly positive. The choice of j is still at our disposal. Since $S_j \uparrow \infty$ there exists some j such that $P[S_j > T + t] > \frac{1}{2}$ (say) and therefore for this value of t, (3.10.3.9) is true.

We next assert

$$(3.10.3.10) \qquad P(A_\infty | \tilde{Y}_0 = t) = 0 \text{ or } 1,$$

which is a direct consequence of Hewitt-Savage applied to the iid (hence the conditioning on $\tilde{Y}_0$) sequence of random vectors $\{(Y_1, \tilde{Y}_1), (Y_2, \tilde{Y}_2), \ldots, \}$ since

$$P[A_\infty | \tilde{Y}_0 = t] = P\{\limsup_{j \to \infty} \cup_{n=1}^\infty [t + \sum_{i=1}^n \tilde{Y}_i \in (S_j, S_j + \delta)]\}.$$

Now let

$$\Lambda = \{t : P(A_\infty | \tilde{Y}_0 = t) = 1\},$$

and we show $F_0(\Lambda) = 1$. If $F_0(\Lambda) < 1$, then a contradiction will ensue: Write

$$PA_0 = \int P(A_0 | \tilde{Y}_0 = t) F_0(dt)$$

and

$$PA_0 = PA_\infty = \int P(A_\infty | \tilde{Y}_0 = t) F_0(dt)$$
$$= \int_\Lambda 1 F_0(dt) + \int_{\Lambda^c} 0 F_0(dt)$$
$$= F_0(\Lambda),$$

since by (3.10.3.10) we have $P(A_\infty | \tilde{Y}_0 = t) = 0$ for $t \in \Lambda^c$. Note $A_0 \supset A_\infty$ so on Λ we have $P(A_0 | \tilde{Y}_0 = t) = 1$, whence

$$F_0(\Lambda) = PA_0 = \int P(A_0 | \tilde{Y}_0 = t) F_0(dt)$$
$$= \int_\Lambda 1 F_0(dt) + \int_{\Lambda^c} P(A_0 | \tilde{Y}_0 = t) F_0(dt).$$

Recall $P[A_0 | \tilde{Y}_0 = t) > 0$ for all t. If $F_0(\Lambda) < 1$ so that $F_0(\Lambda^c) > 0$, then the above gives a contradiction.

Since $F_0(\Lambda) = 1$, we have

$$PA_\infty = \int P(A_\infty | \tilde{Y}_0 = t) F_0(dt)$$
$$= \int_\Lambda P(A_\infty | \tilde{Y}_0 = t) F_0(dt$$
$$= \int_\Lambda 1 F_0(dt) = F_0(\Lambda) = 1.$$

This is the desired conclusion and shows there are plenty of potential switching times.

Step 2: As a switching time we choose the first epoch when renewals from the two processes are close. Define

$$T = \inf\{i \geq 0 : \tilde{B}(S_i) < \delta\}, \tilde{T} = \inf\{j : \tilde{S}_j \geq S_T\}$$

and

$$(S_n^*, n \geq 0) = (0, S_1, S_2, \ldots, S_T, \tilde{S}_{\tilde{T}+1} - \tilde{B}(S_T), \tilde{S}_{\tilde{T}+2} - \tilde{B}(S_T), \ldots).$$

See Figure 3.10.

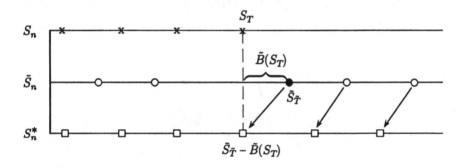

FIGURE 3.10.

Proposition 3.10.5. *We have*

$$\{S_n^*, n \geq 0\} \stackrel{d}{=} \{S_n, n \geq 0\}$$

in R^∞.

Proof. Set $Y_n^* = S_n^* - S_{n-1}^*$, $n \geq 1$, and it suffices to show

$$(Y_n^*, n \geq 1) \stackrel{d}{=} (Y_n, n \geq 1)$$

in R^∞. Observe that for integers n_1, n_2

$$[T = n_1, \tilde{T} = n_2] \in \sigma(S_1, \ldots, S_{n_1}, \tilde{S}_0, \ldots, \tilde{S}_{n_2})$$

(3.10.3.11)
$$= \sigma(Y_1, \ldots, Y_n, \tilde{Y}_0, \tilde{Y}_1, \ldots, \tilde{Y}_{n_2}).$$

Therefore for any k, and $x_i > 0$, $i = 1, \ldots, k$

$$P[Y_i^* \le x_i, i \le k] = \sum_{n_1, n_2} P[Y_i^* \le x_i, i \le k, T = n_1, \tilde{T} = n_2]$$

$$= \sum_{n_1 \le k, n_2 \ge 0} + \sum_{n_1 > k, n_2 \ge 0}.$$

The first sum equals

$$\sum_{n_1 \le k, n_2 \ge 0} P[Y_i \le x_i, i \le n_1, T = n_1, \tilde{T} = n_2,$$

$$\tilde{Y}_{n_2+1} \le x_{n_1+1}, \ldots, \tilde{Y}_{n_2+k-n_1} \le x_k].$$

Now from (3.10.3.11), $[Y_i \le x_i, i \le n_1] \cap [T = n_1, \tilde{T} = n_2]$ is independent of

$$P[\tilde{Y}_{n_2+1} \le x_{n_1+1}, \ldots, \tilde{Y}_{n_2+k-n_1} \le x_k],$$

so the sum is

$$\sum_{n_1 \le k, n_2 \ge 0} P[Y_i \le x_i, i \le n_1, T = n_1,$$

$$\tilde{T} = n_2] P[\tilde{Y}_{n_2+1} \le x_{n_1+1}, \ldots, \tilde{Y}_{n_2+k-n_1} \le x_k].$$

Since $(\tilde{Y}_{n_2+1}, \tilde{Y}_{n_2+2}, \ldots) \overset{d}{=} (Y_{n_1+1}, Y_{n_1+2}, \ldots)$, the above is

$$\sum_{n_1 \le k, n_2 \ge 0} P[Y_i \le x_i, i \le n_1, T = n_1, \tilde{T} = n_2] P[Y_{n_1+1} \le x_{n_1+1}, \ldots, Y_k \le x_k],$$

and applying (3.10.3.11) again yields

$$\sum_{n_1 \le k, n_2 \ge 0} P[Y_i \le x_i, i \le n_1, T = n_1, \tilde{T} = n_2, Y_{n_1+1} \le x_{n_1+1}, \ldots, Y_k \le x_k]$$

$$= P[Y_i \le x_i, i \le k, T \le k].$$

The second sum under consideration is

$$\sum_{n_1 > k, n_2 \ge 0} P[Y_i \le x_i, i \le k, T = n_1, \tilde{T} = n_2] = P[Y_i \le x_i, i \le k, T > k].$$

Combining the two sums gives

$$P[Y_i \le x_i, i \le k] = P[Y_i \le x_i, i \le k, T \le k] + P[Y_i \le x_i, i \le k, T > k]$$

$$= P[Y_i \le x_i, i \le k],$$

as required.

Step 3: We now prove

$$\lim_{t\to\infty} P[B(t) > x] = 1 - F_0(x).$$

Since $(S_n^*, n \geq 0) \overset{d}{=} (S_n, n \geq 0)$, it is equivalent to prove

$$\lim_{t\to\infty} P[B^*(t) > x] = 1 - F_0(x).$$

Toward this end, observe that, for $t > S_T$ (i.e., t is larger than the switching time), we have

$$
\begin{aligned}
[\tilde{B}(t+\delta) > x + \delta] &= [\tilde{N}(t+\delta, t+\delta+x+\delta] = 0] \\
&= [N^*(t+\delta - \tilde{B}(S_T), t+\delta+x+\delta - \tilde{B}(S_T)] = 0] \\
&\subset [N^*(t+\delta, t+\delta+x] = 0],
\end{aligned}
$$

since

$$(t+\delta, t+\delta+x] \subset (t+\delta - \tilde{B}(S_T), t+\delta+x+\delta - \tilde{B}(S_T)].$$

We conclude

(3.10.3.12) $[\tilde{B}(t+\delta) > x + \delta] \subset [B^*(t+\delta) > x]$ on $[t > S_T]$.

Also, we have on $[t > S_T]$

$$
\begin{aligned}
[B^*(t) > x] &= [N^*(t, t+x) = 0] \\
&= [\tilde{N}(t+\tilde{B}(S_T), t+x+\tilde{B}(S_T)] = 0] \\
&\subset [\tilde{N}(t+\delta, t+x] = 0]
\end{aligned}
$$

(since $(t+\tilde{B}(S_T), t+x+\tilde{B}(S_T) \supset (t+\delta, t+x])$

$$= [\tilde{B}(t+\delta) > x - \delta].$$

So again we conclude

(3.10.3.13) $[B^*(t) > x] \subset [\tilde{B}(t+\delta) > x - \delta]$ on $[S_T < t]$.

Therefore, using (3.10.3.13), we obtain

$$
\begin{aligned}
P[B^*(t) > x] &= P[B^*(t) > x, S_T \geq t] + P[B^*(t) > x, S_T < t] \\
&\leq P[S_T \geq t] + P[\tilde{B}(t+\delta) > x - \delta, S_T < t] \\
&\leq o(1) + P(\tilde{B}(t+\delta) > x - \delta),
\end{aligned}
$$

since $P[S_T \geq t] \to 0$ as $t \to \infty$. Recalling that $\tilde{B}$ is stationary and hence $\tilde{B}(t + \delta)$ has distribution F_0 gives

$$\limsup_{t \to \infty} P[B^*(t) > x] \leq 1 - F_0(x - \delta).$$

Likewise, from (3.10.3.12)

$$\begin{aligned}
P[B^*(t + \delta) > x] &\geq P[B^*(t + \delta) > x, t > S_T] \\
&\geq P[\tilde{B}(t + \delta) > x + \delta, t > S_T] \\
&= P[\tilde{B}(t + \delta) > x + \delta] - P[\tilde{B}(t + \delta) > x + \delta, t \leq S_T] \\
&= P[\tilde{B}(t + \delta) > x + \delta] - o(1) \\
&= 1 - F_0(x + \delta) - o(1),
\end{aligned}$$

so that

$$\liminf_{t \to \infty} P[B^*(t) > x] \geq 1 - F_0(x + \delta).$$

Let $\delta \downarrow 0$, and use the continuity of F_0. Combining the lim inf and lim sup statements gives

$$1 - F_0(x) = \lim_{t \to \infty} P[B^*(t) > x] = \lim_{t \to \infty} P[B(t) > x].$$

This completes the proof of the renewal theorems.

3.11. IMPROPER RENEWAL EQUATIONS.

Recall that if $F(\infty) \neq 1$, the equation

$$Z = z + F * Z$$

is called improper. Example 3.5.2, on the age dependent branching process, provides an improper renewal equation: Just choose $m < 1$.

In the case where $z(\infty) := \lim_{t \to \infty} z(t)$ exists and $F(\infty) < 1$, asymptotic analysis of $Z(t) = U * z(t)$ is easy. We have $U(\infty) = \sum_{n=0}^{\infty} F(\infty)^n = (1 - F(\infty))^{-1} < \infty$. Recall that z is always assumed locally bounded, that is, bounded on compact intervals, and the existence of $z(\infty)$ means z is bounded on $[0, \infty)$. Since z is bounded on $[0, \infty)$, from dominated convergence we obtain

$$Z(\infty) =: \lim_{t \to \infty} Z(t) = z(\infty)U(\infty).$$

Frequently, a simple transformation allows the key renewal theorem to refine the above argument.

Proposition 3.11.1. *Suppose there exists $\beta \in R$ such that*

$$\hat{F}(-\beta) = \int_0^\infty e^{\beta x} F(dx) = 1,$$

and suppose Z satisfies $Z = z + F * Z$. Define $Z^\#(t) = e^{\beta t} Z(t)$, $z^\#(t) = e^{\beta t} z(t)$, $F^\#(dt) = e^{\beta t} F(dt)$. Note

$$F^\#(\infty) = \int_0^\infty e^{\beta t} F(dt) = 1;$$

i.e., $F^\#$ is proper. Then $Z^\#$ satisfies

$$Z^\# = z^\# + F^\# * Z^\#,$$

and, if $z^\#$ is dRi,

$$\lim_{t \to \infty} Z^\#(t) = \lim_{t \to \infty} e^{\beta t} Z(t)$$
$$= \frac{\int_0^\infty z^\#(t) dt}{\int_0^\infty t F^\#(dt)}$$
$$= \frac{\int_0^\infty e^{\beta t} z(t) dt}{\int_0^\infty t e^{\beta t} F(dt)}.$$

Proof. Take the equation

$$Z(t) = z(t) + \int_0^t Z(t - s) F(ds)$$

and multiply through by $e^{\beta t}$ to get

$$e^{\beta t} Z(t) = e^{\beta t} z(t) + \int_0^t e^{\beta(t-s)} Z(t - s) e^{\beta s} F(ds)$$

or

$$Z^\#(t) = z^\#(t) + \int_0^t Z^\#(t - s) F^\#(ds).$$

The rest follows in a straightforward way by the key renewal theorem. ∎

Special Case 3.11.2. $F(0) < 1 < F(\infty)$. Note that $\hat{F}(0) = F(\infty) > 1$, $\hat{F}(-\infty) = F(0) < 1$ and $\hat{F}$ is continuous and monotone. Therefore, a

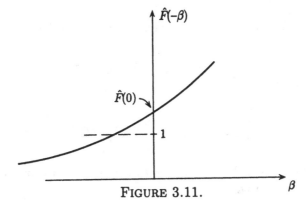

FIGURE 3.11.

solution β to $\hat{F}(-\beta) = 1$ always exists, is unique and $-\infty < \beta < 0$. See Figure 3.11. Furthermore,

$$\mu^{\#} := \int_0^\infty x F^{\#}(dx) = \int_0^\infty x e^{\beta x} F(dx) < \infty,$$

since, as $x \to \infty$, we have $x e^{\beta x} \to 0$ for $\beta < 0$, and therefore $x e^{\beta x}$ is bounded on $[0, \infty)$. If z is dRi then

$$\lim_{t \to \infty} Z^{\#}(t) = \lim_{t \to \infty} e^{\beta t} Z(t) = \int_0^\infty e^{\beta s} z(s) ds / \mu^{\#},$$

so that $Z(t) \sim e^{-\beta t} \int_0^\infty e^{\beta s} z(s) ds / \mu^{\#}$, and Z grows at an exponential rate.

Example 3.5.2 (continued). Consider a supercritical age dependent branching process with population size at time t given by $X(t)$. As usual, we assume $X(0) = 1$ and that G is the life length distribution. Let $\{p_k\}$ be the offspring distribution, with $m = \sum_{k=0}^\infty k p_k \in (1, \infty)$. Finally set $Z(t) = EX(t)$. We found in Example 3.5.2 that whatever the value of m ($m < \infty$, however)

$$Z(t) = 1 - G(t) + \int_0^t Z(t - u) m G(du).$$

Set $F(x) = mG(x)$ so $F(\infty) = m > 1$. We know $\beta < 0$ exists and is the unique solution of $\hat{F}(-\beta) = 1$. Let $\gamma = -\beta$, and

$$m \hat{G}(\gamma) = 1.$$

Observe $z^{\#}(t) = e^{-\gamma t}(1 - G(t))$ is dRi. The reason is that $z^{\#}$ is non-increasing and integrable,

$$\int_0^\infty z^{\#}(t) dt = \int_0^\infty e^{-\gamma t}(1 - G(t)) dt = \gamma^{-1}(1 - \hat{G}(\gamma)) < \infty,$$

so by criterion (3.10.1.3) $z^\#$ is dRi. Hence, by the Key Renewal Theorem,

$$\lim_{t\to\infty} Z^\#(t) = \int_0^\infty z^\#(s)ds/\mu^\#,$$

i.e., as $t \to \infty$,

$$Z(t) \sim e^{\gamma t}\frac{\int_0^\infty e^{-\gamma s}(1 - G(s))ds}{m\int_0^\infty se^{-\gamma s}G(ds)}.$$

This is the famous Malthusian law of exponential growth, and γ is called the *Malthusian parameter*.

Special Case 3.11.3. $F(\infty) < 1$. As discussed in the beginning of this section, if $z(\infty)$ exists, then $Z(t) = U * z(t) \to z(\infty)U(\infty)$, and sometimes the technique summarized in the proposition can be used to obtain a rate of convergence result. However, the existence of a solution to $\hat{F}(-\beta) = 1$ is no longer guaranteed, and, if a solution β does exist, we must have $\beta > 0$, since $\hat{F}(0) = F(\infty) < 1$, $\hat{F}(-\infty) = F(0)$. See Figure 3.12.

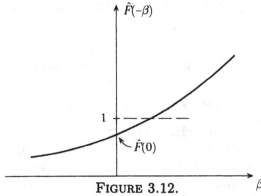

FIGURE 3.12.

Consider the following example: Define

$$F(x) = \begin{cases} 0, & \text{if } x < 2, \\ \frac{1}{2} - x^{-1}, & \text{if } x \geq 2. \end{cases}$$

For every $\beta > 0$

$$\hat{F}(-\beta) = \int_0^\infty e^{\beta t}F(dt) = \int_2^\infty e^{\beta t}t^{-2}dt = \infty,$$

so no solution exists for the equation $\hat{F}(-\beta) = 1$. For existence of a solution, we need

$$F(\infty) - F(t) \leq Ke^{-\delta t}, \quad \delta > 0, K > 0$$

for some $\delta > 0, K > 0$. Also sufficient is that

$$1 \leq \hat{F}(-\delta) = \int_0^\infty e^{\delta t} F(dt) < \infty$$

for some $\delta > 0$.

Now suppose that $F(\infty) < 1$, that there exists $\beta > 0$ such that $\hat{F}(-\beta) = 1$ and that $z(\infty)$ exists. Write $Z_1(t) = Z(\infty) - Z(t)$, where recall $Z(\infty) = z(\infty)U(\infty)$, and set

$$z_1(t) = z(\infty) - z(t) + z(\infty)\left(\frac{F(\infty) - F(t)}{1 - F(\infty)}\right).$$

Then, since $Z = z + F * Z$, it follows that

(3.11.2) $$Z_1 = z_1 + F * Z_1.$$

This can be checked directly (drudgery) or as follows. We seek a function z_1 which satisfies (3.11.2). Treat z_1 as the unknown. For $\lambda > 0$, set

$$\hat{Z}(\lambda) = \int_0^\infty e^{-\lambda s} Z(s)\,ds, \quad \hat{z}(\lambda) = \int_0^\infty e^{-\lambda s} z(s)\,ds,$$

and so on so that

$$\hat{Z} = \hat{z}\hat{U} = \hat{z}/(1 - \hat{F}).$$

Since

$$z_1 * U(t) = Z_1(t) = Z(\infty) - Z(t)$$
$$= z(\infty)U(\infty) - U * z(t),$$

we have

$$\hat{z}_1(\lambda)\hat{U}(\lambda) = z(\infty)U(\infty)\lambda^{-1} - \hat{z}(\lambda)\hat{U}(\lambda).$$

Upon solving, this yields

$$\hat{z}_1(\lambda) = \frac{z(\infty)U(\infty)\lambda^{-1}}{\hat{U}(\lambda)} - \hat{z}(\lambda)$$
$$= z(\infty)U(\infty)\lambda^{-1}(1 - \hat{F}(\lambda)) - \hat{z}(\lambda).$$

Use

$$(F(\infty) - \hat{F}(\lambda))/\lambda = \int_0^\infty e^{-\lambda x}(F(\infty) - F(x))\,dx$$

to get

$$\begin{aligned}
\hat{z}_1(\lambda) &= z(\infty)U(\infty)\lambda^{-1}[1 - F(\infty) + F(\infty) - \hat{F}(\lambda)] - \hat{z}(\lambda) \\
&= \lambda^{-1}z(\infty)U(\infty)(1 - F(\infty)) \\
&\quad + z(\infty)U(\infty)\int_0^\infty e^{-\lambda x}(F(\infty) - F(x))dx - \hat{z}(\lambda) \\
&= \lambda^{-1}z(\infty) + z(\infty)U(\infty)\int_0^\infty e^{-\lambda x}(F(\infty) - F(x))dx - \hat{z}(\lambda).
\end{aligned}$$

Inverting, we obtain

$$z_1(t) = z(\infty) - z(t) + z(\infty)\frac{F(\infty) - F(t)}{1 - F(\infty)},$$

as predicted.

If $e^{\beta t}z_1(t)$ is dRi then

$$\lim_{t\to\infty} e^{\beta t}(Z(\infty) - Z(t)) = \frac{\int_0^\infty e^{\beta s}z_1(s)ds}{\int_0^\infty se^{\beta s}F(ds)}.$$

To unwind the right side, we write

$$\int_0^\infty e^{\beta s}z_1(s)ds$$
$$= \int_0^\infty e^{\beta s}(z(\infty) - z(s))ds + z(\infty)U(\infty)\int_0^\infty e^{\beta s}(F(\infty) - F(s))ds.$$

For the last term note

$$\begin{aligned}
\int_0^\infty e^{\beta s}(F(\infty) - F(s))ds &= \int_0^\infty e^{\beta s}\int_s^\infty F(du)ds \\
&= \int_0^\infty \left(\int_0^u e^{\beta s}ds\right)F(du) \\
&= \int_0^\infty (e^{\beta s} - 1)\beta^{-1}F(du) \\
&= \hat{F}(-\beta) - F(\infty))/\beta = (1 - F(\infty))/\beta,
\end{aligned}$$

since $\hat{F}(-\beta) = 1$. Therefore, we conclude

$$Z(\infty) - Z(t) \sim e^{-\beta t}\frac{\int_0^\infty e^{\beta x}(z(\infty) - z(x))dx + z(\infty)\beta^{-1}}{\int_0^\infty xe^{\beta x}F(dx)}.$$

The Risk Process (continued). Recall the risk process: Non-negative, independent, identically distributed claims $X_1, X_2, \ldots$ arrive according to a Poisson process rate α. Between claims epochs, capital increases at rate c. Let $f(t)$ be the fortune (risk reserve) of the company at time t and set

$$R(x) = P[f(t) > 0 \text{ for all } t | f(0) = x].$$

In Section 3.5.1, we found that if $\alpha E X_1 < c$, then $R(\infty) = 1$, $R(0) = 1 - c^{-1}\alpha E X_1$ and $R(x) = R(0)U(x)$, where

$$U(x) = 1 + \int_0^x \sum_{n=1}^{\infty} g^{n*}(u)du$$

and

$$g(x) = \alpha c^{-1} P[X_1 > x].$$

The previous discussion provides a rate of convergence of $R(x) \to 1$. The parameter β must satisfy

$$1 = \int_0^{\infty} e^{\beta t}g(t)dt = \alpha c^{-1} \int_0^{\infty} P[X_1 > x]e^{\beta x}dx.$$

Since $R(0) = z(x)$, we have $z_1 \equiv 0$, and (3.11.3) becomes

$$1 - R(t) \sim \frac{e^{-\beta t}R(0)\beta^{-1}}{\alpha c^{-1} \int_0^{\infty} xe^{\beta x}P[X_1 > x]dx}$$

$$= \frac{e^{-\beta t}(1 - c^{-1}\alpha E X_1)}{\beta \alpha c^{-1} \int_0^{\infty} xe^{\beta x}P[X_1 > x]dx}.$$

3.12. MORE ON REGENERATIVE PROCESSES.

In this section we discuss more intensively and rigorously regenerative processes and develop more fully some examples from queueing theory.

3.12.1. DEFINITION AND EXAMPLES*.

Consider a stochastic process $\{X(t), t \geq 0\}$ for which there exists an epoch S_1 such that the process beyond S_1 is a probabilistic replica of the process starting from zero. Such processes can be decomposed into cycles and

---

* This section contains advanced material which may be skipped on first reading by beginning readers.

asymptotic behavior studied by means of renewal theory. There are several
proposed definitions which attempt to make this description precise. See
Cinlar (1975), Thorisson (1983), or Asmussen (1987). Suppose the state
space of the process is a Euclidean space E (R, R^k or even the integers)
with $\mathcal{E}$ being the usual σ-algebra of Borel sets. Suppose the process is
defined on a probability space $(\Omega, \mathcal{F}, P)$.

Definition. Let $\{S_n, n \geq 0\}$ be an increasing sequence of finite ran-
dom variables defined on $(\Omega, \mathcal{F})$ such that $0 \leq S_n \uparrow \infty$. Then a process
$\{X(t), t \geq 0\}$ is a regenerative process with regeneration times $\{S_n\}$ if for
every k, $0 \leq t_1 < \cdots < t_k, B \in \mathcal{B}(R_+^k), A \in \mathcal{B}(R_+^\infty)$, for all $n \geq 0$, we have

$$P[(X(S_n + t_i), i = 1, \ldots, k) \in B, \{S_{n+i} - S_n, i \geq 1\} \in A | S_0, \ldots, S_n]$$

(3.12.1.1) $$= P[(X(t_i), i = 1, \ldots, k) \in B, \{S_i - S_0, i \geq 1\} \in A].$$

It will be seen that this definition requires the process past S_n to be
independent of $S_0, \ldots, S_n$. This is a bit more flexible than other definitions
which require the process past S_n to be independent of all information up to
S_n. Also, note that although the index set is assumed to be the continuous
set $[0, \infty)$, obvious modifications would allow us to consider, for example,
regenerative processes with the index set $\{0, 1, \ldots\}$. Finally, the definition as stated requires the renewal process to be non-terminating. The
definition can be easily modified, however, to include terminating renewal
processes.

The following comments explain the definition further.

Remarks. (a) $\{S_n\}$ is a renewal process. Set $Y_0 = S_0, Y_k = S_k - S_{k-1}$,
$k \geq 1$. Use (3.12.1.1) with $B = R_+^k$. For $x_j \in R_+, j = 0, \ldots, i$, we have

$$P[Y_j \leq x_j, 0 \leq j \leq i] = EP[Y_j \leq x_j, 0 \leq j \leq i | S_0, \ldots, S_{i-1}]$$
$$= E1_{[\cap_{j=0}^{i-1} \{Y_j \leq x_j\}]} P[Y_i \leq x_i | S_0, \ldots, S_{i-1}].$$

Applying (3.12.1.1) we have $P[Y_i \leq x_i | S_0, \ldots, S_{i-1}] = P[Y_1 \leq x_i]$, so

$$P[Y_j \leq x_j, 0 \leq j \leq i] = P[Y_1 \leq x_i] P[Y_j \leq x_j, 0 \leq j \leq i - 1]$$

Continuing inductively, this equals

$$= P[Y_1 \leq x_i] P[Y_1 \leq x_{i-1}] \ldots P[Y_1 \leq x_2] P[Y_0 \leq x_0, Y_1 \leq x_1].$$

For the last factor, observe

$$P[Y_0 \leq x_0, Y_1 \leq x_1] = E1_{[Y_0 \leq x_0]} P[Y_1 \leq x_1 | S_0]$$
$$= P[Y_0 \leq x_0] P[Y_1 \leq x_1]$$

from (3.12.1.1).

(b) In (3.12.1.1) set $A = R_+^\infty$ and take expectations on both sides to obtain

$$(X(S_n + t_i), i \le k) \stackrel{d}{=} (X(t_i), i \ge k)$$

in E^k.

(c) The post-S_n process is independent of $\sigma(S_0, \ldots, S_n)$. To check this, take $G \in \sigma(S_0, \ldots, S_n)$. Then, using (3.12.1.1) with $A = R_+^\infty$, we have

$$P\{[(X(S_n + t_i), i \le k) \in B] \cap G]\}$$
$$= E1_G P\{[(X(S_n + t_i), i \le k) \in B] | S_0, \ldots, S_n]\}$$
$$= E1_G P\{(X(t_i), i \le k) \in B\}.$$

Applying (b) above, this is

$$= P(G) P[(X(S_n + t_i), i \le k) \in B].$$

(d) **Smith's construction of a regenerative process using random tours.** Suppose that $\{Y(t), t \ge 0\}$ is a stochastic process on E and that $Y \ge 0$ is R_+-valued and defined on the same probability space. Let

$$(\{Y_j(t), t \ge 0\}, Y_j)$$

be iid copies of $(\{Y(t), t \ge 0\}, Y)$. Define $S_0 = 0$, $S_n = \sum_{i=1}^n Y_i$, $n \ge 1$, and set

$$X(t) := \sum_{j=1}^\infty Y_j(t - S_{j-1}) 1_{[S_{j-1}, S_j)}(t), t \ge 0.$$

Then $X(\cdot)$ is regenerative according to the definition. To see this, note for any n and $t > 0$

$$X(t + S_n) = \sum_{j=1}^\infty Y_j(t + S_n - S_{j-1}) 1_{[S_{j-1}, S_j)}(t + S_n)$$

$$= \sum_{j=n+1}^\infty Y_j(t - (S_{j-1} - S_n)) 1_{[S_{j-1} - S_n, S_j - S_n)}(t)$$

$$= \sum_{j=1}^\infty Y_{j+n}(t - (S_{j+n-1} - S_n)) 1_{[S_{j+n-1} - S_n, S_{j+n} - S_n)}(t).$$

Set $Y_j^*(\cdot) = Y_{j+n}(\cdot)$, $S_j^* = S_{j+n} - S_n, j \ge 1$, and the above becomes

$$\sum_{j=1}^{\infty} Y_j^*(t - S_{j-1}^*) 1_{[S_{j-1}^*, S_j^*)}(t) = X^*(t).$$

Note $X^*(\cdot) \stackrel{d}{=} X(\cdot)$ and $X^*(\cdot)$ is independent of $\sigma(S_0, \ldots, S_n)$. Therefore

$$P[(X(S_n + t_i), i \leq k) \in B, (S_{n+k} - S_n, k \geq 1) \in A | S_0, \ldots, S_n]$$
$$= P[(X^*(t_i), i \leq k) \in B, (S_k^*, k \geq 1) \in A | S_0, \ldots, S_n]$$
$$= P[(X^*(t_i), i \leq k) \in B, (S_k^*, k \geq 1) \in A]$$
$$= P[(X(t_i), i \leq k) \in B, (S_k, k \geq 1) \in A],$$

and since $S_0 = 0$ this verifies (3.12.1.1).

Recall the following terminology: The epochs S_n are called the *regeneration points* for $X(\cdot)$, and the time intervals $[S_{n-1}, S_n)$, $n \geq 1$, are called *cycles*.

Recall also the examples of Section 3.7:

(1) A renewal process $\{S_n, n \geq 0\}$ is regenerative in discrete time. The regeneration epochs are $\{S_n\}$.

(2) An irreducible Markov chain $\{X_n\}$ in discrete time with the integers as state space is regenerative, and for any state j return times to j constitute regeneration epochs. If the process is recurrent, the renewal process is non-terminating. In Chapter 5 this example will be discussed in continuous time.

(3) Storage process: Inputs to a dam occur according to a compound Poisson process $A(t) = \sum_{i=1}^{N(t)} X_i$, where $\{X_i, i \geq 1\}$ are iid, non-negative random variables with common distribution G and $N(t), t \geq 0$, is a Poisson process with rate α and independent of $\{X_i\}$. Between inputs content decreases at rate c. The content process $X(t)$ satisfies

$$dX(t) = dA(t) - cdt,$$

i.e., change in content equals change in input minus change in output.

The content process is regenerative. One set of regeneration points is obtained by letting S_n be the beginning of the nth dry period. See Figure 3.13. Alternatively pick any level $x \geq 0$ and let S_n be the epoch of the nth downcrossing of level x by the process $X(t)$. See Figure 3.14.

(4) Markov processes on general state spaces may be regenerative if they have atoms, for example, singleton states that the process returns to infinitely often. This is the case in the second

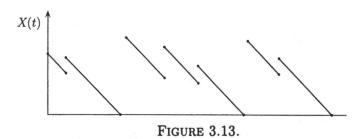

FIGURE 3.13.

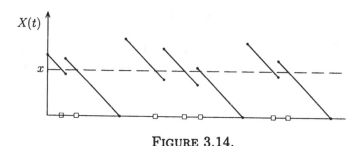

FIGURE 3.14.

and third examples above. Another example is the process
$\{B(t), t \geq 0\}$ of forward recurrence times in renewal theory
where $\{0\}$ is a convenient atom.

3.12.2. THE RENEWAL EQUATION AND SMITH'S THEOREM*.

The informal discussion of regenerative processes in Section 3.7.1 assumed
the state space was a subset of the integers. In this section, we allow more
general state spaces and consider carefully when Smith's theorem holds.

Fix a set $A \in \mathcal{E}$, and define

$$Z(t) = P[X(t) \in A].$$

Suppose for now $S_0 = 0$, and let

$$K(t, A) = P[X(t) \in A, S_1 > t],$$

so that this kernel describes the probability that at t we are in A and the
initial cycle has not ended. Then we have the decomposition

$$Z(t) = P[X(t) \in A, S_1 > t] + P[X(t) \in A, S_1 \leq t].$$

---

* This section contains advanced material which may be skipped on first
reading by beginning readers.

The first term is $K(t, A)$ and the second equals

$$E1_{[S_1 \le t]} P[X(S_1 + t - S_1) \in A|S_1].$$

The idea now is to move the time origin up to S_1, so that the above equals (with F being the distribution of S_1)

$$\int_0^t P[X(t - s) \in A]F(ds).$$

To make this precise, we use the following result from the theory of conditional expectation (Breiman, 1968): Suppose ξ, η are independent random elements in complete, separable metric spaces S, S'. If $\phi : S \times S' \mapsto R_+$ is measurable, then

$$E(\phi(\xi, \eta)|\eta = a) = E\phi(\xi, a) \text{ a.s.}$$

Now let $\xi = X(S_1 + \cdot)$ and $\eta = t - S_1$. Suppose that the paths of $X(\cdot)$ belong to a complete separable metric space S; then we define $\psi : S \times R_+ \mapsto E$ by

$$\psi(x(\cdot), s) = x(s).$$

Then

$$P[X(S_1 + t - S_1) \in A, S_1 \le t]$$
$$= \int_0^t P[X(S_1 + t - S_1) \in A|S_1 = s]F(ds)$$
$$= \int_0^t P[\psi(X(S_1 + \cdot), t - S_1) \in A|S_1 = s]F(ds)$$
$$= \int_0^t E(\phi(X(S_1 + \cdot), t - S_1)|S_1 = s)F(ds),$$

where

$$\phi(x(\cdot), s) = \begin{cases} 1, & \text{if } \psi(x(\cdot), s) \in A \\ 0, & \text{otherwise.} \end{cases}$$

By Remark (c) following (3.12.1.1), we have that $X(S_1 + \cdot)$ is independent of $t - S_1$, so the above integral is

$$\int_0^t E\phi(X(S_1 + \cdot), t - s)F(ds) = \int_0^t E\phi(X(\cdot), t - s)F(ds)$$

(by Remark (b) following (3.12.1.1)

$$= \int_0^t P[X(t - s) \in A]F(ds),$$

as required. Thus

$$
\begin{aligned}
P[S(t) \in A, S_1 \le t] &= EP[X(S_1 + t - S_1) \in A_1, S_1 \le t | \mathcal{F}_{S_1}) \\
&= E1_{S_1 \le t} P[X(S_1 + t - S_1) \in A | \mathcal{F}_{S_1}] \\
&= \int_0^t P[X(t - u) \in A] \cdot F(du)
\end{aligned}
$$

where $P[S_1 \le u] = F(u)$. Therefore,

(3.12.2.1) $$Z(t) = K(t, A) + \int_0^t Z(t - u) F(du),$$

and, solving, we find

(3.12.2.2) $$Z(t) = K(\cdot, A) * U(t).$$

Smith's Theorem 3.12.1. *Suppose $\{X(t)\}$ is a regenerative process with state space E and A is a measurable subset. For fixed A, assume $K(t, A)$ is Riemann integrable. Set $\mu = ES_1$, $S_0 = 0$.*
   a) If $\mu < \infty$, then

$$
\lim_{t \to \infty} P[X(t) \in A] = \mu^{-1} \int_0^\infty K(s, A) ds
$$

$$
= \mu^{-1} E \int_0^{S_1} 1_{[X(s) \in A]} ds
$$

$$
= \frac{E\{ \text{ occupation time in } A \text{ in the first cycle } \}}{E\{ \text{ cycle length } \}}.
$$

   b) If $\mu = \infty$ then
$$
\lim_{t \to \infty} P[X(t) \in A] = 0.
$$

Remarks. (i) Our convention regarding arithmetic distributions is still in force.

(ii) $K(t, A)$ need not be Riemann integrable. An example is given in Miller (1972): Let F put mass 2^{-n} at the point $n^{-1}, n = 1, 2, \ldots$. The regenerative process $X(t)$ is defined by

$$
X(t) = 1_Q(t),
$$

where Q is the set of positive rationals. The kernel K is

$$
K(t, \{1\}) = P[X(t) = 1, S_1 > t.]
$$

For $0 \le t \le 1$, this is $1_Q(t)F(t,1]$, which is not continuous a.e. and hence not Riemann integrable.

    (iii) Smith's theorem is easily seen to hold also in the case when the renewal process is delayed.

Proof. a) We check $K(t,A) =: z(t)$ is dRi. Note $K(t,A)$ is assumed Riemann integrable and

$$K(t,A) = P[X(t) \in A, S_1 > t] \le 1 - F(t),$$

so K is bounded above by $1 - F$, which is monotone and integrable (recall $\mu < \infty$). Therefore, from Remark 3.10.3, $1 - F$ is dRi, and, from Remark 3.10.5, K(t,A) is dRi. Applying the key renewal theorem yields

$$\lim_{t \to \infty} P[X(t) \in A] = \lim_{t \to \infty} Z(t) = \lim_{t \to \infty} K(\cdot, A) * U(t)$$

$$= \mu^{-1} \int_0^\infty K(s, A) ds$$

$$= \mu^{-1} \int_0^\infty E 1_{[X(s) \in A, S_1 > s]} ds$$

which, by Fubini's theorem, is

$$\mu^{-1} E \int_0^\infty 1_{[X(s) \in A, S_1 > s]} ds = \mu^{-1} E \int_0^{S_1} 1_A(X(s)) ds. \quad \blacksquare$$

Remark. For the process in Remark (ii) above, we have

$$P[X(t) = 1] = 1_Q(t),$$

which does not converge as $t \to \infty$.

    We now must consider when $K(t, A)$ is Riemann integrable. Since K is bounded, K is Riemann integrable iff K is continuous a.e., so it is enough to give conditions for the a.e. continuity of K.

    Say the process X has a *fixed discontinuity* at t if it is false that whenever $s_n \to t$, $X(s_n) \to X(t)$ almost surely (Doob 1953, p. 357). The extreme example of a fixed discontinuity is to take any random variable X and define

$$X(t) = X 1_{(1,\infty)}(t),$$

so that 1 is a fixed discontinuity.

    This concept must be distinguished from moving discontinuities such as possessed by the Poisson process $\{N(t)\}$. Each sample path consists of jumps, but

$$P[N(\cdot) \text{ jumps at } t] = P\{\cup_{n=0}^\infty [S_n = t]\} = 0,$$

so that if $s_n \to t$, $P[N(s_n) \to N(t)] = 1$.

Consider the function class $D[0, \infty)$ which is the class of functions $x : [0, \infty) \mapsto R$ such that x is right continuous on $[0, \infty)$ and has finite left limits on $(0, \infty)$. We use the following fact (Billingsley, 1968, p. 124). If $\{X(t), t \geq 0\}$ is a process with $P[X(\cdot) \in D[0, \infty)] = 1$, then X has at most a countable number of fixed discontinuities.

Proposition 3.12.2. *If $X(\cdot)$ is regenerative with at most a countable number of fixed discontinuities, then $K(t, A)$ is a.e. continuous for all open A.*

Corollary 3.12.3. *If $X(\cdot)$ has paths in $D[0, \infty)$ and A is open,*

$$\lim_{t \to \infty} P[X(t) \in A] = \mu^{-1} \int_0^\infty K(s, A)ds.$$

In particular, $X(t) \Rightarrow X(\infty)$ (convergence in distribution), where

$$P[X(\infty) \in A] = \mu^{-1} \int_0^\infty K(s, A)ds$$

for $A \in \mathcal{E}$.

Proof. Define the countable subset Disc of $[0, \infty)$ by

$$\text{Disc} = \{t : F \text{ discontinuous at } t\} \cup \{t : t \text{ is a fixed discontinuity of } X\}.$$

For open sets A, we show $K(\cdot, A)$ is continuous at $t \notin$ Disc and hence $K(\cdot, A)$ is continuous a.e. Fix $t \notin$ Disc. Define

$$\begin{aligned}\Lambda_s &= [S_1 > t, X(t) \in A] \cap [S_1 > s, X(s) \in A]^c \\ &= [S_1 > t, X(t) \in A, S_1 \leq s] \cup [S_1 > t, X(t) \in A, X(s) \notin A].\end{aligned}$$

Hence
(3.12.2.3)
$$P\Lambda_s \leq P[S_1 > t, X(t) \in A, S_1 \leq s] + P[S_1 > t, X(t) \in A, X(s) \notin A].$$

The first term is bounded above by

$$P[t \wedge s < S_1 \leq t \vee s] \to 0$$

as $s \to t$ since F is continuous at t. The second term of (3.12.2.3) is bounded above by $P[X(t) \in A, X(s) \notin A]$. Since t is not a fixed discontinuity of X, we have $X(s) \to X(t)$ as $s \to t$. Since A is open, if $X(t) \in A$, then ultimately we must have $X(s) \in A$ for s sufficiently close to t.

Hence the above converges to zero as well. A similar result is obtained by interchanging t and s, and this leads to

$$P\{[S_1 > t, X(t) \in A]\triangle[S_1 > s, X(s) \in A]\} \to 0,$$

as $s \to t$, which yields

$$K(t, A) - K(s, A) \to 0,$$

and K is continuous at $t \notin \text{Disc.}$ ∎

We end this section with a little result useful for some of the queueing examples.

Proposition 3.12.4. *Suppose* $X(t), t \geq 0$ *is regenerative and* $X(t) \Rightarrow X(\infty)$, *where*

$$(3.12.2.4) \qquad P[X(\infty) \in A] = \mu^{-1} E \int_0^{S_1} 1_A(X(s))ds, \quad \mu < \infty.$$

For any $h \geq 0$, *measurable, and* $h : E \mapsto [0, \infty)$, *we have*

$$(3.12.2.5) \qquad Eh(X(\infty)) = \mu^{-1} E \int_0^{S_1} h(X(s))ds.$$

In particular, if $E = [0, \infty)$ *and* $h(x) = e^{-sx}$, $s > 0$, $x \in E$, *then the Laplace transform of* $X(\infty)$ *satisfies*

$$E \exp\{-\zeta X(\infty)\} = \mu^{-1} E \int_0^{S_1} \exp\{-\zeta X(s)\}ds.$$

Proof. If $h(x) = 1_A(x)$, $A \in \mathcal{E}$, then (3.12.2.5) reduces to (3.12.2.4). Next, suppose $h(x) = \sum_{i=1}^k a_i 1_{A_i}(x)$ where $A_i \in \mathcal{E}$ are disjoint for $i = 1, 2, \ldots, k$ and $a_i \geq 0$. Then

$$Eh(X(\infty)) = \sum_1^k a_i E 1_{A_i}(X_\infty))$$

$$= \sum_1^k a_i P[X(\infty) \in A_i]$$

$$= \sum_1^k a_i \mu^{-1} E \int_0^{S_1} 1_{A_i}(X(s))ds$$

$$= \mu^{-1} E \int_0^{S_1} h(X(s))ds.$$

Finally, if $h \geq 0$ is measurable, there exist simple $h_n \geq 0$ of the form considered in the previous step, and $h_n \uparrow h$ so that

$$Eh(X(\infty)) = \lim_{n \to \infty} \uparrow Eh_n(X(\infty))$$

(monotone convergence)

$$= \lim_{n \to \infty} \mu^{-1} E \int_0^{S_1} h_n(X(s))ds$$

(previous step)

$$= \mu^{-1} E \lim_{n \to \infty} \int_0^{S_1} = \mu^{-1} E \int_0^{S_1} \lim_{n \to \infty} h_n(X(s))ds$$

$$= \mu^{-1} E \int_0^{S_1} h(X(s))ds. \ \blacksquare$$

3.12.3. Queueing Examples.

In what follows we describe (rather than construct) the processes used to study the G/G/1 queue. The symbols stand for general input (arrivals occur according to a renewal process), general service times (service times of successive customers are iid) and one server. An exhaustive discussion of regenerative processes and queueing models is given in Cohen (1976).

In a general one server queue, suppose customer number 0 arrives at time 0. Let σ_{n+1} be the interarrival time between the nth and the $(n+1)$st arriving customer. Assume $\sigma_n, n \geq 1$ are iid with common distribution $A(x) = P[\sigma_1 \leq x]$; set $a^{-1} = E\sigma_1 < \infty$ so that customers arrive at rate a. Let t_k be the time of arrival of customer $k, k \geq 0$, so that $t_0 = 0$, $t_k = \sigma_1 + \cdots + \sigma_k, k \geq 1$. Further define the renewal counting function ν by

$$\nu(t) = \# \text{ of arrivals in } [0, t]$$

$$= \sum_{k=0}^{\infty} 1_{[t_k \leq t]}.$$

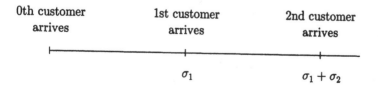

| 0th customer arrives | 1st customer arrives | 2nd customer arrives |

σ_1 $\sigma_1 + \sigma_2$

Arrivals

Focus now on service. Let τ_n be the service time of the nth customer, and suppose $\{\tau_n, n \geq 0\}$ is iid with common distribution $B(x) = P[\tau_0 \leq x]$. Set $b^{-1} = E\tau_0 < \infty$ so that customers are served at rate b. Define the traffic intensity ρ by

$$(3.12.3.1) \qquad \rho = a/b = E\tau_0/E\sigma_1 = (E\sigma_1)^{-1}/(E\tau_0)^{-1},$$

so that ρ is the ratio of the arrival rate to the service rate. If $\rho < 1$, then, on the average, the server is able to cope with his load. Assume $\{\tau_n\}$ and $\{\sigma_n\}$ are independent. (Sometimes it suffices that $\{(\tau_n, \sigma_{n+1}), n \geq 0\}$ be iid.)

We assume that there is one server who serves customers on a first come first served basis. A basic process of interest is W_n, the waiting time of the nth customer. This is the elapsed time between the arrival of the nth customer and the beginning of his service period. A basic recursion for W_n is

$$W_0 = 0, \quad W_{n+1} = (W_n + \tau_n - \sigma_{n+1})^+, \quad n \geq 0,$$

where $x^+ = x$ if $x \geq 0$, and $X^+ = 0$ if $x < 0$. To check this recursion note that there are two possible pictures. For the first, $W_n + \tau_n > \sigma_{n+1}$. The darkened region in Figure 3.15 represents the wait of the $(n + 1)$st customer. The second picture (Figure 3.16) is when $W_n + \tau_n \leq \sigma_{n+1}$. In this case, the $(n + 1)$st customer enters service immediately upon arrival and has no wait.

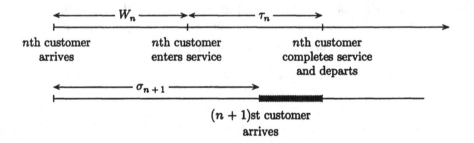

FIGURE 3.15. CUSTOMER WAITS.

For $n \geq 0$ define

$$(3.12.3.2) \qquad X_{n+1} = \tau_n - \sigma_{n+1},$$

so that $\{X_n, n \geq 1\}$ is iid. With this notation,

$$W_{n+1} = (W_n + X_{n+1})^+.$$

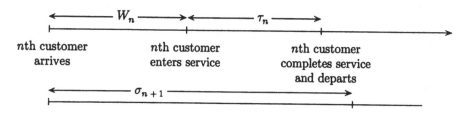

FIGURE 3.16. CUSTOMER DOES NOT WAIT.

It is evident that $\{W_n\}$ is a random walk with a boundary at 0, that is, a sum of iid random variables, which is prevented from going negative. (See also Chapter 7.)

To help us organize information, define the σ-algebras

$$\mathcal{F}_n = \sigma(\tau_0, \ldots, \tau_{n-1}, \sigma_1, \ldots, \sigma_n), \quad n \geq 0,$$

so $\mathcal{F}_n$ is the information obtained by watching until the nth customer arrrives or the $(n-1)$st customer departs—whichever is later. Let $\overline{N}$ be the number of customers served in the initial busy period. The initial busy period starts at time 0 and continues until the server becomes free for the first time and there is nobody waiting for service. Note

$$\overline{N} = \inf\{n \geq 1 : W_n = 0\},$$

since, if $W_{\overline{N}} = 0$, the busy period ended in $(t_{\overline{N}-1}, t_{\overline{N}}]$, and, counting customer number 0, there were $\overline{N}$ served in the initial busy period. Observe too that

$$\overline{N} = \inf\{n \geq 1 : \sum_{i=1}^{n} X_i \leq 0\};$$

this shows why random walk theory, to be explored later in Chapter 7, is relevant. Finally, $\overline{N}$ is a stopping time for $\{\mathcal{F}_n, n \geq 1\}$ since

$$[\overline{N} = n] = [W_1 > 0, \ldots, W_{n-1} > 0, W_n = 0]$$

$$= [\sum_{i=1}^{k} X_i > 0, k \leq n - 1, \sum_{i=1}^{n-1} X_i \leq 0]$$

and $X_n = \tau_{n-1} - \sigma_n$. This merely says the event $[N = n]$ depends only on $\tau_0, \ldots, \tau_{n-1}, \sigma_1, \ldots, \sigma_n$.

Now let C_1 be the length of the initial busy cycle, that is, the combination of the initial busy period and the server's idle period which follows.

Let $\{C_n, n \geq 1\}$ be successive busy cycle lengths, and suppose $\overline{N}_n$ is the number served in the nth busy period. Then $\{C_n, n \geq 1\}$ and $\{\overline{N}_n, n \geq 1\}$ are iid sequences. Note that

$$(3.12.3.3) \qquad\qquad C_1 = \sum_{i=1}^{\overline{N}} \sigma_i = t_{\overline{N}},$$

the initial busy period BP_1 satisfies

$$(3.12.3.4) \qquad\qquad BP_1 = \sum_{i=0}^{\overline{N}-1} \tau_i,$$

and the length of the initial idle period I_1 is

$$I_1 = C_1 - BP_1.$$

To verify that $\{C_n\}$ and $\{\overline{N}_n\}$ are each iid sequences, see that

$$\{(\tau_{\overline{N}-1+k}, \sigma_{\overline{N}+k}), k \geq 1\}$$

is iid, independent of the pre-$\overline{N}$ field and identically distributed as $\{(\tau_{k-1}, \sigma_k), k \geq 1\}$ (cf. Section 1.8.2 and the material on random walks in Chapter 7).

If $\rho < 1$ then $EC_1 < \infty$, $E\overline{N} < \infty$, and the system is stable. This will be proved in Chapter 7 in connection with random walk theory. Note $1 > \rho = a/b$ means $b > a$ and $EX_1 = E\tau_0 - E\sigma_1 = b^{-1} - a^{-1} < 0$ and the drift of the associated random walk $\{\sum_{i=1}^n X_i\}$ is negative. Since $\overline{N}$ is a stopping time, Wald's equation (Section 1.8.1) yields

$$(3.12.3.5) \qquad\qquad EC_1 = E\sum_{i=1}^{\overline{N}} \sigma_i = a^{-1}E\overline{N},$$

so $EC_1 < \infty$ iff $E\overline{N} < \infty$. Henceforth we always assume $\rho < 1$.

Aside from $\{W_n\}$, the following are two processes of interest: $Q(t)$, the number of customers in the system (waiting and in service) at time t, and $V(t)$, the virtual waiting time at t, that is, the wait of a fictitious customer who arrives at time t. $V(t)$ may also be interpreted as the work load of the server at time t.

We now observe that $\{Q(t), V(t), t \geq 0\}$ is regenerative on R_+^2 with regenerations at $\{\sum_{i=1}^n C_i, n \geq 1\}$, and $\{W_n, n \geq 0\}$ is regenerative with regenerations at $\{\overline{N}_k, k \geq 1\}$. So if C_1 has a non-arithmetic distribution (for instance, if σ_1 has a non-arithmetic distribution), then

$$(3.12.3.6) \qquad\qquad (V(t), Q(t)) \Rightarrow (V(\infty), Q(\infty))$$

in R_+^2, where

$$P[V(\infty) \le v, Q(\infty) = n] = (EC_1)^{-1} E \int_0^{C_1} 1_{[V(s) \le v, Q(s)=n]} ds,$$

since the paths of $(V(\cdot), Q(\cdot))$ are in $D[0, \infty)$. Further, provided $\overline{N}_1$ is a non-arithmetic integer valued random variable, we have

(3.12.3.7) $$W_n \Rightarrow W_\infty$$

and

$$P[W_\infty \le w] = (E\overline{N}_1)^{-1} E \left(\sum_{i=0}^{\overline{N}_1 - 1} 1_{[W_i \le w]} \right).$$

A simple corollary of (3.12.3.7) is the following: Let

$$S^{(n)} = W_n + \tau_n$$

represent the total time in the system of the nth customer. Then

(3.12.3.8) $$S^{(n)} \Rightarrow S^{(\infty)},$$

where

$$S^{(\infty)} \overset{d}{=} W_\infty + \tau_\infty$$

and where (W_∞, τ_∞) are independent, with W_∞ specified as in (3.12.3.6) and τ_∞ has distribution B. The proof is easy: Since W_n and τ_n are independent (remember $W_n \in \mathcal{F}_n$) and $\{\tau_n, n \ge 0\}$ are iid, we have

$$P[W_n \le x, \tau_n \le y] = P[W_n \le x]B(y) \to P[W_\infty \le x]B(y),$$

and therefore

$$(W_n, \tau_n) \Rightarrow (W_\infty, \tau_\infty)$$

in R_+^2. The continuous mapping theorem implies that

$$W_n + Z_n \Rightarrow W_\infty + \tau_\infty.$$

The above are existence results. The limit distribution is shown to exist, but little information is given about explicit forms. Several qualitative conclusions emerge, however.

Little's Formula. This is a balance expression which does a bit of accounting to equate certain moments of asymptotic distributions. Observe first that

(3.12.3.9) $$\int_0^{C_1} Q(s)ds = \sum_{j=0}^{\overline{N}_1 - 1} S^{(j)},$$

since

$$\int_0^{C_1} Q(s)ds = \int_0^{C_1} \sum_{j=0}^{\overline{N}_1-1} 1_{[\text{ customer } j \text{ is present in the system at time } s]}ds$$

$$= \sum_{j=0}^{\overline{N}_1-1} \int_0^{C_1} 1_{[\text{ customer } j \text{ is present in the system at time } s]}ds$$

$$= \sum_{j=0}^{\overline{N}_1-1} (\text{ departure time } - \text{ arrival time of } j\text{th customer })$$

$$= \sum_{j=0}^{\overline{N}_1-1} S^{(j)}.$$

We now apply Proposition 3.12.4. Write

$$EQ(\infty) = (EC_1)^{-1} E \int_0^{C_1} Q(s)ds = (EC_1)^{-1} E \sum_{j=0}^{\overline{N}_1-1} S^{(j)}.$$

Since $S^{(j)} = W_j + \tau_j$,

$$E\left(\sum_{j=1}^{\overline{N}_1-1} S^{(j)}\right) = E\left(\sum_{j=1}^{\overline{N}_1-1} W_j\right) + E\left(\sum_{j=0}^{\overline{N}_1-1} \tau_j\right).$$

Applying Proposition 3.12.4 again, (3.12.3.7) gives

$$(E\overline{N}_1)^{-1} E \sum_{j=1}^{\overline{N}_1-1} W_j = EW_\infty.$$

Since $\overline{N}_1$ is a stopping time for $\{\mathcal{F}_n\}$ and $\tau_{j+1} \in \mathcal{F}_{j+1}$, Wald's identity yields

$$E\left(\sum_{j=0}^{\overline{N}_j-1} \tau_j\right) = E(\overline{N}_1)E(\tau_1) = b^{-1}E(\overline{N}_1).$$

Putting the pieces together, we get

$$EQ(\infty) = (E\overline{N}_1 EW_\infty + b^{-1}E\overline{N}_1)/EC_1$$
$$= E\overline{N}_1(EW_\infty + E\tau_\infty)/EC_1$$
$$= \frac{E\overline{N}_1 ES^{(\infty)}}{EC_1}.$$

Applying (3.12.3.5), we obtain Little's formula

(3.12.3.10)
$$EQ(\infty) = aES^{(\infty)},$$

which says for the limit distributions that the expected number in system is equal to the arrival rate times the average sojourn in system.

On the relation between $P[W_\infty \le w]$ **and** $P[V(\infty) \le v]$. (See Lemoine, 1974 and also Takacs, 1962.) Define

(3.12.3.11)
$$V(x) = P[V(\infty) \le x], \quad W(t) = P[W_\infty \le t]$$

and

$$B_0(t) = b \int_0^t (1 - B(s))ds.$$

For $x \ge 0$, we prove

(3.12.3.12)
$$V(x) = 1 - \rho + \rho W * B_0(x),$$

where $\rho = a/b$ is the traffic intensity. (Note the letter V is being used in two ways. Distinguish between $V(\infty)$, the random variable, and $V(x)$, the distribution of the random variable $V(\infty)$.)

Let us proceed using Laplace transforms. In transform notation (3.12.3.12) is ($\zeta > 0$)

(3.12.3.13)
$$\hat{V}(\zeta) = 1 - \rho + \rho \hat{W}(\zeta)(1 - \hat{B}(\zeta))b\zeta^{-1}.$$

Now write

$$\hat{V}(\zeta) = Ee^{-\zeta V(\infty)}.$$

From Proposition 3.12.4 this is

$$(EC_1)^{-1} E \int_0^{C_1} e^{-\zeta V(s)} ds.$$

During the idle period $[BP_1, C_1]$ we have $V(s) = 0$, so the above is

(3.12.3.14)
$$(EC_1)^{-1} \left(E \int_0^{BP_1} e^{-\zeta)v(s)} ds + E \int_{BP_1}^{C_1} 1 ds \right).$$

Now

$$(EC_1)^{-1} E \int_{BP_1}^{C_1} ds = (EC_1 - EBP_1)/EC_1 = 1 - EBP_1/EC_1$$

$$= 1 - \frac{E\left(\sum_{i=0}^{\overline{N}_1 - 1} \tau_i\right)}{E\left(\sum_{i=1}^{\overline{N}_1} \sigma_i\right)},$$

from (3.12.3.3), (3.12.3.4); applying the Wald identity twice gives

$$1 - E\overline{N}_1 b^{-1})/(E\overline{N}_1 a^{-1}) = 1 - \rho.$$

We now focus on the first piece of (3.12.3.14) and write

$$\int_0^{BP_1} e^{-\zeta V(s)} ds = \sum_{k=1}^{\overline{N}_1-1} \int_{t_{k-1}}^{t_k} + \int_{t_{\overline{N}_1-1}}^{BP_1}.$$

On $[t_{k-1}, t_k)$ we have

$$V(s) = W_{k-1} + \tau_{k-1} - (s - t_{k-1}),$$

and on $[t_{\overline{N}_1-1}, BP_1)$ we have

$$V(s) = W_{\overline{N}_1-1} + \tau_{\overline{N}_1-1} - (s - t_{\overline{N}_1-1}).$$

Changing variables gives

$$\int_0^{BP_1} e^{-\zeta V(s)} ds = \sum_{k=1}^{\overline{N}_1-1} \int_0^{\sigma_k} e^{-\zeta(W_{k-1}+\tau_{k-1}-s)} ds$$

$$+ \int_0^{BP_1-t_{\overline{N}_1-1}} e^{-\zeta(W_{\overline{N}_1-1}+\tau_{\overline{N}_1-1}-s)} ds$$

$$= \sum_{k=1}^{\overline{N}_1-1} e^{-\zeta(W_{k-1}+\tau_{k-1})}(e^{\zeta\sigma_k} - 1)\zeta^{-1}$$

(3.12.3.15)

$$+ e^{-\zeta(W_{\overline{N}_1-1}+\tau_{\overline{N}_1-1})}(e^{\zeta(BP_1-t_{\overline{N}_1-1})} - 1)\zeta^{-1}.$$

We now make two statements.
   (a) Observe that

$$W_{\overline{N}_1-1} + \tau_{\overline{N}_1-1} - (BP_1 - t_{\overline{N}_1-1}) = 0,$$

since $t_{\overline{N}_1-1} + W_{\overline{N}_1-1} + \tau_{\overline{N}_1-1}$ adds together for the last customer in the first cycle the arrival time, waiting time, and service time. The sum of these three is the departure time of the last customer and hence signals the end of the busy period. Thus the last bit of (3.3.12.15) equals

$$(1 - \exp\{-\zeta(W_{\overline{N}_1-1} + \tau_{\overline{N}_1-1})\})\zeta^{-1}.$$

(b) For $1 \leq k \leq \overline{N}_1 - 1$

$$W_{k-1} + \tau_{k-1} - \sigma_k = W_k$$

and the first sum in (3.12.3.15)) is

$$\zeta^{-1} \sum_{k=1}^{\overline{N}_1 - 1} (e^{-\zeta W_k} - e^{-\zeta(W_{k-1} + \tau_{k-1})}).$$

Combining the pieces of (3.12.3.15) and remembering that $W_0 = 0$ we get

$$\int_0^{BP_1} e^{-\zeta V(s)} ds = \zeta^{-1} \sum_{k=1}^{\overline{N}_1 - 1} (e^{-\zeta W_k} - e^{-\zeta(W_{k-1} + \tau_{k-1})})$$

$$+ \zeta^{-1}(1 - e^{-\zeta(W_{\overline{N}_1 - 1} + \tau_{\overline{N}_1 - 1})})$$

$$= \zeta^{-1} \left(\sum_{k=0}^{\overline{N}_1 - 1} e^{-\zeta W_k} - \sum_{k=0}^{\overline{N}_1 - 1} e^{-\zeta(W_k + \tau_k)} \right)$$

$$= \zeta^{-1} \sum_{k=0}^{\overline{N}_1 - 1} e^{-\zeta W_k} (1 - e^{-\zeta \tau_k}),$$

so

$$E \int_0^{BP_1} e^{-\zeta V(s)} ds = \zeta^{-1} E \sum_{k=0}^{\overline{N}_1 - 1} e^{-\zeta W_k} (1 - e^{-\zeta \tau_k})$$

$$= \zeta^{-1} E \sum_{k=0}^{\infty} e^{-\zeta W_k} (1 - e^{-\zeta \tau_k}) 1_{[k \leq \overline{N}_1 - 1]}.$$

Now

$$[\overline{N}_1 - 1 \geq k] = [\overline{N} \geq k + 1] = [\overline{N} > k]$$

$$= [N \leq k]^c \in \mathcal{F}_k = \sigma(\tau_0, \ldots, \tau_{k-1}; \sigma_1, \ldots, \sigma_k).$$

Also $W_k \in \mathcal{F}_k$. Since τ_k is independent of $\mathcal{F}_k$, we have the above equal to

$$\zeta^{-1} \sum_{k=0}^{\infty} E \left(e^{-\zeta W_k} 1_{[k \leq \overline{N}_1 - 1]} \right) (1 - e^{-\zeta \tau_k})$$

$$= \zeta^{-1} \sum_{k=0}^{\infty} E e^{-\zeta W_k} 1_{[k \leq \overline{N}_1 - 1]} E(1 - e^{-\zeta \tau_k})$$

$$= E \left(\sum_{k=0}^{\overline{N}_1 - 1} e^{-\zeta W_k} \right) (1 - \hat{B}(\zeta)) \zeta^{-1}$$

$$= E \overline{N}_1 E e^{-\zeta W_\infty} (1 - \hat{B}(\zeta)) \zeta^{-1}$$

(from Proposition 1.8.1). Finally, we have

$$\begin{aligned}
\hat{V}(\zeta) &= (EC_1)^{-1}\{E\overline{N}_1 Ee^{-\zeta W_\infty}(1 - \hat{B}(\zeta))\zeta^{-1}\} + 1 - \rho \\
&= (a^{-1}E\overline{N}_1)^{-1}\{E\overline{N}_1 Ee^{-\zeta W(\infty)}(1 - \hat{B}(\zeta))\zeta^{-1}\} + 1 - \rho \\
&= ab^{-1}\{Ee^{-\zeta W_\infty}(1 - \hat{B}(\zeta))b\zeta^{-1}\} + 1 - \rho \\
&= \rho Ee^{-\zeta W_\infty}\hat{B}_0(\zeta) + 1 - \rho,
\end{aligned}$$

as desired.

A closing remark: Let $x \to 0$ in (3.12.3.12). Then

$$\begin{aligned}
V(0) &= P[V(\infty) = 0] = 1 - \rho + \rho W * B_0(0) \\
&= 1 - \rho.
\end{aligned}$$

So $1 - \rho$ is a measure of the fraction of time the server is idle.

The M/G/1 Queue. The "M" stands for Markovian, which means the input stream to the system is a Poisson process and $P[\sigma_1 > x] = e^{-ax}$. For such a distribution, $P[\sigma_1 > x] = a^{-1}P[\sigma_1 \in dx]$. We now check

(3.12.3.16) $$V(\infty) \stackrel{d}{=} W_\infty.$$

We have that

$$\begin{aligned}
P[V(\infty) \le v] &= (EC_1)^{-1}E\int_0^{C_1} 1_{[V(s)\le v]}ds \\
&= (EC_1)^{-1}E\sum_{k=1}^{\overline{N}_1}\int_{t_{k-1}}^{t_k} 1_{[V(s)\le v]}ds.
\end{aligned}$$

As we have done previously, for $s \in [t_{k-1}, t_k)$, we have

$$V(s) = (W_{k-1} + \tau_{k-1} - (s - t_{k-1}))^+,$$

so

$$\begin{aligned}
\int_{t_{k-1}}^{t_k} 1_{[V(s)\le v]}ds &= \int_{t_{k-1}}^{t_k} 1_{[(W_{k-1}+\tau_{k-1}-(s-t_{k-1}))^+\le v]}ds \\
&= \int_0^{\sigma_k} 1_{[(W_{k-1}+\tau_{k-1}-s)^+\le v]}ds.
\end{aligned}$$

Thus

$$P[V(\infty) \le v] = (EC_1)^{-1} E \sum_{k=1}^{\overline{N}_1} \int_0^{\sigma_k} 1_{[(W_{k-1}+\tau_{k-1}-s)^+ \le v]} ds$$

$$= (EC_1)^{-1} E \sum_{k=1}^{\infty} \int_0^{\infty} 1_{[(W_{k-1}+\tau_{k-1}-s)^+ \le v]} 1_{[k \le \overline{N}_1]} 1_{[s \le \sigma_k]} ds$$

(3.12.3.17)

$$= (EC_1)^{-1} \sum_{k=1}^{\infty} \int_0^{\infty} E 1_{[(W_{k-1}+\tau_{k-1}-s)^+ \le v]} 1_{[k \le \overline{N}_1]} 1_{[s \le \sigma_k]} ds,$$

by Fubini's theorem. Now $[k \le \overline{N}_1] = [\overline{N}_1 \le k-1]^c \in \mathcal{F}_{k-1} = \sigma(\tau_0,\ldots,\tau_{k-2},\sigma_1,\ldots,\sigma_{k-1})$, and also $W_{k-1} \in \mathcal{F}_{k-1}$. Since τ_{k-1} is assumed independent of σ_k, we have σ_k independent of

$$[k \le \overline{N}_1] \cap [(W_{k-1}+\tau_{k-1}-s)^+ \le v],$$

and (3.12.3.17) becomes

$$(EC_1)^{-1} \sum_{k=1}^{\infty} \int_0^{\infty} E 1_{[(W_{k-1}+\tau_{k-1}-s)^+ \le v]} 1_{[k \le \overline{N}_1]} P[\sigma_k > s] ds.$$

Because σ_k has an exponential density, this is

$$(EC_1)^{-1} \sum_{k=1}^{\infty} \int_0^{\infty} 1_{[(W_{k-1}+\tau_{k-1}-s)^+ \le v]} 1_{[k \le \overline{N}_1]} a^{-1} P[\sigma_k \in ds],$$

and, again using the independence of σ_k from

$$[\overline{N}_1 \ge k] \cap [(W_{k-1}+\tau_{k-1}-s)^+ \le v],$$

we get that the above equals

$$a^{-1}(EC_1)^{-1} \sum_{k=1}^{\infty} E 1_{[(W_{k-1}+\tau_{k-1}-\sigma_k)^+ \le v]} 1_{[k \le \overline{N}_1]}$$

$$= a^{-1}(EC_1)^{-1} \sum_{k=1}^{\infty} E 1_{[W_k \le v]} 1_{[k \le \overline{N}_1]}$$

(since $W_k = (W_{k-1}+\tau_{k-1}-\sigma_k)^+$)

$$= a^{-1}(EC_1)^{-1} E \sum_{k=1}^{\overline{N}_1} 1_{[W_k \le v]}.$$

$$= a^{-1}(EC_1)^{-1} E \sum_{k=1}^{\overline{N_1}} 1_{[W_k \le v]}.$$

Now $W_0 = 0 = W_{\overline{N_1}}$, so $1_{[W_0 \le v]} = 1_{[W_{\overline{N_1}} \le v]}$, and thus

$$P[V(\infty) \le v] = a^{-1}(EC_1)^{-1} E \sum_{k=0}^{\overline{N_1}-1} 1_{[W_k \le v]}$$

$$= a^{-1}(E\overline{N_1})(EC_1)^{-1} P[W_\infty \le v]$$

(from Proposition 3.12.2.4)

$$= P[W_\infty \le v],$$

as asserted.

Combining this result with (3.12.3.12) allows us to compute V or W for the M/G/1 queue. From (3.12.3.12),

$$V(t) = 1 - \rho + \rho V * B_0(t).$$

This is a renewal equation with $z(t) \equiv 1 - \rho$, $Z(t) = V(t)$, $F = \rho B_0$ and hence the solution, $U * z$, is

$$V(t) = (1 - \rho) \sum_0^\infty \rho^n B_0^{n*}(t).$$

EXERCISES

3.1. Reliability of the Power Supply. The nuclear reactor supplying power to Harry's restaurant has a certain component with lifetime distribution $F_1(x)$. If the component is operative, the system is in state 1. Upon failure, there is probability p that this component is instantaneously replaced by an attentive maintenance man $(0 < p < 1)$. If not, the system passes into state 2. After a random amount of time with distribution $F_2(x)$, a fail-safe mechanism shuts down the reactor. The system is shut down (state 3) for a random amount of time with distribution $F_3(x)$. Then it passes back into state 1 and the reactor operation proceeds as before. Let $X(t)$ be the state of the system at time t. Distinguish a set of regeneration points, and use Smith's theorem to compute

$$\lim_{t \to \infty} P[X(t) = k], \quad \text{for } k = 1, 2, 3.$$

Assume $\mu_i = \int_0^\infty x dF_i(x) < \infty$ for $i = 1, 2, 3$.

3.2. Harry Copes with the Young Republicans; Counter Models. Members of the Young Democrats (YD) arrive at Harry's restaurant according to a Poisson process rate α. However whenever a Young Republican (YR) spots a Young Democrat (YD) entering the restaurant, the YR's get mad at Harry and declare a boycott of the restaurant during which time no YR's enter. The boycott lasts a random amount of time with distribution G. Boycott periods are iid with the condition that if a YD arrives while a boycott is in effect, a new boycott of random duration G is immediately initiated, regardless of how long the previous boycott had been in place. Suppose a YD arrives at time 0 and L is the total duration of the initial boycott.

(a) Write a renewal equation for

$$Z(t) = P[L > t].$$

(b) Find the Laplace transform $\phi(\lambda)$ for the distribution of L in terms of the Laplace transform of G and α.

(c) Find $E(L)$ in terms of $g = \int_0^\infty x dG(x)$ and α.

(d) Now suppose that a YD arriving during a boycott has no effect on the boycott. It is as if a YD arriving during a boycott is unobserved by the YR's. Show that arrival times of YD's when no boycott is in progress form a renewal process and find the interarrival distribution.

(e) Suppose in (d) that the boycott duration distribution G is known but the rate α is not known. Discuss how one would naively estimate α based on the number of boycott initiated by time t and find some properties of the estimator as $t \to \infty$.

(f) If the renewal equation written in (a) is of the form

$$Z = z + Z * F,$$

show $F(\infty) < 1$. When does there exist β such that

$$1 = \int_0^\infty e^{\beta x} dF(x).$$

When such a β exists, use it to derive the exponential decay of $P[L > t]$.

(g) Now suppose that Young Democrats arrive according to a renewal process with interarrival distribution F. Suppose there is an arrival at time 0 and that the boycott periods are for a fixed duration d. During boycott periods, arrivals have no effect on boycotts. Show that the distribution of the length of time between when one boycott ends and the next begins is

$$\int_0^d (F(d + t - y) - F(d - y))U(dy).$$

What is this distribution in the case F is exponential? In the exponential case, what is the probability that at time t there is no boycott?

(This model can also be used for counters of radioactive particles. Arriving particles block the counter. The scheme in (d) corresponds to a *type I* counter and the scheme in (a), (b), (c) corresponds to a *type II* counter.)

3.3. Find the renewal function corresponding to

(a) $F'(x) = \alpha e^{-\alpha x}$, $\alpha > 0$, $x > 0$. (This is the Poisson process.)

(b) $F'(x) = \alpha^2 x e^{-\alpha x}$, $x > 0$. (This is a gamma density.)

Do the drill without Laplace transforms. In both examples, verify the elementary renewal theorem directly.

3.4. Suppose F is a distribution of a positive random variable and $p_k \geq 0, \sum_{k=0}^{\infty} p_k = 1$. Define $G = \sum_{k=0}^{\infty} p_k F^{k*}$. If $X_1, X_2, \ldots$ are iid with common distribution F and if N is a non-negative integer valued random variable independent of $\{X_n\}$ then check G is the distribution of $X_1 + \cdots + X_N$. Find the Laplace transform of G.

3.5. If $X > 0$ and $Y > 0$ are independent with distributions F and G, show

$$Ee^{-\lambda XY} = \int_0^{\infty} \hat{F}(\lambda y) dG(y) = \int_0^{\infty} \hat{G}(\lambda y) dF(y).$$

3.6. Items of a certain type have a mean lifetime of one day. The standard deviation σ of the lifetime distribution is two hours. There is a unit cost per item, and items are used successively according to a renewal model. Use of items starts at $t = 0$ and continues for a 10-year period (3600 days). If C is the total cost for the 10-year period, give an upper bound α and a lower bound β such that

$$P[\beta \leq C \leq \alpha] \approx .95.$$

Hint: Use the central limit theorem.

3.7. (a) Write a renewal equation for

$$Z(t) = P[A(t) > x, B(t) > y].$$

(b) Solve it.

(c) What is

$$\lim_{t \to \infty} Z(t)?$$

(d) When is the limit distribution a product measure?

3.8. Suppose the interarrival density of a renewal process is

$$F'(x) = \begin{cases} \alpha e^{-\alpha(x-\delta)}, & \text{for } x > \delta \\ 0, & \text{for } x \leq \delta, \end{cases}$$

where $\delta > 0$ is fixed. Find $P[N(t) \geq k]$. (Express as an incomplete gamma integral.)

3.9. Suppose $Z(t)$ satisfies the renewal equation

$$Z = z + F * Z,$$

where $z(t)$ is assumed bounded, non-negative, non-decreasing, so that

$$\lim_{t \to \infty} \uparrow z(t) =: z(\infty) < \infty.$$

Show

$$\lim_{t \to \infty} Z(t)/t = z(\infty)/\mu,$$

where μ is the mean of F.

3.10. Forgetfulness Property. If you have been living in a cave and haven't encountered the "forgetfulness property" of the exponential density, then check it now: If X is a random variable with exponential density, then

$$P[X > t + x | X > t] = P[X > x]$$

for $x > 0$, $t > 0$.

Waiting times in doctors' offices in Ithaca follow exponential waiting times with mean 75 minutes. Given that you have already waited one hour, what is the probability that you will have to wait another 45 minutes to see the doctor?

The Youngest Goes Last. Dana, Tali and Sheli go to the post office where there are two servers. The service requirements D, T and S of the three girls are iid exponentially distributed random variables. Because Sheli is the youngest, Dana and Tali enter service first. Sheli can only enter the service when one of the older girls finishes the service (at time $D \wedge T$). What is the probability that Sheli is not the last to leave the service?

3.11. Optima Street Fire Department. In Harry's neighborhood of Optima Street, calls to the fire department occur according to a Poisson process with a rate of three per day. The fire department must respond to each call (although they know a certain percentage represent false alarms).

(a) A fireman gets a raise after the 300th call. How many days should the fireman expect to wait for the raise?

(b) Of the calls that come into the fire department, on the average one third turn out to be false alarms. In a single day, what is the probability of two false alarms?

(c) Firemen are presently paid $100 a day. A new pay plan is proposed where they will be paid a random amount of money per fire *that they actually fight* (the reward being based on the difficulty of the fire fought). In the new scheme the expected pay per fire fought is $40 per fire. What is the long run reward rate in the new scheme? (To help your thinking, you might wish to represent the payments as random variables $R_1, R_2, \ldots$.) Is the new scheme more or less advantageous than the old one?

(d) If we consider only calls to the fire department which occur after midnight, January 1, 1995, what is the probability that a false alarm occurs before a real alarm is phoned in?

3.12. Whenever machines of a given type are "operative", they stay that way for a negative exponential length of time of mean duration $1/\lambda$. When breakdowns occur, repairs lasting for a negative exponential length of time of mean $1/\mu$ are started immediately. Repairs return the machines again to the operative state.

At time $t = 0$, N identical machines are independently placed in use and all are operative.

(a) Show that the number of machines that are operative at time $t > 0$ has a binomial distribution with parameters N and

$$p = \frac{1}{\lambda + \mu} \left(\mu + \lambda e^{-(\lambda+\mu)t} \right).$$

(b) How large should N be chosen to guarantee that the probability that at least one machine will be operative is at least 0.99?

(c) Give a simple approximation to the number asked for in (b) when t is very large.

(Concentrate on one machine alternating periods of being operative and inoperative. Compute a renewal function. Laplace transform techniques may help. We will see later that this problem has a somewhat simpler solution using continuous time Markov chains.)

3.13. Harry's Customers Like Their Brew. Kegs of beer at Happy Harry's bar hold 24 liters. Only one is kept under the bar at a time; the rest are in the back. Customers arrive according to a renewal process with finite mean interarrival times. Each customer orders half a liter. When a keg is empty, it is instantaneously replaced with a full keg from the back. Let $X(t)$ be the level in the keg under the bar at time t. Assume the state space is $\{.5, 1, 1.5, \ldots, 24\}$.

(a) Is $X(t)$ regenerative? Why or why not?

(b) Compute

$$\lim_{t \to \infty} P[X(t) = k].$$

(c) What is the the long run percentage of time the keg on the bar is full?

3.14. Wildlife at Harry's Restaurant. The cockroach population grows linearly at Happy Harry's with a random growth rate. When the population reaches 5000 the Public Health Service comes in and shuts

Happy Harry's Restaurant for a random amount of time while the place is fumigated and cleaned up. Then the restaurant reopens, and the cockroach population again begins to grow. This happy pattern continues ad nauseam.

Assume cockroach growth rates are $G_1, G_2, \ldots$, which are iid random variables, so that, for instance, if t is before the first time the Public Health Service shuts down the restaurant, we have the cockroach population $G_1 t$. Similarly, the periods when the restaurant is closed are iid random variables $D_1, D_2, \ldots$. Assume $\{D_i\}$ and $\{G_i\}$ are independent, $ED_i = d, EG_i^{-1} = \hat{g} < \infty$.

(a) Let $N(t)$ be the number of times the Public Health authorities have closed Happy Harry's in $[0, t]$. Give an expression for $EN(t)$ in terms of the distribution of G_1 and D_1. What is the long run rate of closures per unit time of happy Harry's? If Harry is fined \$50 per closure, what is the expected total amount Harry has paid in fines by time t? (Clearly state assumptions about the initial configuration of the system at time 0.)

(b) Let $B(t)$ be the time until the next closure of Happy Harry's after time t. Give an expression for the distribution of $B(t)$. What is

$$\lim_{t \to \infty} P[B(t) > x] \ ?$$

(c) Maximum profits are achieved when the cockroach population is between 1000 and 3000. (Anything below 1000 leaves the regular patrons nervous that Harry is serving roaches in the soup, and anything more than 3000 is excessive even for Harry's devoted customers.) Let $C(t)$ be the number of cockroaches at time t. Does $\{C(t), t \geq 0\}$ have any notable structure? Compute

$$\lim_{t \to \infty} P[1000 \leq C(t) \leq 3000].$$

3.15. Age Dependent Branching Process. Let $T(t)$ be the total number of particles born up to time t. Define

$$F(s, t) = Es^{T(t)}.$$

Derive an integral equation for F and then differentiate to get a renewal type equation for $ET(t)$. Give the solution of the equation.

3.16. Plumbing at Happy Harry's Restaurant. In the back of Harry's restaurant is a small building with a tube running down to a huge tank of capacity 100,000 cc. Harry's customers visit the small building in a steady procession according to a renewal process $\{S_n, n \geq 0\}$, with $S_n - S_{n-1} \sim F(x)$ and $E(S_n - S_{n-1}) = 35$ seconds. Upon arrival to the small building, each person instantaneously and discretely deposits 10 cc of liquid down the tube which is collected in the tank. The tank is initially empty. When the tank is full, a red light glows and the tank is replaced by a shiny new, empty tank. Let $X(t)$ be the level in the tank at time t. For $j = 0, \ldots, 10,000 - 1$, compute

$$\lim_{t \to \infty} P[X(t) = j].$$

Why does this limit exist?

3.17. Consider a discrete pure renewal process such that the span of the distribution governing times between renewals is 1, and assume

$$EY_1 = \mu < \infty, \quad \text{Var}(Y_1) = \sigma^2 < \infty.$$

Define

$$q_n = P[Y_1 > n] = \sum_{j=n+1}^{\infty} f_j, \quad r_n = \sum_{j=n+1}^{\infty} q_j.$$

Show that the generating functions $Q(s)$ and $R(s)$ converge for $s = 1$, and prove that

$$\sum_n \left(u_n - \frac{1}{\mu} \right) s^n = \frac{R(s)}{\mu Q(s)},$$

and hence that

$$\sum_n \left(\mu_n - \frac{1}{\mu} \right) = \frac{\sigma^2 - \mu + \mu^2}{2\mu^2}.$$

3.18. Consider a delayed renewal process with initial distribution

$$G(x) = F_0(x) = \mu^{-1} \int_0^x (1 - F(s))\, ds.$$

Let

$$H(t, x) = P\{\cup_{n=0}^{\infty} [S_n \in (t, t+x]]\}.$$

Write a renewal equation for $H(t, x)$. What is $H(t, x)$?

3.19. Given a renewal process with finite mean, suppose the random variables $A(t)$ and $B(t)$ are independent for each $t > 0$. Show the process is Poisson.

(Hints: Use the identity

$$P[A(t) > x, B(t) > y] = P[A(t) > x]P[B(t) > y],$$

and let $t \to \infty$. Recall Exercise 3.7(d). Derive a functional equation for $1 - F_0(x)$. You may want to use the fact that a version of *Cauchy's Equation*

$$f(x + y) = f(x)f(y)$$

has only exponentials as solutions if f is reasonable. Monotone implies reasonable.)

3.20. Suppose you have an alternating renewal process: A new machine begins operating at time 0 and breaks down after a random amount of time whose distribution is $F_1(x)$. It takes a random amount of time whose distribution is $F_2(x)$ to replace the broken machine with a new machine with characteristics identical to the original. Successive periods are independent and alternate between having distributions F_1 and F_2.

Define $f_x(t)$ to be the probability that at time t there is an operating mchine and that it stays operative for at least x more time units. In what follows, x is always fixed.

(a) Write the renewal equation satisfied by $f_x(t)$.
(b) Solve the equation.
(c) Compute $\lim_{t\to\infty} f_x(t)$.

3.21. Consider an ordinary renewal process with counting function $N(t)$.
(a) Show

$$E(N(t))^2 = \sum_{n=1}^{\infty} (2n-1) F^{(n-1)*}(t).$$

(b) Find the Laplace transform of $\sum_{n=1}^{\infty} n F^{(n-1)*}(t)$. Use this to write $\sum_{n=1}^{\infty} n F^{(n-1)*}(t)$ in terms of $U(t)$, the renewal function.
(c) Use (a), (b) to find an expression for $E(N(t))^2$ only in terms of U.

3.22. (a) If X, Y are independent, X has density $\alpha e^{-\alpha x}$, $x > 0$ and Y has density $\beta e^{-\beta x}$, $x > 0$, show the density of $X + Y$ is

$$\alpha\beta \left(\frac{e^{-\alpha x} - e^{-\beta x}}{\beta - \alpha} \right).$$

Do this directly by computing the convolution of the two densities, and then do this a second time using Laplace transforms.

(b) Let $E(\alpha_0), \ldots, E(\alpha_n)$ be independent exponentially distributed with parameters $\alpha_0, \ldots, \alpha_n$. Find the density of $\sum_{i=0}^{n} E(\alpha_i)$ by Laplace transform methods using the following outline:

(b1) The Laplace transform of the sum $\sum_{i=0}^{n} E(\alpha_i)$ is

$$\prod_{i=0}^{n} \frac{\alpha_i}{\alpha_i + \lambda}.$$

(b2) Do a partial fraction expansion on the product by writing

$$\prod_{i=0}^{n} \frac{1}{\alpha_i + \lambda} = \sum_{i=0}^{n} \frac{A_i}{\alpha_i + \lambda}.$$

Multiply through by $\prod_{i=0}^{n} (\alpha_i + \lambda)$ to solve for A_i, $i = 0, \ldots, n$.

(b3) Verify

$$E \exp\{-\lambda \sum_{i=0}^{n} E(\alpha_i)\} = \int_0^\infty e^{-\lambda x} \left(\sum_{i=0}^{n} \left[\frac{\prod_{m=0}^{n} \alpha_m}{\prod_{j \neq i}(\alpha_j - \alpha_i)} \right] e^{-\alpha_i x} \right) dx.$$

3.23. Prove without using transforms that if $V(t) = t/\mu$, then $G = F_0$.

3.24. Replacement Policies. Machines having iid lifetimes with common distribution $F(x)$ of finite mean μ and variance σ^2 are used successively.

(a) Show that if machines are replaced upon failure only, then the limiting average cost per unit of time for replacements is A/μ, where A is the cost of a new machine.

As an alternative, we can replace machines regardless of age at times $0, T, 2T, \ldots$ at a cost B or whenever a failure occurs (at cost A).

(b) Show that the expected cost per time interval $[KT, KT + T]$ for replacements is $B + AU(T)$ where U is the renewal function corresponding to F.

3.25. Let $\hat{U}(\lambda) = \int_0^\infty e^{-\lambda x} U(dx)$ exist for $\lambda > 0$. Then $\hat{U}$ is integrable over $(0,1)$ (respectively $(1,\infty)$) with respect to Lebesgue measure iff x^{-1} is integrable with respect to U over $(1,\infty)$ (respectively $(0,1)$).

3.26. Busy Periods in M/G/1 Queues; the Bullpen Discipline. Customers arrive at a single server queue according to a Poisson process $\{N_A(t), t \geq 0\}$ of rate a. Service times are iid random variables $\{S_n\}$ with common distribution $B(x) = P[S_1 \leq x]$, $x \geq 0$. Assume service times are independent of input to the system. Let $Q(t)$ be the number of people in the system at time t; we are interested in the busy period

$$BP := \inf\{t \geq 0 : Q(t) = 0\},$$

and we seek to calculate the busy period distribution

$$\beta(x) = P[BP \leq x | Q(0) = 1].$$

(a) As a warm-up, first compute the distribution of the number of arrivals during a service period, i.e., compute $P[N_A(S_1) = n]$ and the generating function $Es^{N_A(S_1)}$. Compute $EN_A(S_1)$.

(b) Argue that the busy period discipline does not depend on the queue discipline, i.e., on the order in which the customers are served.

(c) Thus we may consider the bullpen discipline: Let the initial customer enter service and remain in service for a duration S_1. Label the customers who arrive in the interval of length S_1 as $c_1, \ldots, c_{N_A(S_1)}$. Put these customers in the bullpen. At the end of the interval of length S_1, the customer

c_1 is removed from the bullpen and allowed to enter service. He is served, and all the people who arrive during c_1's service period (call these arrivees the descendants of c_1) are served, as are all the people who arrive during the service periods of the descendants, etc. until all customers descending from c_1 are served, and the server is free. The next customer c_2 is released from the bullpen and allowed to enter service. He is served, and all of his descendants are served, etc. until the server becomes free. This process continues until the last customer is led from the bullpen to the server, and he and his descendants are served. Show that this procedure leads to the recursion

$$BP \stackrel{d}{=} S_1 + BP_1 + BP_2 + \ldots BP_{N_A(S_1)}$$

where BP_i can be considered the busy period initiated by c_i. Note $\{BP_i\}$ are iid with $BP_i \stackrel{d}{=} BP$.

(d) Derive *Takacs' Equation* by taking Laplace transforms:

$$\hat{\beta}(\theta) = \hat{B}(a + \theta - a\hat{\beta}(\theta)),$$

where $\hat{\beta}$ is the Laplace transform of the busy period distribution and $\hat{B}$ is the Laplace transform of the service time distribution.

(e) In the special case $B(x) = 1 - e^{-bx}$, $x \geq 0$, what is the form of Takacs' Equation ? Solve for $\hat{\beta}$. (Takacs' Equation leads to two solutions; one needs to be discarded.)

Assume for the rest of the problem that the service time distribution is exponential.

(f) Define the traffic intensity to be $\rho = a/b$. Show

$$P[BP < \infty] = 1 \text{ iff } \rho \leq 1.$$

Compute $P[BP = \infty]$ in case $\rho > 1$.

(g) If $\rho \leq 1$ compute $E(BP|Q(0) = 1)$. Show that if $\rho = 1$, then this expectation is infinite. Compute $\text{Var}(BP)$.

Now consider the random variable N, the number of customers served during a busy period. We seek the transform $g(s) = E(s^N|Q(0) = 1)$.

(h) Argue as in part (c) that N satisfies the recursion

$$N \stackrel{d}{=} 1 + N_1 + \cdots + N_{N_A(S_1)},$$

where $\{N_i\}$ are iid with $N_i \stackrel{d}{=} N$, and N_i represents the number served during the busy period initiated by customer c_i.

(i) From (h) convert to generating functions, and show with a general service distribution B that g satisfies the functional equation

$$g(s) = s\hat{B}(a(1 - g(s))).$$

Specialize this to the case that the service time distribution is exponential with parameter b, and solve for g in this case. (Again, there will be a quadratic equation in the unknown g and thus two roots; one root must be discarded.)

(j) Compute $P[N < \infty|Q(0) = 1]$. Prove this is 1 iff $\rho \le 1$. In this case compute $E(N|Q(0) = 1)$ and $\mathrm{Var}(N|Q(0) = 1)$.

3.27. Bulk Arrivals at Harry's Restaurant. The late shift at Happy Harry's restaurant consists of one chef. Traffic at this time of night arrives at the restaurant in bulk owing to arrivals of busloads of hungry athletes returning home from athletic competitions. Buses arrive at Harry's according to a Poisson process $\{A(t)\}$ with rate a. The number of occupants in each bus is a sequence of iid random variables $\{N_i, i \ge 1\}$ with common distribution

$$P[N_1 = k] = \theta_k, \quad 1 \le k \le 30,$$

generating function

$$\Theta(s) = Es^{N_1} = \sum_{i=1}^{30} \theta_k s^k,$$

and mean

$$m := \Theta'(1) = \sum_k k\theta_k.$$

(Note the support of the distribution contains small numbers to allow for single arrivals by car.) The chef processes the order of each customer in an exponential length of time with parameter b. We seek to analyze the busy period of the chef assuming the busy period is initiated by the arrival of a busload of hungry athletes.

(a) What should be the condition that the busy period is finite?

(b) Under the condition given in (a) find $E(BP)$. Do this by writing an equation among random variables, many of which are iid replicates of BP. Express the answer neatly and succinctly in terms of λ, μ, m. Does the answer depend on the assumption that the chef works at an *exponential* rate?

(c) Let

$$\Psi(\zeta) = E \exp\{-\zeta BP\}$$

be the Laplace transform of the busy period distribution. Write a functional equation for $\Psi(\zeta)$. The equation should involve the quantities $\Psi(\zeta), \hat{B}(\zeta) = b/(b+\zeta), \zeta, \lambda, \Theta(\cdot)$.

(d) Let

$$p = P[BP < \infty].$$

Use (c) to write a recursion for p and to verify your answer to (a).

(e) Use (c) to compute $E(BP)$ when $p = 1$. When is $E(BP) < \infty$?

3.28. If $g \geq 0$ is integrable on $[0, \infty)$, check

$$\int_0^\infty e^{-\lambda x} g(x) dx \int_0^\infty e^{-\lambda x} G(dx)$$

is the Laplace transform of the density $G * g(x)$.

3.29. Check the elementary renewal theorem via Laplace transforms when the process is delayed.

3.30. Formulate and prove a central limit theorem for a renewal reward process.

3.31. Prove for a delayed renewal process

$$EN(t, t + b] = U * G_t(b),$$

where

$$G_t(x) = P[B(t) \leq x].$$

3.32. Let $M = \max\{S_n : S_n < \infty\}$ be time of the last renewal in a terminating ordinary renewal process with defective interarrival distribution F.

(a) Find the distribution of M in terms of F.
(b) What is the distribution of $N(\infty)$, the total number of renewals?
(c) What is EN? EM?
(d) **Large gaps; pedestrian delay:** Harry strains his Achilles tendon in a pick-up basketball game and has to be on crutches while mending. While on crutches, he needs τ time units to cross Optima Street. Assume Optima Street has only one way traffic and that the traffic passes according to a renewal process with interarrival distribution F. Suppose Harry arrives on crutches at $t = 0$ and seeks a gap in traffic sufficient for him to cross. Therefore, he starts to cross when he sees a renewal interval at least τ in length. Let M be the time when he starts to cross so that $M = S_n$ iff $Y_1 \leq \tau, \ldots, Y_n \leq \tau, Y_{n+1} \geq \tau$. Compute the distribution of M and give EM. Specialize to the case that the renewal process is Poisson with parameter α.
(Hint: Let

$$\tilde{Y}_n = \begin{cases} Y_n, & \text{if } Y_n \leq \tau, \\ \infty, & \text{otherwise.} \end{cases}$$

Let $\tilde{S}_n = \tilde{Y}_1 + \cdots + \tilde{Y}_n$. Find the distribution of $\tilde{Y}_n$, and then M is the lifetime of the terminating renewal process $\{\tilde{S}_n\}$.)

3.33. A General Risk Process. Let $r(x), x \geq 0$, satisfy $r(0) = 0$, and r is continuous. Assume

$$S(x) := \int_0^x \frac{1}{r(y)} dy < \infty$$

for $x > 0$ and that $S(\infty) = \infty$. Define $q(x,t) = S^{\leftarrow}(S(x)+t)$ where $S^{\leftarrow}$ is the inverse of the monotone function S. Check that q satisfies

$$q(x,t) = x + \int_0^t r(q(x,s))ds.$$

Suppose at Poisson times $t_1, t_2, \ldots$ iid jumps downward of size $X_1, X_2, \ldots$ are taken. Define for $x \geq 0$

$$
\begin{aligned}
X(t) &= q(x,t), && \text{if } 0 \leq t < t_1, \\
&= (q(x,t_1) - X_1)_+, && \text{if } t = t_1, \\
X(t_1 + t) &= q(X(t_1),t), && \text{if } 0 \leq t < t_2 - t_1, \\
X(t_2) &= (q(X(t_1), t_2 - t_1) - X_2)_+, && \text{if } t = t_2,
\end{aligned}
$$

etc. Analyze the ruin probability.

3.34. Construct a Riemann integrable function on $[0,\infty)$ which is continuous and unbounded.

3.35. If z is dRi, then $F * z$ is dRi for F a distribution function on $[0,\infty)$. (Cinlar, 1975)

3.36. If F is arithmetic with span Δ (in which case regard z as a function on $\{k\Delta\}$), then z is dRi iff $\sum_k z(k\Delta) < \infty$.

3.37. Any finite linear combination of functions which are dRi is itself dRi.

3.38. If z is continuous almost everywhere and

$$\sum_k \sup_{0 \leq t < 1} z(k+t) < \infty,$$

then z is dRi.

3.39. Suppose in a ruin problem, the company invests capital at interest d, so that, in the absence of claims, the fortune obeys the differential equation

$$f'(t) = c + df(t).$$

Show that $R(x)$ is positive for all x no matter what the values of α, c, EX_1.

3.40. Find the limiting behavior of the expected number of all individuals who ever lived before time t (which equals $1+$ the number of births up to time t) in a supercritical ($m > 1$) age-dependent branching process.

3.41. In the G/G/1 model, let $\zeta(t)$ be the residual service time of the customer who is being served at time t, provided $Q(t) > 0$ (meaning there is someone in the system). If the system is empty, $Q(t) = 0$, then define $\zeta(t) = 0$. Show

$$(Q(t), \zeta(t)) \Rightarrow (Q(\infty), \zeta(\infty)),$$

and express the distribution of $(Q(\infty), \zeta(\infty))$ in terms of paths up to C_1.

3.42. (a) Suppose a process $\{X(t), t \geq 0\}$ may be in one of three states 1, 2, 3. Suppose initially the process starts in state 1 and remains there for a random length of time distributed by F_1. After leaving state 1, it enters state 2, where it sojourns for a random length of time distributed according to F_2. Then it goes to state 3, where the sojourn is governed by F_3, and then on to state 1, etc. Successive sojourns are independent. Assuming F_i is non-arithmetic and has finite mean μ_i for $i = 1, 2, 3$, find the limiting state probabilities

$$\lim_{t \to \infty} P[X(t) = i].$$

(b) Customers arrive at the dispatching room of the Optima Street Taxi Service according to a renewal process having mean interarrival time μ. When N customers are present, a taxi departs with the N waiting passengers. The taxi company incurs a cost at the rate of nc dollars per unit time whenever there are n customers waiting. What is the long run average cost incurred?

(c) Replace the taxi company by a railroad company. Imagine customers arrive at the train station according to a Poisson process. When N passengers are waiting, a train is summoned to pick them up. The train takes a random number of time units (with mean k) to arrive. When it arrives, it takes everyone away. Again, assume the train company incurs a cost per unit time of nc when n passengers are waiting in the station. Find the long run average cost incurred by the train company.

3.43. If F^{2*} has a density which is directly Riemann integrable, then $V = U - 1 - F$ has a density v and $v(t) \to 1/\mu$ (Feller, 1971).

3.44. Suppose for $i = 1, 2$

$$Z_i = z_i + Z_i * F.$$

(a) Show that if

$$\lim_{t \to \infty} z_1(t)/z_2(t) = 1,$$

then

$$\lim_{t \to \infty} Z_1(t)/Z_2(t) = 1.$$

(b) Suppose z_2 is the integral of z_1 and $z_1(0) = 0$. Then Z_2 is the integral of Z_1.

(c) Use (b) to show that if $z(t) = t^{n-1}$, then $Z(t) \sim t^n/(n\mu)$ as $t \to \infty$.

(d) If $z_2 = G * z_1$ (where G is a Radon measure, that is, a measure which puts finite mass on finite intervals), then $Z_2 = G * Z_1$. If $G(x) = x^a$ with $a \geq 0$ then

$$z_1(x) \sim x^{a-1} \text{ implies } Z_1(x) \sim x^a/(a\mu).$$

3.45. On Length Biased Sampling, Inspection Paradox, Waiting Time Paradox. The renewal interval covering t tends to be longer than a typical interval with interarrival distribution F. We explore this phenomenon in this exercise.

(a) As a warm-up, let X be uniform on $[0, 1]$. Then

$$[0,1] = [0, X] \cup (X, 1].$$

By symmetry, each subinterval has mean length $1/2$. Now pick one of the subintervals at random in the following way: Let Y be independent of X and uniformly distributed on $[0, 1]$, and pick the subinterval $[0, X]$ or $(X, 1]$ that Y falls in. Let L be the length of the interval chosen so that

$$L = \begin{cases} X, & \text{if } Y \leq X \\ 1 - X, & \text{if } Y > X. \end{cases}$$

Find EL (Taylor and Karlin, 1984).

(b) Consider an ordinary renewal process $\{S_n, n \geq 0\}$ with interarrival distribution F. The lengths of the interarrival intervals are $\{Y_j, j \geq 1\}$. The length of the interval covering t is $S_{N(t)} - S_{N(t)-1} = Y_{N(t)} = A(t) + B(t) =: L_t$. The point of this exercise is that the length of this interval does not have distribution F, since longer intervals tend to cover t.

(1) Show the distribution of L_t is

$$P[L_t \leq x] = \begin{cases} \int_{u=t-x}^{t}(F(x) - F(t-u))dU(u), & \text{if } x \leq t, \\ \int_0^t(F(x) - F(t-u))dU(u), & \text{if } x > t. \end{cases}$$

(2) Evaluate the distribution of L_t in the case when the renewal process is a Poisson process with rate α. Show that L_t converges in distribution to L_∞ and give the distribution of L_∞. What is the mean of L_∞?

(3) Tired of the rigors of city driving and seeking to be ecologically correct, Harry eschews the automobile in favor of the bus. Buses arrive at his stop according to a Poisson process with rate α. Harry arrives at time t and tries to estimate his expected wait until the expected next bus arrives. Harry reasons in two ways: (i) On the one hand, Harry knows that the forward recurrence time for the Poisson process is exponentially distributed with parameter α and hence his expected wait should be α^{-1}. (ii) On the other hand, he reasons that because of symmetry, and in the absence of knowledge about when he arrived relative to bus arrivals, he might as well assume that his arrival is uniformly distributed in the interval. Harry thinks the interval should have expected length

α^{-1} and hence his expected wait should be approximately $\alpha^{-1}/2$. Criticize Harry's reasoning given in (ii), and explain why, if Harry's reasoning in (ii) is fixed up, he has an approximately (as $t \to \infty$) correct method of estimating the expected wait.

(c) Show that
$$P[Y_t > x] \geq P[Y_1 > x].$$

3.46. Find the limiting distribution of the forward recurrence time when the interarrival distribution is

(1) uniform on $[0,1]$,
(2) the triangle density on $[0,2]$
(3) Erlang with k stages with density f
$$f(x) = \alpha(\alpha x)^{k-1}\frac{e^{-\alpha x}}{(k-1)!}.$$

3.47. Consider a discrete renewal process with interarrival distribution which is geometric:
$$P[X_1 = k] = a(1-a)^{k-1}, \quad k = 1,2,\ldots.$$

Compute the discrete density of the forward recurrence time.

3.48. Consider an age-dependent branching process with offspring distribution $\{p_k\}$, $f(s) = \sum_{k=0}^{\infty} p_k s^k$ and life length distribution G. Let $X(t)$ be the population size at time t and $F(s,t) = Es^{X(t)}$; show that $F(s,t)$ satisfies the recursion
$$F(s,t) = s(1 - G(t)) + \int_0^t f(F(s,t-u))dG(u).$$

From this, recover the recursion for $EX(t)$ given in Example 3.5.2.

3.49. Suppose for an ordinary renewal process that $\sigma^2 := \mathrm{Var}(Y_1) < \infty$. Show that
$$\lim_{t\to\infty} \frac{\mathrm{Var}(N(t))}{t} = \frac{\sigma^2}{\mu^3}.$$

3.50. Average Age, Average Forward Recurrence Time. Check
$$\int_0^\infty x dF_0(x) = \frac{\sigma^2 + \mu^2}{2\mu}.$$

Then check that
$$\lim_{t\to\infty} t^{-1}\int_0^t B(s)ds = EB(\infty) = EA(\infty)$$
$$= \int_0^\infty x dF_0(x)$$
$$= \lim_{t\to\infty} t^{-1}\int_0^t A(s)ds,$$

where $B(\infty) \stackrel{d}{=} A(\infty)$ has distribution F_0.

3.51. The lifetime of Harry's computer, which performs essential restaurant management functions, has distribution H. Harry buys a new computer as soon as the old one breaks down or reaches the obsolescence age of T years. A shiny new computer costs C_1 dollars, and an additional cost of C_2 dollars is incurred whenever a computer breaks down and leaves Harry's business without high tech support. A T-year-old antique computer has minimal resale value but can be donated to a school for a tax benefit yielding a gain of $b(T)$ dollars. Find Harry's long run average cost.

(a) If H is uniform on $(2,8)$, $C_1 = 4, C_2 = 1$ and $b(T) = 4 - (T/2)$, then what value of T minimizes Harry's long run average cost?

(b) If H is exponential with mean 5, $C_1 = 3$, $C_2 = 1/2$, $b(T) = 0$, what value of T minimizes the long run average cost? (Ross, 1985.)

3.52. Conserving the Server. Suppose in an M/M/1 queue, the service facility is turned off whenever it is empty. It is turned on again when the amount of work in the system exceeds a set quantity v. Suppose the system is initially empty. The server is activated by customer number $N = n$ if

$$\sum_{i=1}^{n-1} \tau_i \le v < \sum_{i=1}^{n} \tau_i.$$

Once on, the server remains active until the system is empty. Assume the traffic intensity is less than 1 so that the system is stable.

(a) How is $\sum_{i=1}^{N} \tau_i - v$ distributed?

(b) Find $E \sum_{i=1}^{N} \tau_i$ and EN.

(c) What is the expected time from when the server is activated, until the system is empty? (Wolf, 1989.)

3.53. Show by example that $\mu < \infty$ is not necessary for convergence in distribution of a regenerative process. (A simple example, say $X(t) \equiv 0$, suffices.)

3.54. Customers arrive at a train station according to a Poisson process of rate α, and trains arrive according to a renewal process with interarrival distribution F with mean μ. What is the long run fraction of customers who arrive for whom the waiting time for the next train does not exceed y. (Wolf, 1987.)

3.55. (a) If $Y_j = c$ for all j with probability 1, what is F_0?

(b) $F_0 = F$ iff F is an exponential distribution function.

3.56. Let ξ_0 be a random variable with distribution F_0, and suppose Y_1 has distribution F concentrating on $[0, \infty)$. Express the moments of ξ_0 in terms of the moments of Y_1.

3.57. Consider an alternating renewal process $\{X(t), t \geq 0\}$ with states $0, 1$. Sojourn lengths in state i are governed by a continuous distribution F_i for $i = 1, 2$. Let

$$
B_i(t) = \begin{cases} \text{excess life in state } i, \text{ i.e., time until the next transition,} \\ \qquad\qquad\qquad\qquad\qquad\qquad\quad \text{if } X(t) = i, \\ 0, \qquad\qquad\qquad\qquad\qquad\qquad\quad \text{if } X(t) \neq i. \end{cases}
$$

Find the limit distribution of $B_i(t)$ for $i = 1, 2$.

3.58. Restaurant Maintenance. The expresso machine in Harry's restaurant breaks down periodically and is eventually fixed. Times between breakdowns are iid random variables $\{X_i\}$ with mean EX_i. The expresso machine is inspected by Harry periodically to determine whether it is still working adequately. The times between inspections are iid exponentially distributed random variables with mean μ. When an inspection determines a repair is necessary, repairs begin. Repair times are iid random variables $\{Y_i\}$ with mean EY.

(a) What fraction of the time is the machine working?

(b) If we assume the inspections stop when the machine is being repaired, what is the average number of inspections per unit time? (Wolf, 1989.)

3.59. Demand for an item in inventory at a warehouse is a Poisson process at rate α, meaning customers arrive according to a Poisson process, and each customer demands one unit. When the inventory level hits 0, an order of constant size S units is placed with the factory. Delivery time of the entire order from the factory to the warehouse is exponential with mean $1/\mu$. During that time, demand is lost at a cost of \$c per unit. The inventory holding cost is \$h per unit per time, based on the *average* amount of inventory held.

(a) Check that iid cycles $\{C_i\}$ can be discerned.

(b) Suppose the inventory level hits 0 at epoch 0. Let $I(t)$ be the inventory level at epoch t, and define

$$
A = \int_0^{C_1} I(t)dt,
$$

where C_1 is the length of the first cycle. Show that

$$
EA_1 = \frac{S(S+1)}{2\alpha}.
$$

(Hint: Break the integral into rectangles.)

(c) Find the long run cost per unit time of inventory and lost sales (Wolf, 1989).

3.60. More on Counter Models. Suppose particles arrive at a counter according to an ordinary renewal process $\{S_n, n \geq 0\}$, where $S_n - S_{n-1} = Y_n$ has distribution F with mean μ. A registered particle locks the counter for a random duration, during which time no other particles can be registered. Lock periods are iid random variables $\{L_n, n \geq 0\}$ with distribution G and mean ν. Registration times have indices $n_0 = 0$, $n_1 = \inf\{n : S_n > L_0\}, \ldots, n_j = \inf\{n : S_n > S_{n_{j-1}} + L_{j-1}\}$. Times of registration are $\{S_{n_j}, j \geq 0\}$

(a) Define $Z_j = S_{n_j} - S_{n_{j-1}}$, and show $\{Z_k\}$ are iid, so that registration times form a renewal process.

Let $U(t) = \sum_{n=0}^{\infty} F^{n*}(t) = EN(t)$ and let $M(t)$ be the number of registrations by time t, so that $M(t)$ is the counting function of the renewal process with interarrival times $\{Z_j\}$. Set $V(t) = EM(t)$ to be the expected number of registrations up to time t. Finally, let $W_j = Z_j - L_{j-1}$ be the amount of time prior to the jth registration that the counter is free since the last registration.

(b) Prove the $\{W_j\}$ are iid and

$$P[W_j \leq x] = \int_0^{\infty} [\int_{\tau=t}^{t+x} (1 - F(x + t - \tau))dU(\tau)]dG(t).$$

(Hint: Recall the formula for the forward recurrence time distribution for an ordinary renewal process.)

(c) Prove $\{Z_j\}$ are iid and

(1) $H(z) := P[Z_j \leq z] = \int_0^z G(t)(1 - F(z - t))dU(t)$.

Show for $\lambda > 0$ that the Laplace transform $\hat{H}$ satisfies

$$\hat{H}(\lambda) = \frac{\int_0^{\infty} e^{-\lambda s} G(s)dU(s)}{\hat{U}(\lambda)}.$$

(2) $P[Z_j < \infty] = 1$. (Hint: Show $\lim_{\lambda \downarrow 0} \hat{H}(\lambda) = 1$.)

(3)

$$EZ_j = \begin{cases} \mu \int_0^{\infty} U(t)dG(t), & \text{if } \mu < \infty, \nu < \infty, \\ \infty, & \text{otherwise.} \end{cases}$$

(Hint: Wald's equation is one possible method.)

Let

$$B(t) = \frac{U(t)}{V(t)}$$

be the *bias* of the counter.

(d) Show the asymptotic bias of the counter is

$$\lim_{t \to \infty} B(t) = \begin{cases} \infty, & \text{if } \mu = \infty, \\ \int_0^{\infty} U(t)dG(t), & \text{if } \mu < \infty. \end{cases}$$

(Hint: The elementary renewal theorem suffices.)

Let $P_0(t)$ be the probability the counter is locked at time t, and let $P_1(t)$ be the probability the counter is free at time t.

(e) Show that

$$P_0(t) = \int_0^t (1 - G(t - s))dV(s)$$

and that

$$\lim_{t \to \infty} P_0(t) = \frac{\nu}{\mu \int_0^\infty U(t)dG(t)}.$$

3.61. Two-State Semi-Markov Process. Given two distributions F_1 and F_2 and the matrix

$$P = \begin{pmatrix} \alpha & 1 - \alpha \\ 1 - \beta & \beta \end{pmatrix},$$

$0 < \alpha < 1, 0 < \beta < 1$. A stochastic process $\{X(t), t \geq 0\}$ has state space $\{1, 2\}$ and moves according to the following scheme: When the process enters i, it stays in i a random amount of time governed by distribution $F_i(x), 1 = 1, 2$. Then, given that this sojourn time is at an end, the process jumps to state j (possibly $j = i$) with probability p_{ij}, which is the (i, j)th entry of the matrix P. Sojourn times are conditionally independent given that you know the state the process is sojourning in. Use Smith's theorem to calculate

$$\lim_{t \to \infty} P[X(t) = j]$$

for $j = 1, 2$.

CHAPTER 4

Point Processes

W E WANT to build models for a random distribution of points in a space, usually, a subset of R or $[0, \infty)$ or R^d, $d \geq 1$. We have seen several examples of such models already. Renewal processes distribute points on $[0, \infty)$ so that the gaps between points are iid random variables and the Poisson process on $[0, \infty)$ is a renewal process which distributes points so the gaps are iid exponential random variables.

Point processes contribute components to a solution of many varied modelling problems. We may need to model any of the following:

(1) the times of arrivals (departures, service initiations, and so forth) in a queue;

(2) the breakdown times (repair times) of a machine or a group of machines;

(3) the positions and times of earthquakes in the next 50 years;

(4) the location of oil relative to a known deposit;

(5) the location of trees in a forest;

(6) the location of tanks in a battlefield.

First, we consider some basic issues connected with modelling random distributions of points in a space. We wish our models to be general enough that they do not require the nice ordering present in the real line but absent in higher dimensional Euclidean spaces.

4.1. BASICS.

Suppose E is a subset of a Euclidean space; for us it is enough to suppose E is a subset of $[0, \infty), R$ or R^d for some $d \geq 1$. We want to distribute points randomly in E and have a convenient notation for a function which counts the random number of points which fall in a bounded set A. Suppose that

$\{X_n, n \geq 0\}$ are random elements of E which represent random points in the state space E. If we define the discrete measure

$$\epsilon_{X_n}(A) = \begin{cases} 1, & \text{if } X_n \in A, \\ 0, & \text{if } X_n \notin A, \end{cases}$$

then, by summing over n, we get the total number of random points X_n which fall in A. If we define the counting measure N by

$$N(\cdot) := \sum_n \epsilon_{X_n}(\cdot),$$

then

$$N(A) = \sum_n \epsilon_{X_n}(A)$$

is the random number of points which fall in the set A. N is called a *point process* and $\{X_n\}$ are called the *points*. The notation is designed to exhibit explicitly the dependence of the counting measure on its points. Note if the domain of the random elements $\{X_n\}$ is the sample space Ω, then N is also a function whose domain is Ω. Therefore, with $\omega \in \Omega$, we could write

$$N(\omega, A) = \sum_n \epsilon_{X_n(\omega)}(A),$$

but we will almost always suppress the dependence on ω in the notation.

A technical requirement for point processes, which allows us to manage infinities, is that bounded regions A must always contain a finite number of points with probability one; viz for any bounded set A,

$$P[N(A) < \infty] = 1.$$

For instance, for a renewal process, we have $E = [0, \infty)$, and the points are the renewal epochs $\{S_n, n \geq 0\}$. The point process corresponding to the points $\{S_n\}$ is

$$N = \sum_n \epsilon_{S_n}.$$

The property that bounded regions should contain a finite number of points is ensured by the finiteness of the renewal function proven in Chapter 3 (Theorem 3.3.1), since, if A is bounded, there exists a large $t > 0$ such that $A \subset [0, t]$, and, therefore,

$$EN(A) \leq EN([0, t]) = V(t) < \infty,$$

where V is the renewal function. Since $EN(A) < \infty$, we have

$$P[N(A) < \infty] = 1.$$

In Chapter 3, the notation for $N([0, t])$ was $N(t)$; this is a typical and convenient convention when $E = [0, \infty)$.

As another example, consider the modelling of earthquake locations and times. A suitable choice of state space would be $E = [0, \infty) \times R^2$, and the point process could be represented as

$$N = \sum_n \epsilon_{(T_n, (L_{n1}, L_{n2}))},$$

where T_n represents the time of the nth earthquake, and (L_{n1}, L_{n2}) represents its latitude and longitude. For $t > 0$ and $B \subset R^2$, $N([0, t] \times B)$ would be the number of earthquakes occurring in $[0, t]$ whose locations were in region B. If we wanted to add intensity to the model, a suitable state space would be $E = [0, \infty) \times R^2 \times [0, \infty)$, and the point process could be represented as

$$\sum_n \epsilon_{(T_n, (L_{n1}, L_{n2}), I_n)}(\cdot),$$

where I_n is the intensity of the nth earthquake.

The notation is flexible enough that it can be modified slightly if we wish to model multiple points at a single location. As before, regard $\{X_n\}$ as a sequence of random elements of E which will now represent *sites*, and let $\{\xi_n\}$ be non-negative integer valued random variables, where ξ_n represents the random number of points located at random site X_n. Then the point process with multiple points at its sites can be represented as

$$\sum_n \xi_n \epsilon_{X_n}.$$

Then, for a region $A \subset E$,

$$\sum_n \xi_n \epsilon_{X_n}(A)$$

is the number of points in A, since we have counted the number of points at each site in A. This type of set-up is ideal for modelling bulk arrivals. Imagine buses arriving at Harry's restaurant according to a point process with points $\{X_n\}$; the nth bus arrives at time X_n. If we represent the number of passengers in the nth bus as ξ_n, then the total number of customers arriving in $[0, t]$ is

$$\sum_n \xi_n \epsilon_{X_n}([0, t]).$$

An important summary statistic for the point process is the *mean measure*, frequently called the *intensity*, defined to be

$$\mu(A) = EN(A).$$

Thus $\mu(A)$ is the expected number of points in the region A. If N is the renewal process $\sum_{n=0}^{\infty} \epsilon_{S_n}$ then $\mu([0,t]) = EN(t) = V(t)$ where V is the renewal function of the renewal process.

The most tractable and commonly used point process model is the Poisson process and we discuss this in the next section.

4.2. THE POISSON PROCESS.

Let N be a point process with state space E. Suppose $\mathcal{E}$ is a class of reasonable subsets of E. (If you are happy with the adjective *reasonable*, skip the rest of this parenthetical remark: $\mathcal{E}$ should contain sets A for which it would be useful to ask how many points of N are contained in A. For instance, we demand that $\mathcal{E}$ contain all open subsets of E or all open intervals, etc. We also want $\mathcal{E}$ to have desirable closure properties: If $A \in \mathcal{E}$, then $A^c \in \mathcal{E}$, and if $A_n \in \mathcal{E}$ for $n \geq 1$, then $\cup_{n \geq 1} A_n \in \mathcal{E}$.*) Then N is a *Poisson process with mean measure μ* or, synonomously, a *Poisson random measure (PRM(μ))* if

(1)   For $A \in \mathcal{E}$

$$P[N(A) = k] = \begin{cases} \frac{e^{-\mu(A)}(\mu(A))^k}{k!}, & \text{if } \mu(A) < \infty \\ 0 & \text{if } \mu(A) = \infty. \end{cases}$$

(2)   If $A_1, \ldots, A_k$ are disjoint subsets of E in $\mathcal{E}$, then $N(A_1), \ldots, N(A_k)$ are independent random variables.

So N is Poisson if the random number of points in a set A is Poisson distributed with parameter $\mu(A)$ and the number of points in disjoint regions are independent random variables.

Property (2) is called *complete randomness*. When $E = R$ it is called the *independent increments* property, since we have for any $t_1 < t_2 < \cdots < t_k$ that $(N((t_i, t_{i+1}]), i = 1, \ldots, k-1)$ are independent random variables. When the mean measure is a multiple of Lebesgue measure (i.e., length when $E = [0, \infty)$ or R, area when $E = R^2$, volume when $E = R^3$, etc.), we call the process *homogeneous*. Thus, in the homogeneous case, there is

---

*An advanced student will by now have figured out that we have described the Borel σ-algebra in E; i.e., the σ-algebra generated by the open subsets.

a parameter $\alpha > 0$ such that for any A we have $N(A)$ Poisson distributed with mean $EN(A) = \alpha|A|$, where $|A|$ is the Lebesgue measure of A. When $E = [0, \infty)$ the parameter α is called the *rate* of the (homogeneous) Poisson process.

When $E = [0, \infty)$, we asserted in Chapter 3 that epochs of a pure renewal process in $(0, \infty)$ whose interarrival density was exponential deserved to be called a Poisson process. We must check if the earlier renewal theory definition of a Poisson process coincides with this one when $E = [0, \infty)$. This is asserted in the next proposition.

Proposition 4.2.1. Let $\{E_j, j \geq 1\}$ be iid random variables with a unit exponential distribution. Define $\Gamma_n = \sum_{i=1}^{n} E_i$ to be the renewal epochs of the renewal process and set $N = \sum_{n=1}^{\infty} \epsilon_{\Gamma_n}$. Then N is a homogeneous Poisson process on $[0, \infty)$ with unit rate $\alpha = 1$; i.e., N satisfies (1),(2) of this section and the mean measure is $|\cdot|$. Conversely, a homogeneous Poisson process on $[0, \infty)$ of unit rate satisfying (1),(2) has points arranged in increasing order which form a renewal process with interarrival density which is exponential with mean 1.

In Section 4.8 we give a proper proof of Proposition 4.2.1, but let us now explore informally why this equivalence must be true.

Start with a renewal sequence $\{\Gamma_n, n \geq 1\}$ described in Proposition 4.2.1. Why does it satisfy properties (1),(2) listed at the beginning of this section? We know from Section 3.6 that $N((0, t]) := \sum_{n \geq 1} \epsilon_{\Gamma_n}((0, t])$ is Poisson distributed with parameter t. This is property (1) above for sets A of the form $A = (0, t]$. We also need to check the independence property given in (2) above. The independent increment property (slightly weaker than (2)) can be checked by brute force (Doob, 1953) or by using Proposition 1.8.2.

Conversely, given a point process N on $[0, \infty)$ satisfying (1),(2) and such that $EN(A) = |A|$, let $T_1 < T_2 < \ldots$ be the points of N in increasing order. Note first of all that

$$P[T_1 > t] = P[N((0, t]) = 0] = e^{-t}$$

since $|(0, t]| = t$, the length of the interval. T_1 has an exponential density with parameter 1 and thus

$$T_1 \overset{d}{=} \Gamma_1.$$

Next, observe that for $0 < t_1 < t_2$

$$P[T_1 \leq t_1, T_2 \leq t_2] = P[N((0,t_1]) \geq 1, N((0,t_2]) \geq 2]$$

$$= \sum_{j=1}^{\infty} P[N((0,t_1]) = j, N((0,t_2]) \geq 2]$$

$$= \sum_{j=2}^{\infty} P[N((0,t_1]) = j] + P[N((0,t_1]) = 1, N((t_1,t_2]) \geq 1]$$

(since for $j \geq 2$ we have $[N((0,t_1]) = j] \subset [N((0,t_2]) \geq 2]$)

$$= \sum_{j=2}^{\infty} \frac{e^{-t_1} t_1^j}{j!} + t_1 e^{-t_1} (1 - e^{-(t_2-t_1)})$$

(using the complete randomness property (2))

$$= \sum_{j=1}^{\infty} \frac{e^{-t_1} t_1^j}{j!} + t_1 e^{-t_2}$$

$$= 1 - e^{-t_1} - t_1 e^{-t_2}.$$

Also, we have for $0 < t_1 < t_2$

$$P[\Gamma_1 \leq t_1, \Gamma_2 \leq t_2] = \iint_{\{(u,v): u \leq t_1, u+v \leq t_2\}} e^{-u} e^{-v} \, du \, dv$$

$$= \int_0^{t_1} e^{-u} \left(\int_0^{t_2-u} e^{-v} \, dv \right) du$$

$$= \int_0^{t_1} e^{-u} (1 - e^{-(t_2-u)}) \, du$$

$$= 1 - e^{-t_1} - t_1 e^{-t_2}.$$

Thus we see that

$$(T_1, T_2) \stackrel{d}{=} (\Gamma_1, \Gamma_2).$$

A small leap of faith makes it believable that $\{T_n\}$ and $\{\Gamma_n\}$ are distributionally equivalent. This verification can also be approached by noting that $\{N((0,t]), t > 0\}$ is a Markov process and using results from Chapter 5. See, for example, the remarks on the linear birth process in Section 5.1.

We know that to construct a homogeneous Poisson process on $[0, \infty)$ we merely construct a renewal process with exponential interarrival density.

A more general construction of the Poisson process which is applicable to a wide variety of spaces will be given in Section 4.9.

Suppose N is a homogeneous Poisson process on $[0, \infty)$ with rate α. What is the interpretation of α? We know $EN((0,t]) = \alpha|(0,t]| = \alpha t$, so we may interpret α as the expected rate of arrivals of points. Furthermore, from Theorem 3.3.2 we know

$$P[\lim_{t \to \infty} t^{-1}N((0,t]) = \alpha] = 1,$$

so, in fact, we may interpret α as the pathwise arrival rate of points. This also tells us that in a statistical context, if we know N is homogeneous Poisson, we may consistently estimate α by $\hat{\alpha} = N((0,t])/t$, assuming we have observed the process up to time t. (See Exercises 4.40 and 4.47 for more information on estimation.)

Next, let $o(h)$ represent a function, perhaps different with each usage, which has the property

$$\lim_{h \to 0} o(h)/h = 0.$$

From a Taylor expansion of the exponential function, we have $1 - e^{-h} = h + o(h)$. The relevance to the Poisson process is as follows: For $h > 0$

$$P[N((t, t+h]) = 1] = e^{-\alpha h}\alpha h = \alpha h + o(h)$$

and

$$
\begin{aligned}
P[N((t, t+h]) > 1] &= 1 - e^{-\alpha h} - \alpha h e^{-\alpha h} \\
&= \alpha h + o(h) - \alpha h e^{-\alpha h} \\
&= \alpha h (1 - e^{-\alpha h}) + o(h) \\
&= o(h).
\end{aligned}
$$

Therefore, the probability of finding a point of a Poisson process in an interval of length h is roughly proportional to h; the proportionality factor is the rate α.

Example. Arrivals at the Infamous Orthopedist's Office. Limping customers arrive at the infamous orthopedist's office according to a homogeneous Poisson process with rate $\alpha = 1/10$ minutes. Careful observation by Harry's niece reveals that the doctor does not bother admitting patients until at least three patients are in the waiting room. If $\Gamma_n = E_1 + \ldots E_n$, where $\{E_j\}$ are iid unit exponential random variables, then $\{E_j/\alpha\}$ are iid exponentially distributed random variables with parameter α. Thus, if a Poisson process has rate $\alpha = 1/10$, its points can be represented as

$\{\Gamma_n/\alpha\} = \{10\Gamma_n\}$. Therefore the expected waiting time until the first patient is admitted to see the doctor is

$$E(10\Gamma_3) = 10 \times 3 = 30 \text{ minutes.}$$

What is the probability that nobody is admitted in the first hour to see the doctor? This is the probability that at most two patients arrive in the first 60 minutes:

$$
\begin{aligned}
P[N((0,60)) \le 2] &= P[N((0,60)) = 0] + P[N((0,60)) = 1] \\
&\quad + P[N((0,60)) = 2] \\
&= e^{-60/10} + (60/10)e^{-60/10} + \frac{1}{2}(60/10)^2 e^{-60/10} \\
&= e^{-6}(1 + 6 + \frac{1}{2}6^2) = 25e^{-6} \\
&= 0.0620. \quad \blacksquare
\end{aligned}
$$

Example. Location of Competition. Harry wonders who the nearest competitor to his restaurant would be if restaurants were geographically distributed relative to his restaurant as a spatial Poisson process with rate $\alpha = 3$ per square mile. Let R be the distance of the nearest competitor and let $d(r)$ be a disc of radius r centered at Harry's restaurant. Then

$$P[R > r] = P[N(d(r)) = 0] = e^{-\alpha|d(r)|}.$$

Since the area of the disc is

$$|d(r)| = \pi r^2,$$

we get

$$P[R > r] = e^{-3\pi r^2}.$$

This means that R^2 is exponentially distributed with parameter 3π, since for $r > 0$

$$P[R^2 > r] = P[R > \sqrt{r}] = e^{-3\pi\sqrt{r^2}} = e^{-3\pi r}.$$

Also, the expected distance to the nearest competitor is

$$
\begin{aligned}
E(R) &= \int_0^\infty P[R > r]\,dr \\
&= \int_0^\infty e^{-3\pi r^2}\,dr.
\end{aligned}
$$

Setting $3\pi = 1/(2\sigma^2)$, we get

$$= \sqrt{2\pi\sigma^2} \int_0^\infty \frac{e^{-r^2/2\sigma^2}}{\sqrt{2\pi\sigma^2}} dr.$$

Because a normal density integrates to 1, this is

$$= \frac{1}{2}\sqrt{\frac{1}{3}} = 0.2887 \text{ miles. } \blacksquare$$

4.3. TRANSFORMING POISSON PROCESSES.

Some very useful results arise by considering what happens to a Poisson process under various types of transformations. The first result, though elementary, is enormously useful in understanding inhomogeneity. To prepare for this result, suppose $\sum_n \epsilon_{X_n}$ is a Poisson process with state space E and mean measure μ. Suppose T is some transformation with domain E and range E', where E' is another Euclidean space:

$$T : E \mapsto E'.$$

The function T^{-1} defines a set mapping of subsets of E' to subsets of E, defined for $A' \subset E'$ by

$$T^{-1}(A') = \{e \in E : T(e) \in A'\}.$$

Thus $T^{-1}(A')$ is the pre-image of A' under T; it is the set of points of E which T maps into A'. See Figure 4.1.

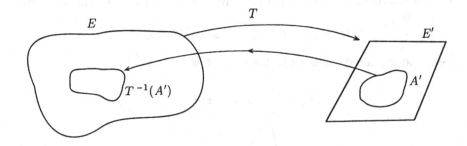

FIGURE 4.1.

As an example, suppose $E = (0, \infty)$, $E' = (-\infty, \infty)$, $T(x) = \log x$. If $a < b$, and $A' = (a, b)$, we have

$$T^{-1}((a,b)) = \{x > 0 : T(x) \in (a, b)\}$$
$$= \{x > 0 : \log x \in (a, b)\}$$
$$= \{x > 0 : x \in (e^a, e^b)\}.$$

Given the measures N, μ defined on subsets of E, we may use T to define induced measures N', μ' on subsets of E'. For $A' \subset E'$ define

$$N'(A') = N(T^{-1}(A')), \quad \mu'(A') = \mu(T^{-1}(A')).$$

To get the measure of A', we map A' back into E and take the measure of the pre-image under T. Also, if N has points $\{X_n\}$, then N' has points $\{X'_n\} = \{T(X_n)\}$, since for $A' \subset E'$

$$N'(A') = N(T^{-1}(A')) = \sum_n \epsilon_{X_n}(T^{-1}(A'))$$

$$= \sum_n 1_{[X_n \in T^{-1}(A')]}$$

$$= \sum_n 1_{[T(X_n) \in A']}$$

$$= \sum_n \epsilon_{T(X_n)}(A').$$

See Figure 4.2.

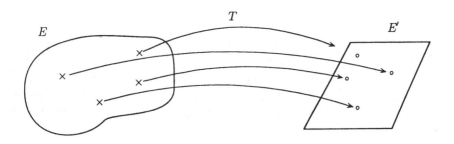

FIGURE 4.2.

The next result asserts that if N is a Poisson process with mean measure μ and points $\{X_n\}$ living in the state space E, then $N' = N(T^{-1}(\cdot))$ is a Poisson process with mean measure μ' and points $\{T(X_n)\}$ living in the state space E'.

Proposition 4.3.1. *Suppose*

$$T : E \mapsto E'$$

is a mapping of one Euclidean space, E, into another, E', with the property that if $B' \subset E'$ is bounded in E', then $T^{-1}B' := \{e \in E : Te \in B'\}$ is bounded in E. If N is $PRM(\mu)$ on E with points $\{X_n\}$, then $N' := N \circ T^{-1}$ is $PRM(\mu')$ on E' with points $\{T(X_n)\}$ and where $\mu' := \mu \circ T^{-1}$.

Remember that if N has the representation

$$N = \sum_n \epsilon_{X_n},$$

then

$$N' = \sum_n \epsilon_{TX_n}.$$

This result says that if you shift the points of a Poisson process around you still have a Poisson process.

Proof. We have

$$P[N'(B') = k] = P[N(T^{-1}(B')) = k] = p(k, \mu(T^{-1}(B'))),$$

so N' has Poisson distributions. It is easy to check the independence property, since, if $B'_1, \ldots, B'_m$ are disjoint, then so are $T^{-1}(B'_1), \ldots, T^{-1}(B_m)$, from which

$$(N'(B'_1), \ldots, N'(B'_m)) = \left(N(T^{-1}(B'_1)), \ldots, N(T^{-1}(B'_m))\right)$$

are independent. Thus (1),(2), which define a Poisson process, are satisfied for N'. ∎

Examples. For examples (1),(2) and (4) below, let $N = \sum_{n=1}^{\infty} \epsilon_{\Gamma_n}$ be a homogeneous Poisson process with rate $\alpha = 1$ on the state space $E = [0, \infty)$. The mean measure μ is Lebesgue measure so that $\mu(A) = |A|$ and, in particular, $\mu([0, t]) = t$. Recall that we can always construct this process by the renewal theory method of summing iid unit exponential random variables.

   (1)   If $Tx = x^2$, then $\sum_n \epsilon_{\Gamma_n^2}$ is PRM, and the mean measure μ' is given by

$$\mu'[0, t] = \mu\{x : Tx \le t\} = \mu[0, \sqrt{t}] = \sqrt{t}.$$

Note that μ' has a density

$$\alpha(t) = \frac{d}{dt}\sqrt{t} = \frac{1}{2}t^{-1/2}.$$

(2) If $T : E \mapsto E \times E$ via $Tx = (x, x^2)$, then $\sum_n \epsilon_{T\Gamma_n} = \sum_n \epsilon_{(\Gamma_n, \Gamma_n^2)}$ is Poisson on $E \times E$. The mean measure concentrates on the graph $\{(x, x^2) : x \geq 0\}$.

(3) Suppose N is homogeneous Poisson on $E = (-\infty, \infty)$, and define $T : E \mapsto (0, \infty)$ via $Tx = e^x$. Then $N' = N(T^{-1}(\cdot))$ is Poisson on $(0, \infty)$ with mean measure μ' given by $(0 < a < b)$

$$\mu'(a, b] = \mu\{x : e^x \in (a, b]\} = \log b - \log a.$$

Again, μ' has a density

$$\alpha(t) = \frac{d}{dt}\log t = \frac{1}{t}.$$

(4) If $\sum_n \epsilon_{\Gamma_n}$ is homogeneous Poisson on $[0, \infty)$, then $\sum_n \epsilon_{\Gamma_n^{-1}}$ is Poisson on $(0, \infty]$ with mean measure μ' given for $t > 0$ by

$$\mu'(t, \infty] = \mu\{s \geq 0 : s^{-1} \geq t\} = \mu[0, t^{-1}) = t^{-1}.$$

μ' has a density

$$\alpha(t) = -\frac{d}{dt}t^{-1} = t^{-2}.$$

Note the density becomes infinite as $t \to 0$, indicating a large density of points about 0.

This example contains a subtlety. We know that for point processes, bounded regions of $E' = (0, \infty]$ must contain a finite number of points. Therefore, $(0, 1]$ had better not be bounded in E', because it contains an infinite number of points $\{\Gamma_n^{-1}\}$, following from the fact that $[1, \infty)$ contains an infinite number of the points $\{\Gamma_n\}$. In E', the bounded sets are those sets bounded away from 0; i.e., the bounded sets are neighborhoods of ∞. The transformation of E to E' in this case essentially interchanges the roles of 0 and ∞. Distance may be measured in E' by the metric $(e'_1, e'_2 \in E')$

$$d'(e'_1, e'_2) = |\frac{1}{e'_1} - \frac{1}{e'_2}|,$$

and we interpret $1/\infty$ as 0.

Poisson processes with mean measure $\mu'(dt) = t^{-2}dt$, $t > 0$, are particularly important in extreme value theory and in the theory of stable processes. We touch briefly on these connections in Section 4.3.1 below.

The non-homogeneous Poisson process. Suppose N is a Poisson process on $[0, \infty)$ with mean measure μ and that μ is absolutely continuous with density $\alpha(t)$. For reasons to be explained later, we call N a *non-homogeneous Poisson process with local intensity* $\alpha(t)$. This process may be obtained as a transformation of the homogeneous Poisson process by the following scheme. Define

$$m(t) = \mu[0, t] = \int_0^t \alpha(s)ds,$$

and define the inverse function

$$m^{\leftarrow}(x) = \inf\{u : m(u) \geq x\}.$$

If we suppose $m(\infty) = \infty$, then the set $\{u : m(u) \geq x\}$ is non-empty for all x, so

$$m^{\leftarrow} : [0, \infty) \mapsto [0, \infty),$$

and, since m is continuous, we have that $m^{\leftarrow}$ is strictly increasing. If $\sum_n \epsilon_{\Gamma_n}$ is homogeneous Poisson, then

$$N' = \sum_n \epsilon_{m^{\leftarrow}(\Gamma_n)}$$

is also a Poisson process. The mean measure is (remember $|A|$ is the Lebesgue measure or length of A)

$$\begin{aligned}
\mu'[0, t] &= |\{x : m^{\leftarrow}(x) \leq t\}| \\
&= |\{x : x \leq m(t)\}| = |[0, m(t)]| \\
&= m(t) = \mu[0, t].
\end{aligned}$$

Thus we have constructed PRM(μ) from a homogeneous Poisson process.

This means that if $\sum_{n=1}^{\infty} \epsilon_{X_n}$ is PRM(μ) on $[0, \infty)$ where $X_1 < X_2 < \ldots$, then

$$\{X_n, n \geq 1\} \stackrel{d}{=} \{m^{\leftarrow}(\Gamma_n), n \geq 1\},$$

so that in R^{∞}

$$\{m(X_n), n \geq 1\} \stackrel{d}{=} \{\Gamma_n, n \geq 1\}.$$

Therefore, $\sum_n \epsilon_{m(X_n)}$ and $\sum_n \epsilon_{\Gamma_n}$ have the same distributions, and thus $\sum_n \epsilon_{m(X_n)}$ is a homogeneous Poisson process with rate 1.

Note that the assumption that μ is absolutely continuous, though traditional, is more than what is needed. Continuity of m and $m(\infty) = \infty$ are adequate assumptions.

If N is a non-homogeneous Poisson process with mean measure μ and local intensity $\alpha(\cdot)$, we have, for any $t > 0$ where $\alpha(\cdot)$ is continuous and small $h > 0$, that

$$
\begin{aligned}
P[N((t, t+h] = 1] &= \mu((t, t+h])e^{-\mu((t,t+h])} \\
&= \mu((t, t+h]) + o(h) \\
&= \int_t^{t+h} \alpha(u)du + o(h) \\
&= \alpha(t)h + o(h).
\end{aligned}
$$

As in the homogeneous case, we may check that

$$
P[N((t, t+h]) \geq 2] = o(h).
$$

The probability of finding a point in a neighborhood of t, therefore, is proportional to the size of the neighborhood. The proportionality factor is the local intensity.

Example 4.3.1. Periodic Demand. To model the periodic nature of customer demand in queueing models, researchers frequently propose a traffic input which is a non-homogeneous Poisson process with a periodic local intensity function. Suppose, for instance, that

$$
\alpha(t) = 2 + \sin t.
$$

Then

$$
m(t) = \int_0^t (2 + \sin s)ds = 2t + 1 - \cos t
$$

gives the expected number of arrivals in $(0, t)$.

4.3.1. MAX-STABLE AND STABLE RANDOM VARIABLES*.

Poisson processes similar to those constructed in the last section are basic to understanding extreme value random variables and stable random variables. Here is a brief treatment.

Extreme value laws. To see the connection between Poisson processes and extreme value theory, we start with a particular Poisson process and construct a random variable which will have a classical extreme value distribution. Proceed as follows: Let $N' = \sum_n \epsilon_{X_n'}$ be a Poisson process

---

* This section may be skipped on first reading without loss of continuity.

on $(0, \infty]$ with mean measure $\mu'(dt) = t^{-2}dt$, $t > 0$. Note that for any $x > 0$, since $\mu'((x, \infty]) = x^{-1} < \infty$, we have

$$P[N'((x, \infty]) < \infty] = 1.$$

There are only a finite number of large points, and we may define

$$Y = \bigvee_n X'_n,$$

where Y is simply the biggest point of the Poisson process. The distribution of Y is easily computed: For $x > 0$

$$
\begin{aligned}
P[Y \le x] &= P[\bigvee_n X'_n \le x] \\
&= P[N'((x, \infty]) = 0] \\
&= e^{-\mu'((x,\infty])} \\
&= e^{-x^{-1}},
\end{aligned}
$$

which is one of the classical extreme value distributions.

Distributions similar to the distribution of Y arise in extreme value theory as approximations to the distribution of extremes of a random sample. Given a random sample $X_1, \ldots, X_n$ of size n from the underlying distribution $F(x)$, the distribution of the maximum $M_n = \max\{X_1, \ldots, X_n\} = \bigvee_{i=1}^n X_i$ is given by

$$P[M_n \le x] = P[X_1 \le x, \ldots, X_n \le x] = F^n(x).$$

In order to obtain an approximation to the distribution of M_n, we hope that there exist a non-degenerate limit distribution function $G(x)$, centering constants $b_n \in R$ and scaling constants $a_n > 0$ such that

(4.3.1.1) $P[a_n^{-1}(M_n - b_n) \le x] = F^n(a_n x + b_n) \to G(x).$

In this case, we call G an *extreme value distribution*, and we say F is in the *domain of attraction* of G. The possible forms of G were worked out by Gnedenko (1943), and good treatments can be found in de Haan (1970), Leadbetter et al. (1983) and Resnick (1987). The distribution of Y given above is one of the extreme value distributions.

A distribution G is called *max-stable* if for every $t > 0$ there exist $\alpha(t) > 0, \beta(t) \in R$ such that

(4.3.1.2) $G(x) = G^t(\alpha(t)x + \beta(t)).$

The class of extreme value distributions coincides with the non-degenerate max-stable distributions; this is discussed in the references above. (One direction of the equivalence is easy since if G is max-stable and (4.3.1.2) holds, we may replace t by n, and then (4.3.1.1) is immediate.)

We can quickly see that the distribution of Y is max-stable. If $G(x) = \exp\{-x^{-1}\}$ for $x > 0$, we have, for $x > 0$,

$$G^t(x) = \exp\{-tx^{-1}\},$$

and thus

$$G^t(tx) = G(x).$$

Stable laws. There are several equivalent definitions of a stable law. Here is Feller's (1971) definition: The distribution G is (sum) *stable* if it is non-degenerate and for every n there exist scaling constants $c_n > 0$ and centering constants $d_n \in R$ such that if $X_1, \dots, X_n$ are iid random variables with common distribution G then

$$\sum_{i=1}^{n} X_i \overset{d}{=} c_n X_1 + d_n.$$

The scaling constants c_n can always be taken to be $c_n = n^{1/\alpha}$ (Feller, 1971, page 171, for example), and in this case we say G is stable with index α.

To see the connection with the Poisson process, it is simplest to look at a variant of the construction discussed in the extreme values section: Let $0 < \alpha < 1$ and define the Poisson process

$$N^\# = \sum_{n=1}^{\infty} \epsilon_{\Gamma_n^{-1/\alpha}}.$$

Let $\mu^\#$ denote the mean measure of $N^\#$, and, since $N^\#$ is constructed from the transformation $T(x) = x^{-1/\alpha}$, we have that

$$\mu^\#((x, \infty]) = x^{-\alpha}.$$

Define the random variable

$$X = \sum_{n=1}^{\infty} \Gamma_n^{-1/\alpha}.$$

We will verify that X is a stable random variable with index α. It is easy to see the series converges, because $\Gamma_n \sim n$ as $n \to \infty$ by the law of large numbers. Since $1/\alpha > 1$, convergence of the series defining X follows by comparison with the series $\sum_n n^{-1/\alpha}$.

We now verify that X is a stable random variable with index α. For $i = 1, 2$, let

$$N_i^\# = \sum_{n=1}^{\infty} \epsilon_{(\Gamma_n^{(i)})^{-1/\alpha}}$$

be two iid Poisson processes each having mean measure $\mu^\#((x, \infty]) = x^{-\alpha}$. The superposition $N_1^\# + N_2^\#$ is a Poisson process with mean measure $2\mu^\#$, and thus

$$(4.3.1.3) \qquad N_1^\# + N_2^\# \overset{d}{=} \sum_{n=1}^{\infty} \epsilon_{2^{1/\alpha}\Gamma_n^{-1/\alpha}},$$

since the process on the right side of (4.3.1.3) has the correct mean measure. (As usual, $\{\Gamma_n\}$ is the renewal sequence corresponding to interarrival distribution which is exponential with parameter 1.) Summing the points, we get, with obvious notation,

$$X_1 + X_2 \overset{d}{=} 2^{1/\alpha} X$$

where $X_1 \overset{d}{=} X_2 \overset{d}{=} X$ and X_1 is independent of X_2. This argument is easily generalized from two summands to n summands and shows X is a stable random variable with index α.

A similar construction can be used to define stable random variables with indices $1 \le \alpha < 2$, but the terms $\Gamma_n^{-1/\alpha}$ must be centered before summing in order to guarantee convergence of the infinite series of random variables.

4.4. MORE TRANSFORMATION THEORY; MARKING AND THINNING.

Given a Poisson process, under certain circumstances it is possible to enlarge the dimension of the points and retain the Poisson structure. One way to do this was given in example (2) of Section 4.3, but the enlargement of dimension was illusory since the points concentrated on a graph $\{(x, x^2) : x > 0\}$. The result presented here allows independent components to be added to the points of the Poisson process. This proves very useful in a variety of applications. We present the result in two parts; the first is simpler. The second is presented in Section 4.10 and is used in Section 4.11 in connection with records.

Proposition 4.4.1. *Suppose $\{X_n\}$ are random elements of a Euclidean space E_1 such that*

$$\sum_n \epsilon_{X_n}$$

is PRM(μ). Suppose $\{J_n\}$ are iid random elements of a second Euclidean space E_2 with common probability distribution F and suppose the Poisson process and the sequence $\{J_n\}$ are defined on the same probability space and are independent. Then the point process

$$\sum_n \epsilon_{(X_n, J_n)},$$

on $E_1 \times E_2$ is PRM with mean measure $\mu \times F$, meaning that if $A_i \subset E_i$, $i = 1, 2$, then

$$\mu \times F(A_1 \times A_2) = \mu \times F(\{(e_1, e_2) : e_1 \in A_1, e_2 \in A_2\}) = \mu(A_1)F(A_2).$$

Often this procedure is described by saying we give to point X_n the *mark J_n*. See Figure 4.3 for a picture of this construction. The points of the original Poisson process $\{X_n\}$ appear on the horizontal axis, and the marked points appear in the $E_1 \times E_2$ plane.

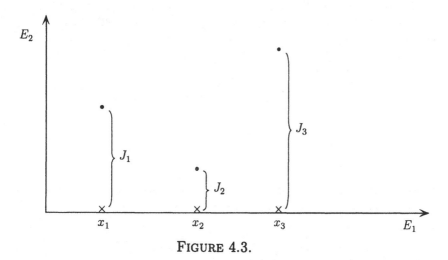

FIGURE 4.3.

The proof is deferred to Section 4.10, by which time some machinery will have been developed which makes the proof extremely simple. For now, note the mean measure is correct, since, for a nice set of the form

$A_1 \times A_2 = \{(e_1, e_2) : e_1 \in A_1 \subset E_1, e_2 \in A_2 \subset E_2\}$, we have

$$E \sum_n \epsilon_{(X_n, J_n)}(A_1 \times A_2) = \sum_n P[(X_n, J_n) \in A_1 \times A_2]$$
$$= \sum_n P[X_n \in A_1]P[J_n \in A_2]$$

because $\{J_n\}$ is independent of the Poisson process. Since $\{J_n\}$ are iid random variables this is the same as

$$= \sum_n P[X_n \in A_1]P[J_1 \in A_2]$$
$$= E(\sum_n \epsilon_{X_n}(A_1))P[J_1 \in A_2]$$
$$= \mu(A_1)P[J_1 \in A_2].$$

Example 4.4.1. The Stationary M/G/∞ Queue. Calls arrive to a telephone exchange at time points which constitute a Poisson process on R with mean measure μ. The lengths of the calls do not depend on the time when they are initiated and are iid random variables $\{J_n\}$ with common distribution F. We will check that the times when calls terminate and free up lines also constitute a Poisson process.

To analyze this situation, suppose the point process of arriving calls is represented by the point process $\sum_n \epsilon_{X_n}$ which is PRM(μ). Then, by Proposition 4.4.1,

$$\sum_n \epsilon_{(X_n, J_n)}$$

is also PRM with mean measure $\mu \times F$. From Proposition 4.3.1,

$$\sum_n \epsilon_{X_n + J_n},$$

which represents the point process of times when calls terminate, is also Poisson with mean measure μ_1 where $(a < b)$

$$\mu_1(a, b] = \mu \times F\{(x, y) : x + y \in (a, b]\}$$
$$= \iint_{\{(x,y):x+y\in(a,b]\}} \mu(dx)F(dy)$$
$$= \int_R \mu(dx)\left(F(b - x) - F(a - x)\right).$$

Note that if the original Poisson process on R is assumed homogeneous, so that μ is Lebesgue measure or length, then

$$\mu_1(a,b] = \int_0^\infty |(a-y,b-y]|F(dy) = \int_0^\infty (b-a)F(dy) = b-a,$$

so that μ_1 is also Lebesgue measure. Thus, if the input (times of call initiations) is homogeneous Poisson, then so is the output (times of call terminations). Therefore, randomly shifting the points of a homogeneous Poisson process results in a process which is still homogeneous Poisson.

Example 4.4.2. How to Construct a Homogeneous Planar Poisson Process. Suppose we know that $\sum_n \epsilon_{(X_n,Y_n)}$ is a homogeneous planar Poisson process. What happens if we transform to polar coordinates? Let $T(x,y) = (r,\theta) = (\sqrt{x^2+y^2}, \arctan y/x)$ be the polar coordinate transformation. Then according to Proposition 4.3.1,

$$\sum_n \epsilon_{T(X_n,Y_n)} =: \sum_n \epsilon_{(R_n,\Theta_n)}$$

is a two-dimensional Poisson process. If we call the mean measure μ', we have

$$\mu'([0,r] \times [0,\theta]) = \iint_{\{(x,y):T(x,y)\in[0,r]\times[0,\theta]\}} dx\,dy.$$

Switching to polar coordinates to do the integration yields

$$= \iint_{\{s\leq r, \eta\leq\theta\}} s\,ds\,d\eta$$

$$= \frac{1}{2}r^2\theta = \pi r^2 \frac{\theta}{2\pi}.$$

From Section 4.3, the non-homogeneous Poisson process with local intensity $\alpha(r) = 2\pi r$ can be represented as

$$\sum_n \epsilon_{\Gamma_n^{1/2}/\sqrt{2\pi}},$$

where, as usual, $\{\Gamma_n\}$ is a sequence of partial sums of iid unit exponential random variables. If $\{U_n\}$ are iid random variables, uniformly distributed on $[0,2\pi)$ and independent of $\{\Gamma_n\}$, then by Proposition 4.4.1 we get that

$$\sum_n \epsilon_{(\Gamma_n^{1/2}/\sqrt{2\pi}, U_n)}$$

is a Poisson process with the same mean measure as $\sum_n \epsilon_{(R_n, \Theta_n)}$.

This gives us a method for constructing or simulating a homogeneous planar Poisson process: Construct a one dimensional planar Poisson process $\{\Gamma_n\}$ and an independent iid family $\{U_n\}$ where each U_n is uniform on $[0, 2\pi)$. With these ingredients, build a point process on $[0, \infty) \times [0, 2\pi)$ with points

$$(\frac{\Gamma_n^{1/2}}{\sqrt{2\pi}}, U_n).$$

Transform these points by applying the map from polar coordinates to cartesian coordinates:

$$(r, \theta) \mapsto (x, y) = (r \cos \theta, r \sin \theta).$$

The resulting points will be the points of a homogeneous planar Poisson process.

Thinning a Poisson process. Suppose $\sum_n \epsilon_{X_n}$ is a Poisson process on the state space E with mean measure μ. Suppose we inspect each point independently of others and decide with probability p to retain the point and with probability $1 - p = q$ to delete the point. Let N_r be the point process of retained points and N_d be the point process of deleted points. We will show that N_r, N_d are independent Poisson processes with mean measures $p\mu$ and $q\mu$ respectively.

To analyze this, let $\{B_i\}$ be iid Bernoulli random variables independent of the points of the Poisson process $\{X_n\}$ so that

$$P[B_1 = 1] = p = 1 - P[B_1 = -1].$$

Then we know by Proposition 4.4.1 that

$$\sum_n \epsilon_{(X_n, B_n)}$$

is PRM on $E \times \{-1, 1\}$ with mean measure $\mu \times P[B_1 = \cdot]$. We think of the points $\{X_n : B_n = 1\}$ as the retained points; similarly, $\{X_n : B_n = -1\}$ is the collection of deleted points. Then we have by the complete randomness property (2) in Section 4.2 that

$$N_r(\cdot) = \sum_n \epsilon_{(X_n, B_n)}((\cdot) \times \{1\}) = \sum_{n: B_n = 1} \epsilon_{X_n}$$

and

$$N_d(\cdot) = \sum_n \epsilon_{(X_n, B_n)}(\cdot \times \{-1\}) = \sum_{n: B_n = -1} \epsilon_{X_n}$$

are independent processes since they are based on the values of $\sum_n \epsilon_{(X_n, B_n)}$ on two disjoint regions of the state space.

This result could, of course, be generalized. Previously we categorized or marked the points in two ways: retained or deleted. We could just as well randomly assign the points to any of k categories. The B's of the discussion in the previous paragraph would be replaced by multinomial random variables with k cells. The resulting point processes $N_1, \ldots, N_k$ would still be independent Poisson processes.

Results like these are used to justify splitting a Poisson input stream to a queue into independent Poisson substreams.

Example. Rush Hour at Harry's Restaurant. Harry's restaurant is well known for serving great food, but it is filthy and hence not for those with weak stomachs. During rush hour, customers arrive at the restaurant according to a Poisson process of rate a. Customers peek in the door and with probability q they decide the filth is not for them and depart; with probability p they enter and eat.

What is the distribution of the waiting times between entrances of customers into the restaurant? What is the mean and variance of this waiting time?

The stream of customers who enter is independent of the stream that departs in disgust. Customers are thinned and those that actually enter constitute a Poisson process of rate pa. (The mean measure of the process before thinning is $a|\cdot|$ where $|\cdot|$ is Lebesgue measure. After thinning, the mean measure is $pa|\cdot|$ from which the rate is pa.) We know that points of a Poisson process have the distribution of sums of iid exponentially distributed random variables and hence the times between entrances into the restaurant are iid exponential with parameter ap. Therefore, the mean time between entrances is $1/(ap)$ and the variance is $1/(ap)^2$. ∎

4.5. THE ORDER STATISTIC PROPERTY.

As we shall see, one way to construct a homogeneous Poisson process on a bounded region A is to sprinkle a Poisson number of iid random variables on the region where the random variables have a uniform distribution over A. Thus, conditional on there being n points in the bounded region A these points are distributed as n iid uniformly distributed random elements of A. Consequently, in statistics, the homogeneous Poisson process is often used as a null hypothesis of no interaction between the points, since the Poisson process can be thought of as a model for points distributed independently and at random in a region.

In this section, we suppose the state space E is $[0, \infty)$ and explore this result and its uses. This result is examined in a more sophisticated way in Section 4.9 and in the exercises.

We begin by reviewing the concept of order statistics and proving some useful connections between order statistics of the uniform distribution and the points of a homogeneous Poisson process.

Suppose $X_1, \ldots, X_n$ are iid random variables defined on the sample space Ω and having the common distribution F. If we suppose $F(x)$ is a continuous distribution function, then ties among the X's occur only with probability 0 and may thus be neglected. We define new random variables $X_{(1)}, \ldots, X_{(n)}$ with domain Ω, called order statistics, as follows: For $\omega \in \Omega$ define

$$X_{(1)}(\omega) = \min\{X_1(\omega), \ldots, X_n(\omega)\}$$
$$X_{(2)}(\omega) = \text{second smallest of } \{X_1(\omega), \ldots, X_n(\omega)\}$$
$$\vdots$$
$$X_{(n-1)}(\omega) = \text{second largest of } \{X_1(\omega), \ldots, X_n(\omega)\}$$
$$X_{(n)}(\omega) = \max\{X_1(\omega), \ldots, X_n(\omega)\},$$

so that with probability 1

$$X_{(1)} < \cdots < X_{(n)}.$$

We now derive the joint density of the order statistics when F is the uniform distribution on $(0, t)$.

Lemma 4.5.1. (a) Suppose $U_1, \ldots, U_n$ are iid uniformly distributed on $(0, t)$ and that $U_{(1)} < \cdots < U_{(n)}$ are the order statistics. Then the joint density of the order statistics is given by

$$f_{U_{(1)}, \ldots, U_{(n)}}(u_1, \ldots, u_n) = \begin{cases} \frac{n!}{t^n}, & \text{if } 0 < u_1 < \cdots < u_n < t \\ 0, & \text{otherwise.} \end{cases}$$

(b) Suppose, as usual, that $\{E_n\}$ are iid unit exponential random variables and that for $n \geq 1$ we define $\Gamma_n = E_1 + \cdots + E_n$. Then conditional on $\Gamma_{n+1} = t$ the joint density of $\Gamma_1, \ldots, \Gamma_n$ is

$$f_{\Gamma_1, \ldots, \Gamma_n | \Gamma_{n+1} = t}(u_1, \ldots, u_n) = \begin{cases} \frac{n!}{t^n}, & \text{if } 0 < u_1 < \cdots < u_n < t \\ 0, & \text{otherwise .} \end{cases}$$

Therefore the conditional distribution of $\Gamma_1, \ldots, \Gamma_n$ *given* $\Gamma_{n+1} = t$ *is the same as that of the order statistics from the uniform distribution on* $(0, t)$.

Part (b) is not needed until Section 4.8, where we discuss fully the reason why the renewal theory construction with exponentially distributed interarrival times yields a Poisson process satisfying (1),(2) of Section 4.2.

Proof. (a) Let Π be the collection of $n!$ permutations of the integers $\{1, \ldots, n\}$. Then for $\pi \in \Pi$ we have $(U_{(1)}, \ldots, U_{(n)}) = (U_{\pi(1)}, \ldots, U_{\pi(n)})$ on the set $[U_{\pi(1)} < \cdots < U_{\pi(n)}]$. Thus, for any bounded function $g(u_1, \ldots, u_n)$, we have

$$Eg(U_{(1)}, \ldots, U_{(n)}) = \sum_{\pi \in \Pi} Eg(U_{\pi(1)}, \ldots, U_{\pi(n)}) 1_{[U_{\pi(1)} < \cdots < U_{\pi(n)}]}.$$

Since the joint density of $(U_{\pi(1)}, \ldots, U_{\pi(n)})$ is

$$f_{U_{\pi(1)}, \ldots, U_{\pi(n)}}(u_1, \ldots, u_n) = f_{U_1, \ldots, U_n}(u_1, \ldots, u_n),$$

we get

$$f_{U_{\pi(1)}, \ldots, U_{\pi(n)}}(u_1, \ldots, u_n) = \begin{cases} t^{-n}, & \text{if } (u_1, \ldots, u_n) \in [0, 1]^n \\ 0, & \text{otherwise,} \end{cases}$$

and, therefore,

$$Eg(U_{(1)}, \ldots, U_{(n)})$$
$$= \sum_{\pi \in \Pi} \int_{[0 < u_1 \cdots < u_n < t]} g(u_1, \ldots, u_n) t^{-n} du_1 \ldots du_n$$
$$= \int_{[0,1]^n} g(u_1, \ldots, u_n) \left(\frac{n!}{t^n} 1_{[u_1 < \cdots < u_n]}(u_1, \ldots, u_n) \right) du_1 \ldots du_n.$$

This shows that the density of $(U_{(1)}, \ldots, U_{(n)})$ is

$$\frac{n!}{t^n} 1_{[u_1 < \cdots < u_n]}(u_1, \ldots, u_n),$$

as was to be proved.

(b) Since $E_1, \ldots, E_{n+1}$ are independent the joint density is

$$f_{E_1, \ldots, E_{n+1}}(x_1, \ldots, x_{n+1}) = \prod_{i=1}^{n+1} e^{-x_i}$$
$$= e^{-\sum_{i=1}^{n+1} x_i}, \quad \text{for } x_i > 0, \ 1 \le i \le n+1.$$

From this density we get the density of $(\Gamma_1, \ldots, \Gamma_{n+1})$ by change of variables: For $i = 1, \ldots, n+1$ define

$$s_i = \sum_{j=1}^{i} x_i,$$

so that the inverse transformation is

$$x_i = s_i - s_{i-1}, \quad 1 \le i \le n+1; \quad s_0 = 0.$$

The Jacobean of the inverse transform is

$$\left| \det\left(\frac{\partial x_i}{\partial s_j}\right) \right| = \left| \det \begin{pmatrix} 1 & 0 & \cdots & 0 & 0 \\ -1 & 1 & 0 & \cdots & 0 \\ 0 & -1 & 1 & \cdots & 0 \\ \vdots & \vdots & \ddots & \ddots & \vdots \\ 0 & 0 & \cdots & -1 & 1 \end{pmatrix} \right| = 1,$$

and, therefore, the joint density of $\Gamma_1, \ldots, \Gamma_{n+1}$ is

$$f_{\Gamma_1, \ldots, \Gamma_{n+1}}(s_1, \ldots, s_{n+1}) = e^{-s_{n+1}}, 0 \le s_1 < s_2 < \ldots < s_{n+1}.$$

Since Γ_{n+1} has a gamma density

$$f_{\Gamma_{n+1}}(t) = \frac{e^{-t} t^n}{n!}, \quad t \ge 0,$$

we get the conditional density of $\Gamma_1, \ldots, \Gamma_n$ given Γ_{n+1} by dividing:

$$f_{\Gamma_1, \ldots, \Gamma_n | \Gamma_{n+1} = t}(s_1, \ldots, s_n) = \frac{f_{\Gamma_1, \ldots, \Gamma_{n+1}}(s_1, \ldots, s_n, t)}{f_{\Gamma_{n+1}}(t)}$$

$$= \frac{n!}{t^n}, \quad 0 < s_1 < \ldots < s_n < s.$$

The latter density is the density of the order statistics of n iid random variables which are uniformly distributed on $[0, t]$. ∎

We now prove a homogeneous Poisson process on $[0, \infty)$ has the *order statistic property*; namely that, conditional on there being n points in $(0, t]$, the locations of these points are distributed like the order statistics from a sample of size n from the uniform distribution on $(0, t)$.

Theorem 4.5.2. *If N is a homogeneous Poisson process on $[0, \infty)$ with rate α, then, conditional on*

$$[N((0, t]) = n],$$

the points of N in $[0, t]$ in increasing order are distributed as the order statistics from a sample of size n from the uniform distribution $U(0, t)$ on $[0, t]$; i.e.,

$$(\Gamma_1, \dots, \Gamma_n | N[0, t] = n) \stackrel{d}{=} (U_{(1)}, \dots, U_{(n)}).$$

Thus, if the non-negative function $g(x_1, \dots, x_n)$ is a symmetric function of its arguments, meaning, for any $\pi \in \Pi$,

$$g(x_1, \dots, x_n) = g(x_{\pi(1)}, \dots, x_{\pi(n)}),$$

we have the following equality in distribution for the conditional distribution of $g(\Gamma_1, \dots, \Gamma_n)$ given n points in $(0, t]$:

$$(4.5.1) \quad (g(\Gamma_1, \dots, \Gamma_n) | N[0, t] = n) \stackrel{d}{=} g(U_{(1)}, \dots, U_{(n)}) = g(U_1, \dots, U_n).$$

Remark. *If $n = 2$, g symmetric means $g(x_1, x_2) = g(x_2, x_1)$. For general n, the typical application is when g is of the form*

$$g(x_1, \dots, x_n) = \sum_{i=1}^{n} h(x_i)$$

for some nice function $h(x)$.

Proof. A short non-computational proof is presented in Section 4.9. Assuming we believe the conditional density exists, we can derive it in an elementary manner as follows: Suppose the homogeneous Poisson process has points $\{\Gamma_n\}$. Then suppose $a_1 < b_1 < a_2 < b_2 < \cdots a_n < b_n < t$, and for the conditional distribution of $(\Gamma_1, \dots, \Gamma_n)$ given $N(0, t] = n$ we have

$$P[\Gamma_i \in (a_i, b_i], \ i = 1, \dots, n | N((0, t]) = n]$$
$$= P[a_i < \Gamma_i \le b_i, \ i = 1, \dots, n, N((0, t]) = n] / P[N((0, t]) = n]$$

For the numerator we have

$$P[N((0, a_1] = 0, N((a_1, b_1]) = 1,$$
$$\cap_{i=1}^{n-1} [N((b_i, a_{i+1}]) = 0, N((a_{i+1}, b_{i+1}]) = 1], N((b_n, t]) = 0]$$
$$= e^{-\alpha a_1} \alpha(b_1 - a_1) e^{-\alpha(b_1 - a_1)}$$
$$\prod_{i=1}^{n-1} e^{-\alpha(a_{i+1} - b_i)} \alpha(b_{i+1} - a_{i+1}) e^{-\alpha(b_{i+1} - a_{i+1})} e^{-\alpha(t - b_n)}$$

and therefore

$$P[\Gamma_i \in (a_i, b_i], i = 1, \ldots, n | N((0, t]) = n]$$

$$= \frac{n!}{t^n} \prod_{i=1}^{n} (b_i - a_i).$$

Dividing through by $\prod_{i=1}^{n} (b_i - a_i)$ and letting $b_i \downarrow a_i$ for $i = 1, \ldots, n$, the left side becomes the conditional density given $[N((0, t]) = n]$:

$$f_{\Gamma_1, \ldots, \Gamma_n | N((0, t]) = n}(a_1, \ldots, a_n) = \frac{n!}{t^n},$$

as required. ∎

Example 4.5.1. Shot Noise Processes. This class of processes is a superposition of iid random impulses. Assume electrons arrive according to a homogeneous Poisson process with rate α on $[0, \infty)$. An arriving electron produces an electrical current whose intensity t time units after arrival is $w(t)$. Typical choices for w are exponential functions:

$$w(t) = \exp\{-\theta t\}, \quad \theta > 0.$$

If arrivals occur at $\{\Gamma_n\}$, then the total current at time t is

$$X(t) = \sum_{i=1}^{N((0, t])} w(t - \Gamma_i),$$

and we may use the order statistic property to determine the distribution of $X(t)$ for each fixed t. The Laplace transform is

$$E \exp\{-\lambda X(t)\} = E \exp\left\{-\lambda \sum_{i=1}^{N((0, t])} w(t - \Gamma_i)\right\}$$

$$= \sum_{n=0}^{\infty} E\left(\exp\left\{-\lambda \sum_{i=1}^{n} w(t - \Gamma_i)\right\} | N((0, t]) = n\right) P[N((0, t]) = n]$$

$$= \sum_{n=0}^{\infty} E\left(\exp\left\{-\lambda \sum_{i=1}^{n} w(t - U_{(i)})\right\}\right) P[N((0, t]) = n]$$

$$= \sum_{n=0}^{\infty} E\left(\exp\left\{-\lambda \sum_{i=1}^{n} w(t - U_i)\right\}\right) P[N((0, t]) = n]$$

(where we have used (4.5.1) twice). Letting U be a $U(0, 1)$ random variable, and using the fact that $U \overset{d}{=} 1 - U$ and $tU \overset{d}{=} U_1$, we have the Laplace transform equal to

$$= \sum_{n=0}^{\infty} \left(E \exp\left\{ -\lambda w(t - tU) \right\} \right)^n P\left[N[0, t] = n \right]$$

$$= \sum_{n=0}^{\infty} \left(E \exp\left\{ -\lambda w(tU) \right\} \right)^n P\left[N[0, t] = n \right].$$

Recalling the generating function of a Poisson random variable, this is

$$= \exp\left\{ \alpha t \left(E e^{-\lambda w(tU)} - 1 \right) \right\}$$

$$= \exp\left\{ \alpha t \int_0^1 \left(e^{-\lambda w(tu)} - 1 \right) du \right\}.$$

Thus we conclude the Laplace transform of the shot-noise process is

$$(4.5.2) \qquad E\exp\{-\lambda X(t)\} = \exp\left\{ \alpha \int_0^t \left(e^{-\lambda w(v)} - 1 \right) dv \right\}.$$

From (4.4.2), moments of $X(t)$ can readily be found by differentiating, setting $\lambda = 0$ and inserting a minus sign. For example, we get

$$EX(t) = \alpha \int_0^t w(v)dv.$$

4.6. VARIANTS OF THE POISSON PROCESS.

Many variants of the Poisson process have been proposed. We discuss several briefly and give some simple properties.

• **Mixed Poisson.** Suppose Λ is a random variable with $P[\Lambda > 0] = 1$, and suppose N is a Poisson process on the state space $E = [0, \infty)$ independent of Λ. Then $\{N((0, \Lambda t]), t > 0\}$ is a mixed Poisson process. Because of the random time change induced by Λ, the mixed Poisson process does not in general have independent increments. If N is homogeneous, the mixed Poisson process still has the order statistic property (which shows that the order statistic property does not characterize the Poisson process), as can be seen by conditioning on Λ and repeating the argument of the previous section. (In fact, with the correct formulation of the order statistic property, the class of mixed Poisson processes is the exact class to possess the order statistic property.)

To obtain the marginal distribution of $N((0, \Lambda t])$, we may easily compute the Laplace transform: Suppose for simplicity that N is homogeneous with rate 1, and let the distribution function of Λ be G so that $P[\Lambda \leq s] = G(s)$. Then for $\lambda > 0$ we condition on Λ to get

$$Ee^{-\lambda N((0,\Lambda t])} = \int_0^\infty Ee^{-\lambda N((0,\alpha t])} dG(\alpha),$$

and, because $N((0, \alpha t])$ is a Poisson random variable with parameter αt, this is

$$= \int_0^\infty e^{\alpha t(e^{-\lambda} - 1)} dG(\alpha)$$
$$= \hat{G}(t(1 - e^{-\lambda}))$$

where $\hat{G}$ is the Laplace transform of G.

From this the mean can easily be computed. We have

$$EN((0, \Lambda t]) = \hat{G}'(t(1 - e^{-\lambda}))te^{-\lambda}|_{\lambda=0}$$
$$= -\hat{G}'(0)t$$
$$= E(\Lambda)t,$$

as expected.

The scheme describing the mixed Poisson process in one dimension can easily be generalized to cases where the state space E is of dimension higher than 1. The relationship of the mixed Poisson process and the linear birth process is explored in Section 5.11.

• **Doubly stochastic Poisson process.** Suppose N is a Poisson process on $[0, \infty)$, and suppose $\{\Lambda(t), t \geq 0\}$ is a stochastic process independent of N with non-decreasing paths. Suppose also that $\Lambda(0) \geq 0$. Then N^* defined by

$$N^*((0, t]) = N((0, \Lambda(t)])$$

is called a *doubly stochastic Poisson process*. We think of $\Lambda(t)$ as a random transformation of the time scale.

As an example, think of an alternating renewal process (Example (iv) of Section 3.1 and Problem 3.12). Imagine that there are two independent sequences $\{U_n\}$ and $\{D_n\}$ and that each sequence consists of independent, identically distributed random variables. Define $S_0 = 0, S_1 = U_1, S_2 = U_1 + D_1, S_3 = U_1 + D_1 + U_2, S_4 = U_1 + D_1 + U_2 + D_2$, etc. Think of a transmission node in a telephone network which has alternating periods of being operable (up) and inoperable (down). The operable periods have lengths $\{U_n\}$, and the inoperable periods have lengths $\{D_n\}$. Let $\xi(t)$ be 1 if the node is operable at time t and 0 otherwise:

$$\xi(t) = \begin{cases} 1, & \text{if } t \in \cup_{n=0}^{\infty}[S_{2n}, S_{2n+1}), \\ 0, & \text{if } t \in \cup_{n=0}^{\infty}[S_{2n+1}, S_{2n+2}). \end{cases}$$

Define $\Lambda(t)$ to be the amount of time in $[0, t]$ that the node has been operable:

$$\Lambda(t) = \int_0^t \xi(s)ds.$$

Now suppose there is a Poisson stream N of arriving calls to the node which is independent of whether the node is operable or inoperable. Calls arriving when the node is operable are instantaneously processed and routed while calls arriving when the node is down are lost. The number of processed calls in $[0, t]$ should have the distribution of $N^*((0, t]) = N((0, \Lambda(t)])$, while the number of lost calls at the node should have the distribution of $N((0, t - \Lambda(t)])$.

The properties of the doubly stochastic Poisson process are discovered by conditioning on the process $\Lambda(\cdot)$, since this conditioning allows the process to be analyzed as a Poisson process. If N is homogeneous Poisson, then, after conditioning on $\{\Lambda(t), t \geq 0\}$, the process N^* is Poisson with mean measure of the interval $(a, b]$ equal to $\Lambda(b) - \Lambda(a)$.

Particular cases of the doubly stochastic Poisson process include the mixed Poisson process ($\Lambda(t) \equiv \Lambda$) and the *Markov modulated Poisson process* discussed in Exercise 5.3. In the Markov modulated Poisson process there is an environmental, continuous time Markov chain $\{X(t), t \geq 0\}$ which is independent of a homogeneous Poisson process. The Markov chain evolves among m states. While the Markov chain is in state j ($1 \leq j \leq m$), we run the Poisson process at rate $\alpha(j)$. Thus, conditional on the Markov chain $\{X(t)\}$, the Markov modulated Poisson process behaves like a non-homogeneous Poisson process with local intensity function $\alpha(X(t))$ and the conditional expected number of points in $(0, t]$ is

$$\Lambda(t) =: \int_0^t \alpha(X(s))ds.$$

The Markov modulated Poisson model is extensively utilized in queueing and teletraffic models because of its algorithmic properties and because it forms a flexible statistical parametric model which can be fit to a wide variety of data sets.

- **Compound Poisson processes.** This model is closely related to the renewal reward models of Section 3.4. Suppose N is a homogeneous

Poisson process with state space $E = [0, \infty)$ and independent of the iid sequence $\{D_n, n \geq 1\}$. Define

$$C(t) = \begin{cases} \sum_{i=1}^{N((0,t])} D_i, & \text{if } N((0,t]) > 0, \\ 0, & \text{otherwise.} \end{cases}$$

Imagine insurance claims arriving at an insurance company according to a homogeneous Poisson process with the nth claim being D_n. Then $C(t)$ is the total amount of claims made up to time t.

The process $\{C(t)\}$ is a simple example of a process with *stationary, independent increments*, or, synonomously, it is a simple example of a *Lévy process*. This means, for any $0 \leq s < t$, $C(t) - C(s)$ has a distribution equal to $C(t - s)$ and that, for any k and $0 \leq t_0 < t_1 < \cdots < t_k$, the random variables $C(t_i) - C(t_{i-1}), 1 \leq i \leq k$, are independent. To verify the independent increment property, suppose $N = \sum_n \epsilon_{X_n}$. From the result on marking (Theorem 4.4.1), we then know

$$\sum_n \epsilon_{(X_n, D_n)}$$

is also Poisson. Because of the complete randomness property of the Poisson process, the following are independent Poisson processes:

$$\sum_n 1_{[X_n \in (t_{i-1}, t_i]]} \epsilon_{(X_n, D_n)}, \quad i = 1, \ldots, k,$$

since the different counting processes look for points in disjoint regions. Therefore the sums

$$C(t_i) - C(t_{i-1}) = \sum_{\{j : X_j \in (t_{j-1}, t_j]\}} D_j, \quad i = 1, \ldots, k.$$

must be independent.

To show the distribution of $C(t) - C(s)$ depends only on $t - s$ we can proceed in a similar fashion. The point process

$$\sum_i 1_{[X_i \in (0, t-s]]} \epsilon_{(X_i, D_i)}$$

is a Poisson process with state space $(0, t - s] \times R$ with mean measure $|\cdot| \times R$ restricted to $(0, t-s] \times R$. By the transformation theory of Section 4.3 we have

$$\sum_i 1_{[X_i \in (0, t-s]]} \epsilon_{(X_i + s, D_i)}$$

is a Poisson process on the state space $(s, t] \times R$ with mean measure $|\cdot| \times R$ restricted to $(s, t] \times R$. However, the process

$$\sum_i 1_{[X_i \in (s,t]]} \epsilon_{(X_i, D_i)}$$

is also a Poisson process on the state space $(s, t] \times R$ with mean measure $|\cdot| \times R$ restricted to $(s, t] \times R$. Since the two Poisson processes have the same mean measure, they are equivalent in distribution:

$$\sum_i 1_{[X_i \in (0,t-s]]} \epsilon_{(X_i+s, D_i)} \overset{d}{=} \sum_i 1_{[X_i \in (s,t]]} \epsilon_{(X_i, D_i)},$$

and, therefore,

$$C(t-s) = \sum_{\{i : X_i \in (0, t-s]\}} D_i$$

$$= \sum_{\{i : X_i + s \in (s, t]\}} D_i$$

$$\overset{d}{=} \sum_{\{j : X_j \in (s, t]\}} D_j = C(t) - C(s).$$

We may easily compute the transform of the distribution of $C(t)$. If the D_n variables are non-negative, it is convenient to compute the Laplace transform. (In cases where the random variables $\{D_n\}$ can be positive or negative valued, it is necessary to use the transform called the characteristic function.) Let the Laplace transform of D_1 be $\phi(\lambda)$. We have, for $\lambda > 0$,

$$Ee^{-\lambda C(t)} = Ee^{-\lambda \Sigma_{i=1}^{N((0,t])} D_i}$$

$$= \sum_{n=0}^{\infty} Ee^{-\lambda \Sigma_{i=1}^{n} D_i} P[N((0, t]) = n]$$

$$= \sum_{n=0}^{\infty} \phi^n(\lambda) P[N((0, t]) = n].$$

Supposing the rate of the Poisson process is α and recognizing that the previous line is the generating function of $N((0, t])$ at the point $s = \phi(\lambda)$, we get

$$= e^{\alpha t(\phi(\lambda) - 1)}$$

and, if G is the distribution of D_1, this can be expressed as

$$= \exp\{-\alpha t \int_0^\infty (1 - e^{-\lambda x}) dG(x)\}.$$

Note

$$EC(t) = -e^{\alpha t(\phi(\lambda)-1)}\alpha t\phi'(\lambda)|_{\lambda=0}$$
$$= \alpha t(-\phi'(0))$$
$$= \alpha t ED_1 = EN((0,t])ED_1,$$

as expected.

• **Cluster Processes.** The simplest kind of cluster process based on the Poisson process is obtained by treating the Poisson points as *sites* and locating at each site a random number of points. Let $\{\xi_n\}$ be independent, identically distributed non-negative integer valued random variables which are independent of the Poisson process $\sum_n \epsilon_{X_n}$ with state space E and points $\{X_n\}$. The simple cluster process, similar to the scheme used to define the compound Poisson model, is

$$N_{CL} = \sum_n \xi_n \epsilon_{X_n},$$

so that, for a region $A \subset E$, the number of points in A is

$$N_{CL}(A) = \sum_{\{n:X_n \in A\}} \xi_n.$$

Thus, to compute the number of points in A, we see which Poisson sites X_n fall in A and sum the associated ξ_n's.

To help fix ideas when $E = [0, \infty)$, think of bulk arrivals to a service facility; for example, imagine cars arrive at a roadside restaurant at Poisson times points $\{X_n\}$ and that each car contains a random number ξ_n of customers. For a spatial version when $E = R^2$, imagine an epidemiological study to determine the effect on a region of proximity to a low-level nuclear waste dump. The null hypothesis is that houses which contain cancer victims are distributed like a spatially homogeneous Poisson process and that the number of victims in each house is random.

Note that N_{CL} has the complete randomness property: If $A_1, \ldots, A_k$ are disjoint regions, the random variables $N_{CL}(A_1), \ldots, N_{CL}(A_k)$ are independent. The proof is the same as the one which proved the independent increment property for the compound Poisson process. Furthermore, the distribution of $N_{CL}(A)$ has a Laplace transform given by the formula for the Laplace transform of a compound Poisson process.

A more sophisticated type of cluster process, sometimes called the *center–satellite process*, uses Poisson points as *centers* or *parents*. At each center, independent of other centers, a random number of *satellites* are generated. The number of satellites per center is given by independent, identically distributed non-negative random variables. Also, each satellite is displaced from the center according to some dispersal distribution. Thus, let E be the state space and let the E-valued random elements $\{X_n, n \geq 0\}$ represent the Poisson points of centers, and suppose ξ_n represents the number of satellites about the nth center X_n. Let the displacement of the jth satellite from X_n ($1 \leq j \leq \xi_n$) be given by Y_{nj} so that each Y_{nj} has range E. Assume $\{X_n\}, \{\xi_n\}$ and $\{Y_{nj}\}$ are independent of each other and that both $\{\xi_n\}$ and $\{Y_{nj}\}$ are iid sequences. The cluster process with displacements is then given by

$$N_{CLD} = \sum_n \sum_{j=1}^{\xi_n} \epsilon_{X_n + Y_{nj}},$$

so the points of N_{CLD} are

$$\{X_n + Y_{nj}, 1 \leq j \leq \xi_n, n \geq 0\}.$$

Such processes have been used as models for dispersals of stars in galaxies and to model stands of trees in forests. Alternate schemes for dispersing satellites about centers when $E = [0, \infty)$ have been proposed. For example, when $E = [0, \infty)$, we may suppose that from each center point a renewal process generates satellites. Such cluster models have been used as models of machine repair. Consider primary failures occurring according to a Poisson process. Due to imperfect repair, each primary failure generates a renewal process of secondary failures.

For a description of the *batch Markovian arrival process* which has recently received some attention in the teletraffic literature, see Exercise 5.72.

4.7. TECHNICAL BASICS*.

In this section we discuss in more detail the sample space of our models. It is important to realize the sample space comes equipped with distinguished sets which determine the distribution of the point process.

---

* This section contains advanced material which may be skipped on first reading by beginning readers.

Suppose E is the space where the points of our model live. This will be a subset of Euclidean space R^d ($d = 1$ is the most important case, but $d > 1$ is also very useful), and we suppose E comes with a σ-field $\mathcal{E}$ which can be the σ-field generated by the open sets or, equivalently, the rectangles of E. So the important sets in $\mathcal{E}$ are built up from rectangles. How can we model a random distribution of points in E? One way is to specify random elements in E, say $\{X_n\}$, and then to say that a stochastic point process is the counting function whose value at the region $A \in \mathcal{E}$ is the number of the random elements $\{X_n\}$ which fall in A. This procedure has some technical drawbacks, and it is mathematically preferable to focus directly on counting functions rather than points.

Accordingly, define a point measure m on E as follows. For $x \in E$ and $A \in \mathcal{E}$, define the delta measure

$$\epsilon_x(A) = \begin{cases} 1, & \text{if } x \in A \\ 0, & \text{if } x \notin A. \end{cases}$$

A point measure m on E is a measure of the form

$$m = \sum_i \epsilon_{x_i},$$

where $x_i \in E$ and any bounded region A contains only a finite number of x_i. Then $m(A)$ is the number of x_i which fall in region A. As an example, let $E = [0, \infty)$, and let $\{S_n\}$ be a renewal process. The counting process associated with renewal epochs $\{S_n\}$ is

$$N = \sum_{n=0}^{\infty} \epsilon_{S_n},$$

so that $N[0, t]$ is the number of renewals up to time t.

Let the set of all point measures on E be denoted $M_p = M_p(E)$. Important subsets of M_p are of the form ($k \geq 0$)

$$\{m \in M_p : m(I) = k\},$$

where I is a bounded rectangle in E. The smallest σ-algebra containing such important sets (let I and k vary) is denoted by $\mathcal{M}_p = \mathcal{M}_p(E)$. Then we have specified the class of all point measures M_p along with a distinguished σ-field of subsets of M_p, which is denoted $\mathcal{M}_p$. (We may also make M_p into a metric space. The metric would make two point measures m_1, m_2 close if the values of these two measures are close on a lot of bounded sets. See Kallenberg, 1983, or Resnick, 1987.) We may now define a *point process* as a random element of $(M_p, \mathcal{M}_p)$; there is a probability space $(\Omega, \mathcal{F}, P)$ where $\mathcal{F}$ is the σ-field of events, and N is a *point process* if it is a measurable map from $(\Omega, \mathcal{F})$ to $(M_p, \mathcal{M}_p)$, which means that for $\Lambda \in \mathcal{M}_p$, we have $N^{-1}(\Lambda) := \{\omega \in \Omega : N(\omega) \in \Lambda\} \in \mathcal{F}$, so that, in particular,

$$[N(I) = k] = N^{-1}\{m \in M_p : m(I) = k\} \in \mathcal{F}.$$

Since $[N(I) = k]$ is an event, it is meaningful to talk of $P[N(I) = k]$.

Instead of picking points at random as initially suggested, we pick point measures at random. If you specify $\omega \in \Omega$ and then plug into the map N, you get a point measure. However, if $\{X_n\}$ is a sequence of random elements in E, it is not hard to show (but we will not dwell on this point) that $\sum_n \epsilon_{X_n}$ satisfies the definition of a point process, provided bounded regions contain finitely many points with probability 1.

Suppose N is a point process

$$N : (\Omega, \mathcal{F}) \mapsto (M_p, \mathcal{M}_p),$$

and P is the probability measure on $(\Omega, \mathcal{F})$. The *distribution of N* is the measure $P \circ N^{-1} = P[N \in \cdot]$ on $(M_p, \mathcal{M}_p)$. Therefore, the probability of any event depending on N can be specified if we know the distribution of N. The *finite dimensional distributions* of N are the collection of multivariate mass functions indexed by bounded rectangles:

$$p_{I_1,\dots,I_k}(n_1,\dots,n_k) := P[N(I_1) = n_1,\dots,N(I_k) = n_k],$$

where $I_j, 1 \le j \le k$ are bounded rectangles and $n_1,\dots,n_k$, are nonnegative integers.

Proposition 4.7.1. *The finite dimensional distributions of the point process N uniquely determine the distribution $P \circ N^{-1}$ of N.*

Proof. Let $\mathcal{G}$ be the class of finite intersections of sets of the form

$$\{m \in M_p : m(I) = k\}, \quad k \ge 0, \quad I \text{ is a bounded rectangle.}$$

Then $\mathcal{G}$ is closed under intersections and generates $\mathcal{M}_p$, and hence any measure on $\mathcal{M}_p$ is uniquely determined by its values on $\mathcal{G}$ (cf. Section 1.7 and Billingsley, 1986, p. 38). In particular, $P \circ N^{-1}$ is a measure defined on $\mathcal{M}_p$ and its values on $\mathcal{G}$ are given by the finite dimensional distributions. ∎

4.7.1. THE LAPLACE FUNCTIONAL*.

A transform technique is useful in manipulating distributions of point processes. Let $\mathcal{B}_+$ be the non-negative and bounded functions on E, and for $m \in M_p$ and $f \in \mathcal{B}_+$ define

$$m(f) = \int_{x \in E} f(x) dm(x) = \sum_i f(x_i),$$

where $\{x_i\}$ are the points of the point measure m. If we think of f ranging over all of $\mathcal{B}_+$, we get $m(f)$ to yield all the information contained in m; certainly we learn about the value of m on each set $A \in \mathcal{E}$ since we can always set $f = 1_A$. Therefore, integrals of measures with respect to arbitrary test functions contain as much information as evaluating the measures on arbitrary sets. This will be a guiding principle.

The Laplace functional of the point process N is the non-negative function on $\mathcal{B}_+$ given by

$$\Psi_N(f) = E \exp\{-N(f)\} = \int_\Omega \exp\{-N(\omega, f)\} dP(\omega)$$

$$= \int_{M_p} \exp\{-m(f)\} P \circ N^{-1}(dm).$$

Proposition 4.7.2. *The Laplace functional of N uniquely determines the distribution of N.*

Proof. For bounded rectangles $I_1, \ldots, I_k$ let f be the simple function

$$f(x) = \sum_{i=1}^k \lambda_i 1_{I_i}(x), \quad x \in E,$$

for $\lambda_i \geq 0, i = 1, \ldots, k$. Then

$$\Psi_N(f) = E \exp\{-\sum_{i=1}^k \lambda_i N(I_i)\},$$

the joint Laplace transform of the random vector $(N(I_1), \ldots, N(I_k))$. Its transform uniquely determines the distribution of this random vector, and, by Proposition 4.7.1, we know that the Laplace functional determines the finite dimensional distributions of N and therefore the distribution of N. ■

---

   * This section, as well as Section 4.8, contains advanced material which may be skipped on the first reading by beginning readers. The definition of the Laplace functional can be understood by most readers.

4.8. More on the Poisson Process.

We begin by repeating the definition.

Let μ be a measure on $(E, \mathcal{E})$ which is finite on bounded sets. A point process N is a *Poisson process with mean measure* μ or, synonomously, a *Poisson random measure (PRM(μ))* if

(1)   For $A \in \mathcal{E}$

$$P[N(A) = k] = \begin{cases} \frac{e^{-\mu(A)}(\mu(A))^k}{k!}, & \text{if } \mu(A) < \infty \\ 0 & , \text{ if } \mu(A) = \infty. \end{cases}$$

(2)   If $A_1, \ldots, A_k$ are disjoint sets in $\mathcal{E}$ then $N(A_1), \ldots, N(A_k)$ are independent random variables.

Therefore, N is Poisson if the random number of points in a set A is Poisson distributed with parameter $\mu(A)$ and the number of points in disjoint regions are independent random variables.

Recall that we must check if the Chapter 3 renewal theory definition of a Poisson process in $[0, \infty)$ being generated by sums of iid exponentially distributed random variables coincides with this one.

Proposition 4.8.1. *Let $\{E_j, j \geq 1\}$ be iid random variables with a unit exponential distribution. Define $\Gamma_n = \sum_{i=1}^{n} E_i$ to be the renewal epochs of the renewal process, and set $N = \sum_{n=1}^{\infty} \epsilon_{\Gamma_n}$. Then N satisfies (1),(2) above and is therefore PRM on $[0, \infty)$ with Lebesgue measure as the mean measure.*

We defer the proof of this result until after the characterization of the Laplace functional of the Poisson process given in the next result.

Theorem 4.8.2. *The distribution of PRM(μ) is uniquely determined by (1),(2) in the definition. Furthermore, the point process N is PRM(μ) iff its Laplace functional is of the form*

(4.8.1) $$\Psi_N(f) = \exp\{-\int_E (1 - e^{-f(x)})\mu(dx)\}, \quad f \in \mathcal{B}_+.$$

PRM(μ) can be identified by the characteristic form of its Laplace functional.

Proof. We first show (1) and (2) imply (4.8.1). If $f = \lambda 1_A$ where $\lambda > 0$, then, because $N(f) = \lambda N(A)$, and $N(A)$ is Poisson with parameter $\mu(A)$

we get

$$\Psi_N(f) = E e^{-\lambda N(A)} = \exp\{(e^{-\lambda} - 1)\mu(A)\}$$

$$= \exp\{-\int_E (1 - e^{-\lambda}) 1_A(x)\mu(dx)\}$$

$$= \exp\{-\int_E (1 - e^{-\lambda 1_A(x)})\mu(dx)\}$$

$$= \exp\{-\int_E (1 - e^{-f(x)})\mu(dx)\},$$

which is the correct form given in (4.8.1)

Next suppose f has a somewhat more complex form,

$$f = \sum_{i=1}^{k} \lambda_i 1_{A_i},$$

where $\lambda_i \geq 0, A_i \in \mathcal{E}, 1 \leq i \leq k$, and $A_1, \ldots, A_k$ are disjoint. Then

$$\Psi_N(f) = E \exp\left\{-\sum_{i=1}^{k} \lambda_i N(A_i)\right\}$$

$$= \prod_{i=1}^{k} E \exp\{-\lambda_i N(A_i)\} \quad \text{from independence,}$$

$$= \prod_{i=1}^{k} \exp\left\{-\int_E (1 - e^{-\lambda_i 1_{A_i}(x)})\mu(dx)\right\} \quad \text{from the previous step,}$$

$$= \exp\left\{-\int_E \sum_{i=1}^{k}(1 - e^{-\lambda_i 1_{A_i}(x)})\mu(dx)\right\}$$

$$= \exp\left\{-\int_E (1 - e^{-\sum_{i=1}^{k} \lambda_i 1_{A_i}(x)})\mu(dx)\right\}$$

$$= \exp\left\{-\int_E (1 - e^{-f(x)})\mu(dx)\right\},$$

which again verifies (4.8.1). The last step is to take general $f \in \mathcal{B}_+$ and verify (4.8.1) for such f. We may approximate f from below by simple f_n of the form just considered. We may take, for instance,

$$f_n(x) = \sum_{i=1}^{n2^n} \frac{i-1}{2^n} 1_{[\frac{i-1}{2^n}, \frac{i}{2^n})}(f(x)) + n 1_{[n,\infty)}(f(x)),$$

so that
$$0 \le f_n(x) \uparrow f(x).$$

By monotone convergence $N(f_n) \uparrow N(f)$, and, since $e^{-f} \le 1$, dominated convergence yields
$$\Psi_N(f) = \lim_{n \to \infty} \Psi_N(f_n).$$

We have from the previous step that
$$\Psi_N(f_n) = \exp\left\{-\int_E (1 - e^{-f_n(x)})\mu(dx)\right\}.$$

Since
$$1 - e^{-f_n} \uparrow 1 - e^{-f},$$

by monotone convergence we conclude that
$$\int_E (1 - e^{-f_n(x)})\mu(dx) \uparrow \int_E (1 - e^{-f(x)})\mu(dx),$$

and thus we conclude (4.8.1) holds for any $f \in \mathcal{B}_+$. Since the distribution of N is uniquely determined by Ψ_N, we have shown that (i) and (ii) determine the distribution of N.

Conversely, if the Laplace functional of N is given by (4.8.1), then $N(A)$ must be Poisson distributed with parameter $\mu(A)$ for any $A \in \mathcal{E}$. This is readily checked by substituting $f = \lambda 1_A$ in (4.8.1) to get a Laplace transform of a Poisson distribution. Furthermore, if $A_1, \ldots, A_k$ are disjoint sets in $\mathcal{E}$ and $f = \sum_{i=1}^k \lambda_i 1_{A_i}$, then substituting in (4.8.1) gives

$$
\begin{aligned}
Ee^{-\sum_{i=1}^k \lambda_i N(A_i)} &= \exp\left\{-\int_E (1 - e^{-\sum_{i=1}^k \lambda_i 1_{A_i}})d\mu\right\} \\
&= \exp\left\{-\int_E \sum_{i=1}^k (1 - e^{-\lambda_i 1_{A_i}})d\mu\right\} \\
&= \prod_{i=1}^k \exp\{-(1 - e^{-\lambda_i})\mu(A_i)\} \\
&= \prod_{i=1}^k Ee^{-\lambda_i N(A_i)}.
\end{aligned}
$$

Then the joint Laplace transform of $(N(A_i), 1 \le i \le k)$ factors into a product of Laplace transforms, and this shows independence. ∎

Proof of Proposition 4.8.1. We show the Laplace functional of $N = \sum_n \epsilon_{\Gamma_n}$ is given by (4.8.1) with $E = [0, \infty)$ and $\mu(dx) = dx$ being Lebesgue measure. We have

$$\Psi_N(f) = \lim_{m \to \infty} E e^{-\sum_{i=1}^m f(\Gamma_i)}.$$

Conditioning on the value of Γ_{m+1}, iterating expectations (Law of Total Probability) yields that the expectation on the right is

$$(4.8.2) \qquad \int_0^\infty E\left(e^{-\sum_{i=1}^m f(\Gamma_i)} \Big| \Gamma_{m+1} = s\right) P[\Gamma_{m+1} \in ds].$$

We know from Lemma 4.5.1(b) that if $U_1, \ldots, U_m$ are iid uniformly distributed random variables on $(0, 1)$ and $U_{(1)}, \ldots, U_{(m)}$ are the order statistics in increasing order, then the conditional distribution of the Γ's is

$$(\Gamma_1, \ldots, \Gamma_m | \Gamma_{m+1} = s) \stackrel{d}{=} (sU_{(1)}, \ldots, sU_{(m)}).$$

Therefore

$$\left(\sum_{i=1}^m f(\Gamma_i) | \Gamma_{m+1} = s\right) \stackrel{d}{=} \left(\sum_{i=1}^m f(sU_{(i)})\right) = \left(\sum_{i=1}^m f(sU_i)\right),$$

because of symmetry.

We are now in position to compute the Laplace functional of N. From (4.8.2) we have

$$\Psi_N(f) = \lim_{m \to \infty} \int_0^\infty E e^{-\sum_{i=1}^m f(sU_i)} P[\Gamma_{m+1} \in ds]$$

$$= \lim_{m \to \infty} \int_0^\infty \left(E e^{-f(sU_1)}\right)^m P[\Gamma_{m+1} \in ds]$$

$$= \lim_{m \to \infty} \int_0^\infty \left(\int_0^s e^{-f(x)} \frac{dx}{s}\right)^m P[\Gamma_{m+1} \in ds]$$

$$= \lim_{m \to \infty} E\left(\int_0^{\Gamma_{m+1}} e^{-f(x)} \frac{dx}{\Gamma_{m+1}}\right)^m$$

$$= \lim_{m \to \infty} E\left(1 - \frac{\int_0^{\Gamma_{m+1}}(1 - e^{-f(x)})dx}{\Gamma_{m+1}}\right)^{\Gamma_{m+1} \cdot \frac{m}{\Gamma_{m+1}}}.$$

Since $\Gamma_{m+1} \to \infty$ and $\Gamma_{m+1}/m \to 1$, by the strong law of large numbers,

$$\left(1 - \frac{\int_0^{\Gamma_{m+1}}(1 - e^{-f(x)})dx}{\Gamma_{m+1}}\right)^{\Gamma_{m+1} \cdot \frac{m}{\Gamma_{m+1}}} \to \exp\left\{-\int_0^\infty (1 - e^{-f(x)})dx\right\},$$

and dominated convergence finishes the result. ∎

4.9. A GENERAL CONSTRUCTION OF THE POISSON PROCESS; A SIMPLE DERIVATION OF THE ORDER STATISTIC PROPERTY*.

When $E = [0, \infty)$ we know the renewal theory construction yields a Poisson process with Lebesgue measure as the mean measure. Here is a general scheme for constructing a Poisson process with mean measure μ.

Start by supposing that $\mu(E) < \infty$ and define the probability measure $F(dx) = \mu(dx)/\mu(E)$. Let $\{X_n, n \geq 1\}$ be iid random elements of E with common distribution F and let ν be independent of $\{X_n\}$ with a Poisson distribution with parameter $\mu(E)$. Define

$$N = \begin{cases} \sum_{i=1}^{\nu} \epsilon_{X_i}, & \text{if } \nu \geq 1 \\ 0, & \text{if } \nu = 0. \end{cases}$$

We claim N is $\mathrm{PRM}(\mu)$ and this is easy to check by computing the Laplace functional of N:

$$\begin{aligned} \Psi_N(f) &= E e^{-\sum_{i=1}^{\nu} f(X_i)} \\ &= \sum_{j=0}^{\infty} E e^{-\sum_{i=1}^{j} f(X_i)} P[\nu = j] \\ &= \sum_{j=0}^{\infty} \left(E e^{-f(X_1)} \right)^j P[\nu = j] \\ &= \exp\left\{ \mu(E)(E e^{-f(X_1)} - 1) \right\} \\ &= \exp\left\{ -\mu(E) \left(1 - \int_E e^{-f(x)} \frac{\mu(dx)}{\mu(E)} \right) \right\} \\ &= \exp\left\{ -\int_E \left(1 - e^{-f(x)} \right) \mu(dx) \right\}. \end{aligned}$$

The Laplace functional has the correct form, so N is indeed $\mathrm{PRM}(\mu)$. Note that what the construction does is to toss points at random into E according to distribution $\mu(dx)/\mu(E)$; the number of points tossed is Poisson with parameter $\mu(E)$.

When the condition $\mu(E) < \infty$ fails, we proceed as follows to make a minor modification in the foregoing construction: Decompose E into disjoint sets $E_1, E_2, \ldots$ such that $E = \cup_i E_i$ and $\mu(E_i) < \infty$ for each i. Let

---

* This section contains advanced material which may be skipped on first reading by beginning readers.

$\mu_i(dx) = \mu(dx)1_{E_i}(x)$, and let N_i be PRM(μ_i) on E_i (do the construction just previously outlined). Arrange things so the collection $\{N_i\}$ is independent. Define $N := \sum_i N_i$, and N is PRM(μ) since

$$\Psi_N(f) = \prod_i \Psi_{N_i}(f)$$

$$= \prod_i \exp\left\{-\int_{E_i} \left(1 - e^{-f(x)}\right) \mu_i(dx)\right\}$$

$$= \exp\left\{-\sum_i \int_E \left(1 - e^{-f(x)}\right) \mu_i(dx)\right\}$$

$$= \exp\left\{-\int_E \left(1 - e^{-f(x)}\right) \sum_i \mu_i(dx)\right\}$$

$$= \exp\left\{-\int_E \left(1 - e^{-f(x)}\right) \mu(dx)\right\},$$

since $\sum_i \mu_i = \mu$. This completes the construction.

The construction just completed tells us PRM(μ) exists, but it also gives us information about the distribution of the points: Conditional on there being n points in a bounded region A, these points are distributed as n iid random elements of A with common distribution $F(dx) = \mu(dx)/\mu(A)$. When $E = [0, \infty)$, this yields the order statistic property for a homogeneous Poisson process which is now restated from Section 4.5.

Theorem 4.5.2. *If N is a homogeneous Poisson process on $[0, \infty)$ with rate α, then conditional on*

$$[N((0,t]) = n]$$

the points of N in $[0, t]$ in increasing order are distributed as the order statistics from a sample of size n from the uniform distribution $U(0, t)$ on $[0, t]$; i.e.,

$$(\Gamma_1, \dots, \Gamma_n | N[0, t] = n) \stackrel{d}{=} (U_{(1)}, \dots, U_{(n)}).$$

Proof. Let $N' = N(\cdot \cap [0, t])$ be the restriction of N to the bounded set $[0, t]$. We know from the general construction that as random elements of $M_p[0, t]$

$$N' \stackrel{d}{=} \sum_{i=1}^{\tau} \epsilon_{U_i},$$

where $\{U_i\}$ are iid random elements with distribution dx/t (which is identified as $U(0, t)$) and τ is independent of $\{U_i\}$ and Poisson distributed with parameter t. Then, if $\{\Gamma_n\}$ are the points of N, we have

$$(\Gamma_1, \dots, \Gamma_n | N[0, t] = n) \stackrel{d}{=} (U_{(1)}, \dots, U_{(n)} | \tau = n) =^d (U_{(1)}, \dots, U_{(n)}). \blacksquare$$

4.10. MORE TRANSFORMATION THEORY;
LOCATION DEPENDENT THINNING*.

Given a Poisson process, recall that one can enlarge the dimension of the points by appending independent marks and that this retains the Poisson structure. The result presented here allows independent (or almost independent) components to be added.

We present the result in two parts; the first is simpler and arises more frequently.

Proposition 4.10.1. *(a) Suppose $\{X_n\}$ are random elements of a Euclidean space E_1 such that*

$$\sum_n \epsilon_{X_n}$$

is PRM(μ). Suppose $\{J_n\}$ are iid random elements of a second Euclidean space E_2 with common probability distribution F, and suppose the Poisson process and the sequence $\{J_n\}$ are defined on the same probability space and are independent. Then the point process on $E_1 \times E_2$

$$\sum_n \epsilon_{(X_n, J_n)}$$

is PRM with mean measure $\mu \times F$.

(b) Suppose $\{X_n\}$ are random elements of a Euclidean space E_1 such that

$$\sum_n \epsilon_{X_n}$$

is PRM(μ). Suppose we have a second Euclidean space $(E_2, \mathcal{E}_2)$ and that $K : E_1 \times \mathcal{E}_2 \mapsto [0,1]$ is a transition function. Then $K(\cdot, A_2)$ is a measurable function of the first variable for every fixed $A_2 \in \mathcal{E}_2$, and, for every $x \in E_1$, we have $K(x, \cdot)$ is a probability measure on $\mathcal{E}_2$. Let $\{J_i\}$ be random elements of E_2 which are conditionally independent given $\{X_n\}$:

(4.10.1) $P[J_i \in A_2 | X_i = x, \{X_j, j \neq i\}, \{J_j, j \neq i\}] = K(x, A_2).$

Then the point process on $E_1 \times E_2$

$$\sum_n \epsilon_{(X_n, J_n)}$$

---

* This section contains advanced material which may be skipped on first reading by beginning readers.

is PRM with mean measure

$$\mu_1(dx, dy) = \mu(dx)K(x, dy).$$

It is not necessary for $\{J_n\}$ to be independent of $\{X_n\}$; conditional independence will do. If the distribution of J_n depends on the $\{X_i\}$, it must do so only through X_n and not the other X's.

Proof. (a) Start by assuming that μ is finite. From the construction in Section 4.9 we may, without loss of generality, assume that the PRM(μ) is of the form

$$\sum_{i=1}^{\nu} \epsilon_{Y_i},$$

where ν, $\{Y_n\}$, and $\{J_n\}$ are independent and

$$\nu \sim p(k, \mu(E_1))$$

and $\{Y_n\}$ are iid with common distribution $\mu(dx)/\mu(E_1)$. It follows that $\{(Y_n, J_n)\}$ is iid in $E_1 \times E_2$ with common distribution $\mu(\cdot)/\mu(E_1) \times F$, so from the construction of Section 4.9,

$$\sum_n \epsilon_{(X_n, J_n)} \overset{d}{=} \sum_{i=1}^{\nu} \epsilon_{(Y_i, J_i)}$$

is PRM with mean measure $\mu(E_1)\mu(\cdot)/\mu(E_1) \times F = \mu \times F$ as required.

If μ is not finite, then, as in the construction of Section 4.9, we patch things together by repeating the argument of the previous paragraph on patches of E_1 where μ is finite. We need at most a countable number of patches which are disjoint and exhaust E_1.

(b) Write

$$K(x, A_2) = P[J_1 \in A_2 | X_1 = x]$$

for the conditional distribution of J_1. It is always possible to realize a distribution as a function of a uniform random variable (e.g., Billingsley, 1971). That is, there exists a function, say $g(x, u)$, such that

$$K(x, A_2) = P[g(x, U_1) \in A_2],$$

where we suppose $\{U_n\}$ are iid $U(0, 1)$ random variables, independent of $\{X_n\}$. (If $E_2 = R$ so that $K(x, (-\infty, y]) = P[J_1 \leq y | X_1 = x]$, then we take $g(x, u) = K^{\leftarrow}(x, u)$, where $K^{\leftarrow}(x, u)$ is the inverse function for fixed x of the y-function $K(x, (-\infty, y])$ so that

$$P[g(x, U_1) \leq z] = P[K^{\leftarrow}(x, U_1) \leq z]$$
$$= P[U_1 \leq K(x, (-\infty, z]) = K(x, (-\infty, z]).$$

If E_2 has dimension higher than 1, a discrete approximation must be used as discussed in Billingsley, 1971.)

The impact of this transformation is that

$$\{(X_n, J_n)\} \stackrel{d}{=} \{(X_n, g(X_n, U_n))\}.$$

We know

$$\sum_n \epsilon_{(X_n, U_n)}$$

is PRM with mean measure $\mu \times L(dy)1_{[0,1]}(y)$. ($L$ is Lebesgue measure.) Therefore, from Proposition 4.3.1, we get that

$$\sum_n \epsilon_{(X_n, J_n)} \stackrel{d}{=} \sum_n \epsilon_{(X_n, g(X_n, U_n))}$$

is PRM(μ_1). To compute μ_1, define $T : E_1 \times [0, 1] \mapsto E_1 \times E_2$ via $T(x, u) = (x, g(x, u))$. Then the mean measure μ_1 is $(A_1 \in \mathcal{E}_1, A_2 \in \mathcal{E}_2)$

$$\mu_1(A_1 \times A_2) = (\mu \times L) \circ T^{-1}(A_1 \times A_2)$$

$$= \int_{[x \in A_1]} \mu(dx) L\{u \in [0, 1] : g(x, u) \in A_2\}$$

$$= \int_{[x \in A_1]} \mu(dx) P[g(x, U_1) \in A_2]$$

$$= \int_{A_1} K(x, A_2)\mu(dx). \blacksquare$$

Example 4.10.1. Thinning Dependent on Location. Suppose we have a homogeneous Poisson process on $[0, \infty)$ with rate a. If a point is at t, we delete it with probability $q(t)$ and retain it with probability $p(t)$ where $p(t) + q(t) = 1$, $0 \le p(t) \le 1$. The point process of retained points is non-homogeneous Poisson with local intensity $ap(t)$.

As in the discussion of thinning in Section 4.4, we imagine having Bernoulli random variables $\{B_i\}$ with values $\{-1, 1\}$. If the homogeneous PRM is $\sum_n \epsilon_{X_n}$ on $[0, \infty)$, then we have

$$P[B_i = 1 | X_i = t] = p(t).$$

The process

$$\sum_n \epsilon_{(X_n, B_n)}$$

is PRM with mean measure of region $dt \times \{1\}$

$$\mu_1(dt \times \{1\}) = \mu(dt) K(t, \{1\}) = ap(t)dt.$$

Note that the process of retained points is still independent of the process of deleted points.

4.11. RECORDS*.

Suppose $\{X_n, n \geq 1\}$ is an iid sequence of random variables with common distribution F which we suppose to be continuous.

Define

$$L(1) = 1$$

and, for $k \geq 1$,

$$L(k+1) = \inf\{m > L(k) : X_m > X_{L(k)}\}.$$

The sequence $\{X_{L(k)}, k \geq 1\}$ is called the *record value* sequence and the sequence $\{L(k), k \geq 1\}$ is the sequence of *record times*. Thus, $X_{L(k)}$ is the kth record which means it is greater than all preceding $L(k) - 1$ values in the sequence. Another way to think about this is to set $M_n = \vee_{i=1}^n X_i$; then $\{X_{L(k)}\}$ are the distinct states visited by the stochastic process $\{M_n, n \geq 1\}$, and the times when the monotone sequence $\{M_n\}$ jumps are the record times $\{L(k)\}$.

There are many connections between the Poisson process and records.

Proposition 4.11.1. (a) *Suppose* $F(x) = 1 - e^{-x}, x > 0$, *is the exponential distribution. Then the records* $\{X_{L(k)}, k \geq 1\}$ *are the points of a homogeneous Poisson process on* $[0, \infty)$. *This means that in* R^∞

$$\{X_{L(k)}, k \geq 1\} \stackrel{d}{=} \{\Gamma_n, n \geq 1\},$$

where $\Gamma_n = E_1 + \cdots + E_n$ *and* $\{E_n, n \geq 1\}$ *are iid exponentially distributed random variables with parameter 1.*

(b) *For general continuous* F, *the records* $\{X_{L(k)}, k \geq 1\}$ *are the points of a Poisson process with mean measure of* $(a, b]$ *equal to* $R(b) - R(a)$ *where* $R(x) = -\log(1 - F(x))$. *The Poisson process has all its points in the interval*

$$(x_l, x_r) := (\inf\{x : F(x) > 0\}, \sup\{x : F(x) < 1\}).$$

Proof. (a) We proceed in a slightly informal way; the procedure about to be given can be used in a more rigorous manner to compute the Laplace functional of the counting function of the sequence $\{X_{L(k)}\}$ or, alternatively, Markov process theory can be used (Resnick, 1987, page 166).

---

* This section can be skipped without loss of continuity in the same sense that dessert can be skipped when eating dinner.

We compute the density directly for $(X_{L(1)}, X_{L(2)})$: For $0 < x_1 < x_2$ we have

$$P[X_{L(1)} \in dx_1, X_{L(2)} \in dx_2] = \sum_{l=1}^{\infty} P[X_{L(1)} \in dx_1, X_{L(2)} \in dx_2, L(2) = l]$$

$$= \sum_{l=1}^{\infty} P[X_{L(1)} \in dx_1, X_{L(2)} \in dx_2, \max\{X_2, \ldots, X_{l-1} < X_1\}]$$

(since in order for a record to occur at index l, we need $X_2, \ldots, X_{l-1}$ not to exceed the previous record X_1)

$$= \sum_{l=1}^{\infty} e^{-x_1} dx_1 (1 - e^{x_1})^{l-1} e^{-x_2} dx_2$$

$$= e^{-x_1} dx_1 \frac{e^{-x_2} dx_2}{1 - (1 - e^{-x_1})}$$

$$= e^{-x_2} dx_1 dx_2$$

$$= P[\Gamma_1 \in dx_1, \Gamma_2 \in dx_2].$$

This shows

$$(X_{L(1)}, X_{L(2)}) \stackrel{d}{=} (\Gamma_1, \Gamma_2).$$

The generalization from two variables to an arbitrary number is not hard and simply requires more bookkeeping.

(b) Suppose $\{X_k, k \geq 1\}$ is an iid sequence with common distribution function $F(x)$. Remember, we suppose that F is continuous. Define $R(x) = -\log(1 - F(x))$ and its inverse

$$R^{\leftarrow}(x) = \inf\{u : R(u) \geq x\}.$$

If $\{E_k, k \geq 1\}$ is a sequence of iid unit exponential random variables then for $i \geq 1$

$$X_i \stackrel{d}{=} R^{\leftarrow}(E_i)$$

since

$$P[R^{\leftarrow}(E_i) \leq x] = P[E_i \leq R(x)]$$

$$= 1 - e^{-R(x)} = F(x).$$

Thus in R^{∞}

$$\{X_k, k \geq 1\} \stackrel{d}{=} \{R^{\leftarrow}(E_k), k \geq 1\}.$$

Since F is continuous, $R^{\leftarrow}$ is strictly increasing, and hence

$$\{R^{\leftarrow}(E_{L(k)}), k \geq 1\} \stackrel{d}{=} \{X_{L(k)}, k \geq 1\}.$$

(We have abused notation a bit; on the left $L(k)$ refers to the kth record time of the $\{E_n\}$ sequence, and on the right $L(k)$ refers to the kth record time of the $\{X_n\}$ sequence.) From part (a) we know that $\{E_{L(k)}\}$ are the points of a homogeneous Poisson process, and from the transformation theory (Proposition 4.3.1) we get that $\{R^{\leftarrow}(E_{L(k)})\}$ are also the points of a Poisson process with mean measure determined by R.

Proposition 4.10.1(b) allows a more sophisticated conclusion:

Proposition 4.11.2. *Suppose F is continuous with*

$$(x_l, x_r) := (\inf\{x : F(x) > 0\}, \sup\{x : F(x) < 1\}).$$

Then

$$\{X_{L(k)}, L(k+1) - L(k), k \geq 1\}$$

are the points of a two-dimensional Poisson process with state space $(x_l, x_r) \times \{1, 2, \dots\}$ and mean measure

$$\mu^*((a, b] \times \{j\}) = (F^j(b) - F^j(a))/j$$

for $x_l < a < b < x_r$, $j \geq 1$.

Proof. We know that $\{X_{L(k)}, k \geq 1\}$ are the points of a Poisson process whose mean measure is determined by R. If we hope to apply Proposition 4.10.1(b), we need to check that $\{L(k+1) - L(k), k \geq 1\}$ are conditionally independent given $\{X_{L(k)}, k \geq 1\}$. This is easily checked by a simple calculation: For positive integers $m, n_1, \dots, n_m$ and real numbers $x_1 < \cdots < x_m$

$$P[L(k+1) - L(k) = n_k, k = 1, \dots, m | X_{L(1)} = x_1, \dots, X_{L(m)} = x_m]$$

$$= \prod_{k=1}^{m} F(x_k)^{n_k - 1}(1 - F(x_k))$$

$$= \prod_{k=1}^{m} P[L(k+1) - L(k) = n_k | X_{L(1)} = x_1, \dots, X_{L(m)} = x_m,$$

$$L(j+1) - L(j) = n_j, j \neq k].$$

Proposition 4.10.1(b) applies, and the mean measure is

$$\mu^*((a,b] \times \{j\}) = \int_a^b F(x)^{j-1}(1-F(x))R(dx)$$

$$= \int_a^b F(x)^{j-1}(1-F(x))\frac{F(dx)}{1-F(x)}$$

$$= \int_{F(a)}^{F(b)} y^{j-1}dy$$

$$= (F^j(b) - F^j(a))/j.$$

EXERCISES

4.1. Electrical pulses have iid amplitudes $\{\xi_n, n \geq 1\}$ and arrive at a detector at random times $\{\Gamma_n, n \geq 1\}$ according to a homogeneous Poisson process rate with α. Assume $\{\xi_n\}$ is independent of the Poisson process. The detector output for the kth pulse at time t is

$$\theta_k(t) = \begin{cases} 0, & \text{if } t < \Gamma_k \\ \xi_k \exp\{-\beta(t-\Gamma_k)\}, & \text{if } t \geq \Gamma_k. \end{cases}$$

Assume the detector is additive, so the output at time t is

$$Z(t) = \sum_{k=1}^{N(t)} \theta_k(t),$$

where $N(t)$ is the number of Poisson points in $[0,t]$. Find the Laplace transform of $Z(t)$ and $E(Z(t))$.

4.2. The Renyi Traffic Model. At time 0 cars are positioned along an infinite highway according to a homogeneous Poisson process with rate α. Assume the initial position of the nth car is X_n. Each car chooses a velocity independently of all other cars and proceeds to travel at that fixed velocity. Call the velocities $\{V_n\}$, and assume these are iid random variables with state space $(-\infty, \infty)$. A negative velocity means the car is travelling to the left and a positive velocity means travel to the right. Assume collisions are impossible; if necessary, cars pass through each other rather than collide. Let $N_0(\cdot)$ be the Poisson process of initial positions and let $N_t(\cdot)$ be the point process describing positions at time t. What sort of process is $N_t(\cdot)$?

Let T_n be the time the nth car passes through 0, so that T_n satisfies $X_n + T_n V_n = 0$. Show $\sum_n \epsilon_{T_n}$ is Poisson, and find its mean measure. (Assume $E|V_1| < \infty$ and consider the point process with points $\{(X_n, V_n)\}$. Proceed by transformations.) (Renyi, 1964.)

4.3. For a homogeneous Poisson process $\{N(t), t \geq 0\}$ of rate 1, we know from renewal theory that

$$\lim_{t \to \infty} N(t)/t = 1.$$

If N' is non-homogeneous Poisson with local intensity $1/t$, what is the asymptotic form of $N'(1, t]$? Find a function $\beta(t)$ such that

$$\lim_{t \to \infty} N'(1, t]/\beta(t)$$

exists and is finite. (This gives the asymptotic rate at which points are coming.) Hint: Use the transformation theory to compare N' with N.

4.4. Harry Combats Drugs. Incidents of drug usage occur in the bathroom of Harry's restaurant. The times of these incidents constitute a Poisson process of rate two per hour. Harry is concerned about this because of the possible effect on business and because he worries about how effective the government's war on drugs will be. Consequently, Harry visits the bathroom at time points which constitute a Poisson process of rate one per hour. Assume this Poisson process is independent of the Poisson process of drug incidents. Assume also that if a drug incident has occurred, Harry detects it.

(a) On his first visit, what is the probability that Harry finds evidence of drug use; i.e., what is the probability that a drug incident occurs before Harry checks the bathroom for the first time?

(b) What is the expected time until Harry detects evidence of drug use?

4.5. Harry's Stressful Life. Due to the stress of coping with business, Harry begins to experience migraine headaches of random severities. The times when headaches occur follow a Poisson process of rate λ. Headache severities are independent of times of occurrences and are independent, identically distributed random variables $\{H_i\}$ with common exponential distribution

$$P[H_i \leq x] = 1 - \exp\{-x\}.$$

(Assume headaches are instantaneous and have duration zero.)

Harry decides he will commit himself to the hospital if a headache of severity greater than $c > 0$ occurs in the time period $[0, t]$. Compute

$$P\{\text{Harry commits himself in } [0, t]\}.$$

(Hint: Compute the probability of the complementary event that Harry won't commit himself in $[0, t]$. Remember that $\max\{a, b\} \le c$ if and only if both $a \le c$ and $b \le c$.)

4.6. Suppose you are a historiometrician (one who uses statistics to aid in the study of history—if there isn't a word like this, there will be eventually), and you need a Poisson process model to explain the archeological findings of coins on the winding road between London and Canterbury. Explain how you would construct such a model.

4.7. M/G/∞ Queue. Times of call initiations are homogeneous Poisson time points on $(0, \infty)$ (not $(-\infty, +\infty)$ as in Section 4.4), and call durations are iid with common distribution G and independent of call initiation times. Let $N(t)$ be the number of calls in progress at time t. Mark an incoming call according to whether or not it is still clogging a line at time t. Show that $N(t)$ has a Poisson distribution.

4.8. Bulk Arrivals. Buses arrive at Harry's on Friday nights according to a Poisson process with rate α. Each bus contains a random number of hungry athletes, and we assume the number per bus constitute iid non-negative integer valued random variables. Expecting Friday night traffic will be heavy, Harry has added many servers; the number is infinite. Each hungry customer spends a random amount of time ordering and waiting for his/her food, and we assume the waiting times are iid and independent of other random variables. Let $X(t)$ be the number of customers who arrived in $[0, t]$ and by time t have still not commenced eating.
(a) Find $EX(t)$.
(b) What is the distribution of $X(t)$? (Think compound Poisson.)
(c) If we assume customers only come to Harry's to buy food, but intend to eat it on the bus rather than in the restaurant, describe the departure process of customers from Harry's.

4.9. If $\{E_n\}$ are iid exponentially distributed with parameter α, N is geometric with
$$P[N = n] = pq^{n-1}, \quad n \ge 1,$$
and N is independent of $\{E_n\}$, show that
$$\sum_{n=1}^{N} E_n$$
has an exponential distribution by computing the Laplace transform. What is the parameter? What is the relation of this result to thinning? Construct a second probabilistic proof using thinning of a Poisson process.

4.10. A critical submarine component has a lifetime which is exponentially distributed with mean 0.5 years. Upon failure, replacement with a new

component of identical characteristics occurs. What is the smallest number
of spare components that the submarine should stock if it is leaving for a
one year tour and wishes the probability of an inoperative unit caused by
failure exceeding the spare inventory to be less than 0.02? (Taylor and
Karlin, 1984.)

4.11. Let

$$N = \sum_{i=1}^{\infty} \epsilon_{X_i}$$

be a non-homogeneous Poisson process with state space $[0, \infty)$ and with
local intensity $\alpha(t)$ and points $\{X_i\}$, with $X_1 < X_2 < \ldots$ Let $T_1 = X_1, T_2 = X_2 - X_1, \ldots$ be the interpoint distances.

(a) Are $\{T_i\}$ independent?

(b) Are $\{T_i\}$ iid?

(c) Compute the joint distribution of T_1, T_2 and X_1, X_2.

4.12. Bulk Arrivals. Customers arrive for Friday night amateur night
at Harry's by bus. The buses arrive according to a Poisson process of rate
α, and the number on the kth bus is a random variable A_k (independent
of numbers on other buses) with generating function $P(s)$. If the Poisson
process is represented by $\sum_n \epsilon_{\Gamma_n}$, then we may represent the process of
bulk arrivals by

$$N = \sum_n A_n \epsilon_{\Gamma_n}.$$

Compute the Laplace functional of N, and find the mean measure

$$\mu(\cdot) = EN(\cdot).$$

(Assume $E(A_1) < \infty$.) Hint: You may wish to condition first on $\{\Gamma_n\}$,
which will give you the Laplace functional of the Poisson process at a
different function than the one you started with. You can then capitalize
on the known form of the Laplace functional for the Poisson process.

4.13. Suppose the input to an M/G/∞ queue (infinite server queue) after
time 0 is a non-homogeneous Poisson process with local intensity $\alpha(t)$. Suppose durations of calls are iid random variables with common distribution
G. Describe the times of terminations of calls.

**4.14. The Order Statistic Property Characterizes the Mixed
Poisson Process.** Suppose $N = \sum_i \epsilon_{T_k}$ is a point process on $(0, \infty)$
with $0 < T_1 < T_2 < \ldots$ and set $N(t) = N((0, t])$. Assume N has the order
statistic property, namely that

$$(T_1, \ldots, T_k | N(t) = k) \stackrel{d}{=} (U_{(1)}, \ldots, U_{(k)}),$$

where $(U_{(1)}, \ldots, U_{(k)})$ are the order statistics from a sample of size k from the uniform distribution on $(0, t)$. Show there exists a random variable $W \geq 0$ such that

$$\{N(t), t > 0\} = \{N^*(Wt), t > 0\},$$

where N^* is a homogeneous Poisson process with unit rate and independent of the random variable W.

This result may be derived by following these simple steps:

(a) Let $e_k = T_k/T_{k-1}$ for $k \geq 2$. Conditional on $N(t) = k$ show that $e_2, \ldots, e_k, T_k$ are independent and that $\log e_j$ is exponentially distributed with mean $1/(j-1)$.

(b) For any $k \geq 2$ show $(e_2, \ldots, e_k, T_k)$ are independent by computing

$$P[e_2 \leq x_2, \ldots, e_k \leq x_k, T_k \leq x]$$
$$= \sum_{j=k}^{\infty} P[e_2 \leq x_2, \ldots, e_k \leq x_k, T_k \leq x | N(t) = j] P[N(t) = j].$$

(c) Since the homogeneous Poisson process with points $\{\Gamma_n, n \geq 1\}$ has the order statistic property,

$$(e_j, j \geq 2) = (\frac{\Gamma_j}{\Gamma_{j-1}}, j \geq 2).$$

(d) By the law of large numbers, for any j,

$$\lim_{n \to \infty} \frac{\Gamma_n}{n\Gamma_j} = \frac{1}{\Gamma_j}.$$

Thus, there exists T_j^* such that

$$\lim_{n \to \infty} \frac{T_n}{nT_j} = \frac{1}{T_j^*},$$

since

$$\lim_{n \to \infty} \frac{T_n}{nT_j} = \lim_{n \to \infty} \frac{T_n}{nT_{n-1}} \frac{T_{n-1}}{T_{n-2}} \cdots \frac{T_{j+1}}{T_j}$$
$$= \lim_{n \to \infty} \frac{1}{n} e_n e_{n-1} \cdots e_{j+1}$$
$$\overset{d}{=} \lim_{n \to \infty} \frac{\Gamma_n}{n\Gamma_{n-1}} \frac{\Gamma_{n-1}}{\Gamma_{n-2}} \cdots \frac{\Gamma_{j+1}}{\Gamma_j}$$
$$= \lim_{n \to \infty} \frac{\Gamma_n}{n\Gamma_j} = \frac{1}{\Gamma_j}.$$

(e) Check $\{T_j\} \stackrel{d}{=} \{\Gamma_j\}$; there exists V such that $T_n/n \to V > 0$.

(f) Since $(e_j, j \le k)$ is independent of T_k, conclude V is independent of $(e_j, j \ge 2)$ and of $\{T_j^*\}$.

(g) On $[V < \infty]$ we have a.s.

$$V^{-1}(T_1, T_2, \dots) = (T_1^*, T_2^*, \dots).$$

Set $W = V^{-1}$. (This outline is from Feigin, 1979; the more general characterization is nicely described in Neveu, 1976.)

4.15. Consider the Poisson process $N = \sum_k \epsilon_{j_k}$ on $(0, \infty]$ with points $\{j_k\}$ and mean measure $\mu(dx) = \alpha x^{-\alpha-1} dx$ for $x > 0$. Let $Y_1 = \sup_k j_k$ be the largest point, and let Y_2 be the second largest point. Compute the joint distribution function of Y_1, Y_2.

4.16. Let $\{\Gamma_n\}$ be as usual the renewal sequence corresponding to the exponential distribution. Suppose ν is a measure on R with the property that $Q(x) := \nu(x, \infty) < \infty$ for any $x \in R$. For $y > 0$, define

$$Q^{\leftarrow}(y) = (1/Q)^{\leftarrow}(y^{-1}),$$

where, as usual, we define the inverse of the non-decreasing function by

$$(1/Q)^{\leftarrow}(u) = \inf\{v : 1/Q(v) \ge u\}.$$

Show

$$\sum_n \epsilon_{Q^{\leftarrow}(\Gamma_n)},$$

is Poisson on R with mean measure ν.

4.17. Compute the Laplace functional of the simple cluster process

$$\sum_n \xi_n \epsilon_{X_n},$$

where $\{X_n\}$ are Poisson points independent of the iid non-negative integer valued sequence $\{\xi_n\}$. Use the form of the Laplace functional to verify that the simple cluster process is completely random.

4.18. Lévy Decomposition of a Poisson Process.* (a) Let N be a Poisson process on $[0, \infty)$ with mean measure μ. Show

$$N = N_f + N_c,$$

---

* This problem is for advanced students.

where N_f, N_c are independent, where $\{N((0,t]), t > 0\}$ is a stochastically continuous Poisson process, and where $\{N((0,t]), t > 0\}$ is a process with independent increments with only fixed discontinuities. (Hint: The fixed discontinuities consist of $\{t > 0 : \mu(\{t\}) > 0.)$

(b) Arrivals at a barber shop form a non-stationary and non-stochastically continuous Poisson process. On a particular day, there are appointments made for 12:00, 12:20, 1:00, 3:40, 4:20, 4:40. Past experience indicates that an appointment is kept with probability 2/3. Customers without appointments arrive at the rate of one per hour during the first three hours, at the rate of 0.4 per hour during the next two hours and at the rate of 0.2 per hour during the last hour. Discuss the structure of the mean measure. Compute

  (a) $P[N_f((0,4]) = k]$,
  (b) $P[N_c((0,4]) = k]$,
  (3) $P[N((0,4]) = k]$.

(Note the number of fixed discontinuities is not Poisson distributed.)

4.19.* Let N be a point process with mean measure μ. Suppose $f \geq 0$ is bounded. Show for any measurable set A

$$E \int_A f(x)dN(x) = \int_A f(x)\mu(dx).$$

4.20. The Mixed Poisson Process and the Order Statistic Property.

(a) Let $N^*((0,t]) = N((0, \Lambda t])$ be a mixed Poisson process. Here Λ and N are independent and N is a homogeneous Poisson process on $[0, \infty)$. Let G be the distribution of Λ, and let $\hat{G}$ be the Laplace transform. Show

$$P[N^*((t, t + \tau]) = n] = (-\tau)^n \hat{G}^{(n)}(\tau)/n!,$$

where $\hat{G}^{(n)}(\tau)$ is the nth derivative of $\hat{G}$ at the point τ.

(b) More generally, show that if the intervals $(t_i, t_i + \tau_i], i = 1, \ldots, r$, are disjoint

$$P[N^*((t_i, t_i + \tau_i]) = n_i, i = 1 \ldots, r] = \frac{\prod_{i=1}^r \tau_i^{n_i}}{\prod_{i=1}^r n_i!} \hat{G}^{(\Sigma_{i=1}^r n_i)}\left(\sum_{i=1}^r \tau_i\right).$$

(c) Show if the intervals $(t_i, t_i + h_i], i = 1, \ldots, n$, are disjoint and $t_i + h_i < \tau$ for each i, then

$$P[N^*((t_i, t_i + h_i]) = 1, i = 1, \ldots, n;$$

$$N^*((0, \tau] \setminus \cup_{i=1}^n (t_i, t_i + h_i]) = 0 | N((0, \tau]) = n] = \frac{n!}{\tau^n} \prod_{i=1}^n h_i,$$

---

* This problem is suitable for more advanced students.

and therefore the mixed Poisson process has the order statistic property.

(This problem can be done independently of Problem 4.14).

4.21. Mixed Poisson, continued.* Let N be a stationary point process on $[0, \infty)$ with points $\{X_n\}$ arranged in increasing order. Suppose $EN((0, t]) < \infty$ for each t and, for each n,

$$((X_1, \ldots, X_n) | N((0, t]) = n) \stackrel{d}{=} (U_{(1)}, \ldots, U_{(n)})$$

where $U_1, \ldots, U_n$ are iid uniform random variables on $(0, t)$ and the order statistics are $U_{(1)}, \ldots, U_{(n)}$. Also, N is Poisson if $\{N((n, n+1])\}$ is a stationary, ergodic sequence, and N is mixed Poisson if $\{N((n, n+1])\}$ is not ergodic. In the latter case,

$$\Lambda = \lim_{n \to \infty} N((0, n])/n$$

almost surely.

4.22. Constructions of the Poisson Process on R. Show that the following constructions lead to a homogeneous Poisson process on R:

(a) Let $\{E_n, -\infty < n < \infty\}$ be iid unit exponentially distributed random variables. Define

$$X_0 = E_0, X_1 = E_0 + E_1, \ldots,$$
$$X_{-1} = -E_{-1}, X_{-2} = -(E_{-1} + E_{-2}), \ldots.$$

Show $\sum_n \epsilon_{X_n}$ is a homogeneous Poisson process.

(b) Suppose $\{E_n, n \neq 0\}$ are iid unit exponentially distributed random variables independent of the uniformly distributed on $(0, 1)$ random variable W. Also suppose E_0 is independent of $\{E_n, n \neq 0\}$ and W and E_0 has density

$$f_{E_0}(x) = xe^{-x}, \quad x > 0.$$

Prove WE_0 and $(1 - W)E_0$ are iid unit exponentially distributed random variables. Define

$$t_0 = -WE_0, \quad t_1 = (1 - W)E_0$$

and

$$t_i = \begin{cases} t_1 + E_1 + \cdots + E_{i-1}, & \text{for } i > 1 \\ t_0 - E_{-1} - \cdots - E_{-i}, & \text{for } i < 0. \end{cases}$$

Show $\sum_i \epsilon_{t_i}$ is a homogeneous Poisson process on R.

---

* This problem is suitable for more advanced students.

4.23. For the mixed Poisson process N^*, show, using, say Theorem 3.3.2, that

$$\lim_{t \to \infty} \frac{N^*((0,t])}{t} = \Lambda.$$

4.24. Doubly Stochastic Poisson Process and Thinning. Let N_m be Poisson process with mean measure m on E. Let μ be a random element of the space $M_+(E)$ of measures on E which are finite on bounded sets, and suppose the random element μ is independent of the family $\{N_m, m \in M_+(E)\}$.

(a) Compute the Laplace functional of N_μ.

(b) For any point process N, let D_pN be the thinned point process where we inspect the points of N independently and discard each with probability $1 - p$ and retain them with probability p. Show the Laplace functionals of N and D_pN are related by

$$\Psi_{D_pN}(f) = \Psi_n(-\log(1 - p + pe^{-f})).$$

(c) If N_μ is a doubly stochastic Poisson process (i.e., N_m is PRM(m) for each m), then D_pN_μ is doubly stochastic Poisson also and

$$D_pN_\mu \stackrel{d}{=} N_{p\mu}.$$

(d) To each doubly stochastic Poisson process N_μ corresponds another doubly stochastic Poisson process $N^{(p)}$ such that

$$N = D_pN^{(p)}.$$

(The converse of (d) is true as well and characterizes the class of doubly stochastic Poisson processes.)

4.25. Let N be a homogeneous Poisson process on $[0, \infty)$. Suppose the non-negative stochastic process $\{V(t), t > 0\}$ is stationary and that it can be integrated. Define

$$\Lambda(t) = \int_0^t V(s)ds.$$

Show that the doubly stochastic Poisson process $\{N((0, \Lambda(t)]), t > 0\}$ is a stationary point process. (If necessary, review the definition of a stationary point process from Section 3.9.)

4.26. Consider random variables X, Y such that $X \geq 0$ and has distribution F (with $F(0) = 0$) and the conditional distribution of Y given X is uniform on $(0, X)$. Consider the point process

$$N = \sum_{n=-\infty}^{\infty} \epsilon_{nX+Y}.$$

Compute $P[N((0,h]) = 0]$, and evaluate this explicitly in the case where F is the uniform distribution on $(0,1)$. In this special case compute

$$\lim_{h\to 0} \frac{P[N((0,h]) > 0]}{h}.$$

In contrast to the behavior of the process presented above, compute this last limit for the mixed Poisson process.

4.27. The Mixed Poisson Process and Independent Increments. Show that a mixed Poisson process has independent increments iff it is Poisson.

4.28. Let $\{S_n, n \geq 0\}$ be a renewal sequence with $S_0 = 0$. Let $A(t)$ and $B(t)$ be the age and forward recurrence times at t. If the interarrival distribution is exponential, we know $B(t)$ has an exponential distribution. Prove if $EB(t) < \infty$ is independent of t, then the renewal process is a Poisson process. (Hint: Let U be the renewal function. Express, as usual, the probability $P[B(t) > x]$ in terms of the renewal function. Integrate over x to get $EB(t)$. Suppose the latter is constant, and take Laplace transforms in t.) (Cinlar and Jagers, 1973.)

4.29. Harry and the Hailstorm. Harry's restaurant has a flat roof which is subject to damage during a hailstorm. When a hail stone strikes the roof, there is damage due to the primary impact of the hail, as well as secondary impacts caused by the hail bouncing and restriking the surface of the roof. Both Harry and the insurance adjuster believe the positions of primary impacts form a Poisson process and assume this to be true. Harry believes the total collection of impact points on his roof form a spatial Poisson process. The insurance adjuster vehemently disagrees. Who is right?

4.30. Assume N is a point process on $[0,\infty)$, and let $N(t) = N([0,t])$. Suppose $\{N(t), t \geq 0\}$ satisfies $N(0) = 0$, N has independent increments and

$$P[N((t, t+h]) = 1] = \lambda(t)h + o(h)$$
$$P[N((t, t+h]) \geq 2] = o(h).$$

Prove N is a nonhomogeneous Poisson process with local intensity $\lambda(t)$. (Hint: Let $P_t(s) = Es^{N(t)}$, and use the postulates to show

$$\frac{d}{dt}P_t(s) = \lambda(t)(s-1)P_t(s).$$

To do this, consider the difference quotient

$$\frac{P_{t+h}(s) - P_t(s)}{h}.$$

Solve the differential equation—it's easy—for $P_t(s)$, and invert or identify the generating function.)

4.31. Consider a non-homogeneous Poisson process on $[0,\infty)$ with local intensity

$$\lambda(t) = \begin{cases} \lambda, & \text{if } 2m \le t < 2m+1, \\ 0, & \text{otherwise,} \end{cases}$$

for $m = 0, 1, \ldots$. A sampling interval of unit length is selected at random by first choosing an integer m at random according to some discrete distribution $\{p_k\}$ and initiating the sampling interval at a point uniformly chosen in the interval $[m, m+1)$. Compute the probability that the sampling interval contains n points, and compute the mean and variance of the number in the sampling interval.

4.32. Consider a homogeneous planar Poisson process, and let $\|x\|$ be the Euclidean norm. Suppose the points of the process are $\{X_n\}$ arranged so that

$$\|X_0\| < \|X_1\| < \|X_2\| < \cdots.$$

What type of process is $\{\|X_n\|\}$? Let

$$A_n = \pi\|X_n\|^2$$

be the area of the sphere with radius $\|X_n\|$. What type of process is $\{A_n\}$?

4.33. As a simple example of a cluster process, consider the following scheme: Let $\sum_n \epsilon_{X_n}$ be homogeneous Poisson on $[0,\infty)$. Suppose $\{Y_n\}$ are iid displacements. Retain the old point X_n, and add a new point at $X_n + Y_n$, so the new point process is

$$N_{CL} = \sum_n \epsilon_{X_n} + \sum_n \epsilon_{X_n+Y_n}.$$

Compute the Laplace transform of $N_{CL}(A)$ or compute the Laplace functional. Is this process ever Poisson?

4.34. The Order Statistic Property in Higher Dimensions. Let N be a PRM(μ) on a state space E. Prove that for a bounded region A, the conditional distribution of the points in A, given $N(A) = n$, is the same as the distribution of n iid random variables from the distribution $\mu(dx)/\mu(A)$.

4.35. Moran's (1967) Example—Poisson Marginals but Not a Poisson Process. Take a two-dimensional density of the form

$$f(x,y) = e^{-(x+y)} + f_\delta(x,y),$$

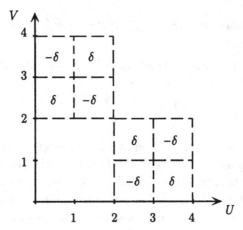

FIGURE 4.4. MORAN'S EXAMPLE

where $f_\delta(x, y)$ takes the values $\pm\delta$ $(\delta < e^{-6})$ in the regions shown in the figure and is 0 elsewhere.

Note that if the random vector (U, V) has the density $f(x, y)$, then the marginal densities of both U and V are unit exponential, and, furthermore, $U + V$ has density $xe^{-x}1_{(0,\infty)}(x)$, which is the density of the sum of two iid unit exponentially distributed random variables.

Suppose $\{(U_i, V_i)\}$ are iid copies of the vector (U, V). Use these iid vectors to construct interpoint distances of a point process N' (where will the origin go?) which has the properties

(1) The number of points of N' in any interval is Poisson distributed with a mean proportional to the length of the interval.
(2) The interpoint distances of points of N' are marginally exponentially distributed.
(3) N' is *not* a Poisson process.

4.36.* Let $N = \sum_n \epsilon_{X_n}$ be a mixed Poisson process on R. Let $\{Y_n\}$ be iid random variables independent of $\{X_n\}$. Show

$$N \overset{d}{=} \sum_n \epsilon_{X_n + Y_n}.$$

Conversely, if a point process $N = \sum_n \epsilon_{X_n}$ on R has the property that

$$N \overset{d}{=} \sum_n \epsilon_{X_n + Y_n}$$

for any sequence $\{Y_n\}$ independent of $\{X_n\}$, show N is mixed Poisson. (Goldman, 1967.)

---

* This is a challenging problem suitable for advanced students.

4.37. EARMA and Friends. As an alternative to iid interpoint distances of the homogeneous Poisson process on $[0, \infty)$, consider the following schemes inspired by time series constructions: Let $\{E_n, n \geq 0\}$ be iid exponentially distributed random variables with parameter ρ, and let $\{U_n\}$ and $\{V_n\}$ be independent sequences of iid Bernoulli variables with

$$\alpha = P[U_n = 0], \quad \beta = P[V_n = 0].$$

Define the following three sequences which may serve as interpoint distances of point processes on $[0, \infty)$:

(1) *Exponential autoregressive process of order 1 (EAR 1):*

$$X_0 = E_0, \quad X_n = \alpha X_{n-1} + U_n E_n, \ n \geq 1;$$

(2) *Exponential moving average process of order 1 (EMA 1):*

$$Y_n = \beta E_n + V_n E_{n-1}, \ n \geq 1;$$

(3) *Exponential autoregressive moving average process (EARMA (1,1)):*

$$Z_n = \beta E_n + V_n X_{n-1}, \ n \geq 1,$$

where X_n is given in (1).

(a) Verify that (1) and (2) are special cases of (3) and that if $\alpha = \beta = 0$ the point process generated by these interpoint distances is Poisson.

(b) Show that the Z_n are marginally exponentially distributed. (Compute the Laplace transform of $U_n E_n$ and then a recursion for the Laplace transform of X_n; solve the recursion. Then compute the Laplace transform of Z_n.)

(c) Compute the autocovariances

$$\mathrm{Cov}(Z_n, Z_{n+k}).$$

Verify that this is independent of n; thus $\{Z_n\}$ is (covariance) stationary, and we may set

$$\gamma(k) = \mathrm{Cov}(Z_n, Z_{n+k}).$$

(Jacobs and Lewis, 1977; Lawrance and Lewis, 1977.)

4.38. Let N be a non-homogeneous Poisson process on $[0, \infty)$ with points $0 < T_1 < T_2 < \ldots$. Given $T_n = t$, what is the distribution of $(T_1, \ldots, T_{n-1})$?

4.39. Consider a mixed Poisson $\{N((0, \Lambda t]), t \geq 0\}$. Work out the distribution of $N((0, \Lambda t])$ when Λ has the following distributions: (a) Gamma, (b) Poisson.

Consider the simple cluster process $N_{CL} = \sum_n \xi_n \epsilon_{X_n}$ based on the homogeneous Poisson process with points $\{X_n\}$ and in which there are no translation satellites about centers. Compute the distribution of $N_{CL}((0, t])$ when ξ_n has the following distributions: (a) Geometric, (b) Poisson, (c) Binomial.

4.40. Estimating the Rate of a Homogeneous Poisson Process. Suppose it is known that N is a homogeneous Poisson process but the rate α is unknown. Suppose up to time t, the process is observed; we observe $N((0, t]) = n$ and points $X_1 = x_1 < X_2 = x_2 < \cdots < X_n = x_n \leq t$.

(a) Show that the likelihood function is

$$e^{-\alpha t} \alpha^n.$$

(b) For t fixed, show that $N((0, t])$ is a sufficient statistic for α.
(c) Show

$$\hat{\alpha} =: \frac{N((0, t])}{t}$$

is the maximum likelihood estimator of α. Show that this estimator is consistent and asymptotically normal. Is it also a uniform minimum variance unbiased estimator? What is an exact 95% confidence interval for α? Use the normal approximation, and give an approximate 95% confidence interval.

(d) Consider testing the null hypothesis

$$H_0 : \alpha = \alpha_0$$

against the alternative

$$H_1 : \alpha > \alpha_0.$$

Using the sufficient statistic as a test statistic, when would you reject the null hypothesis? What is the exact significance level of the test. Compare the exact significance level with the normal approximation if $t = 15$ and $N((0, t]) = 12$.

Change the alternative hypothesis to $H_1 : \alpha < \alpha_0$ and answer the questions in the previous paragraph. (Cox and Lewis, 1966.)

4.41. Local Poissonification. Neuts et al. (1992) describe a transformation of a stationary point process called *local poissonification* designed to test how critical local characteristics are in determining global properties. The procedure is as follows: Suppose N is a stationary point process on the state space R. Fix a length h, and decompose R as a union of intervals of length h:

$$R = \bigcup_{n=-\infty}^{\infty} ((n-1)h, nh] = \bigcup_{n=-\infty}^{\infty} I_n(h).$$

The *poissonified* version of N, call it N_{PO}, is obtained by looking at each interval $I_n(h)$. If $N(I_n(h)) = m$, we replace the m points of N by m random variables uniformly distributed on $I_n(h)$.

(a) Express the Laplace functional of N_{PO} in terms of the Laplace functional of N. Alternatively, if you skipped Section 4.7.1 on Laplace functionals, express the Laplace transform of $N_{PO}(I_n(h))$ in terms of that of N.

(b) If N is either a Poisson process or a mixed Poisson process, show

$$N \overset{d}{=} N_{PO};$$

i.e., show N_{PO} is a Poisson process of the same rate or a mixed Poisson process.

4.42. Suppose the arrival process of customers to the bus stop is a homogeneous Poisson process with rate α and that the arrival process of buses is an independent renewal process with interarrival distribution F. Find the long run percentage of customers who wait for a bus less than x units of time.

4.43. Under what circumstances would you be willing to model the following phenomena as Poisson processes? Which Poisson processes?

(a) Telephone calls arriving to a central exchange.

(b) Emissions from a radioactive souce arriving to a counter. Does the counter have a significant block time after a registration?

(c) Stops of a machine due to either breakdowns or lack of work.

4.44. Superposition of Poisson Processes. Suppose $N_1, \ldots, N_k$ are independent Poisson processes on a state space E, and suppose the mean measure of N_i is μ_i. Verify that $\sum_{i=1}^{k} N_i$ is a Poisson random measure also with mean measure $\sum_{i=1}^{k} \mu_i$. If you wish, think about this when the state space E is $[0, \infty)$.

4.45. Who Will Be First?

(a) Suppose E_1, E_2 are independent exponentially distributed random variables with parameters α_1, α_2. Show the events

$$[E_1 < E_2] \text{ and } [E_1 \wedge E_2 > x]$$

are independent. (See Proposition 5.1.1 for help.)

(b) Entry of yuppies and computer nerds into Harry's restaurant after opening each day constitute independent Poisson processes of rates α_1, α_2. What is the probability the first person to enter after opening is a yuppie? Show that the event that the first person to enter is a yuppie is independent of the entry time of the first customer.

4.46. The arrival of customers at a self-service market is a Poisson process of rate α. Customers have iid shopping times S_j which have common distribution $G(t)$. After shopping, each customer checks himself out. The checkout times C_j are iid and independent of shopping times and have common distribution $H(t)$. Assume customers begin checking out as soon as their shopping is completed.

For each $t > 0$, let $X(t)$ be the number of customers shopping at time t, and let $Y(t)$ be the number of customers who are checking out.

(a) Find the joint distribution of $(X(t), Y(t))$.

(b) Now suppose that $\{(S_j, C_j)\}$ are iid random vectors but that for each j we might have S_j and C_j dependent. Suppose we know the joint distribution of (S_1, C_1). Show $(X(t), Y(t))$ are independent, and compute the joint distribution.

(c) Take the joint distribution found in (b) and let $t \to \infty$ to show a limiting joint mass function exists. Identify it. (Wolf, 1989.)

4.47. More on Estimating the Parameter of a Poisson Process. Suppose we know

$$N = \sum_{n=1}^{\infty} \epsilon_{X_n}$$

is a homogeneous Poisson process on $[0, \infty)$ with unknown rate α where the points $\{X_n\}$ are assumed to satisfy

$$X_1 < X_2 < \ldots .$$

For some fixed n, we get to observe the process up until the time of the nth point and thus we observe until X_n.

(a) Show that the likelihood function is the joint density of $X_1, \ldots, X_n$, namely

$$f_{X_1, \ldots, X_n}(x_1, \ldots, x_n) = \alpha^n e^{-\alpha x_n}, \quad x_1 < x_2 < \cdots < x_n.$$

(b) Show X_n is a sufficient statistic for α. What is the maximum likelihood estimator of α? What is the UMVUE (uniform minimum variance unbiased estimator)?

(c) What is the density of $2\alpha X_n$? Does it depend on α? If not, use it as a pivotal quantity to get a 95% confidence interval for α.

(d) Explain how to test the hypothesis $H_0 : \alpha = \alpha_0$ against the alternative $H_1 : \alpha < \alpha_0$. (Cox and Lewis, 1966.)

4.48. Motor vehicles arrive at a bridge toll gate according to a Poisson process with rate $\alpha = 2$ vehicles per minute. The drivers pay tolls of \$1, \$2 or \$5 depending on which of three weight classes their vehicles belong. Assuming that the vehicles arriving at the gate belong to classes 1, 2 and 3 with probabilities 1/2, 1/3, and 1/6, respectively, find

(a) The mean and variance of the amount in dollars collected in any given hour.

(b) The probability that exactly \$3 is collected in each of three consecutive 1-minute intervals (which are specified).

(c) The probability that the waiting time between the first two \$5 vehicles is more than 10 minutes.

4.49. Customers arrive to a bus depot according to the mixed Poisson process $\{N((0, \xi t]), t \geq 0\}$ where N is homogeneous Poisson with rate α and ξ is a positive random variable independent of N. The first bus to leave after $t = 0$ leaves at $t = 1$. Compute the expected total wait of all customers arriving in $(0, 1)$.

4.50. Let $\{N(t) = N((0, t]), t \geq 0\}$ be a non-homogeneous Poisson process with $N(0) = 0$ and local intensity $\alpha(t) = \theta t$, where $\theta > 0$ is a constant.

(a) Compute $EN(1)N(2)$ and $P[N(1) = 2 | N(2) = 6]$.

(b) Let $W_n = \inf\{t > 0 : N(t) = n\}$, for $n = 1, 2, \ldots$. Find first the distribution of W_n and then the joint distribution of $W_1, W_2 - W_1$.

(c) Suppose θ is unknown and needs to be estimated.

(1) If the process is observed only at $t = 5$ and $t = 10$ with $N(5) = 1, N(10) = 2$, find the maximum likelihood estimate of θ.

(2) If the process is observed over the entire time interval $[0, 10]$ with $N(10) = 2, W_1 = 5, W_2 = 8$, find the maximum likelihood estimate of θ.

(d) Define $Y(t) = N(t^2)$. Determine the local intensity for the point process $Y(t)$ in terms of θ.

4.51. Extremal Processes. Consider a two dimensional Poisson process

$$N = \sum_k \epsilon_{(t_k, j_k)}$$

on $[0, \infty) \times (0, \infty]$ with mean measure determined by

$$\mu((0, t] \times (x, \infty]) = t(-\log F(x)),$$

where F is a distribution function on R. Define

$$Y(t) = \sup\{j_k : t_k \leq t\}$$

to be the largest second component whose first component is prior to t. $\{Y(t)\}$ is called an *extremal process*. (Dwass, 1964; Resnick, 1987.)

(a) Compute $P[Y(t) \leq x]$.

(b) For any k and $x_1, \ldots, x_k$ real and $0 < t_1 < \cdots < t_k$, compute

$$P[Y(t_1) \leq x_1, \ldots, Y(t_k) \leq x_k].$$

4.52. Consider $\{X_n, n \geq 1\}$ iid with continuous distribution F and

$$(x_l, x_r) = (\sup\{x : F(x) > 0\}, \inf\{x : F(x) < 1\}).$$

Set $M_n = \bigvee_{i=1}^n X_i$, and define

$$\eta(t) = \inf\{n : M_n > t\}$$

for $x_l < t < x_r$. Prove $\{\eta(t)\}$ is a process with independent increments, and for $x_l < a < b < x_r$,

$$P[\eta(b) = k] = F^{k-1}(b)(1 - F(b)), \quad k \geq 1$$

$$P[\eta(b) - \eta(a) = 0] = \frac{1 - F(b)}{1 - F(a)}.$$

Also, for $n \geq 1$

$$P[\eta(b) - \eta(a) = n] = \left(\frac{1 - F(b)}{1 - F(a)} \right) F(a, b] F^{n-1}(b).$$

(Hint: Observe that

$$\eta(b) - \eta(a) = \sum_n 1_{[M_n \in (a,b]]}$$

$$= \sum_{k=1}^{\infty} (L(k+1) - L(k)) \epsilon_{X_{L(k)}}(a, b]$$

and use Proposition 4.11.2.)

CHAPTER 5

Continuous Time Markov Chains

W E TURN now to the continuous time version of the Markov property. Some of the simplicity of Chapter 2 is retained, because we assume the state space S is discrete. Usually we can suppose that $S = \{0, 1, \dots\}$. The succession of states visited still follows a discrete parameter Markov chain but now the flow of time is perturbed by exponentially distributed holding times in each state. An easy generalization of the dissection argument of Chapter 2 shows that the process regenerates at return times to a fixed reference state, so renewal theory and regenerative processes are useful.

Our approach is to construct the process starting from some basic ingredients: a discrete parameter Markov chain and a supply of iid exponentially distributed random variables. The discrete parameter Markov chain controls movements through the state space, and the exponential random variables control how rapidly these movements take place.

In Chapter 2, useful descriptive quantities such as absorption probabilities and expected absorption times were calculated as solutions of difference equations; in this chapter, however, we must solve systems of differential equations.

There is a large class of queueing models which are analyzed as Markov chains in continuous time. Other applications include models for population growth, material transfer, reliability of mechanical systems, and epidemics.

5.1. DEFINITIONS AND CONSTRUCTION.

We consider a process $\{X(t), t \geq 0\}$ with index set $[0, \infty)$ and a state space which is a subset of the integers. Given that the process is in state i, the holding time in that state will be exponentially distributed with parameter $\lambda(i)$. The succession of states visited will follow a discrete time Markov chain. Given a knowledge of the states visited, holding times are independent random variables.

For example, think of the number of patrons in Harry's restaurant as fluctuating with time. For this to be modelled as a Markov chain, we must assume that when there are n patrons present, the time until the number present changes is exponentially distributed with a parameter which depends on n, the current number of patrons.

We want the process to have the *Markov property*; i.e., for any $m \geq 0$, time points $t_1 < t_2 < \ldots < t_m$ and states $k_1, \ldots, k_{m-2}, i, j$, we want

$$P[X(t_m) = j | X(t_1) = k_1, \ldots, X(t_{m-2}) = k_{m-2}, X(t_{m-1}) = i]$$
$$\text{(5.1.1)} \qquad = P[X(t_m) = j | X(t_{m-1}) = i] =: P_{ij}(t_m - t_{m-1}).$$

We also want the process to have stationary transition probabilities, meaning for $s, t \geq 0$,

$$P[X(t + s) = j | X(t) = i] =: P_{ij}(s).$$

Furthermore, we wish this process to have the continuous time analogues of property (2.1.4), namely, for any collection of paths B (allow this to be vague), we have

$$P[\{X(s), s \geq t\} \in B | X(t_1) = k_1, \ldots, X(t_{k-1}) = i]$$
$$\text{(5.1.2)} \qquad\qquad = P[\{X(s), s \geq t\} \in B | X(t_{k-1}) = i].$$

We defer consideration of (5.1.1) and (5.1.2) and concentrate on the construction of the process $\{X(t)\}$. Suffice it to say here that the existence of the Markov property is crucially dependent on the forgetfulness property of the exponential distributions used for the holding times in a state. Conditioning on the past up to time t means knowing the current state and elapsed time since the last transition. The exponential distributions allow us to pretend a transition has just been made, so prior history becomes irrelevant. In Section 5.2.1 we will prove that the process we construct satisfies (5.1.1).

We construct the process as follows. Start with a discrete time Markov chain $\{X_n, n \geq 0\}$ with state space S, initial distribution $\{a_k\}$ and transition matrix $Q = (Q_{ij})$. We assume $Q_{ii} = 0$ for all $i \in S$. Suppose we have a sequence $\{E_n, n \geq 0\}$ of iid exponentially distributed random variables with unit mean, and suppose the exponential random variables are independent of $\{X_n\}$. We are given a function $\{\lambda(i), i \in S\}$ which will govern holding times. This function satisfies $\lambda(i) > 0$ for all i. We now construct the continuous time Markov chain $\{X(t)\}$. (Distinguish in your mind between the discrete time Markov chain $\{X_n\}$ and the about to be constructed continuous time Markov chain $\{X(t)\}$ which uses $\{X_n\}$ as an ingredient in the construction.) Let $T_0 = 0$, and define $W(0) = E_0/\lambda(X_0)$, which implies

$$P[W(0) > x|X_0 = i] = \exp\{-\lambda(i)x\}, \quad x > 0,$$

so that, given the initial state of the discrete time Markov chain, $W(0)$ has an exponential distribution with parameter $\lambda(X_0)$. Remember, we assume that $\{X_n\}$ and $\{E_m\}$ are independent. Now define

$$T_1 = T_0 + W(0)$$

and

$$X(t) = X_0, \quad \text{for } T_0 \le t < T_1;$$

define $W(T_1) = E_1/\lambda(X_1)$ so that

$$P[W(T_1) > x|X_1 = i] = \exp\{-\lambda(i)x\};$$

define

$$T_2 = T_1 + W(T_1)$$

and

$$X(t) = X_1, \quad \text{for } T_1 \le t < T_2.$$

Continuing in this way, suppose $\{W(T_m), m \le n-1\}, \{T_m, 0 \le m \le n\}$ and $\{X(s), 0 \le s < T_n\}$ have been defined. We define $W(T_n) = E_n/\lambda(X_n)$,

$$T_{n+1} = T_n + W(T_n)$$

and

$$X(t) = X_n \quad \text{for } T_n \le t < T_{n+1}.$$

Setting $T_\infty = \lim_{n\to\infty} \uparrow T_n$, we have the process defined on $[0, T_\infty)$, and for $t < T_\infty$,

(5.1.3) $$X(t) = \sum_{n=0}^{\infty} X_n 1_{[T_n, T_{n+1})}(t),$$

and furthermore,

(5.1.4) $$T_{n+1} - T_n = W(T_n) = E_n/\lambda(X_n).$$

Two important properties of the sequence $\{X_n, T_n\}$ now follow. The first is that $\{T_m - T_{m-1}, m \ge 1\}$ are conditionally independent and exponentially distributed given $\{X_n\}$: For $u_m > 0, m = 1, \ldots, n$,

$$P[T_m - T_{m-1} > u_m, 1 \le m \le n|X_0 = i_0, \ldots, X_{n-1} = i_{n-1}]$$
$$= P[E_{m-1}/\lambda(X_{m-1}) > u_m, 1 \le m \le n|X_0 = i_0, \ldots, X_{n-1} = i_{n-1}]$$

(5.1.5) $$= P[E_{m-1}/\lambda(i_{m-1}) > u_m, 1 \le m \le n] = \prod_{m=1}^{n} e^{-\lambda(i_{m-1})u_m}.$$

The second property is the distributional structure of $\{X_n, T_n\}$. For $u > 0, i_0, \ldots, i_{n-1}, i \in S$, we have

$$P[X_{n+1} = j, T_{n+1} - T_n > u | X_0 = i_0, \ldots, X_n = i, T_0, \ldots, T_n]$$
$$= P[X_{n+1} = j, E_n/\lambda(X_n) > u | X_0 = i_0, \ldots, X_n = i, T_m - T_{m-1},$$
$$1 \le m \le n]$$
$$= P[X_{n+1} = j, E_n/\lambda(i) > u | X_0 = i_0, \ldots, X_n = i, E_{m-1}/\lambda(i_{m-1}),$$
$$1 \le m \le n]$$

and, since $\{X_n\}$ and $\{E_m\}$ are independent, we conclude

$$P[X_{n+1} = j, T_{n+1} - T_n > u | X_0, \ldots, X_n, T_0, \ldots, T_n] = Q_{X_n j} e^{-\lambda(X_n)u}$$
$$= P[X_{n+1} = j, T_{n+1} - T_n > u | X_n].$$
(5.1.6)

Note that because we assume $Q_{ii} = 0$ for all $i \in S$, $\{T_n\}$ are precisely the epochs when the process can be observed to change states. It is conceptually simpler to construct the Markov chain when we assume the discrete time Markov chain cannot make transitions from i to i, for any $i \in S$. There is also a statistical reason for desiring $Q_{ii} = 0$, since otherwise, from observing the continuous time process $\{X(t)\}$ transitions from state i back to i would not be identifiable. Note, however, if you are dying to construct a continuous time model based on a transition matrix $\{Q_{ij}\}$ where $Q_{ii} > 0$ for some i, then things can still be arranged neatly. Replace the original Markov chain $\{X_n\}$, governed by a Q which has zeros on the main diagonal, by a new chain $\{\hat{X}_n\}$, recognizing only transitions of $\{X_n\}$ to new states and ignoring the geometrically distributed number of visits back to the same state. See Section 5.10 and Exercise 5.8

Example 5.1.1. The Pure Birth Process. Consider the deterministically monotone Markov chain $\{X_n, n \ge 0\}$ starting from state 1 with $Q_{i,i+1} = 1$ so that the transition from i to $i+1$ is sure. Then there is no randomness in the $\{X_n\}$ sequence, and thus there is no difference between the distribution of $\{T_{n+1} - T_n, n \ge 0\}$ conditional on $\{X_n\}$ and the unconditional distribution. We conclude for this process that the holding times $\{T_{n+1} - T_n, n \ge 0\}$ are independent exponentially distributed random variables. The process $\{X(t)\}$ represents the population size at time t for a growing population. The case where $\lambda(i) = \lambda i$ and $\lambda > 0$ is called the *linear birth process* or the *Yule process*.

Before proceeding with the discussion of the dynamics of the birth process, it is wise to stop and collect some facts about the exponential distribution.

Proposition 5.1.1. *Let* $E(a), E(b)$ *be independent with exponential distributions with parameters* a, b *respectively where* $a > 0, b > 0$. *Then*

(i) Forgetfulness property: For $s, t > 0$

$$P[E(a) > t + s | E(a) > s] = P[E(a) > t] = e^{-at}.$$

(ii) Denote the minimum of two numbers x, y *by* $x \wedge y$. *Then* $E(a) \wedge E(b)$ *is exponentially distributed with parameter* $a + b$:

$$P[E(a) \wedge E(b) > x] = e^{-(a+b)x}, \quad x > 0.$$

(iii) We have
$$P[E(a) > E(b)] = b/(a + b).$$

(iv) Suppose $\{E_n\}$ *are independent exponentially distributed random variables*
$$P[E_n > x] = e^{-\lambda(n)x}, \quad x > 0, \quad \lambda(n) > 0.$$

Then the sum

$$\sum_n E_n < \infty \text{ a.s. iff } \sum_n \frac{1}{\lambda(n)} < \infty.$$

Proof. For (i) we calculate

$$P[E(a) > t + s | E(a) > s] = \frac{P[E(a) > t + s, E(a) > s]}{P[E(a) > s]}$$
$$= \frac{P[E(a) > t + s]}{P[E(a) > s]}$$
$$= e^{-a(t+s)}/e^{-as} = e^{-at} = P[E(a) > t].$$

For (ii) we have

$$P[E(a) \wedge E(b) > x] = P[E(a) > x, E(b) > x] = e^{-ax}e^{-bx} = e^{-(a+b)x}.$$

For (iii) we integrate

$$P[E(a) > E(b)] = \int \int_{\{(u,v):u>v>0\}} ae^{-au}be^{-bv}\,du\,dv$$
$$= \int_{v=0}^{\infty} be^{-bv}\,dv \int_{u=v}^{\infty} ae^{-au}\,du$$
$$= \int_{v=0}^{\infty} be^{-bv}e^{-av}\,dv$$
$$= b/(a+b).$$

For (iv) note that if $\sum_n 1/\lambda(n) < \infty$, then $\sum_n \mathbf{E}(E_n) = \mathbf{E}(\sum_n E_n) < \infty$, and, because $\sum_n E_n \geq 0$, we have $\sum_n E_n < \infty$. For the converse, note that if $\sum_n E_n < \infty$ then for all $s > 0$

$$0 < \mathbf{E}\exp\{-s\sum_n E_n\} = \prod_n \mathbf{E}\exp\{-sE_n\}$$

$$= \prod_n \frac{\lambda(n)}{\lambda(n) + s}.$$

Then, from Lemma 2.9.1 we find

$$\sum_n \left(1 - \frac{\lambda(n)}{\lambda(n) + s}\right) = \sum_n \frac{s}{s + \lambda(n)} < \infty.$$

This means $s/(\lambda(n) + s) \to 0$ as $n \to \infty$ so that $\lambda(n) \to \infty$. Since $n \to \infty$, we have

$$\frac{s}{\lambda(n) + s} \sim \frac{s}{\lambda(n)},$$

and therefore

$$\sum_n \frac{s}{s + \lambda(n)} < \infty \text{ iff } \sum_n \frac{1}{\lambda(n)} < \infty,$$

as desired. ∎

Remark. For the linear birth process on the states $\{1, 2, \ldots\}$, we have the holding time in state n, namely $T_n - T_{n-1}$, distributed as an exponential random variable with parameter λn. Therefore

$$T_n - T_{n-1} \stackrel{d}{=} \wedge_{i=1}^n \lambda^{-1} E_i,$$

where $\{E_i\}$ are iid unit exponential random variables. This follows from Proposition 5.1.1 (iii).

If $X(t)$ is the number in the population at time t, then the dynamics of the process are as follows: Particles in a population live for random lengths of time independent of each other. Each lifetime is exponentially distributed with parameter λ. When a particle dies, it is replaced by two other particles. Given that the population is n, the waiting time until the population is $n + 1$ is the minimum of n exponential random variables. At this random time, there are now, because of the forgetfulness property, $n + 1$ lifetimes which are iid with exponential distribution. The minimum of these $n + 1$ gives the sojourn in state $n + 1$, and so on.

Example 5.1.2. Birth–Death Process. A birth–death process is a continuous time Markov chain whose embedded discrete time Markov chain $\{X_n\}$ is a simple random walk on $\{0, 1, \dots\}$ with

$$Q_{i,i+1} = p_i, \quad Q_{i,i-1} = q_i = 1 - p_i, \quad i \geq 1,$$

and

$$Q_{01} = 1.$$

Given the holding time parameters $\{\lambda(i), i \geq 1\}$, we define

$$\lambda_i = p_i \lambda(i), \quad \mu_i = q_i \lambda(i).$$

(Be careful to distinguish $\lambda(i)$ and λ_i, which are very different. Tradition demands that the quantity $p_i \lambda(i)$ be labelled λ_i.) For historical reasons and for reasons which will be clearer later, the numbers $\{\lambda_i\}$ are called *birth rates* and the $\{\mu_i\}$ are called *death rates*. Note $\mu_0 = 0$, the holding time parameters are

$$\lambda(i) = \lambda_i + \mu_i, \quad i \geq 0,$$

and the probabilities governing the discrete time random walk are

$$p_i = \frac{\lambda_i}{\lambda_i + \mu_i}, \quad i \geq 0,$$

where we assume for all i that $\lambda_i + \mu_i > 0$.

A way to think about the birth–death process is as follows. When in state i, imagine there are two independent random variables $B(i)$ and $D(i)$ which are exponentially distributed with parameters λ_i and μ_i respectively. A transition from i to $i + 1$ is made if $B(i) \leq D(i)$, which occurs with probability

$$P[B(i) \leq D(i)] = \frac{\lambda_i}{\lambda_i + \mu_i} = Q_{i,i+1}$$

(which follows from Proposition 5.1.1). The holding time in state i is $B(i) \wedge D(i)$, which is exponential with parameter $\lambda_i + \mu_i$ (also from Proposition 5.1.1). We may think of $B(i)$ as the time until a birth when the population is i and similarly $D(i)$ is the time until a death when the population size is i. The population increases by one if a birth occurs prior to a death; otherwise the population decreases by one.

When the birth and death parameters are of the form

$$\lambda_i = \lambda i, \quad \mu_i = \mu i, \quad i \geq 0,$$

we call the process a *linear birth-death process*.

As a particular case of the birth-death process consider the M/M/1 queue, which is a model having arrivals according to a Poisson process of rate a (a stands for arrival) and where arriving customers are served in turn by a single server who serves at rate b (b is the letter which follows a).* This means service times are iid exponentially distributed with parameter b and independent of the arrival process. A birth corresponds to an arrival and a death corresponds to a service completion with a consequent departure; therefore $\lambda_i = a$ and $\mu_i = b$. The process $\{X(t)\}$ gives the number in the system at time t. The sojourn time in state i is $B(i) \wedge D(i)$, which is exponentially distributed with parameter $\lambda(i) = \lambda_i + \mu_i = a + b$ if $i \geq 1$ and $\lambda(0) = \lambda_0 = a$. The number in the system goes from i to $i + 1$ if $B(i) \leq D(i)$, i.e., if there is an arrival before a service completion. Therefore $Q_{i,i+1} = a/(a + b)$, for $i \geq 1$ and $Q_{01} = 1$. Thus the matrix Q is of the form

$$Q = \begin{pmatrix} 0 & 1 & 0 & 0 & \cdots \\ \frac{b}{a+b} & 0 & \frac{a}{a+b} & 0 & \cdots \\ 0 & \frac{b}{a+b} & 0 & \frac{a}{a+b} & \cdots \\ \vdots & & & \ddots & \end{pmatrix}.$$

Example 5.1.3. Uniformizable Chains. This is discussed more fully in Section 5.10, where the reason for the name also becomes obvious. For a brief introduction, suppose $\{X_n\}$ is a Markov chain in discrete time with transition matrix Q, and suppose $\lambda(i) = \lambda > 0$ independently of i. Then

$$T_{n+1} - T_n = E_n/\lambda,$$

so the sojourn times are iid exponentially distributed random variables with parameter λ and independent of $\{X_n\}$. Thus

$$\{T_n, n \geq 1\}$$

are the points of a Poisson process. Let the counting function of this Poisson process be $N(0, t]$, so that the continuous time and discrete time Markov chains are related by the formula

$$X(t) = X_{N(0,t]}.$$

The transition matrix function of the constructed process $\{X(t)\}$ can be

---

*Another way to remember this is to keep in mind that $a =$ ankomst (arrival in Danish) and $b =$ betjening (service in Danish).

computed as

$$P_{ij}(t) := P[X(t) = j | X(0) = i] = P[X_{N(0,t]} = j | X_0 = i]$$

$$= \sum_{n=0}^{\infty} P[X_{N(0,t]} = j, N(0,t] = n | X_0 = i]$$

$$= \sum_{n=0}^{\infty} P[X_n = j | X_0 = i] P[N(0,t] = n]$$

$$= \sum_{n=0}^{\infty} \frac{e^{-\lambda t}(\lambda t)^n}{n!} Q_{ij}^{(n)}.$$

In this last example, it was possible to compute $\{P_{ij}(t)\}$ in closed form. This is frequently not the case. However, from the description of the reality being modelled, it is usually comparatively easy to obtain the sojourn parameters $\{\lambda(i)\}$ and the transition probabilities $\{Q_{ij}\}$ of the discrete time Markov chain. These quantities control the local movement of the continuous time Markov chain, and in subsequent sections we will see what conclusions can be drawn about $\{X(t)\}$ from knowledge of these local movements. Usually the analytic challenges in applying Markov chain theory come from relating local movements to global ones.

5.2. STABILITY AND EXPLOSIONS.

The previous definitions have assumed $\lambda(i) > 0$. The construction could be modified to allow $\lambda(i) = 0$, and, since the mean sojourn time in i is $\lambda(i)^{-1}$, this would be a natural way to model absorbing states. Similarly, if we allowed $\lambda(i) = \infty$, the mean sojourn time in state i would be 0, and such states would be visited and exited instantaneously. (Such states are termed *instantaneous* states.) While such a property has potential applications (for example, to model bulk arrivals in queues), we will assume, unless explicitly stated to the contrary, that $0 < \lambda(i) < \infty$; i.e., that all states are *stable*.

The construction given in Section 5.1 defines $\{X(t)\}$ only up to time $T_\infty := \lim_n \uparrow T_n$. If $T_\infty < \infty$, we say an explosion occurs, because an infinite number of transitions have taken place in finite time. When

$$P_i[T_\infty = \infty] = 1, \quad \forall\, i \in S,$$

we say the process is *regular*. (Recall $P_i(A) = P[A | X(0) = i]$.)

Using Proposition 5.1.1 and the birth process, it is easy to construct examples. The linear birth process is regular since $\sum_n (\lambda n)^{-1} = \infty$, and hence

$$T_\infty = \sum_n E_n/(\lambda n) = \infty \text{ a.s.}$$

On the other hand, if we set $\lambda(i) = \lambda i^2$, then we get an explosive process.

The main criterion for regularity now follows.

Proposition 5.2.1. *For any $i \in S$,*

$$P_i[T_\infty < \infty] = P_i[\sum_n \frac{1}{\lambda(X_n)} < \infty],$$

and thus the continuous time Markov chain is regular iff

$$\sum_n \frac{1}{\lambda(X_n)} = \infty$$

P_i-*almost surely for every $i \in S$. In particular,*

(1) *If $\lambda(i) \leq c$ for all $i \in S$ for some $c > 0$, then the chain is regular.*

(2) *If S is a finite set, then the chain is regular.*

(3) *If $T \subset S$ are the transient states of $\{X_n\}$ and if*

$$P_i[X_n \in T, \forall\, n] = 0$$

for every $i \in S$ and if there are no instantaneous states (our standing assumption), then the chain is regular.

Note for a regular pure birth process every state in the discrete time Markov chain (the deterministically monotone chain) is transient, so

$$P[X_n \in T, \forall\, n] = 1.$$

We may have regularity, therefore, even if condition (3) fails.

Proof. We have

$$T_\infty = \sum_n (T_{n+1} - T_n) = \sum_n E_n/\lambda(X_n).$$

Conditional on the sequence of states $\{X_n\}$, the quantity T_∞ is a sum of independent exponential random variables with parameters $\{\lambda(X_n)\}$. From Proposition 5.1.1(iv), we have, by conditioning on $\{X_n\}$, that a.s.

$$P[T_\infty < \infty | \{X_n\}] = \begin{cases} 1, & \text{if } \sum_n \frac{1}{\lambda(X_n)} < \infty \\ 0, & \text{if } \sum_n \frac{1}{\lambda(X_n)} = \infty. \end{cases}$$

Thus

$$P[T_\infty < \infty | \{X_n\}] = 1_{[\Sigma_n \frac{1}{\lambda(X_n)} < \infty]} \text{ a.s.}$$

Taking expectations yields

$$P_i[T_\infty < \infty] = P_i[\sum_n \frac{1}{\lambda(X_n)} < \infty].$$

For (1), we observe that, if $\lambda(i) < c$ for all $i \in S$, then

$$\sum_n 1/\lambda(X_n) \geq \sum_n 1/c = \infty,$$

which yields regularity. If S is finite, then $\lambda(i) \leq \vee_j \lambda(j) = c < \infty$ and (1) applies. If $P[X_n \in T \ \forall \ n] = 0$, then entry into T^c is sure, and then there must be a state, say $i \in T^c$, which is hit infinitely often. Let the infinite sequence of random indices when $X_n = i$ be $\{n_j\}$. Then

$$\sum_n 1/\lambda(X_n) \geq \sum_j 1/\lambda(X_{n_j}) = \sum_j 1/\lambda(i) = \infty. \ \blacksquare$$

We henceforth assume the chain is regular.

5.2.1. THE MARKOV PROPERTY*.

Consider a Markov chain as constructed in Section 5.1 and suppose for simplicity it is regular. (A more sophisticated discussion without the assumption of regularity is contained in Asmussen, 1987, pages 30-31.) Why is the process that we constructed Markov? Recall the notation

$$P_{ij}(t) = P_i[X(t) = j]$$

and that whether or not a process is Markov can be determined from the form of the finite dimensional distributions (Proposition 2.1.1). We see that for the process in Section 5.1 to be Markov, it is necessary and sufficient that, for any $0 < t_1 < \cdots < t_k$ and states $i, j_1, \ldots, j_k$,

(5.2.1.1)

$$P_i[X(t_1) = j_1, \ldots, X(t_k) = j_k]$$
$$= P_{i,j_1}(t_1) P_{j_1,j_2}(t_2 - t_1) \ldots P_{j_{k-1},j_k}(t_k - t_{k-1}).$$

---

* This section may be skipped without loss of continuity provided one is willing to accept the fact that the process constructed in Section 5.1 is a Markov process.

We first derive a form for $P_{ij}(t)$ for convenient reference. We have

$$P_i[X(t) = j] = \sum_{n=0}^{\infty} P_i[X_n = j, T_n \leq t < T_{n+1}]$$

$$= \sum_{n=0}^{\infty} \sum_{i_1,\ldots,i_{n-1}} P_i[T_n \leq t < T_{n+1}|X_1 = i_1, \ldots, X_{n-1} = i_{n-1}, X_n = j]$$

$$\cdot P_i[X_1 = i_1, \ldots, X_{n-1} = i_{n-1}, X_n = j]$$

$$= \sum_{n=0}^{\infty} \sum_{i_1,\ldots,i_{n-1}} P_i[\sum_{l=0}^{n-1} \frac{E_l}{\lambda(i_l)} \leq t < \sum_{l=0}^{n-1} \frac{E_l}{\lambda(i_l)} + \frac{E_n}{\lambda(j)}]$$

$$\cdot P_i[X_1 = i_1, \ldots, X_{n-1} = i_{n-1}, X_n = j].$$

Since $\{E_j\}$ are iid exponentially distributed random variables, we obtain

$$P_{ij}(t) = \sum_{n=0}^{\infty} \sum_{i_1,\ldots,i_{n-1}} \int_{u_1=0}^{t} P_i[\sum_{l=0}^{n-1} \frac{E_l}{\lambda(i_l)} \in du_1]e^{-\lambda(j)(t-u_1)}$$

$$(5.2.1.2) \qquad\qquad \cdot P_i[X_1 = i_1, \ldots, X_{n-1} = i_{n-1}, X_n = j].$$

We now check (5.2.1.1) when $k = 2$; the proof for any $k > 2$ is similar but notationally cumbersome. The main trick is to remember that, conditional on knowing the succession of states, $\{T_n - T_{n-1}, n \geq 1\}$ is a sequence of independent exponentially distributed random variables.

For $0 < t_1 < t_2$ and states i and $j_1 \neq j_2$, we have

$$P_i[X(t_1) = j_1, X(t_2) = j_2]$$

$$= \sum_{n_1 \leq n_2} P_i[X_{n_l} = j_l, T_{n_l} \leq t_l < T_{n_l} + \frac{E_{n_l}}{\lambda(j_l)}, l = 1, 2].$$

Conditioning on the succession of states yields

$$= \sum_{\substack{i_1,\ldots,i_{n_1-1}, \\ i_{n_1+1},\ldots,i_{n_2-1} \\ i_{n_1}=j_1, i_{n_2}=j_2}} \sum_{n_1 \leq n_2} P_i[T_{n_l} \leq t_l < T_{n_l} + \frac{E_{n_l}}{\lambda(j_l)}, l = 1, 2|X_\alpha = i_\alpha,$$

$$\alpha = 1, \ldots, n_2] \cdot P_i[X_\alpha = i_\alpha, \alpha = 1, \ldots, n_2].$$

Since $E_{n_1}/\lambda(j_1)$ is exponentially distributed with parameter $\lambda(j_1)$ this becomes

$$= \sum_{\substack{i_1,\ldots,i_{n_1-1}, \\ i_{n_1+1},\ldots,i_{n_2-1},i_{n_1}=j_1,i_{n_2}=j_2}} \sum_{n_1\leq n_2} \iint_{\substack{u_1\leq t_1, \\ t_1<u_1+u_2<t_2}} P[\sum_{l=0}^{n_1-1}\frac{E_l}{\lambda(i_l)}\in du_1]$$

$$\cdot \lambda(j_1)e^{-\lambda(j_1)u_2}du_2$$

$$\cdot P_i[u_1+u_2+\sum_{l=n_1+1}^{n_2-1}\frac{E_l}{\lambda(i_l)}\leq t_2<u_1+u_2+\sum_{l=n_1+1}^{n_2}\frac{E_l}{\lambda(i_l)}]$$

$$\cdot P_i[X_\alpha=i_\alpha,\alpha=1,\ldots,n_2].$$

Make the change of variable $s_1=u_1$, $s_2=u_1+u_2-t_1$ in the integral, and use the Markov property for $\{X_n\}$ to obtain

$$= \sum_{\substack{i_1,\ldots,i_{n_1-1}, \\ i_{n_1+1},\ldots,i_{n_2-1},i_{n_1}=j_1,i_{n_2}=j_2}} \sum_{n_1\leq n_2} \iint_{s_1\leq t_1,0<s_2<t_2-t_1} P[\sum_{l=0}^{n_1-1}\frac{E_l}{\lambda(i_l)}\in ds_1]$$

$$\cdot \lambda(j_1)e^{-\lambda(j_1)(s_2-s_1+t_1)}ds_2$$

$$\cdot P_i[s_2+\sum_{l=n_1+1}^{n_2-1}\frac{E_l}{\lambda(i_l)}\leq t_2-t_1<s_2+\sum_{l=n_1+1}^{n_2}\frac{E_l}{\lambda(i_l)}]$$

$$\cdot P_i[X_\alpha=i_\alpha,\alpha=1,\ldots,n_2]$$

$$= \sum_{n_1\leq n_2} \sum_{\substack{i_1,\ldots,i_{n_1-1}, \\ i_{n_1}=j_1}} \int_{s_1=0}^{t_1} e^{-\lambda(j_1)(t_1-s_1)} P[\sum_{l=0}^{n_1-1}\frac{E_l}{\lambda(i_l)}\in ds_1]$$

$$\cdot P_i[X_1=i_1,\ldots,X_{n_1}=i_{n-1}]$$

$$\sum_{\substack{i_{n_1+1},\ldots,i_{n_2-1} \\ i_{n_2}=j_2}} \int_{s_2=0}^{t_2-t_1} P[s_2+\sum_{l=n_1+1}^{n_2-1}\frac{E_l}{\lambda(i_l)}\leq t_2-t_1<s_2+\sum_{l=n_1+1}^{n_2}\frac{E_l}{\lambda(i_l)}]$$

$$\lambda(j_1)e^{-\lambda(j_1)s_2}ds_2 P_{j_1}[X_1=i_{n_1+1},\ldots,X_{n_2-n_1}=i_{n_2}].$$

Applying (5.2.1.2), and setting $m=n_2-n_1$, we get

$$= P_i[X(t_1)=j_1]$$

$$\sum_m \sum_{i_{n_1+1},\ldots,i_{m+n_1-1}i_{m+n_1}=j_2} P_{j_1}[X_1=i_{n_1+1},\ldots,X_m=i_{m+n_1}]$$

$$\cdot P[\frac{E_{n_1}}{\lambda(j_1)}+\sum_{l=n_1+1}^{m+n_1-1}\frac{E_l}{\lambda(i_l)}\leq t_2-t_1<\sum_{l=n_1}^{m+n_1-1}\frac{E_l}{\lambda(i_l)}]$$

$$= P_i[X(t_1)=j_1]P_{j_1}[X(t_2-t_1)=j_2],$$

as was to be verified.

5.3. DISSECTION.

As in discrete time, we may decompose the continuous time Markov chain into independent cycles. Suppose i is a recurrent state of $\{X_n\}$. As in Section 2.5 define $\tau_i(0) = 0$ and

$$\tau_i(1) = \inf\{m \geq 1 : X_m = i\}$$
$$\tau_i(2) = \inf\{m \geq \tau_i(1) : X_m = i\}$$
$$\vdots$$
$$\tau_i(n+1) = \inf\{m \geq \tau_i(n) : X_m = i\}.$$

The epochs $\{T_{\tau_i(n)}, n \geq 1\}$ are the epochs when $\{X(t)\}$ hits i in continuous time, and the path behaviors between these times are part of independent cycles. Suppose that the process starts in state i, so that $P[X_0 = i] = 1$. The precise statement of the dissection principle in continuous time is that the blocks

$$\left\{ \left(X_{\tau_i(m)+1}, T_{\tau_i(m)+1} - T_{\tau_i(m)}, \ldots, X_{\tau_i(m+1)}, T_{\tau_i(m+1)} - T_{\tau_i(m+1)-1} \right) \right.$$

(5.3.1.) $$\left. m \geq 0 \right\}$$

are iid, and hence

(5.3.2) $$\left\{ \left((X(t), T_{\tau_i(m)} < t \leq T_{\tau_i(m+1)}), T_{\tau_i(m+1)} - T_{\tau_i(m)} \right), m \geq 0 \right\}$$

is an iid collection of pieces of path. These are the *excursions* between visits to state i.

As in discrete time, if the process does not start in state i, then, provided it hits i with probability 1, we have a similar result. After the first hit of state i there are iid cycles which are independent of the initial wanderings of the process prior to the first hit.

These rem♣·s should make it clear that a continuous time Markov chain is an example of a regenerative process. The regeneration cycles can be taken to be excursions between visits to a recurrent state i.

5.3.1. MORE DETAIL ON DISSECTION*.

Without loss of generality, suppose $\{X_n\}$ is the "simulated Markov chain" given in Section 2.1 by construction from uniform random variables $\{U_n\}$, and suppose $\{U_n\}$ is independent of $\{E_n\}$. Recall that there is a function $f : S \times [0,1] \mapsto S$ such that

(5.3.1.1.) $$X_{n+1} = f(X_n, U_{n+1}).$$

---

* This section contains harder material which may be skipped on first reading by beginning readers or readers with only average patience.

For $n \geq 1$, define the σ-algebras

$$\mathcal{F}_n = \sigma\left((U_k, E_{k-1}), 1 \leq k \leq n\right).$$

Since we have, for $n \geq 1$,

$$[\tau_i(1) = n] \in \sigma(X_1, \ldots, X_n)$$
$$\subset \sigma(U_1, \ldots, U_n)$$
$$\subset \sigma\left((U_k, E_{k-1}), 1 \leq k \leq n\right) = \mathcal{F}_n,$$

we see that $\tau_i(1)$ is a stopping time of the sequence of σ-algebras $\mathcal{F}_n$.

Now we apply the material from Section 1.8.2 on splitting iid random elements at a stopping time. We have

(5.3.1.2) $\qquad (U_{\tau_i(1)+k}, E_{\tau_i(1)+k-1}, k \geq 1) \stackrel{d}{=} (U_n, E_{n-1}, n \geq 1).$

Furthermore, the post-$\tau_i(1)$ process on the left of (5.3.1.2) is independent of $\mathcal{F}_{\tau_i(1)}$ and therefore independent of the block of variables $(U_m, E_{m-1}, 0 < m \leq \tau_i(1))$. Recalling (5.3.1.1) we have from (5.3.1.2) that

$$\left(X_{\tau_i(1)+m}, m \geq 1, E_{\tau_i(1)+k-1}, k \geq 1\right)$$
$$= \left(f(i, U_{\tau_i(1)+1}), f(f(i, U_{\tau_i(1)+1}), U_{\tau_i(1)+2}), \ldots, E_{\tau_i(1)+k-1}, k \geq 1\right)$$
$$\stackrel{d}{=} \left(f(i, U_1), f(f(i, U_1), U_2), \ldots, E_{k-1}, k \geq 1\right)$$
$$= (X_n, E_{n-1}, n \geq 1)$$

and the left side of the previous equation is independent of

$$(X_n, E_{n-1}, 0 < n \leq \tau_i(1)).$$

Thus we find

$$\left(X_{\tau_i(1)+m}, \frac{E_{\tau_i(1)+m-1}}{\lambda(X_{\tau_i(1)+m-1})}, m \geq 1\right) \stackrel{d}{=} \left(X_m, \frac{E_{m-1}}{\lambda(X_{m-1})}, m \geq 1\right),$$

and the left side is independent of

$$\left(X_n, \frac{E_{n-1}}{\lambda(X_{n-1})}, 0 < n \leq \tau_i(1)\right).$$

Finally, we may conclude that

$$\left(X_{\tau_i(1)+k}, T_{\tau_i(1)+k} - T_{\tau_i(1)+k-1}, k \geq 1\right) \stackrel{d}{=} (X_n, T_n - T_{n-1}, n \geq 1),$$

and the left side is independent of

$$(X_n, T_n - T_{n-1}, 0 < n \leq \tau_i(1)).$$

If we employ this same procedure starting with the post-$\tau_i(1)$ process and splitting at $\tau_i(2)$, then employ this procedure with the post-$\tau_i(2)$ process, etc., we find that the blocks

$$\left\{\left(X_{\tau_i(m)+1}, T_{\tau_i(m)+1} - T_{\tau_i(m)}, \ldots, X_{\tau_i(m+1)}, T_{\tau_i(m+1)} - T_{\tau_i(m+1)-1}\right),\right.$$
(5.3.1.3) $\qquad\qquad\qquad\qquad\qquad\qquad\qquad\qquad\qquad\qquad m \geq 0\}$

are iid.

5.4. THE BACKWARD EQUATION AND THE GENERATOR MATRIX.

Given the ingredients $\{X_n\}$ and $\{E_n\}$. Section 5.1 shows how to construct a continuous time Markov chain. Assume we believe the construction yields a Markov process satisfying (5.1.1) and (5.1.2). We have as the basic parameters of the model the holding time parameters $\lambda(i), i \in S$, and the transition matrix Q of the discrete time Markov chain. How do these parameters, which control local movements of the process, determine the global probabilities of the form

$$P_{ij}(t) = P[X(t) = j | X(0) = i]?$$

We assume the chain is regular. We also assume all states are stable; modest changes are necessary to handle absorbing states.

We start by deriving the backward integral equation by conditioning on the first jump place and time.

Proposition 5.4.1. *For all $i, j \in S$ and $t > 0$, we have*

$$P_{ij}(t) = \delta_{ij}e^{-\lambda(i)t} + \int_0^t \lambda(i)e^{-\lambda(i)s} \sum_{k \neq i} Q_{ik}P_{kj}(t - s)ds,$$

where as usual

$$\delta_{ij} = \begin{cases} 1, & \text{if } i = j \\ 0, & \text{if } i \neq j. \end{cases}$$

Proof. We decompose according to the value of the first jump T_1. We have

$$P[X(t) = j | X(0) = i] = P[X(t) = j, T_1 > t | X(0) = i]$$
$$+ P[X(t) = j, T_1 \leq t | X(0) = i] =: I + II.$$

For I, a first transition has not been made up to time t, so, if $i = j$, we have

$$I = P_i[T_1 > t] = e^{-\lambda(i)t}.$$

For analyzing II, the idea is as follows: Suppose the first jump occurs at $(s, s + ds)$, which occurs with probability $\lambda(i)e^{-\lambda(i)s}ds$. Then there must be a jump to some intermediate state k which has probability Q_{ik}. If we start at k, the Markov property allows us to neglect the history of how we got to k. Now the process must make its way in time $t - s$ from state k to state j; the probability of this occurrence is $P_{kj}(t - s)$. Finally, the law of total probability requires us to sum over intermediate states k and over intermediate time points s.

The more precise analysis of II which follows may be considered optional reading for novices. We have

$$X(t) = \sum_{n=0}^{\infty} X_n 1_{[T_n, T_{n+1})}(t),$$

and, for $n \geq 1$,

$$T_n = \sum_{j=0}^{n-1} \frac{E_j}{\lambda(X_j)}.$$

Therefore, if the process starts from i and $t \geq T_1$, we have

$$X(t) = \sum_{n=1}^{\infty} X_n 1_{[\Sigma_{j=0}^{n-1} E_j/\lambda(X_j), \Sigma_{j=0}^{n} E_j/\lambda(X_j))}(t)$$

(5.4.1)
$$= \sum_{n=1}^{\infty} X_n 1_{[\Sigma_{j=1}^{n-1} E_j/\lambda(X_j), \Sigma_{j=1}^{n} E_j/\lambda(X_j))}(t - E_0/\lambda(i)).$$

Therefore, we have for II

$$P_i[X(t) = j, T_1 \leq t] = \sum_{k \neq i} P_i[X(t) = j, E_0/\lambda(i) \leq t, X_1 = k]$$

$$= \sum_{k \neq i} \int_{s=0}^{t} P_i[X(t) = j | E_0/\lambda(i) = s, X_1 = k]$$

$$P_i[X_1 = k, E_0/\lambda(i) \in ds].$$

From (5.4.1) we get

$$P_i[X(t) = j | E_0/\lambda(i) = s, X_1 = k]$$

$$= P_i[\sum_{n=1}^{\infty} X_n 1_{[\Sigma_{j=1}^{n-1} E_j/\lambda(X_j), \Sigma_{j=1}^{n} E_j/\lambda(X_j))}(t - s) = j | X_1 = k]$$

$$= P_k[\sum_{n=1}^{\infty} X_{n-1} 1_{[\Sigma_{j=1}^{n-1} E_j/\lambda(X_{j-1}), \Sigma_{j=1}^{n} E_j/\lambda(X_{j-1}))}(t - s) = j],$$

where we have used the Markov property (2.1.4) of $\{X_n\}$. Now set $m = n - 1, h = j - 1, E_h' = E_{h+1}$ and the foregoing expression becomes

$$P_k[\sum_{m=0}^{\infty} X_m 1_{[\Sigma_{h=0}^{m-1} E_h'/\lambda(X_h), \Sigma_{h=0}^{m} E_h'/\lambda(X_h))}(t - s) = j]$$

$$= P_k[X(t - s) = j] = P_{kj}(t - s).$$

Since

$$P_i[X_1 = k, E_0/\lambda(i) \in ds] = P[X_1 = k]P_i[E_0/\lambda(i) \in ds] = Q_{ik}\lambda(i)e^{-\lambda(i)s}ds,$$

the required expression for II follows. ∎

We now convert the integral equation into a differential equation. Define the *generator matrix* A by

$$A_{ij} = \begin{cases} -\lambda(i), & \text{if } i = j \\ \lambda(i)Q_{ij}, & \text{if } i \neq j. \end{cases}$$

A is obtained by writing the λ's on the main diagonal with a minus sign and writing $\lambda(i)Q_{ij}$ off the main diagonal. Off the main diagonal, the entries are non-negative, and on the main diagonal the entries are non-positive. Furthermore, for any i,

$$\sum_j A_{ij} = -\lambda(i) + \sum_{j \neq i} \lambda(i)Q_{ij} = -\lambda(i) + \lambda(i) = 0,$$

so the row sums are always equal 0. (This sometimes saves a bit of work when computing the generator matrix, since, if the number of states is m, we need only compute $m - 1$ entries.)

Note that $\lambda(i), i \in S$, and Q determine A and vice versa. Given A, we obtain λ by negating the main diagonal entries, and, for $i \neq j$,

$$Q_{ij} = -\frac{A_{ij}}{A_{ii}}.$$

Before proceeding, we prove a property for the matrix $P(t) = \{P_{ij}(t)\}$ which says $P(t)$ is *standard*.

Lemma 5.4.2. *We have*

$$\lim_{t \downarrow 0} P(t) = I,$$

the identity matrix; i.e.,

$$\lim_{t \downarrow 0} P_{ij}(t) = \delta_{ij}.$$

Proof. This can readily be checked from Proposition 5.4.1. The following alternative verification is also instructive: If the process starts from state i then

$$X(t) = i, \qquad 0 \leq t < T_1,$$

and, since $T_1 = E_0/\lambda(i) > 0$, this shows

$$\lim_{t \downarrow 0} X(t) = i, \qquad P_i - \text{almost surely.}$$

Almost sure convergence implies convergence in distribution, so

$$\lim_{t \downarrow 0} P_i[X(t) = j] = \delta_{ij},$$

as required. ∎

For the following, define the matrix

$$P'(t) = \{\frac{d}{dt} P_{ij}(t)\}.$$

Proposition 5.4.3. *For all* $i, j \in S$ *we have* $P_{ij}(t)$ *differentiable and the derivative is continuous. At* $t = 0$, *the derivative is*

(5.4.2) $$P'(0) = A,$$

i.e.,

$$\frac{d}{dt} P_{ij}(0) = A_{ij} = \begin{cases} -\lambda(i), & \text{if } i = j \\ \lambda(i)Q_{ij}, & \text{if } i \neq j. \end{cases}$$

Also the **backward differential equation** *holds:*

(5.4.3) $$P'(t) = AP(t),$$

i.e.,

$$\frac{d}{dt} P_{ij}(t) = \sum_k A_{ik} P_{kj}(t).$$

Remark. An interpretation of (5.4.2) in terms of *flow rates of probability* is as follows: We have

$$-\lambda(i) = P'_{ii}(0) = \lim_{t \downarrow 0} \frac{P_{ii}(t) - P_{ii}(0)}{t}$$

$$= \lim_{t \downarrow 0} \frac{P_{ii}(t) - 1}{t}.$$

Therefore

$$1 - P_{ii}(t) = \lambda(i)t + o(t).$$

Since $1 - P_{ii}(t)$ is almost (modulo the term $o(t)$) a linear function of t, we have

$$\lambda(i)t \approx \text{ probability the system leaves } i \text{ before } t,$$

and hence we interpret

$$\lambda(i) \approx \text{ flow rate for the probability the system leaves } i \text{ before } t.$$

Similarly, for $i \neq j$,

$$\lambda(i)Q_{ij} = \lim_{t \downarrow 0} \frac{P_{ij}(t) - P_{ij}(0)}{t} = \lim_{t \downarrow 0} \frac{P_{ij}(t)}{t},$$

so that

$$P_{ij}(t) = \lambda(i)Q_{ij}t + o(t).$$

Therefore, since the right side is approximately a linear function of t, we interpret $\lambda(i)Q_{ij}$ as the flow rate of probability leaving i and heading toward j. Note the flow rate of probability out of i is $\lambda(i)$, and that $100Q_{ij}\%$ is directed towards state j.

Recall the birth–death process introduced in Section 5.1. From the above discussion on flow rates

$$\begin{aligned} P_{i,i+1}(\delta) &= A_{i,i+1}\delta + o(\delta) \\ &= p_i\lambda(i)\delta + o(\delta) \\ &= \lambda_i\delta + o(\delta), \end{aligned}$$

and, similarly,

$$\begin{aligned} P_{i,i-1}(\delta) &= A_{i,i-1}\delta + o(\delta) \\ &= q_i\lambda(i)\delta + o(\delta) \\ &= \mu_i\delta + o(\delta). \end{aligned}$$

Thus the probability of a birth in a small interval of length δ is approximately $\lambda_i\delta$, which explains why λ_i is called a *birth rate*. Similarly, the probability of a death in a small interval of length δ is approximately $\mu_i\delta$ and explains why the μ_i are called *death rates*.

Proof of Proposition 5.4.3. In the recursion of Proposition 5.4.1, make the change of variable $u = t - s$ inside the integral to get

$$P_{ij}(t) = \delta_{ij}e^{-\lambda(i)t} + e^{-\lambda(i)t}\int_0^t \lambda(i)e^{\lambda(i)u}\sum_{k \neq i}Q_{ik}P_{kj}(u)du$$

(5.4.4) $$= e^{-\lambda(i)t}\left[\delta_{ij} + \int_0^t \lambda(i)e^{\lambda(i)u}\sum_{k \neq i}Q_{ik}P_{kj}(u)du\right].$$

Observe that what sits inside the integral is a locally bounded function, and thus the integral is a continuous function of t. Hence $P_{ij}(t)$ must be a continuous function of t for all $i, j \in S$, since it has an expression in terms of this integral and an exponential function. If we consider the integral in (5.4.4) once more, we see that the integrand is not only bounded on finite intervals, but it is also continuous, and hence the integral is a continuously differentiable function of t. This shows that $P_{ij}(t)$ is absolutely continuous (it is the product of the exponential and the integral of a bounded function) and continuously differentiable. Differentiating in (5.4.4) we find

$$(5.4.5) \quad P'_{ij}(t) = -[\ ]\lambda(i)e^{-\lambda(i)t} + e^{-\lambda(i)t}\left\{\lambda(i)e^{\lambda(i)t}\sum_{k\neq i}Q_{ik}P_{kj}(t)\right\},$$

and, since (5.4.4) yields

$$[\ ] = P_{ij}(t)e^{\lambda(i)t},$$

we conclude from (5.4.5) that

$$P'_{ij}(t) = -\lambda(i)P_{ij}(t) + \lambda(i)\sum_{k\neq i}Q_{ik}P_{kj}(t)$$

$$= \sum_{k\in S}\left(-\lambda(i)\delta_{ik} + \lambda(i)Q_{ik}\right)P_{kj}(t)$$

$$= \sum_{k\in S}A_{ik}P_{kj}(t),$$

yielding the backward equation (5.4.3) as desired. Note for the derivative at 0, for $i \neq j$,

$$P'_{ij}(0) = \lim_{t\to 0}\frac{P_{ij}(t) - P_{ij}(0)}{t}$$

$$= \lim_{t\to 0}\frac{P_{ij}(t)}{t}$$

$$= \lim_{t\to 0}t^{-1}\int_0^t \lambda(i)e^{-\lambda(i)s}\sum_k Q_{ik}P_{kj}(t-s)ds$$

$$= \lambda(i)\sum_k Q_{ik}\delta_{kj} = \lambda(i)Q_{ij}$$

$$= A_{ij}.$$

Similarly, we can show $P'_{ii}(0) = -\lambda(i) = A_{ii}$. ∎

Note that just as $\{X_n\}, \{E_n\}$ carry the same information as $\{X(t)\}$, we have that $\{P(t), t > 0\}$ carries the same analytic information as $Q, \{\lambda(i), i \in S\}$. Given $\{P(t), t > 0\}$, we have $A = P'(0)$. Given $Q, \{\lambda(i), i \in S\}$, we compute A and then solve the backward equation to get $P(t)$.

There is a companion equation, called the forward equation, which is obtained by conditioning on the last jump before time t and proceeding in a manner very similar to Proposition 5.4.1. For regular systems, such a last jump exists, but if explosions are possible the last jump may fail to exist. (If $T_\infty < \infty$, what is the last jump before T_∞?) The backward equations are more fundamental than the forward equations, but the forward equations are often easier to solve. The forward equations are obtained formally by noting the Chapman-Kolmogorov equation $(t > 0, s > 0)$

$$P(t + s) = P(t)P(s)$$

(the continuous time expression of the Markov property) and then differentiating with respect to s to get

$$P'(t + s) = P(t)P'(s);$$

setting $s = 0$ yields

$$P'(t) = P(t)A.$$

When the state space S is finite, both the backward and forward equations have the formal solution

$$P(t) = e^{At},$$

where the matrix exponential function is defined by

$$e^{At} = \sum_{n=0}^{\infty} \frac{t^n A^n}{n!}.$$

It is usually relatively easy to specify the *infinitesimal parameters* $Q, \{\lambda(i), i \in S\}$ giving the local motion of the system. It is frequently analytically unreasonable to expect to solve for the transition function $P_{ij}(t)$ except in toy problems. Either a system of differential (or integral) equations must be solved, or all the powers of the A matrix must be computed. Generating function techniques sometimes convert the system of differential equations into a partial differential equation, but, in general, an explicit solution can be very difficult to obtain. The following very simple example illustrates two techniques for solving these equations. A third technique, using Laplace transforms, will be given later. A fourth technique using renewal theory was suggested in Exercise 3.12.

Example 5.4.1. On-Off System. Whenever a machine of a given type is "operative," it stays that way for a random length of time with exponential density with mean $1/\lambda$. When a breakdown occurs, repairs lasting for a random length of time with exponential density with mean $1/\mu$ are started immediately. Repairs return the machine again to the operative state. Let $X(t)$ be the state of the machine at time t.

Suppose the state space is $S = \{0,1\}$ with 0 corresponding to the operative state and 1 corresponding to the system being under repair. We have that $X(t)$ is a Markov chain with state space S. We have already seen a renewal theory solution for the transition probabilities; now we give Markov chain solutions for $P(t)$. We present several approaches.

Approach 1. *The backward equation*: Since states 0 and 1 alternate, the matrix Q is

$$Q = \begin{matrix} 0 \\ 1 \end{matrix} \begin{pmatrix} 0 & 1 \\ 1 & 0 \end{pmatrix},$$

and $\lambda(0) = \lambda$, $\lambda(1) = \mu$. Thus the generator matrix is

$$A = \begin{pmatrix} -\lambda & \lambda \\ \mu & -\mu \end{pmatrix},$$

and the backward equations

$$P'(t) = AP(t)$$

become

(5.4.6) $$P_{00}'(t) = \lambda\left[P_{10}(t) - P_{00}(t)\right]$$

(5.4.7) $$P_{10}'(t) = \mu\left[P_{00}(t) - P_{10}(t)\right].$$

Take $\mu(5.4.6) + \lambda(5.4.7)$, and we have

$$\mu P_{00}'(t) + \lambda P_{10}'(t) = 0.$$

If we integrate this equation, we get

$$\mu P_{00}(t) + \lambda P_{10}(t) = c$$

for some constant c. Since we want $P(0) = I$, we let $t = 0$ so that

$$\mu P_{00}(0) + \lambda P_{10}(0) = \mu + 0 = c,$$

from which

$$\mu P_{00}(t) + \lambda P_{10}(t) = \mu.$$

Solve for $\lambda P_{10}(t)$ and put this into (5.4.6) to obtain

$$
\begin{aligned}
P'_{00}(t) &= \mu\left(1 - P_{00}(t)\right) - \lambda P_{00}(t) \\
&= \mu - (\mu + \lambda)P_{00}(t).
\end{aligned}
$$

(5.4.8)

If we set

$$
g(t) = P_{00}(t) - \frac{\mu}{\mu + \lambda},
$$

then we find from (5.4.8) that g satisfies the differential equation

$$
g'(t) = -(\mu + \lambda)g(t).
$$

This is easy to solve since

$$
\frac{g'(t)}{g(t)} = \frac{d}{dt}\log g(t) = -(\mu + \lambda),
$$

which yields

$$
g(t) = ce^{-(\mu+\lambda)t}
$$

for some constant c. Thus

$$
P_{00}(t) = ce^{-(\mu+\lambda)t} + \frac{\mu}{\mu + \lambda}.
$$

Since for $t = 0$ we have

$$
P_{00}(0) = 1 = c + \frac{\mu}{\mu + \lambda},
$$

we finally conclude

$$
P_{00}(t) = \frac{\lambda}{\mu + \lambda}e^{-(\mu+\lambda)t} + \frac{\mu}{\mu + \lambda}.
$$

The remaining entries of the matrix $P(t)$ may now be computed.

Approach 2. *The forward equations*: If we solve the system

$$
P'(t) = P(t)A,
$$

we get

$$
P'_{00}(t) = -P_{00}(t)\lambda + P_{01}(t)\mu
$$
$$
P'_{10}(t) = -P_{10}(t)\lambda + P_{11}(t)\mu.
$$

Because there are only two states in S, upon setting $P_{01}(t) = 1 - P_{00}(t)$, the first equation becomes

$$P_{00}'(t) = -P_{00}(t)(\lambda + \mu) + \mu,$$

which is the same as (5.4.8). The solution now proceeds as before.

Note that the forward equations, as is frequently the case, are easier to solve than the backward equations.

Approach 3. *Matrix methods*: Since A is so simple, there is hope of finding all powers and computing $\exp\{At\}$. The matrix A has eigenvalues $0, -\lambda - \mu$:

$$A \begin{pmatrix} 1 \\ 1 \end{pmatrix} = 0 \begin{pmatrix} 1 \\ 1 \end{pmatrix}$$

and

$$A \begin{pmatrix} \lambda \\ -\mu \end{pmatrix} = (-\lambda - \mu) \begin{pmatrix} \lambda \\ -\mu \end{pmatrix}.$$

Thus we have the decomposition

$$A = B \begin{pmatrix} 0 & 0 \\ 0 & -\lambda - \mu \end{pmatrix} B^{-1},$$

where

$$B = \begin{pmatrix} 1 & \lambda \\ 1 & -\mu \end{pmatrix}, \qquad B^{-1} = \frac{1}{\lambda + \mu} \begin{pmatrix} \mu & \lambda \\ 1 & -1 \end{pmatrix}.$$

Now we can evaluate $P(t)$:

$$
\begin{aligned}
P(t) = e^{At} &= \sum_{k=0}^{\infty} \frac{t^k A^k}{k!} \\
&= \sum_{k=0}^{\infty} \frac{t^k}{k!} B \begin{pmatrix} 0^k & 0 \\ 0 & (-\lambda - \mu)^k \end{pmatrix} B^{-1} \\
&= B \begin{pmatrix} 1 & 0 \\ 0 & e^{-(\lambda + \mu)t} \end{pmatrix} B^{-1} \\
&= \frac{1}{\lambda + \mu} \begin{pmatrix} \mu + \lambda e^{-(\lambda + \mu)t} & \lambda - \lambda e^{-(\lambda + \mu)t} \\ \mu - \mu e^{-(\lambda + \mu)t} & \lambda + \mu e^{-(\lambda + \mu)t} \end{pmatrix}.
\end{aligned}
$$

5.5. STATIONARY AND LIMITING DISTRIBUTIONS.

A measure $\eta = \{\eta_j, j \in S\}$ on S is called *invariant* if (just as in the discrete time case) for any $t > 0$

$$(5.5.1) \qquad\qquad \eta' P(t) = \eta'.$$

(Note that since S is discrete, the term *measure* merely denotes a family of non-negative numbers indexed by S.) If this measure is a probability distribution, then it is called a *stationary distribution*. Paralleling the discrete time Markov chain case, we have the following proposition.

Proposition 5.5.1. *If the initial distribution of the Markov chain* $\{X(t)\}$ *is* η, *i.e.,*

$$P[X(0) = j] = \eta_j, \quad j \in S,$$

and η *is a stationary distribution, then* $\{X(t), t \geq 0\}$ *is a stationary process. Thus for any* $k \geq 1$ *and* $s > 0, 0 < t_1 < \ldots < t_k$

$$(X(t_i), 1 \leq i \leq k) \overset{d}{=} (X(t_i + s), 1 \leq i \leq k).$$

In particular , for any $t \geq 0, j \in S,$

$$P[X(t) = j] = \eta_j$$

is independent of t.

Proof. Copy the proof of Proposition 2.12.1 for the discrete time case. ∎

    A probability distribution $\{L = L_i, i \in S\}$ on S is called a limit distribution if for all $i, j \in S$

$$(5.5.2) \qquad \lim_{t \to \infty} P[X(t) = j | X(0) = i] = \lim_{t \to \infty} P_{ij}(t) = L_j.$$

This definition is formulated under the assumption that the embedded discrete Markov chain $\{X_n\}$ is irreducible. As in discrete time we have the following:

Proposition 5.5.2. *A limit distribution is a stationary distribution.*

Proof. Replicate the proof of Proposition 2.13.1. ∎

    Since the matrix transition function $P(t)$ can be difficult to compute, it is essential to have a means of computing a stationary distribution or a limit distribution from the ingredients Q and $\lambda(i), i \in S$. The following result thoughtfully provides this for us.

Theorem 5.5.3. *Suppose $\{X_n\}$ is irreducible and recurrent. Then $\{X(t)\}$ has an invariant measure η which is unique up to multiplicative factors and can be found as the unique (up to multiplicative factors) solution of the equation*

$$(5.5.3) \qquad\qquad \eta' A = 0.$$

Also, η satisfies

$$(5.5.4) \qquad\qquad 0 < \eta_j < \infty, \quad \forall\, j \in S.$$

A stationary distribution exists for $\{X(t)\}$ iff

$$\sum_{i \in S} \eta_i < \infty,$$

in which case

$$\left\{ \frac{\eta_i}{\sum_{k \in S} \eta_k}, i \in S \right\}$$

is the stationary distribution.

Proof. In this section we only check that (5.5.3) has a solution unique up to multiplicative factors; novices may wish to content themselves with reading only this part of the proof. The remainder of the proof is discussed in Section 5.5.1. Note that (5.5.3) holds iff

$$\sum_{j \in S} \eta_j A_{jk} = \sum_{j \neq k} \eta_j \lambda(j) Q_{jk} - \eta_k \lambda(k) = 0,$$

so that (5.5.3) holds iff

$$\eta_k \lambda(k) = \sum_{j \neq k} \eta_j \lambda(j) Q_{jk},$$

which says that $\{\eta_k \lambda(k), k \in S\}$ is invariant for the Markov matrix Q. From discrete time results (Proposition 2.12.3) we may conclude that (5.5.3) indeed has a solution satisfying (5.5.4) which is unique up to multiplicative constants. If you are curious why this solution is in fact an invariant measure for the continuous time Markov chain, go now to Section 5.5.1.

We may express the solution of (5.5.3) as an occupation time as follows: If $\{\eta_j \lambda(j), j \in S\}$ is invariant for Q, then by Proposition 2.12.2 and 2.12.3 we have for some $c > 0$ and $i \in S$

$$\eta_j \lambda(j) = c \mathbf{E}_i \sum_{n=0}^{\tau_i(1)-1} 1_{[X_n = j]}.$$

Without loss of generality we may and do suppose $c = 1$, since otherwise we could consider $\{\eta_j/c\}$ which would also be a solution of (5.5.3). So if $\{\eta_j\}$ satisfies (5.5.3) then

$$
\eta_j = \frac{1}{\lambda(j)} \mathbf{E}_i \sum_{n=0}^{\tau_i(1)-1} 1_{[X_n=j]} = \frac{1}{\lambda(j)} \mathbf{E}_i \sum_{n=0}^{\infty} 1_{[X_n=j,\, n<\tau_i(1)]}
$$

$$
= \sum_{n=0}^{\infty} \mathbf{E}_i \left(\frac{E_n}{\lambda(X_n)} \right) 1_{[X_n=j,\, n<\tau_i(1)]}
$$

$$
= \sum_{n=0}^{\infty} \mathbf{E}_i (T_{n+1} - T_n) 1_{[X_n=j,\, n<\tau_i(1)]}
$$

$$
= \mathbf{E}_i \sum_{n=0}^{\tau_i(1)-1} (T_{n+1} - T_n) 1_{[X_n=j]},
$$

and since between T_n and T_{n+1} the process $X(s)$ does not change state, we have this equal to

$$
= \mathbf{E}_i \sum_{n=0}^{\tau_i(1)-1} \int_{T_n}^{T_{n+1}} 1_{[X(s)=j]} ds
$$

$$
= \mathbf{E}_i \int_0^{T_{\tau_i(1)}} 1_{[X(s)=j]} ds.
$$

Therefore, η_j may be considered to be the expected occupation time of state j in one cycle. This is to be expected based on our experience in discrete time. See Proposition 2.11.2.

Note that, according to Theorem 5.5.3, a stationary distribution exists iff

$$
\infty > \sum_{k \in S} \eta_k = \sum_{k \in S} \mathbf{E}_i \int_0^{T_{\tau_i(1)}} 1_{[X(u)=k]} du = \mathbf{E}_i T_{\tau_i(1)},
$$

i.e., iff the expected cycle length is finite.

So when $\{X_n\}$ is recurrent and irreducible, any of the following schemes will produce an invariant measure for $P(t)$:

(1)   Solve $\eta' A = 0$.

(2)   Compute expected occupation times

$$
\eta_j = \mathbf{E}_i \int_0^{T_{\tau_i(1)}} 1_{[X(s)=j]} ds, \quad j \in S.
$$

(3)   Find the invariant measure $\{\nu_j\}$ for $\{X_n\}$. Then

$$
\eta_j = \nu_j / \lambda(j), \quad j \in S
$$

is invariant for $P(t)$. (Note the weighting of ν_j by $1/\lambda(j)$ accounts for the random sojourn times of $(X(t))$.)

Ergodicity: Call the regular Markov chain $\{X(t), t \geq 0\}$ ergodic if $\{X_n\}$ is recurrent and irreducible and a stationary distribution exists for $\{X(t)\}$.

Corollary 5.5.4. *Suppose $\{X_n\}$ is recurrent and irreducible. Then the following are equivalent:*

(1)   *The expected cycle lengths are finite:*

(5.5.5) $$\mathbf{E}_i(T_{\tau_i(1)}) < \infty.$$

(2)   *The system*
$$\eta' A = 0$$

has a probability distribution as solution.

(3) $\{X(t)\}$ *is ergodic.*

Proof. As just shown, (5.5.5) holds iff $\sum \eta_j < \infty$ where $\eta' A = 0$. Therefore, (1) and (2) are equivalent and imply ergodicity. Conversely, if the process is ergodic there exists a stationary distribution $\{\eta_j\}$. Since $\{X(t)\}$ is regenerative, Smith's theorem yields

$$\eta_j \equiv P_\eta[X(t) = j] \to \frac{\mathbf{E}_i \int_0^{T_{\tau_i(1)}} 1_{[X(s)=j]} ds}{\mathbf{E}_i T_{\tau_i(1)}},$$

where P_η means the initial distribution is η. Since the limit is non-zero, we conclude $\mathbf{E}_i(T_{\tau_i(1)}) < \infty$ as required. ∎

Corollary 5.5.5. *If $\{X(t)\}$ is ergodic, the limit distribution exists. For any $k, j \in S$*

$$\lim_{t \to \infty} P_k[X(t) = j] = \frac{\mathbf{E}_i \int_0^{T_{\tau_i(1)}} 1_{[X(s)=j]} ds}{\mathbf{E}_i T_{\tau_i(1)}}.$$

Remark. Postive recurrence of $\{X_n\}$ is not enough to ensure that $\{X(t)\}$ is ergodic, since long sojourn times could prevail. As an example, consider the modified success run chain $\{X_n\}$ with state space $\{0, 1, \ldots\}$ and transition matrix

$$Q = \begin{pmatrix} 0 & 1 & 0 & 0 & \cdots \\ q & 0 & p & & \\ q & 0 & 0 & p & \\ \vdots & & & \ddots & \end{pmatrix}.$$

We have for $n \geq 2$

$$P_0[\tau_0(1) = n] = p^{n-2}q,$$

so

$$E_0(\tau_0(1)) < \infty,$$

ensuring that $\{X_n\}$ is positive recurrent. Let $\{X(t), t \geq 0\}$ have this embedded discrete time Markov chain, and suppose the holding time parameters $\{\lambda(i), i \geq 0\}$ are given. Then $\{X(t)\}$ is regular by Proposition 5.2.1(3). Observe that

$$E_0 T_{\tau_0(1)} = E_0 \sum_{m=0}^{\tau_0(1)-1} \frac{E_m}{\lambda(X_m)}$$

$$= E_0 \sum_{m=0}^{\tau_0(1)-1} \frac{E_m}{\lambda(m)}$$

$$= E_0 \sum_{m=0}^{\infty} \frac{E_m}{\lambda(m)} 1_{[m < \tau_0(1)]}.$$

Since $\{X_n\}$ and $\{E_m\}$ are independent, this is

$$= \sum_{m=0}^{\infty} \frac{1}{\lambda(m)} P_0[m < \tau_0(1)]$$

$$= \sum_{m=1}^{\infty} \frac{1}{\lambda(m)} p^{m-1} + \frac{1}{\lambda(0)}.$$

If we let $\lambda(m) = p^m$, for instance, then

$$E_0(T_{\tau_0(1)}) = \infty.$$

Therefore, $\{X_n\}$ is positive recurrent, but $\{X(t)\}$ is not ergodic.

Time averages: Consider an ergodic Markov chain $\{X(t)\}$. Suppose that when the process is in state j, a reward is earned at rate $f(j)$, so that the total reward earned up to time t is

$$\int_0^t f(X(s))ds.$$

Using dissection, we may break this up into iid pieces

$$\sum_{m=1}^{N(t)} \int_{T_{\tau_i(m-1)}}^{T_{\tau_i(m)}} f(X(s))ds + J,$$

where $N(t) = \sup\{m : T_{\tau_i(m)} \leq t\}$ and J is the junk left over by this decomposition of $\int_0^t$. Following the procedure in Proposition 2.11.1, we find that if f is reasonable (non-negative or bounded) that a long run reward rate exists:

$$\lim_{t\to\infty} t^{-1} \int_0^t f(X(s))ds = \eta(f) = \sum_{j\in S} f(j)\eta_j$$

$$= \mathbf{E}_i \int_0^{T_{\tau_i(1)}} f(X(s))ds / \mathbf{E}_i T_{\tau_i(1)}$$

since

$$\sum_j f(j)\mathbf{E}_i \int_0^{T_{\tau_i(1)}} 1_{[X(s)=j]}ds / \mathbf{E}_i(T_{\tau_i(1)})$$

$$= \mathbf{E}_i \int_0^{T_{\tau_i(1)}} \sum_j f(j)1_{[X(s)=j]}ds / \mathbf{E}_i(T_{\tau_i(1)})$$

$$= \mathbf{E}_i \int_0^{T_{\tau_i(1)}} \sum_j f(X(s))1_{[X(s)=j]}ds / \mathbf{E}_i(T_{\tau_i(1)})$$

$$= \mathbf{E}_i \int_0^{T_{\tau_i(1)}} f(X(s))ds / \mathbf{E}_i(T_{\tau_i(1)}).$$

Example. Harry's Basketball Injuries. Harry's illustrious semi-pro basketball career was marred by the fact that he was injury prone. During his playing career he fluctuated between three states: 0 (fit), 1 (minor injury not preventing competition), 2 (major injury requiring complete abstinence from the court). The team doctor, upon observing the transitions between states, concluded these transitions could be modelled by a Markov chain with transition matrix

$$Q = \begin{pmatrix} 0 & \frac{1}{3} & \frac{2}{3} \\ \frac{1}{3} & 0 & \frac{2}{3} \\ 1 & 0 & 0 \end{pmatrix}.$$

Holding times in states 0,1,2 were exponentially distributed with parameters $\frac{1}{3}, \frac{1}{3}, \frac{1}{6}$ respectively.

Let $X(t)$ be the continuous time Markov chain describing the state of health of our hero at time t. (Assume the time units are days.)

(1) What is the generator matrix?
(2) What is the long run proportion of time that our hero had to abstain from playing?

(3) Harry was paid \$40/day when fit, \$30/day when able to play injured and \$20/day when he was injured and unable to play. What was the long run earning rate of our hero?

We see from Q and $\lambda(0) = \frac{1}{3}, \lambda(1) = \frac{1}{3}, \lambda(2) = \frac{1}{6}$ that the generator matrix A is

$$A = \begin{pmatrix} -\frac{1}{3} & \frac{1}{9} & \frac{2}{9} \\ \frac{1}{9} & -\frac{1}{3} & \frac{2}{9} \\ \frac{1}{6} & 0 & -\frac{1}{6} \end{pmatrix}.$$

Solving the system $\eta' A = 0, \eta_0 = 1$, yields the equations

$$\frac{1}{9}\eta_0 - \frac{1}{3}\eta_1 = 0$$

$$-\frac{1}{3}\eta_0 + \frac{1}{9}\eta_1 + \frac{1}{6}\eta_2 = 0,$$

which yield the numbers

$$\eta_0 = 1, \eta_1 = 1/3, \eta_2 = 16/9.$$

Normalizing by dividing by $\eta_0 + \eta_1 + \eta_2$ yields the probabilities

$$(\eta_0, \eta_1, \eta_2) = (\frac{9}{28}, \frac{3}{28}, \frac{16}{28}),$$

so the long run percentage of time Harry abstained was $\eta_2 = 16/28 = 4/7$. Harry's long run earning rate was

$$40\eta_0 + 30\eta_1 + 20\eta_2 = 27.5. \quad \blacksquare$$

When the state space is larger than three elements, finding the solution of (5.5.3) by hand is tedious and undesirable. A method of solution suitable for the computer is discussed in Section 5.7.

5.5.1. MORE ON INVARIANT MEASURES*.

Here we continue the discussion of Theorem 5.5.3. It remains to show that a solution of (5.5.3) is invariant. We may now check that $\{\eta_j\}$, written in the form of an occupation time, is invariant for $P(t)$.

Write $T := T_{\tau_i(1)}$. Note that regardless of whether $t \leq T$, we have, by the usual rules of integration,

$$\int_0^T 1_{[X(u)=j]}du = \int_0^t 1_{[X(u)=j]}du + \int_t^T 1_{[X(u)=j]}du,$$

---

* This section contains harder material requiring above average patience to read. It may be skipped by beginning readers.

and therefore

$$\eta_j = \mathbf{E}_i \int_0^T 1_{[X(u)=j]} du = \mathbf{E}_i \int_0^t 1_{[X(u)=j]} du + \mathbf{E}_i \int_t^T 1_{[X(u)=j]} du.$$

By the dissection principle,

$$\mathbf{E}_i \int_0^t 1_{[X(u)=j]} du = \mathbf{E}_i \int_T^{T+t} 1_{[X(u)=j]} du,$$

so

$$\eta_j = \mathbf{E}_i \int_T^{T+t} 1_{[X(u)=j]} du + \mathbf{E}_i \int_t^T 1_{[X(u)=j]} du$$

$$= \mathbf{E}_i \int_t^{T+t} 1_{[X(u)=j]} du = \mathbf{E}_i \int_0^T 1_{[X(u+t)=j]} du$$

$$= \mathbf{E}_i \int_0^\infty 1_{[X(u+t)=j, u<T]} du$$

$$(5.5.1.1) \qquad = \int_0^\infty P_i[X(u+t)=j, u<T] du.$$

Observe that on $[T_n \le u < T_{n+1}]$ we have

$$(5.5.1.2) \qquad X(u+t) = \sum_{m=n}^\infty X_m 1_{[T_m, T_{m+1})}(t+u),$$

and, for $m > n$,

$$T_m = u + (T_{n+1} - u) + T_m - T_{n+1}$$

$$= u + E_n'/\lambda(X_n) + \sum_{k=n+1}^{m-1} E_k/\lambda(X_k)$$

where $E_n'/\lambda(X_n) = T_{n+1} - u$, and $\sum_{k=n+1}^n = 0$.

FIGURE 5.1. TIME LINE

Therefore $(m > n)$,

$$t + u \in [T_m, T_{m+1})$$

iff

$t \in [T_m - u, T_{m+1} - u)$
(5.5.1.3)
$$= [E'_n/\lambda(X_n) + \sum_{k=1}^{m-1-n} E_{k+n}/\lambda(X_{k+n}), E'_n/\lambda(X_n) + \sum_{k=1}^{m-n} E_{k+n}/\lambda(X_{k+n})).$$

Now pick up the thread from (5.5.1.1). We have

$$\eta_j = \int_0^\infty P_i[X(u+t) = j, u < T]du$$

$$= \int_0^\infty \sum_{n=0}^\infty P_i[X(u+t) = j, T_n \le u < T_{n+1}, n < \tau_i(1)]du.$$

Condition on $X_1 = i_1, \ldots, X_n = i_n$ (observe that $i_1, \ldots, i_n \neq i$ implies $n < \tau_i(1)$) and $T_n, T_{n+1} - T_n > u - v$ to get

$$\eta_j = \int_0^\infty \sum_{n=0}^\infty \sum_{i_1, \ldots, i_n \neq i} \int_{v=0}^u$$
$P_i[X(u+t) = j | X_1 = i_1, \ldots, X_n = i_n, T_n = v, T_{n+1} - T_n > u - v]$
(5.5.1.4)
$$\cdot P[X_1 = i_1, \ldots, X_n = i_n, T_n \in dv, T_{n+1} - T_n > u - v]du.$$

From (5.5.1.2), (5.5.1.3), (5.5.1.4),

$$P_i[X(u+t) = j | X_1 = i_1, \ldots, X_n = i_n, T_n = v, T_{n+1} - T_n > u - v]$$
$$= P_i[X_n 1_{[0, E'_n/\lambda(i_n))}(t)$$

$$+ \sum_{m=n+1}^\infty X_m 1_{[E'_n/\lambda(i_n) + \sum_{k=1}^{m-1-n} E_{k+n}/\lambda(X_{k+n}), E'_n/\lambda(i_n) + \sum_{k=1}^{m-n} E_{k+n}/\lambda(X_{k+n}))}(t)$$

$$= j | X_1 = i_1, \ldots, X_n = i_n, T_n = v, T_{n+1} - T_n > u - v].$$
(5.5.1.5)

Since $E'_n/\lambda(X_n) = T_{n+1} - u$ and

$$P[T_{n+1} - u > x | T_n = v, T_{n+1} - T_n > u - v, X_n = i_n]$$
$$= P[T_{n+1} - T_n + v - u > x | T_n = v, T_{n+1} - T_n > u - v, X_n = i_n]$$
$$= P[E_n/\lambda(i_n) > x + (u - v) | E_n/\lambda(i_n) > u - v]$$
$$= P[E_n/\lambda(i_n) > x],$$

by the forgetfulness property of the exponential distribution and from the Markov property (2.1.4) for $\{X_n\}$, (5.5.1.5) equals

$$P_{i_n}[X_0 1_{[0, E_0/\lambda(i_n))}(t)$$

$$+ \sum_{m=1}^{\infty} X_m 1_{[E_0/\lambda(i_n) + \sum_{k=1}^{m-1-n} E_k/\lambda(X_k), E_0/\lambda(i_n) + \sum_{k=1}^{m-n} E_k/\lambda(X_k))}(t) = j]$$
$$= P_{i_n}[X(t) = j] = P_{i_n, j}(t).$$

Inserting this into (5.5.1.4) yields

$$\eta_j = \int_0^{\infty} \sum_{n=0}^{\infty} \sum_{i_1, \dots, i_n \neq i} \int_{v=0}^u P_{i_n, j}(t)$$
$$P_i[X_1 = i_1, \dots, X_n = i_n, T_n \in dv, T_{n+1} - T_n > u - v] du$$

$$= \int_0^{\infty} \sum_{n=0}^{\infty} \sum_{k \in S} P_{kj}(t) \int_{v=0}^u$$
$$P_i[X_n = k, T_n \in dv, T_{n+1} - T_n > u - v, n < \tau_i(1)] du$$

$$= \int_0^{\infty} \sum_{n=0}^{\infty} \sum_{k \in S} P_{kj}(t) P_i[X_n = k, T_n \leq u < T_{n+1}, n < \tau_i(1)] du$$

$$= \sum_{k \in S} P_{kj}(t) \mathbf{E}_i \sum_{n=0}^{\infty} \int_{T_n}^{T_{n+1}} 1_{[X(u)=k, T_n \leq u < T_{n+1}, n < \tau_i(1)]} du$$

$$= \sum_{k \in S} P_{kj}(t) \mathbf{E}_i \sum_{n=0}^{\infty} \int_{T_n}^{T_{n+1}} 1_{[X(u)=k, n < \tau_i(1)]} du$$

$$= \sum_{k \in S} P_{kj}(t) \mathbf{E}_i \sum_{n=0}^{\tau_i(1)-1} \int_{T_n}^{T_{n+1}} 1_{[X(u)=k]} du$$

$$= \sum_{k \in S} P_{kj}(t) \mathbf{E}_i \int_0^{T_{\tau_i(1)}} 1_{[X(u)=k]} du$$

$$= \sum_{k \in S} \eta_k P_{kj}(t).$$

This shows that $\{\eta_k\}$ is invariant for $P(t)$, which completes the proof of Theorem 5.5.3.

5.6. LAPLACE TRANSFORM METHODS.

Since computers can now do symbolic algebraic computations, the following can be helpful when solving for $P(t)$. Suppose the state space is finite, and set $(\alpha > 0)$

$$\hat{P}_{ij}(\alpha) = \int_0^\infty e^{-\alpha t} P_{ij}(t)dt.$$

Then we have

$$\hat{P}_{ij}(\alpha) = \int_0^\infty e^{-\alpha t} P_{ij}(t)dt$$

$$= \int_0^\infty e^{-\alpha t} \left(P_{ij}(t) - P_{ij}(0) \right) dt + \alpha^{-1}\delta_{ij}$$

$$= \int_0^\infty e^{-\alpha t} \left(\int_0^t P'_{ij}(s)ds \right) dt + \alpha^{-1}\delta_{ij}.$$

Reversing the order of integration yields

$$= \int_{s=0}^\infty P'_{ij}(s) \left(\int_{t=s}^\infty e^{-\alpha t} dt \right) + \alpha^{-1}\delta_{ij}$$

$$= \int_0^\infty P'_{ij}(s)\alpha^{-1}e^{-\alpha s}ds + \alpha^{-1}\delta_{ij},$$

and, recalling the backward equation $P'(t) = AP(t)$, we get

$$= \sum_k \int_0^\infty A_{ik}P_{kj}(s)\alpha^{-1}e^{-\alpha s}ds + \alpha^{-1}\delta_{ij}$$

$$= \alpha^{-1}\sum_k A_{ik}\hat{P}_{kj}(\alpha) + \alpha^{-1}\delta_{ij}.$$

Writing this in matrix notation, we obtain

$$\alpha\hat{P}(\alpha) = A\hat{P}(\alpha) + I,$$

so that

$$(\alpha I - A)\hat{P}(\alpha) = I,$$

and, finally,

(5.6.1) $$\hat{P}(\alpha) = (\alpha I - A)^{-1}$$

provided the inverse exists. To verify the existence of the inverse, it suffices to show for any $\alpha > 0$ that $(\alpha I - A)x = 0$ implies $x = 0$; i.e., we need to show $Ax = \alpha x$ and $\alpha > 0$ implies $x = 0$. To accomplish this goal, we use the forward equation $P'(t) = AP(t)$, which frequently holds; for instance, the forward equation is always valid when the state space is finite. Observe

$$\left(e^{-\alpha t}P(t)x\right)' = e^{-\alpha t}P'(t)x - P(t)x\alpha e^{-\alpha t},$$

which becomes

$$= e^{-\alpha t}P(t)Ax - P(t)x\alpha e^{-\alpha t}$$
$$= 0$$

upon using the forward equation, since $Ax = \alpha x$. For all $t > 0$ we have

$$\left(e^{-\alpha t}P(t)x\right)' = 0,$$

and if we integrate from 0 to s we get

$$0 = \int_0^s \left(e^{-\alpha t}P(t)x\right)' dt = e^{-\alpha s}P(s)x - x.$$

Therefore we conclude that, for all $s > 0$,

$$e^{-\alpha s}P(s)x = x.$$

Let $s \to \infty$. Since $e^{-\alpha s} \to 0$ and $P(s)x$ is a bounded vector (for instance, recalling S is finite we have $|P(s)x| \le \sum_{i\in S}|x_i|$), we see that $x = 0$, as required.

The relation (5.6.1) in the transform domain provides a justification for the formula

$$P(t) = \exp\{At\}$$

since

$$\hat{P}(\alpha) = (\alpha I - A)^{-1} = \alpha^{-1}(I - \alpha^{-1}A)^{-1}$$

$$= \alpha^{-1}\sum_{n=0}^{\infty}(\alpha^{-1}A)^n = \alpha^{-1}\sum_{n=0}^{\infty}A^n\alpha^{-n}$$

$$= \sum_{n=0}^{\infty}\frac{A^n}{n!}\int_0^{\infty}e^{-\alpha t}t^n dt$$

$$= \int_0^{\infty}e^{-\alpha t}\sum_{n=0}^{\infty}\frac{A^nt^n}{n!}dt,$$

and therefore

$$P(t) = \sum_{n=0}^{\infty} A^n t^n / n!.$$

Example 5.6.1. On-Off System (continued). Recall the generator matrix

$$A = \begin{pmatrix} -\lambda & \lambda \\ \mu & -\mu \end{pmatrix};$$

therefore,

$$\alpha I - A = \begin{pmatrix} \alpha + \lambda & -\lambda \\ -\mu & \alpha + \mu \end{pmatrix}.$$

Inverting, we find

$$(\alpha I - A)^{-1} = \frac{1}{\det(\alpha I - A)} \begin{pmatrix} \alpha + \mu & \lambda \\ \mu & \alpha + \lambda \end{pmatrix}$$

$$= \frac{1}{\alpha(\alpha + \mu + \lambda)} \begin{pmatrix} \alpha + \mu & \lambda \\ \mu & \alpha + \lambda \end{pmatrix}.$$

Since

$$\frac{1}{\alpha(\alpha + \mu + \lambda)} = \frac{1}{\lambda + \mu} \left(\frac{1}{\alpha} - \frac{1}{\alpha + \lambda + \mu} \right),$$

we obtain

$$(\alpha I - A)^{-1} = \frac{1}{\lambda + \mu} \begin{pmatrix} \frac{\alpha+\mu}{\alpha} - \frac{\alpha+\mu}{\alpha+\lambda+\mu} & \frac{\lambda}{\alpha} - \frac{\lambda}{\alpha+\lambda+\mu} \\ \frac{\mu}{\alpha} - \frac{\mu}{\alpha+\lambda+\mu} & \frac{\alpha+\lambda}{\alpha} - \frac{\alpha+\lambda}{\alpha+\lambda+\mu} \end{pmatrix}$$

$$= \frac{1}{\lambda + \mu} \begin{pmatrix} 1 + \frac{\mu}{\alpha} - \frac{\alpha+\mu}{\alpha+\lambda+\mu} & \lambda \left(\frac{1}{\alpha} - \frac{1}{\alpha+\lambda+\mu} \right) \\ \mu \left(\frac{1}{\alpha} - \frac{1}{\alpha+\lambda+\mu} \right) & 1 + \frac{\lambda}{\alpha} - \frac{\alpha+\lambda}{\alpha+\lambda+\mu} \end{pmatrix}$$

$$= \frac{1}{\lambda + \mu} \begin{pmatrix} \frac{\mu}{\alpha} + \frac{\lambda}{\alpha+\lambda+\mu} & \lambda \left(\frac{1}{\alpha} - \frac{1}{\alpha+\lambda+\mu} \right) \\ \mu \left(\frac{1}{\alpha} - \frac{1}{\alpha+\lambda+\mu} \right) & \frac{\lambda}{\alpha} + \frac{\mu}{\alpha+\lambda+\mu} \end{pmatrix}$$

$$= \frac{1}{\lambda + \mu} \int_0^{\infty} e^{-\alpha t} \begin{pmatrix} \lambda e^{-(\lambda+\mu)t} + \mu & \lambda(1 - e^{-(\lambda+\mu)t}) \\ \mu(1 - e^{-(\lambda+\mu)t}) & \mu e^{-(\lambda+\mu)t} + \lambda \end{pmatrix} dt$$

$$= \int_0^{\infty} e^{-\alpha t} P(t) dt$$

where we interpret the integral of a matrix as the matrix of integrals of the components of the matrix.

Example 5.6.2. M/M/1 Queue with a Finite Waiting Room. Suppose arrivals occur according to a Poisson process rate a and service lengths are iid, independent of the arrivals with distribution $1 - e^{-bx}$, $x > 0$.

The wrinkle here is that an arriving customer who finds two customers in the system departs immediately without waiting for service, and the interpretation is that the capacity of the system is two or the waiting room only seats one.

Let $Q(t)$ be the number in the system at time t, so that $S = \{0, 1, 2\}$. First we compute $\lambda(i), 0 \le i \le 2$. Since $\lambda(0)$, the holding time parameter of state 0, is the parameter of the time until an arrival, we have $\lambda(0) = a$. For $\lambda(1)$ we consider the holding time in state 1, which is the minimum of the exponential waiting time until an arrival and the independent exponential waiting time until a departure. From Proposition 5.1.1(ii), we have $\lambda(1) = a + b$. Finally, for $\lambda(2)$ we note the system is blocked until a service completion, so $\lambda(2) = b$.

For the Q matrix, we have

$$Q_{10} = P[E(b) < E(a)] = b/(a + b),$$

since Q_{10} is the probability a departure occurs before an arrival. Similarly,

$$Q_{12} = P[E(a) < E(b)] = a/(a + b).$$

This is enough to write the generator matrix:

$$A = \begin{matrix} 0 \\ 1 \\ 2 \end{matrix} \begin{pmatrix} -a & a & 0 \\ b & -(a+b) & a \\ 0 & b & -b \end{pmatrix}.$$

Now $\hat{P}(\alpha) = (\alpha I - A)^{-1}$. We have

$$\alpha I - A = \begin{pmatrix} \alpha + a & -a & 0 \\ -b & \alpha + a + b & -a \\ 0 & -b & \alpha + b \end{pmatrix}.$$

Inverting by means of, for example, *Mathematica*, we get an expression which is not particularly heartwarming: Define

$$d = \alpha \left(\alpha^2 + 2(a+b)\alpha a^2 + ab + b^2 \right),$$

and

$$(\alpha I - A)^{-1}$$
$$= d^{-1} \begin{pmatrix} \alpha^2 + (2b+a)\alpha + b^2 & a(\alpha+b) & a^2 \\ b(b+b\alpha) & \alpha^2 + \alpha(a+b) + ab & a(a+\alpha) \\ b^2 & b(a+\alpha) & \alpha^2 + \alpha(b+2a) + a^2 \end{pmatrix}.$$

To continue further, one requires considerable analytic persistence. To demonstrate on an easy term, pick $\hat{P}_{20}(\alpha) = a^2/d$. Note that

$$d = \alpha(\alpha - \alpha_1)(\alpha - \alpha_2),$$

where

$$(\alpha_1, \alpha_2) = (-(a+b) + \sqrt{ab}, -(a+b) - \sqrt{ab}).$$

Consulting a table of inverse Laplace transforms (Abramowitz and Stegun, 1970, page 1022), we have

$$P_{20}(t) = \frac{2\sqrt{ab} + e^{-(a+b)t}\left((a+b-2\sqrt{ab}e^{-\sqrt{ab}t} - (a+b+\sqrt{ab})e^{\sqrt{ab}t}\right)}{2\sqrt{ab}\left((a+b)^2 - ab\right)}.$$

Perhaps this is a good illustration of the assertion that solutions of the backward differential equations are usually difficult to obtain.

5.7. CALCULATIONS AND EXAMPLES.

In this section we discuss some matrix techniques and consider some examples where the stationary distribution is computed.

For a finite state ergodic Markov chain, there is a matrix method to compute the stationary distribution which parallels formula (2.14.1) in discrete time.

Proposition 5.7.1. *Suppose $\{X(t), t \geq 0\}$ is an ergodic m–state Markov chain with state space $S = \{1, \ldots, m\}$ and that A is the generator matrix. Then we can compute the stationary probability vector for $\{X(t)\}$ by the formula*

$$(5.7.1) \qquad\qquad \eta' = (1, \ldots, 1)(A + ONE)^{-1},$$

where ONE is the $m \times m$ matrix all of whose entries are 1.

Proof. Suppose we know $A + ONE$ is invertible. From Theorem 5.5.3 we know $\eta'A = 0$, and, since η is a probability vector, we know $\eta'1 = 1$, where 1 is a column vector of one's. Thus $\eta'ONE = 1'$, and

$$\eta'(A + ONE) = 1',$$

from which, solving for η', we get (5.7.1).

It remains only to check the invertibility of $A + $ ONE. It suffices to assume that, for $x \in R^m$,

$$(A + \text{ONE})\mathbf{x} = \mathbf{0}$$

and prove this implies $\mathbf{x} = \mathbf{0}$. But if $(A+\text{ONE})\mathbf{x} = \mathbf{0}$ then $\eta'(A+\text{ONE})\mathbf{x} = 0$, so since $\eta'A = 0$ we have $\eta'\text{ONE}\mathbf{x} = 0$. Since η' is a probability vector we have $\eta'\text{ONE} = \text{ONE}$, so we conclude $\text{ONE}\mathbf{x} = \mathbf{0}$. Thus $A\mathbf{x} = \mathbf{0}$, and therefore $P(t)A\mathbf{x} = \mathbf{0}$. From the forward equations we get $P(t)A = P'(t)$, so it follows that $P'(t)\mathbf{x} = \mathbf{0}$. Integrating for any $T > 0$ yields

$$0 = \int_0^T P'(t)\mathbf{x}dt = P(T)\mathbf{x} - I\mathbf{x},$$

so that

$$P(T)\mathbf{x} = I\mathbf{x} = \mathbf{x}.$$

Let $T \to \infty$ so that

$$P(T) \to L$$

where

$$L = \begin{pmatrix} \eta_1 & \cdots & \eta_m \\ \vdots & \ddots & \vdots \\ \eta_1 & \cdots & \eta_m \end{pmatrix}.$$

Thus $L\mathbf{x} = \mathbf{x}$, or, in component form,

$$\sum_{j=1}^m x_j\eta_j = x_i, \quad i = 1,\ldots,m,$$

and $\mathbf{x} = c\mathbf{1}$ for some constant c. Since $\text{ONE}\mathbf{x} = \mathbf{0}$, we realize that

$$0 = (1,\ldots,1)\mathbf{x} = \mathbf{1}'c\mathbf{1} = cm,$$

which requires $c = 0$, and this makes $\mathbf{x} = c\mathbf{1} = \mathbf{0}$. Therefore, $(A+\text{ONE})\mathbf{x} = \mathbf{0}$ implies $\mathbf{x} = \mathbf{0}$, which makes $A + \text{ONE}$ invertible. ∎

As a short example, consider again the example of Section 5.5 describing Harry's injury prone basketball career. There were three states: 0 (fit), 1 (minor injury) and 2 (major injury), and the fluctuation between the states followed a continuous time Markov chain with generator matrix

$$A = \begin{pmatrix} -\frac{1}{3} & \frac{1}{9} & \frac{2}{9} \\ \frac{1}{9} & -\frac{1}{3} & \frac{2}{9} \\ \frac{1}{6} & 0 & -\frac{1}{6} \end{pmatrix}.$$

With the help of a package such as Minitab, we quickly perform the following calculations: First we find

$$A + \text{ONE} = \begin{pmatrix} .66667 & 1.11111 & 1.22222 \\ 1.11111 & .66667 & 1.22222 \\ 1.16667 & 1.00000 & .83333 \end{pmatrix}.$$

Inverting yields

$$(A + \text{ONE})^{-1} = \begin{pmatrix} -1.28572 & .57143 & 1.04262 \\ 0.96429 & -1.67857 & 1.04762 \\ 0.64286 & 1.21429 & -1.52381 \end{pmatrix}.$$

Multiplying by a row vector of ones, we get the stationary distribution

$$\begin{aligned} (\eta_0, \eta_1, \eta_2) &= (1,1,1)(A + \text{ONE})^{-1} \\ &= (.321429, .107143, .571428), \end{aligned}$$

which agrees with the hand calculations of Section 5.5.

Next, we comment on matrix methods for calculating absorption probabilities and expected times until absorption in a finite state Markov chain. For absorption probabilities, the time scale does not matter, so the discussion of Section 2.11 is still applicable. Starting from transient state $i \in S$, the probability u_{ik} of being absorbed in recurrent state k is obtained from the embedded discrete time chain $\{X_n\}$ and its transition matrix Q. For expected absorption times, where the time scale is crucial, we have the following representative result.

Proposition 5.7.2. *Suppose the associated discrete time Markov chain* $\{X_n\}$ *is irreducible. Pick a reference state* $j \in S$, *and let*

$$\mathbf{t} = \{E_i T_{\tau_j(1)}, i \neq j\}'$$

be the column vector of expected hitting times of state j *starting from initial state* $i \neq j$. *Define the* $(m-1) \times (m-1)$ *matrix*

$$B = (A_{ik}, i \neq j, k \neq j)$$

to be the restriction of the matrix A *to the set* $S \backslash \{j\}$. *Then* $-B$ *is invertible, and*

(5.7.2) $$\mathbf{t} = (-B)^{-1}\mathbf{1}$$

where **1** *is a column vector of 1's.*

Proof. Use first jump analysis to verify that, for $i \neq j$,

$$E_i T_{\tau_j(1)} = \lambda(i)^{-1} + \sum_{k \neq i, k \neq j} Q_{ik} E_k T_{\tau_j(1)}$$

$$= \lambda(i)^{-1} + \sum_{k \neq i, k \neq j} \frac{A_{ik}}{\lambda(i)} E_k T_{\tau_j(1)}.$$

Multiply through by $\lambda(i)$ to get

$$t_i \lambda(i) = 1 + \sum_{k \neq i, k \neq j} A_{ik} E_k T_{\tau_j(1)},$$

or, equivalently,

$$t_i \lambda(i) - \sum_{k \neq i, k \neq j} A_{ik} t_k = 1.$$

Then, for all $i \neq j$,

$$-t_i A_{ii} - \sum_{k \neq i, k \neq j} A_{ik} t_k = 1,$$

which in matrix form is

$$(-B)\mathbf{t} = \mathbf{1},$$

so that if $(-B)$ is invertible we get (5.7.2).

To check this invertibility, let $D = \text{diag}(\lambda(1), \ldots, \lambda(m))$; i.e.,

$$D = \begin{pmatrix} \lambda(1) & 0 & 0 & \cdots & 0 \\ 0 & \lambda(2) & 0 & \cdots & 0 \\ \vdots & & \ddots & & \vdots \\ 0 & \cdots & 0 & 0 & \lambda(m) \end{pmatrix}.$$

Check that $A = D(I - Q)$ so $B = D^-(I - Q)^-$ where the minus means delete the jth row and column; i.e., for a $m \times m$ matrix M,

$$M^- = (M_{ik}, i \neq j, k \neq j).$$

From discrete time theory we know that $(I - Q)^-$ is invertible since it is the fundamental matrix (see Section 2.11). The diagonal matrix D^- is invertible since the λ's are assumed positive, and therefore, $(-B)$ must be invertible. ∎

As another quick numerical example, consider again Harry's basketball career, which fluctuated between states 0,1,2, and suppose we and Harry's coach wish to compute the expected time until Harry is felled by a major injury. The generator matrix is

$$A = \begin{pmatrix} -\frac{1}{3} & \frac{1}{9} & \frac{2}{9} \\ \frac{1}{9} & -\frac{1}{3} & \frac{2}{9} \\ \frac{1}{6} & 0 & -\frac{1}{6} \end{pmatrix}.$$

Eliminating the third row and column and multiplying remaining entries by -1 yields

$$-B = \begin{pmatrix} \frac{1}{3} & -\frac{1}{9} \\ -\frac{1}{9} & \frac{1}{3} \end{pmatrix}.$$

Inverting, we find

$$(-B)^{-1} = \begin{pmatrix} \frac{27}{8} & \frac{9}{8} \\ \frac{9}{8} & \frac{27}{8} \end{pmatrix}.$$

Post multiplying by a column vector of ones, we get the expected absorption times starting from states 0,1:

$$(E_0 T_{\tau_2(1)}, E_1 T_{\tau_2(1)})' = (t_0, t_1)' = (-B)^{-1} \mathbf{1}$$
$$= (\frac{27}{8} + \frac{9}{8}, \frac{9}{8} + \frac{27}{8})' = (\frac{36}{8}, \frac{36}{8})'.$$

Additional practice with the formulas given in Propositions 5.7.1 and 5.7.2 can be obtained by trying some of the exercises at the end of the chapter. Problem 5.1 is a good numerical exercise.

The distribution of the time until a state of a continuous time Markov chain is hit for the first time is an example of a *phase distribution*. (See Neuts, 1981, 1989.) Such distributions have been important for modelling complex stochastic networks in a Markovian way. Any distribution can be approximated by a phase distribution. By replacing a distribution in a model by a phase approximation, one allows for simpler Markovian analysis. The basic information about such distributions is explored in Problems 5.5 and 5.50.

We now discuss significant examples where the stationary distribution is of interest.

Example 5.7.1. Birth-Death Processes. Suppose birth rates are $\{\lambda_i\}$ and death rates are $\{\mu_i\}$ with $\mu_0 = 0$ and $S = \{0, 1, \dots\}$, so that

$$\lambda(i) = \lambda_i + \mu_i, \quad i \geq 0,$$

and

$$Q_{i,i+1} = \frac{\lambda_i}{\lambda_i + \mu_i}, \quad Q_{i,i-1} = \frac{\mu_i}{\lambda_i + \mu_i}, \quad i \geq 1,$$

and $Q_{01} = 1$. Then

$$A = \begin{pmatrix} -\lambda_0 & \lambda_0 & 0 & 0 & \cdots \\ \mu_1 & -(\lambda_1 + \mu_1) & \lambda_1 & 0 & \\ 0 & \mu_2 & -(\lambda_2 + \mu_2) & \lambda_2 & \\ & & \ddots & \ddots & \ddots \end{pmatrix}.$$

Note that for δ small,

$$P_{i,i+1}(\delta) = A_{i,i+1}\delta + o(\delta) = \lambda_i\delta + o(\delta)$$
$$P_{i,i-1}(\delta) = A_{i,i-1}\delta_0(\delta) = \mu_i\delta + o(\delta), \ i \geq 1,$$

which helps to explain why λ_i, μ_i are referred to as the birth and death rates respectively.

For regularity, we need, for any starting state $i \in S$,

$$\infty = \sum_{N=0}^{\infty} \frac{1}{\lambda(X_N)} = \sum_{j \in S} \frac{\sum_{n=0}^{\infty} 1_{[X_n=j]}}{\lambda_j + \mu_j}.$$

Setting $N^{(j)} = \sum_{n=1}^{\infty} 1_{[X_n=j]}$ for the number of visits to state j, we find we need

$$\infty = \sum_{j \in S} \frac{N^{(j)}}{\lambda_j + \mu_j}.$$

An obvious sufficient condition is that

$$\sum_j \frac{1}{\lambda_j} = \infty.$$

For instance, in the linear birth-death process where $\lambda_j = \lambda j$ and $\mu_j = \mu j$, we quickly conclude regularity holds.

Supposing that the process is regular, let us analyze $\eta' A = 0$, assuming $\eta_0 = 1$. We have

$$-\lambda_0 + \mu_1\eta_1 = 0$$
$$\lambda_0 - (\lambda_1 + \mu_1)\eta_1 + \mu_2\eta_2 = 0$$
$$\lambda_1\eta_1 - (\lambda_2 + \mu_2)\eta_2 + \mu_3\eta_3 = 0.$$

In general, for $n \geq 0$,

$$\lambda_n\eta_n - (\lambda_{n+1} + \mu_{n+1})\eta_{n+1} + \mu_{n+2}\eta_{n+2} = 0,$$

so

$$\mu_1 \eta_1 = \lambda_0$$

(recall the right side is also $(\lambda_0 + \mu_0)\eta_0$)

$$\lambda_0 + \mu_2 \eta_2 = (\lambda_1 + \mu_1)\eta_1$$

$$\vdots$$

$$\lambda_n \eta_n + \mu_{n+2} \eta_{n+2} = (\lambda_{n+1} + \mu_{n+1})\eta_{n+1}, \quad n \geq 0.$$

Remembering $\eta_0 = 1, \mu_0 = 0$, after summing from 0 to $n-2$, we have

$$\sum_{i=0}^{n-2} \lambda_i \eta_i + \sum_{i=0}^{n} \mu_i \eta_i = \sum_{i=0}^{n-1} \lambda_i \eta_i + \sum_{i=0}^{n-1} \mu_i \eta_i,$$

from which

$$\mu_n \eta_n = \lambda_{n-1} \eta_{n-1}, \quad n \geq 1,$$

and

$$
\begin{aligned}
\eta_n &= \frac{\lambda_{n-1}}{\mu_n} \eta_{n-1} \\
&= \frac{\lambda_{n-1}\lambda_{n-2}}{\mu_n \mu_{n-1}} \eta_{n-2} \\
&= \ldots = \frac{\lambda_{n-1}\lambda_{n-2}\ldots\lambda_0}{\mu_n \mu_{n-1}\ldots\mu_1} \eta_0 \\
&= \frac{\prod_{i=0}^{n-1} \lambda_i}{\prod_{i=1}^{n} \mu_i}, \quad n \geq 1.
\end{aligned}
$$

(5.7.1)

Therefore, a stationary distribution exists iff $\sum \eta_n < \infty$ iff

(5.7.2)
$$\sum_{n=1}^{\infty} \frac{\prod_{i=0}^{n-1} \lambda_i}{\prod_{i=1}^{n} \mu_i} < \infty.$$

Example 5.7.2.(a). M/M/1 Queue. Recall customers arrive according to a Poisson process with rate a. There is one server who serves at rate b; i.e., service times are iid exponential random variables with parameter b independent of the input. We have

$$Q_{i,i+1} = \frac{a}{a+b}, \quad Q_{i,i-1} = \frac{b}{a+b}, \quad i \geq 1,$$

and $(\lambda(0), \lambda(1), \dots) = (a, a+b, a+b, \dots)$, so $\lambda_i = a$, $i \geq 0$, $\mu_i = b$, $i \geq 1$, $\mu_0 = 1$. Then, from (5.7.1),

$$\eta_n = \left(\frac{a}{b}\right)^n, \quad n \geq 0.$$

Let $\rho = \frac{a}{b}$, which is called the traffic intensity. We have $\sum \eta_n < \infty$ iff $\rho < 1$, in which case, by normalizing, the stationary distribution is geometric:

$$\left\{\frac{\eta_n}{\sum_{k=0}^{\infty} \eta_k}, n \geq 0\right\} = \{(1-\rho)\rho^n, n \geq 0\}.$$

Example 5.7.2(b). M/M/s Queue. Customers arrive according to a Poisson process, but now there are s servers ($s > 1$), each of whom serves at rate b.

The holding time in state 0 is the waiting time for an arrival, and is therefore exponential with parameter a. Thus $\lambda(0) = a$. For $1 \leq i \leq s$, the holding time in i has the distribution of

$$E(a) \wedge (\wedge_{k=1}^{i} E_k(b)),$$

where these are independent exponential random variables, $E(a)$ represents the wait until a new arrival and $E_k(b)$ represents the required service time of the customer at server number k. (Recall the notation that $E(parameter)$ represents a random variable with exponential distribution of parameter=$parameter$.) Thus

$$\lambda(i) = a + ib, \quad 1 \leq i \leq s.$$

For $i > s$, the holding time is of the form

$$E(a) \wedge (\wedge_{k=1}^{s} E_k(b)),$$

so $\lambda(i) = a + sb$, $i > s$. Also, for $i \geq 0$,

$$Q_{i,i+1} = \begin{cases} P[E(a) < \wedge_{k=1}^{i} E_k(b)] = \frac{a}{a+ib}, & \text{if } 0 \leq i \leq s \\ P[E(a) < \wedge_{k=1}^{s} E_k(b)] = \frac{a}{a+sb}, & \text{if } i > s. \end{cases}$$

We conclude

$$\lambda_i = a, \quad i \geq 0,$$

$$\mu_i = \begin{cases} ib, & \text{if } 0 \leq i \leq s, \\ sb, & \text{if } i > s. \end{cases}$$

A stationary distribution exists iff (5.7.2) holds. Condition (5.7.2) is

$$1 + \sum_{n=1}^{s} \frac{a^n}{\prod_{i=1}^{n} b i} + \sum_{n=s+1}^{\infty} \frac{a^n}{\prod_{i=1}^{s} b i \prod_{i=s+1}^{n} b s} < \infty,$$

iff the last sum is finite. The last piece is

$$\frac{a^s}{b^s s!} \sum_{n=s+1}^{\infty} \frac{a^{n-s}}{(bs)^{n-s}},$$

and this is finite iff

$$\frac{a}{bs} < 1,$$

i.e., $a < bs$. Therefore, the queue is stable if the overall service rate exceeds the arrival rate.

Example 5.7.2(c). Machine Repair. Suppose there are m machines and one repairman. Machines operate for periods that are exponentially distributed with parameter a. Repair periods are exponential with parameter b. All operative and repair periods are independent. Let $X(t)$ be the number of machines inoperable at time t. The holding time in 0 is $\wedge_{k=1}^{m} E_k(a)$, so $\lambda(0) = am$. For $1 \leq i \leq m$, the holding time in i is

$$\wedge_{k=1}^{m-i} E_k(a) \wedge E(b)$$

(since $m - i$ machines are operating and one is in service), so $\lambda(i) = a(m - i) + b$. Then

$$\lambda_i = \begin{cases} a(m - i), & \text{if } 0 \leq i \leq m, \\ 0, & \text{if } i > m, \end{cases}$$

$$\mu_i = \begin{cases} 0, & \text{if } i = 0, \\ b, & \text{if } 1 \leq i \leq m, \\ 0, & \text{if } i > m. \end{cases}$$

Therefore, the state space is finite, consisting of the states $\{0, 1, \ldots, m\}$, and a stationary distribution will always exist. For $1 \leq n \leq m$, we have

$$\eta_n = \prod_{i=0}^{n-1} \lambda_i / \prod_{i=1}^{n} \mu_i$$

$$= \frac{\prod_{i=0}^{n-1} a(m - i)}{b^n} = \left(\frac{a}{b}\right)^n (m(m - 1) \ldots m - n + 1)$$

$$= \left(\frac{a}{b}\right)^n (m)_n,$$

and the stationary distribution is obtained by normalization.

Many other examples of queues modelled by means of Markov chains are explored in the exercises. Also, a significant set of applications to queueing networks is discussed in the next subsection.

5.7.1. QUEUEING NETWORKS.

In this section we consider important examples of Markov chain analysis applied to queueing networks. A queueing network is an interconnected grid of service facilities called nodes. Each node may have its own service mechanism and queueing discipline. For each node there is a probabilistic mechanism for routing traffic either to other nodes or out of the system.

Networks may be either *closed* or *open*. A closed network has a fixed number k of nodes and a fixed number m of customers. It is assumed that no traffic either enters into or departs from the system. To fix ideas, imagine a company has m trucks. The trucks are either in the field operating (node 1) or in one of $k-1$ service bays being serviced and maintained. In the other important case, the network is open. Arrivals and departures are permitted. Imagine data packets being routed around the terminals of a computer network or telephone traffic being routed in the grid of exchanges controlled by the phone company.

The quantity of interest is the queue length process

$$\mathbf{Q}(t) = (Q_1(t), \ldots, Q_k(t)),$$

where $Q_i(t)$ represents the number of customers at the ith node at time t. The state space for the queue length process is either

$$S = \{\mathbf{n} = (n_1, \ldots, n_k) : n_i \geq 0, \sum_{i=1}^{k} n_i = m\}$$

in the closed network case or

$$S = \{\mathbf{n} = (n_1, \ldots, n_k) : n_i \geq 0, i = 1, \ldots, k\}$$

in the open network case. We will make sufficient assumptions to assure that $\{\mathbf{Q}(t), t \geq 0\}$ is a Markov chain with state space S. We are interested in the stationary distribution, and we will find that it has a product form.

A closed queueing network: There are m customers who move among k nodes (service facilities). Entry into and exit from the system are not permitted. There is a routing matrix $P = (p_{ij})$ which has the interpretation that an item departing node i next goes to node j with probability $p_{ij}, 1 \leq i, j \leq k$, where we assume for all i, j that $p_{ij} \geq 0, \sum_{j=1}^{k} p_{ij} = 1$ and that P is the transition matrix of an irreducible and, hence, a positive recurrent Markov chain. Note we allow $p_{jj} > 0$ as this allows for feedback to node j, as would be common when reworking is necessary in manufacturing situations. Suppose $\pi' = (\pi_1, \ldots, \pi_k)$ is the stationary distribution of P, so that $\pi' P = \pi'$ and $\pi' \mathbf{1} = 1$. Assume the server at node j serves at rate b_j $(1 \leq j \leq m)$, so that service times at node j are iid with exponential density of parameter b_j. Service times at different nodes are independent.

The process of interest is the queue length process

$$\mathbf{Q}(t) = (Q_1(t), \ldots, Q_k(t)),$$

where for $1 \leq i \leq k$, we define $Q_i(t)$ to be the number waiting or in service at node i at time t. The state space of this process is

$$S = \{(n_1, \ldots, n_k) : 0 \leq n_i \leq m, \sum_{i=1}^{k} n_i = m, 1 \leq i \leq k\}.$$

The stationary distribution of $\mathbf{Q}(t)$ turns out to depend on the stationary distribution π' of P and the service rates b_j. Why should π' be a component of the solution? It is reasonable to expect the stationary distribution of $\mathbf{Q}(t)$ to depend on the *throughput rates* at each node where we may think of the throughput rate r_j at node j as the long run number of customers leaving node j per unit time. How would we compute the throughput rates? These should satisfy the system of equations (called the *traffic equations*) which says that if the system is in equilibrium, the long run rate at which customers enter node j should equal the long run rate r_j that customers leave node j. The long run rate of arrivals to node j is the sum of the arrival rates to j from all the nodes; i.e., $\sum_{l=1}^{k} r_l p_{lj}$, since the proportion of the output of node l which goes to j is p_{lj}. Thus the throughput rates should satisfy

$$r_j = \sum_{l=1}^{k} r_l p_{lj}, \quad j = 1, \ldots, k.$$

The vector $\mathbf{r}'$ of throughput rates must be a constant multiple of the stationary distribution π' of the matrix P, which explains why we may expect the stationary dsitribution of $\mathbf{Q}(t)$ to be dependent on π'.

We now compute the infinitesimal parameters $\{\lambda(\mathbf{n}), \mathbf{n} \in S\}, Q, A$. If the process is in state $\mathbf{n} \in S$, the sojourn time is determined by the minimum of the service times at nodes where customers are being served. Thus the sojourn time in $\mathbf{n} \in S$ is of the form

$$\wedge_{i:n_i > 0} E(b_i),$$

and therefore

$$\lambda(\mathbf{n}) = \lambda(n_1, \ldots, n_k) = \sum_{i:n_i > 0} b_i = \sum_{i=1}^{k} b_i(1 - \delta_{0,n_i}).$$

The **Q**-process changes state at service completion times. If service is completed at node i, then the number at node i drops by 1. If the departing item moves to node j, the number at j increases by 1. Then for $1 \leq i, j \leq k$, we have the following transition within S:

$$(n_1, \ldots, n_i + 1, \ldots, n_j - 1, \ldots, n_k) \to (n_1, \ldots, n_i, \ldots, n_j, \ldots, n_k)$$

(where we assume $m > n_j - 1 \geq 0$). To avoid going crazy with the notation, define the transfer function within S: If $\mathbf{n} \in S, n_i > 0$,

$$T(i,j)\mathbf{n} = (n_1, \ldots, n_i - 1, \ldots, n_j + 1, \ldots, n_k),$$

so that an item is transferred from node i to node j. (It is not required that $i < j$.) Note $T(j,i)T(i,j)\mathbf{n} = \mathbf{n}$ if $n_i > 0$, and $\mathbf{m} = T(i,j)\mathbf{n}$ iff $T(j,i)\mathbf{m} = \mathbf{n}$ if $n_i > 0, m_j > 0$.

If $n_i > 0$,

$$Q_{\mathbf{n}, T(i,j)\mathbf{n}} = P[E(b_i) < \wedge_{l \neq i: n_l > 0} E(b_l)] p_{ij}$$

$$= \frac{b_i}{\sum_{l=1}^{k} b_l (1 - \delta_{0, n_l})} p_{ij}$$

$$= \frac{b_i}{\lambda(\mathbf{n})} p_{ij}$$

since in order for there to be a transition $\mathbf{n} \to T(i,j)\mathbf{n}$, service at node i must finish and an item must transfer from i to j. We find, for the generator matrix A,

$$A_{\mathbf{n}, \mathbf{n}} = -\lambda(\mathbf{n}) = -\sum_{l=1}^{k} b_l (1 - \delta_{0, n_l}),$$

and, if $n_i > 0$,

$$A_{\mathbf{n}, T(i,j)\mathbf{n}} = \lambda(\mathbf{n}) Q_{\mathbf{n}, T(i,j)\mathbf{n}} = b_i p_{ij}.$$

We now check that $(\mathbf{n} \in S)$

$$\eta_{\mathbf{n}} = \prod_{l=1}^{k} \left(\frac{\pi_l}{b_l} \right)^{n_l}$$

is an invariant distribution. To do this, we verify $\sum_{\mathbf{n} \in S} \eta_{\mathbf{n}} A_{\mathbf{n}, \mathbf{m}} = 0$ for $\mathbf{m} \in S$. Observe that if $m_j > 0$

$$\eta_{T(j,i)\mathbf{m}} = \prod_{l=1}^{k} \left(\frac{\pi_l}{b_l} \right)^{T(j,i)m_l}$$

$$= \prod_{l \neq i, l \neq j} \left(\frac{\pi_l}{b_l} \right)^{m_l} \left(\frac{\pi_i}{b_i} \right)^{m_i + 1} \left(\frac{\pi_j}{b_j} \right)^{m_j - 1}$$

$$= \eta_{\mathbf{m}} \frac{\pi_i / b_i}{\pi_j / b_j}.$$

We have that

$$\sum_{n \neq m} \eta_n A_{n,m} = \sum_{\{n:m=T(i,j)n, \text{ for some } i,j\}} \eta_n A_{n,m}$$

$$= \sum_{\{n:T(j,i)m=n \text{ for some } i,j\}} \eta_n A_{n,m}$$

$$= \sum_{i,j:m_j>0} \eta_{T(j,i)m} A_{T(j,i)m,m},$$

and since $A_{T(j,i)m,m} = A_{T(j,i)m,T(i,j)(T(j,i)m)}$ we have

$$= \sum_{i,j:m_j>0} \left(\prod_{l=1}^{k} \left(\frac{\pi_l}{b_l}\right)^{m_l} \frac{\pi_i/b_i}{\pi_j/b_j} \right) b_i p_{ij}$$

$$= \sum_{j:m_j>0} \prod_{l=1}^{k} \left(\frac{\pi_l}{b_l}\right)^{m_l} \frac{\sum_i \pi_i p_{ij}}{\pi_j/b_j}$$

$$= \sum_{j:m_j>0} \prod_{l=1}^{k} \left(\frac{\pi_l}{b_l}\right)^{m_l} \frac{\pi_j}{\pi_j/b_j}$$

$$= \prod_{l=1}^{k} \left(\frac{\pi_l}{b_l}\right)^{m_l} \sum_{j=1}^{k} b_j (1 - \delta_{0,m_j})$$

$$= \eta_m \lambda(m).$$

Therefore

$$\sum_{n \in S} \eta_n A_{n,m} = \sum_{n \neq m} \eta_n A_{n,m} - \eta_m \lambda(m) = \eta_m \lambda(m) - \eta_m \lambda(m) = 0,$$

showing $\{\eta_n, n \in S\}$ is invariant.

Because the state space is finite, the stationary distribution exists and equals

$$\frac{\eta_n}{\sum_{l \in S} \eta_l} = \beta_m \prod_{j=1}^{k} \left(\frac{\pi_j}{b_j}\right)^{n_j},$$

where

$$\beta_m = \left(\sum_{l \in S} \prod_{j=1}^{k} \left(\frac{\pi_j}{b_j}\right)^{l_j} \right)^{-1}.$$

Computing the constant β_m can be quite difficult, as the state space S can be quite large. One scheme for accomplishing this, suggested in Kelly

(1979), is discussed in Problem 5.49. Other schemes are discussed in Walrand (1988).

Note that although the stationary distribution has a product form, the stationary distribution is not the distribution of a random vector with independent components. This is because of the constraint on the state space that $\mathbf{n} \in S$ implies $\sum_{i=1}^{k} n_i = m$.

Bottlenecks and clogging: Define for $j = 1, \ldots, k$ and a multiplicative constant $c > 0$ to be specified,

$$\theta_j = c \frac{\pi_j}{b_j},$$

and call this the *clogging factor* for node j. Bottlenecks will occur at the node with the biggest clogging factor, and this will be where customers tend to be located. We explore how to make this precise by means of some limit results.

We know $\mathbf{Q}(t)$ converges in distribution and the limiting distribution is the stationary distribution, which we denote by $\eta_\mathbf{n}, \mathbf{n} \in S$. For $\mathbf{n} \in S$, we will write

$$P[\mathbf{Q}(\infty) = \mathbf{n}] = \eta_\mathbf{n}.$$

To be specific, suppose that node 1 has the biggest clogging factor, so that for $j = 2, \ldots, k$,

$$\theta_j < \theta_1.$$

Take

$$c = \left(\frac{\pi_1}{b_1} \right)^{-1},$$

so that

$$\theta_1 = 1, \quad \theta_j < 1, \ j = 2, \ldots, k.$$

Write $\delta = \vee_{j=2}^{k} \theta_j$. We may re-express the stationary distribution $\eta_\mathbf{n}$ as

$$\eta_\mathbf{n} = \frac{\prod_{j=1}^{k} (\frac{\pi_j}{b_j})^{n_j}}{\sum_{\mathbf{l} \in S} \prod_{j=1}^{k} (\frac{\pi_j}{b_j})^{l_j}}.$$

Multiply numerator and denominator by c^m, and the above is

$$= \frac{\prod_{j=1}^{k} (\frac{c\pi_j}{b_j})^{n_j}}{\sum_{\mathbf{l} \in S} \prod_{j=1}^{k} (\frac{c\pi_j}{b_j})^{l_j}}$$

$$= \frac{\prod_{j=2}^{k} \theta_j^{n_j}}{\sum_{\mathbf{l} \in S} \prod_{j=2}^{k} \theta_j^{l_j}}.$$

Call the denominator in the last expression DEN and we note that as the number of customers m goes to infinity a limit results:

$$DEN = \sum_{\{(l_1,\ldots,l_k):\Sigma_{j=1}^{k}l_j=m\}} \prod_{j=2}^{k} \theta_j^{l_j}$$

$$= \sum_{\{(l_2,\ldots,l_k):\Sigma_{j=2}^{k}l_j\leq m\}} \prod_{j=2}^{k} \theta_j^{l_j}$$

$$\to \sum_{l_2=0}^{\infty}\cdots\sum_{l_k=0}^{\infty}\prod_{j=2}^{k} \theta_j^{l_j}$$

$$= \prod_{j=2}^{k}(1-\theta_j)^{-1}.$$

Thus, as $m \to \infty$

$$P[Q_2(\infty) = n_2,\ldots,Q_k(\infty) = n_k] = \frac{\prod_{j=2}^{k}\theta_j^{n_j}}{DEN}$$

$$\to \frac{\prod_{j=2}^{k}\theta_j^{n_j}}{\prod_{j=2}^{k}(1-\theta_j)^{-1}}$$

$$= \prod_{j=2}^{k}(1-\theta_j)\theta_j^{n_j}.$$

We get a product form limit distribution for nodes which do not have maximal clogging factors. As we will see, the sort of product form for the stationary distribution obtained here is characteristic of open networks. Since behavior in nodes $2,\ldots,k$ is stable as $m \to \infty$, we expect lots of customers to be caught in the bottleneck node 1 with maximal clogging factor. We will check that for any integer q

(5.7.1.1) $$P[Q_1(\infty) \leq q] \to 0,$$

so that with high probability the number in node 1 is very large and enough traffic is clogged at node 1 that nodes $2,\ldots,k$ operate as a stable, open system.

To check (5.7.1.1) we use brute force: For any integer q

$$P[Q_1(\infty) = q] = \sum_{\{(n_2,\ldots,n_k):\Sigma_{i=2}^k n_i = m-q\}} \frac{\prod_{j=2}^k \theta_j^{n_j}}{DEN}$$

$$\leq \sum_{\{(n_2,\ldots,n_k):\Sigma_{i=2}^k n_i = m-q\}} \frac{\prod_{j=2}^k \delta^{n_j}}{DEN}$$

$$= \sum_{\{(n_2,\ldots,n_k):\Sigma_{i=2}^k n_i = m-q\}} \frac{\delta^{\Sigma_{j=2}^k n_j}}{DEN}$$

$$= \frac{\delta^{m-q}}{DEN} \times (\text{the number of ways to assign } m - q \text{ customers}$$
$$\text{to } k - 1 \text{ nodes }).$$

Computing the number of ways to assign $m - q$ customers to $k - 1$ nodes is a classical combinatorial problem (e.g., Feller, 1968, Volume 1, Chapter 2) and equals

$$\binom{m + k - q - 2}{m - q} = \binom{m + k - q - 2}{k - 2}.$$

We have that DEN converges to a non-zero limit as $m \to \infty$ and

$$\delta^{m-q} \binom{m + k - q - 2}{m - q} = \delta^{m-q} \frac{(m - q + k - 2)!}{(m - q)!(k - 2)!}$$

$$= \delta^{m-q} \frac{(m - q + k - 2)\ldots(m - q + 1)}{(k - 2)!}$$

$$= \frac{\delta^{m-q}}{(k - 2)!} m^{k-2}(1 - \frac{(q - k + 2)}{m})\ldots(1 - \frac{(q - 1)}{m})$$

$$\leq c'\delta^m m^k \to 0$$

as $m \to \infty$ for some $c' > 0$. Therefore

$$P[Q_1(\infty) = q] \to 0;$$

thus (5.7.1.1) follows.

Example 5.7.3. A Closed Cyclic Queue. A simple example is constructed by assuming there are k nodes and traffic cycles through the nodes according to the scheme $1 \to 2 \to \cdots \to k \to 1$, so that the routing matrix is given by $p_{i,i+1} = 1$ for $i = 1, \ldots, k - 1$ and also $p_{k1} = 1$. The routing matrix is doubly stochastic, so the stationary probability vector for P is automatically $\pi_j = 1/k$ for $j = 1, \ldots, k$. The stationary distribution for $\{Q(t)\}$ is thus

$$\beta_m \prod_{j=1}^{k} (\frac{1}{b_j})^{n_j},$$

and there is a bottleneck at node j if

$$\frac{1}{b_j} > \frac{1}{b_i}, \quad j \neq i;$$

that is, if $b_j < b_i$ for all $j \neq i$. As expected, the bottleneck occurs at the node with the slowest service rate.

An open queueing network. As before there are k nodes, but now customers can enter and leave the system. In order to guarantee that $\{Q(t)\}$ is Markov, we assume that customers from outside the network enter node i according to a Poisson process of rate a_i and that the arrival streams to the different nodes are independent. Also, we assume node i has a single server serving at rate b_i; service times at node i are iid with exponential distribution with parameter b_i. There is a routing matrix $(p_{ij}, 1 \leq i, j \leq k)$ where p_{ij} represents the probability that a customer leaving node i goes next to node j. In contrast to the closed network, we assume that at least for some i ($1 \leq i \leq k$) we have

$$\sum_{j=1}^{k} p_{ij} < 1,$$

and we interpret

$$q_i := 1 - \sum_{j=1}^{k} p_{ij} > 0$$

as the probability that a customer departing from node i leaves the system. Think of there being a state Δ which corresponds to "leave the system," and define the $(k+1) \times (k+1)$ matrix $\tilde{P}$ by

$$\tilde{P}_{ij} = \begin{cases} p_{ij}, & \text{if } 1 \leq i, j \leq k \\ 1, & \text{if } i = j = \Delta \\ 0, & \text{if } i = \Delta, j \neq \Delta \\ q_i := 1 - \sum_{l=1}^{k} p_{il}, & \text{if } i \neq \Delta, j = \Delta. \end{cases}$$

If we think of $\tilde{P}$ governing a discrete time Markov chain, then states $1, \ldots, k$ are transient, Δ is absorbing and $(I - P)^{-1}$ is the fundamental matrix corresponding to transient states. A customer leaving node i will visit node j a finite number of times before leaving the system $(1 \leq i, j \leq k)$.

As before, the stationary distribution (when it exists) should depend on throughput rates, so suppose r_j is the rate customers leave node j $(1 \leq j \leq k)$. In equilibrium, the rate r_j at which customers leave node j should balance the rate at which they arrive to node j. The rate at which customers leave node i is r_i, and the proportion of the customers who depart from node i and migrate to j is p_{ij}. Furthermore, customers enter the system externally as input to node j at rate a_j. Thus the input rates should satisfy the system of equations

$$r_j = a_j + \sum_{i=1}^{k} r_i p_{ij}, \quad 1 \leq j \leq k,$$

which we again call the traffic equations. If we let $\mathbf{a} = (a_1, \ldots, a_k)'$, $\mathbf{r} = (r_1, \ldots, r_k)'$, then in matrix form the traffic equations for the open system are

$$\mathbf{r}' = \mathbf{a}' + \mathbf{r}'P,$$

or

$$\mathbf{r}'(I - P) = \mathbf{a}',$$

and the solution is

$$\mathbf{r}' = \mathbf{a}'(I - P)^{-1} = \mathbf{a}' \sum_{n=0}^{\infty} P^n.$$

We still need to have the transfer functions $T(i, j)$ defined on the state space S. When $1 \leq i, j \leq k$ define $T(i, j)$ as we did for the closed system. In addition, for $\mathbf{n} = (n_1, \ldots, n_k) \in S$, define

$$T(0, j)\mathbf{n} = (n_1, \ldots, n_j + 1, \ldots, n_k),$$

which means there is an external arrival to node j, and for $n_j \geq 1$, define

$$T(j, 0)\mathbf{n} = (n_1, \ldots, n_j - 1, \ldots, n_k),$$

which means there is a departure from the system at node j.

The generator matrix is now of the form ($\mathbf{n} = (n_1, \ldots, n_k) \in S$)

$$A_{\mathbf{n},\mathbf{n}} = -\lambda(\mathbf{n}) = -\left(\sum_{j=1}^{k} a_j + \sum_{j:n_j>0} b_j \right)$$

$$A_{\mathbf{n},T(0,j)\mathbf{n}} = a_j, \quad 1 \leq j \leq k,$$
$$A_{\mathbf{n},T(j,0)\mathbf{n}} = b_j q_j, \quad 1 \leq j \leq k, \ n_j > 0$$
$$A_{\mathbf{n},T(i,j)\mathbf{n}} = b_i p_{ij}, \quad 1 \leq i, j \leq k, \ n_i > 0.$$

We will show that the invariant measure $\eta_{\mathbf{n}}, \mathbf{n} \in S$ has the product form

(5.7.1.2)
$$\eta_{\mathbf{n}} = \prod_{j=1}^{k} (\frac{r_j}{b_j})^{n_j},$$

and therefore a stationary probability distribution exists iff $\sum_{\mathbf{n} \in S} \eta_{\mathbf{n}} < \infty$. This holds iff

$$\sum_{m=0}^{\infty} (\frac{r_j}{b_j})^m < \infty, \quad j = 1, \ldots, k,$$

and this condition is equivalent to

$$r_j < b_j, \quad j = 1, \ldots, k,$$

which says that, in order for equilibrium to hold, the service rate must be adequate to handle the rate at which customers pass through each node.

To verify that the product form in (5.7.1.2) yields an invariant measure, we make the following observations:

(5.7.1.3)
$$\eta_{T(j,0)\mathbf{n}} = \eta_{\mathbf{n}} \frac{b_j}{r_j};$$

(5.7.1.4)
$$\eta_{T(0,j)\mathbf{n}} = \eta_{\mathbf{n}} \frac{r_j}{b_j};$$

(5.7.1.5)
$$\eta_{T(j,i)\mathbf{n}} = \frac{\eta_{\mathbf{n}} b_j r_i}{b_i r_j};$$

(5.7.1.6)
$$\mathbf{r}' q = \mathbf{a}' \mathbf{1}.$$

The first three observations are readily verified from the form given in (5.7.1.2), and for the last observation we note that multiplying the traffic equations by the column vector of ones yields

$$\mathbf{r}' \mathbf{1} = \mathbf{a}' \mathbf{1} + \mathbf{a}' P \mathbf{1},$$

so that

$$\mathbf{r}'(\mathbf{1} - P\mathbf{1}) = \mathbf{a}' \mathbf{1},$$

and recalling that $\mathbf{1} - P\mathbf{1} = q$ yields the result.

Now we must show that

$$0 = \sum_{n \in S} \eta_n A_{nm}$$

or

$$\sum_{n \neq m} \eta_n A_{nm} = -\eta_m A_{mm}$$

(5.7.1.7)
$$= \eta_m \left(\sum_{i=1}^{k} a_i + \sum_{j:m_j>0} b_j \right).$$

We break $\sum_{n \neq m} \eta_n A_{nm}$ into three pieces corresponding to transitions which yield internal swaps of position, transitions resulting from departures and transitions resulting from arrivals. Using the first observation (5.7.1.3), for arrivals, we have

$$\sum_{j:m_j>0} \eta_{T(j,0)m} A_{T(j,0)m,m} = \eta_m \sum_{j:m_j>0} \frac{b_j a_j}{r_j}.$$

Using the second observation (5.7.1.4), for departures, we have

$$\sum_{j=1}^{k} \eta_{T(0,j)m} A_{T(0,j)m,m} = \sum_{j=1}^{k} \eta_m \frac{r_j}{b_j} b_j q_j$$

$$= \eta_m \sum_{j=1}^{k} r_j q_j.$$

Corresponding to internal swaps of position, we have the terms

$$\sum_{(i,j):m_j>0} \eta_{T(j,i)m} A_{T(j,i)m,m} = \sum_{(i,j):m_j>0} \eta_m \frac{b_j}{b_i} \frac{r_i}{r_j} b_i p_{ij}$$

(from the third observation (5.7.1.5))

$$= \eta_m \sum_{j:m_j>0} \frac{b_j}{r_j} \sum_{i=1}^{k} r_i p_{ij}$$

$$= \eta_m \sum_{j:m_j>0} \frac{b_j}{r_j} (r_j - a_j)$$

(from the traffic equations)

$$= \eta_m \sum_{j:m_j>0} b_j \left(1 - \frac{a_j}{r_j}\right).$$

If we now add terms corresponding to arrivals, departures and internal swaps of positions and use the fourth observation, we verify (5.7.1.7), which verifies the form of the product solution.

Observe that when a stationary probability distribution exists, the distribution is a product of geometric distributions. Each node acts as an M/M/1 queue with the jth node having effective input rate r_j and service rate b_j.

Example 5.7.4.　Queues in Series. As a simple example, suppose there are k nodes in a series so that customers must traverse the network according to the path $1 \to 2 \to \cdots \to k$. After node k, a customer departs. We may suppose $a_1 = a$ corresponding to a Poisson input of rate a to the network at node 1 and $a_2 = \cdots = a_k = 0$, which says customers only enter at node 1. Assume that the service rate at node j is b_j and that service times are exponentially distributed. The routing matrix is $p_{i,i+1} = 1$, $i = 1, \ldots, k-1$ and $p_{k,\Delta} = 1$, the last probability indicating that from node k the customer leaves the system. For the throughputs, we guess that $r_j = r$, and, in fact, it seems obvious that $r_j = a$. It is easy to verify that this is indeed the solution of the traffic equations. Thus the invariant measure is

$$\eta_{\mathbf{n}} = \prod_{j=1}^{k} \left(\frac{r_j}{b_j}\right)^{n_j} = \prod_{j=1}^{k} (\rho_j)^{n_j},$$

where $\rho_j = a/b_j$. When $\rho_j < 1$ for all j, we get a stationary probability distribution, which is a product of geometric distributions, namely,

$$\lim_{t \to \infty} P[\mathbf{Q}(t) = \mathbf{n}] = \prod_{j=1}^{k} (1 - \rho_j)\rho_j^{n_j}.$$

5.8. TIME DEPENDENT SOLUTIONS*.

Some methods of solving forward or backward equations to obtain the matrix of transition functions $P = (p_{ij}(t))$ were discussed in Section 5.4, and Laplace transform techniques were discussed and illustrated in Section 5.6. In this section we illustrate a generating function technique

---

* This section contains material requiring some familiarity with partial differential equations. A rapid review of the technique needed for solving partial differential equations is included. Beginning readers may skip this section on the first reading.

which converts the linked system of differential equations represented by the forward equations into a partial differential equation. We illustrate this technique with the pure birth process whose time-dependent solution by this method presents some challenges but is still fairly simple. A completely different probabilistic analysis of the pure birth process considered as a mixed Poisson process is described in Section 5.11.

Consider then, the pure birth process $\{X(t), t \geq 0\}$, and assume the initial state is 1. The parameters are given for $i \geq 1$ by $\lambda(i) = \lambda i$, and $Q_{i,i+1} = 1$. Thus the generator matrix is

$$A = \begin{pmatrix} 0 & -\lambda & \lambda & 0 & 0 & \cdots \\ 0 & 0 & -2\lambda & 2\lambda & 0 & \cdots \\ 0 & 0 & 0 & -3\lambda & 3\lambda & \cdots \\ \vdots & & \ddots & & \vdots & \end{pmatrix},$$

and the forward equations $P'(t) = P(t)A$ become, starting from 1,

$$P'_{1,n+1}(t) = \sum_{j=0}^{\infty} P_{1j}(t)A_{j,n+1}$$
$$= P_{1n}(t)A_{n,n+1} + P_{1,n+1}(t)A_{n+1,n+1}$$
$$= P_{1n}(t)n\lambda - P_{1,n+1}(t)(n+1)\lambda.$$

Let $q_n(t) = P_{1,n+1}(t)$ be the probability that starting from 1 there are n births up to time t. Then for $n \geq 1$ we have the following system of differential equations:

(5.8.1) $$q'_n(t) = q_{n-1}(t)n\lambda - q_n(t)(n+1)\lambda,$$

with side conditions

$$q_0(0) = 1, \quad q_n(0) = 0, \quad n \geq 1,$$

and

(5.8.2) $$q'_0(t) = p'_{11}(t) = p_{11}(t)A_{11}$$
$$= -q_0(t)\lambda.$$

Define the generating function

$$G(s,t) = \sum_{n=0}^{\infty} q_n(t)s^n.$$

Multiply equation (5.8.1) by s^n and sum over n from 1 to ∞ to get

$$\sum_{n=1}^{\infty} q'_n(t)s^n = \sum_{n=1}^{n} q_{n-1}(t)n\lambda s^n - \sum_{n=1}^{\infty} q_n(t)(n+1)\lambda s^n.$$

On the left, if the sum started at $n = 0$ we would have $\frac{\partial}{\partial t}G(s,t)$. By changing dummy indices on the right, we obtain

$$\frac{\partial}{\partial t}G(s,t) - q'_0(t) = \lambda \sum_{j=0}^{\infty} q_j(t)(j+1)s^{j+1} - \lambda \sum_{j=1}^{\infty} q_j(t)(j+1)s^j$$

$$= \lambda s q_0(t) + \lambda s \sum_{j=1}^{\infty} q_j(t)(j+1)s^j - \lambda \sum_{j=1}^{\infty} q_j(t)(j+1)s^j$$

$$= \lambda s q_0(t) + \lambda(s-1)\sum_{j=1}^{\infty} q_j(t)(j+1)s^j.$$

Using (5.8.2), this becomes

$$\frac{\partial}{\partial t}G(s,t) + q_0(t) = \lambda s q_0(t) + \lambda(s-1)\frac{\partial}{\partial s}(\sum_{j=1}^{\infty} q_j(t)s^{j+1})$$

$$= \lambda s q_0(t) + \lambda(s-1)\frac{\partial}{\partial s}(s(G(s,t) - q_0(t))$$

$$= \lambda s q_0(t) + \lambda(s-1)(G(s,t) - q_0(t) + s\frac{\partial}{\partial s}G(s,t)),$$

and we obtain the partial differential equation

$$(5.8.3) \qquad \frac{\partial}{\partial t}G(s,t) = \lambda(s-1)\left(s\frac{\partial}{\partial s}G(s,t) + G(s,t)\right).$$

We now attempt to solve the partial differential equation (5.8.3) for G and then invert G to obtain $q_n(t)$.

Here is a review of a method for solving some partial differential equations which are likely to arise with Markov processes. (The methods are standard and are explained in standard books on differential equations; the reference on my shelf is Cohen, 1933, page 250ff.) Consider the problem of finding a solution $f(x_1, x_2)$ of an equation of the form

$$(5.8.4) \qquad P_1(x_1, x_2)\frac{\partial f}{\partial x_1} + P_2(x_1, x_2)\frac{\partial f}{\partial x_2} = R(x_1, x_2),$$

where P_1, P_2, R are nice functions of x_1, x_2. We must solve the auxiliary system

(5.8.5)
$$\frac{dx_1}{P_1} = \frac{dx_2}{P_2} = \frac{df}{R},$$

whose general solution is of the form

$$u_1(x_1, x_2) = c_1, \quad u_2(x_1, x_2) = c_2.$$

The general solution of the partial differential equation (5.8.4) is of the form

$$\phi(u_1, u_2) = 0,$$

for ϕ an arbitrary function which must be determined by boundary conditions. Alternatively, the general solution can be expressed as

$$u_2 = \psi(u_1),$$

for ψ an arbitrary function which must be determined.

Let us apply this method to our example (5.8.3) coming from the pure birth process. In this case $f = G, x_1 = s, x_2 = t, P_1 = -\lambda(s-1)s, P_2 = 1,$ and $R = \lambda(s-1)G$. The auxiliary equations (5.8.5) become

$$\frac{dt}{1} = \frac{ds}{\lambda(1-s)s} = \frac{dG}{-\lambda(1-s)G}.$$

The first yields
$$\frac{dt}{1} = \frac{ds}{\lambda(1-s)s},$$

and, since
$$\frac{1}{(1-s)s} = \frac{1}{s} + \frac{1}{1-s},$$

we have for some constant c

$$\lambda t + c = \log s - \log(1-s) = \log(\frac{s}{1-s}),$$

so that
$$u_1(s,t) = -\lambda t + \log(\frac{s}{1-s}).$$

A second equation,
$$\frac{ds}{\lambda(1-s)s} = \frac{dG}{-\lambda(1-s)G},$$

or, equivalently,

$$\frac{ds}{s} = \frac{dG}{-G},$$

yields (treat G simply as a variable)

$$\log s + c = -\log G$$

so that

$$\log sG = u_2 = c.$$

Therefore the general solution is of the form

(5.8.6) $$u_2 = \log sG = \psi(u_1) = \psi(-\lambda t + \log(\frac{s}{1-s})).$$

To determine ψ we use the boundary condition $G(s, 0) = 1$. Substituting $t = 0$ in the previous equation, we obtain

$$\log s = \psi(\log(\frac{s}{1-s})),$$

which allows us to determine ψ. Set

$$y = \log(\frac{s}{1-s}),$$

so that

$$s = \frac{1}{1+e^{-y}},$$

and thus

$$\psi(y) = \log\left(\frac{1}{1+e^{-y}}\right).$$

Using this in (5.8.6) yields

$$sG = \frac{1}{1+e^{\lambda t}\frac{1-s}{s}}$$

$$= \frac{s}{s+e^{\lambda t}(1-s)}$$

$$= \frac{e^{-\lambda t}s}{1-s(1-e^{-\lambda t})},$$

so

$$G = \frac{e^{-\lambda t}}{1-s(1-e^{-\lambda t})}$$

$$= e^{-\lambda t}\sum_{n=0}^{\infty}(1-e^{-\lambda t})^n s^n$$

$$= \sum_{n=0}^{\infty}q_n(t)s^n.$$

This allows us to identify

$$q_n(t) = e^{-\lambda t}(1-e^{-\lambda t})^n, \quad n \geq 0.$$

5.9. REVERSIBILITY.

Reversibility is a concept with many applications in stochastic modelling. It is present in many birth–death processes. It makes the structure of the output processes of many queueing models transparent, which is important since the output of one queue is sometimes needed as the input to a different service facility. Since the publication of Kelly's book (1979), reversibility has assumed ever greater prominence.

We start by considering a discrete time Markov chain $\{X_n, -\infty < n < \infty\}$ with discrete state space S. Note the index set of the process is all integers including the negative ones. This is a convenience when we reverse time. If the index set were $\{0, 1, \dots\}$, then for the reversed time process there would be an awkward barrier at 0 to be dealt with; better to remove the barrier and consider the process on all integers.

We assume $\{X_n\}$ has a transition matrix P, that it is stationary, and that the stationary distribution is $\pi' = (\pi_j, j \in S)$. Thus the finite dimensional joint distributions which determine the distribution of the process are of the form $(k \geq 1, i_1, \dots, i_k \in S)$

$$(5.9.1) \qquad P[X_{n+1} = i_1, \dots, X_{n+k} = i_k] = \pi_{i_1} p_{i_1 i_2} \cdots p_{i_{k-1} i_k}.$$

If you have trouble assimilating the idea of a process with index set $\{\dots, -1, 0, 1, \dots\}$, think of a Markov chain ξ with transition matrix P being started a long time ago. Assume $p_{ij}^{(n)} \to \pi_j$ as $n \to \infty$, and consider the ξ-process started in the remote past at time $-N$. Call this process $\{\xi_j, j \geq -N\}$. If this process has some initial distribution $\mathbf{a}$, then, as $N \to \infty$, the finite dimensional distributions of $\{\xi_j\}$ converge to those of $\{X_n\}$, since for any fixed n as $N \to \infty$

$$P[\xi_{n+1} = i_1, \dots, \xi_{n+k} = i_k] = \sum_{j \in S} a_j p_{ji_1}^{(n+N+2)} p_{i_1 i_2} \cdots p_{i_{k-1} i_k}$$

$$\to \pi_{i_1} p_{i_1 i_2} \cdots p_{i_{k-1} i_k}$$

$$= P[X_{n+1} = i_1, \dots, X_{n+k} = i_k].$$

Sometimes this procedure of letting the starting point drift to the remote past is termed *letting the process reach equilibrium*, but this is very vague language.

Set $Y_n = X_{-n}$ and consider the reversed time process

$$\{Y_n, -\infty < n < \infty\}.$$

Observe first that it is a Markov chain, since

$$P[Y_{n+k+1} = i_{k+1}|Y_{n+1} = i_1, \ldots, Y_{n+k} = i_k]$$
$$= P[X_{-(n+k+1)} = i_{k+1}|X_{-(n+1)} = i_1, \ldots, X_{-(n+k)} = i_k]$$
$$= \frac{P[X_{-(n+k+1)} = i_{k+1}, X_{-(n+1)} = i_1, \ldots, X_{-(n+k)} = i_k]}{P[X_{-(n+1)} = i_1, \ldots, X_{-(n+k)} = i_k]}$$
$$= \frac{\pi_{i_{k+1}} p_{i_{k+1} i_k} \cdots p_{i_2 i_1}}{\pi_{i_k} p_{i_k i_{k-1}} \cdots p_{i_2 i_1}}$$
$$= \frac{\pi_{i_{k+1}}}{\pi_{i_k}} p_{i_{k+1} i_k}$$
$$= P[Y_{n+k+1} = i_{k+1}|Y_{n+k} = i_k],$$

which shows that $\{Y_n\}$ is a Markov chain with stationary transition probabilities and transition matrix

(5.9.2)
$$P[Y_1 = j|Y_0 = i] = \frac{\pi_j}{\pi_i} p_{ji}.$$

Note also that the reversed process $\{Y_n\}$ is stationary. This follows since if the distribution of the vector $(X_{n+1}, \ldots, X_{n+k})$ is independent of n for any k, then the same is true of the vector with Y's replacing the X's. Furthermore $\{Y_n\}$ has stationary probabilites π, since

$$\sum_i \pi_i P[Y_1 = j|Y_0 = i] = \sum_i \pi_i \frac{\pi_j}{\pi_i} p_{ji}$$
$$= \sum_i \pi_j p_{ji} = \pi_j.$$

Next we seek a condition for the process $\{X_n\}$ to be reversible; we want both $\{X_n\}$ and $\{Y_n\} = \{X_{-n}\}$ to have the same finite dimensional distributions, a property which we write as

$$\{X_n\} \overset{d}{=} \{Y_n\}.$$

We know both processes have the same stationary distribution, so examining the form of the finite dimensional distributions in (5.9.1) yields the conclusion that reversibility will hold if the transition probabilities for both the process and the time reversed process are the same. From (5.9.1), we need

$$p_{ij} = P[X_1 = j|X_0 = i] = P[Y_1 = j|Y_0 = i] = \frac{\pi_j}{\pi_i} p_{ji},$$

and we see that a stationary Markov chain indexed by all the integers is time reversible if and only iff

(5.9.3)
$$\pi_j p_{ji} = \pi_i p_{ij}$$

for all $i, j \in S$.

Now consider a continuous time regular Markov chain $\{X(t), -\infty < t < \infty\}$ with countable state space S and transition matrix function $P(t) = \{P_{ij}(t), i, j \in S\}$. Suppose this process is stationary with stationary distribution η so that $(k \geq 1, 0 \leq t_1 \leq \cdots \leq t_k)$

(5.9.4)
$$P[X(t_1) = i_1, \ldots, X(t_k) = i_k] = \eta_{i_1} P_{i_1 i_2}(t_2 - t_1) \ldots P_{i_{k-1} i_k}(t_k - t_{k-1}).$$

As in the discrete case we find that the reversed process $Y(t) = X(-t)$ is also a Markov chain with stationary distribution η and transition matrix function

(5.9.5)
$$P[Y(t + s) = j | Y(t) = i] = \frac{\eta_j}{\eta_i} P_{ji}(s).$$

Furthermore, $\{X(t)\}$ is reversible, that is, the process X and the process Y have the same finite dimensional distributions, if and only if

(5.9.6)
$$\eta_j P_{ji}(s) = \eta_i P_{ij}(s) \quad \forall i, j \in S, \ s \geq 0.$$

Condition (5.9.6) is impractical, since it depends on the transition matrix function $P(t)$ which is frequently unknown, so we wish to recast this condition in terms of the generator matrix A of the Markov chain. If we divide by s in (5.9.6), let $s \to 0$ and use Proposition 5.4.3, we see

(5.9.7)
$$\eta_j A_{ji} = \eta_i A_{ij}, \quad i \neq j.$$

Conversely, assume (5.9.7). The distribution of the regular Markov chain $\{Y(t)\}$ is determined by η and the generator matrix, so to show $\{X(t)\} \overset{d}{=} \{Y(t)\}$ it suffices to show the generator matrix of Y is also A. Again from Proposition 5.4.2 and (5.9.5), we have

$$\lim_{s \to 0} \frac{P[Y(s) = j | Y(0) = i]}{s} = \lim_{s \to 0} \frac{\eta_j}{\eta_i} \frac{P_{ji}(s)}{s}$$
$$= \frac{\eta_j}{\eta_i} A_{ji}$$

Applying (5.9.7)

$$= \frac{\eta_i}{\eta_i} A_{ij} = A_{ij}.$$

Thus we have verified the following:

Proposition 5.9.1. *A stationary Markov chain* $\{X(t), -\infty < t < \infty\}$ *is reversible, i.e.,*

$$\{X(t)\} \overset{d}{=} \{X(-t)\},$$

iff the detailed balance condition (5.9.7) holds.

Fortunately, a large and useful class of processes can be immediately identified as reversible.

Proposition 5.9.2. *A stationary birth-death process on the state space* $\{0, 1, \dots\}$ *is time reversible.*

Proof. Recall the form of the stationary distribution

$$\eta_0 = c, \quad \eta_j = c\frac{\prod_{i=0}^{j-1} \lambda_i}{\prod_{i=1}^{j} \mu_i}, \quad j \geq 1,$$

where c makes $\sum_j \eta_j = 1$. For instance, we have the following case of (5.9.7) for $j \geq 1$,

$$\eta_j A_{j,j+1} = c\frac{\prod_{i=0}^{j-1} \lambda_i}{\prod_{i=1}^{j} \mu_i}\lambda_j$$
$$= \eta_{j+1} A_{j+1,j}.$$

The case $j = 0$ is easily checked, and there are no other non-trivial cases. ∎

This has an immediate application to the output process of certain Markovian queues.

Proposition 5.9.3. *The output process from a stationary M/M/s queue is Poisson with parameter* a. *Let* $Q(t)$ *be the number in the system at time* t, *and let* $D(s, t]$ *be the number of departures from the system in the time interval* $(s, t]$. *Then for any* t_0,

$$\{D(s, t_0], s < t_0\} \text{ and } Q(t_0)$$

are independent.

Proof. A stationary M/M/s queueing model is a stationary birth-death process and hence is time reversible, so $\{Q(t)\} \overset{d}{=} \{Q(-t)\}$. This means that the times of upward jumps of $\{Q(t)\}$, which is a Poisson process of rate a, are equal in distribution to the times of upward jumps of $\{Q(-t)\}$. But the times of upward jumps of $\{Q(-t)\}$ are the times of downward jumps of $\{Q(t)\}$ due to the effect of time reversal. So the times of downward jumps (which correspond to departures resulting in the number in the system dropping) also form a Poisson process of rate a. What comes in must go out in equilibrium.

Now we extend this argument slightly. The departure process prior to t_0 is the times of downward jumps in the path $\{Q(s), s < t_0\}$. Set $v = -s$, and the departure process is equal to the times of upward jumps in the path of the reversed process $\{Q(-v), v > -t_0\}$. Because the reversed process has the same distribution as the original process, these Poisson upward jumps are the input to a M/M/s queueing model. Since these upward jumps represent future arrivals for the reversed process, they are independent of $Q(-(-t_0)) = Q(t_0)$. ∎

Consider the application of these ideas to tandem queues. These are queues in series, arranged so that the output from one queue becomes the input to the next queue. For simplicity, assume we have two queueing systems, each with a single server, arranged in series. Assume the first queue has a Poisson input of rate a_1 and that the server serves at an exponential rate b_1. The second queue has the output from the first queue as its input, and the server serves at an exponential rate b_2. We suppose the stochastic process $Q_1(\cdot)$, representing the number in the first queue, is stationary, so that the output from the first queue (identical to the input for the second queue) is Poisson, parameter a_1. Assume $\rho_i = a_1/b_i < 1$ for $i = 1, 2$ to assure stability. Let $Q_i(t)$ be the number in the ith queue at time t, $i = 1, 2$. The bivariate process $\{\mathbf{Q}(t) = (Q_1(t), Q_2(t)), t \geq 0\}$ is a Markov chain. We have the following result about the series of queues in equilibrium.

Proposition 5.9.4. *The stationary distribution of the Markov chain* $\{\mathbf{Q}(t), t \geq 0\} = \{(Q_1(t), Q_2(t)), t \geq 0\}$ *is the product distribution*

$$\eta_{(n,m)} = (1 - \rho_1)\rho_1^n (1 - \rho_2)\rho_2^m,$$

so that

$$\lim_{t \to \infty} P[Q_1(t) = n, Q_2(t) = m] = \lim_{t \to \infty} P[Q_1(t) = n] P[Q_2(t) = m]$$

$$= \eta_{(n,m)}.$$

This does not say that the two processes $\{Q_1(t)\}$ and $\{Q_2(t)\}$ are independent. Under the stationary distribution, for a fixed t, the two random variables $Q_1(t)$ and $Q_2(t)$ are independent.

Proof. To verify η is the stationary distribution of $\{\mathbf{Q}(t)\}$, it suffices to start the process $\mathbf{Q}$ with distribution η and show that, for any t, η is also the distribution of $\mathbf{Q}(t)$. Take the process $\mathbf{Q}$ and start it off at time 0 with initial distribution η. Since η has a product form we have $Q_1(0)$

independent of $Q_2(0)$. Since the Markov chain Q_1 has stationary distribution $(1 - \rho_1)\rho_1^n$, we have started Q_1 with its stationary distribution, and therefore $\{Q_1(t), t \geq 0\}$ is stationary. We know from Proposition 5.9.3 that, for any t, the departure process $D_1(\cdot)$ from the first queue is Poisson rate a_1 and $\{D_1(s,t], 0 \leq s < t\}$ is independent of $Q_1(t)$. However, the queue length $Q_2(t)$ in the second queue is dependent on its input $\{D_1(s,t], 0 \leq s < t\}$ (which is independent of $Q_1(t)$) and upon the actions of the servers in the second service facility which are independent of the first server. Therefore, $Q_1(t)$ and $Q_2(t)$ are independent.

The Markov chain Q_1 was started with its stationary distribution and hence is stationary, so $Q_1(0) \stackrel{d}{=} Q_1(t)$. The second queue has Poisson input of rate a_1 (being the output from the first queue, which is in equilibrium) and an initial distribution which is geometric with parameter ρ_2. Therefore, the second process Q_2 is also a stationary queueing model with $Q_2(0) \stackrel{d}{=} Q_2(t)$. Hence

$$P_\eta[Q_1(t) = n, Q_2(t) = m] = P_\eta[Q_1(t) = n]P_\eta[Q_2(t) = m],$$

since $Q_1(t)$ and $Q_2(t)$ are independent,

$$= (1 - \rho_1)\rho_1^n(1 - \rho_2)\rho_2^m$$

since $Q_1(0) \stackrel{d}{=} Q_1(t)$ and $Q_2(0) \stackrel{d}{=} Q_2(t)$

$$= \eta_{(n,m)}. \blacksquare$$

This result has an obvious generalization to r queues in series.

5.10. Uniformizable chains.

Recall the example at the end of Section 5.1 of a uniformizable chain which consisted of taking a discrete time Markov chain and allowing it to jump at the times of an independent Poisson process. We now review this construction. Suppose $\{X_n\}$ is the discrete time Markov chain with transition matrix $K = (K_{ij})$ and $\{N(t), t \geq 0\}$ is a homogeneous Poisson process of rate α independent of $\{X_n\}$. Define

$$X^{\#}(t) = X_{N(t)}.$$

The transition matrix function of this process is

$$(5.10.1) \qquad P_{ij}(t) = \sum_{n=0}^{\infty} \frac{e^{-\alpha t}(\alpha t)^n}{n!} K_{ij}^{(n)}.$$

What is the generator of this process $X^\#$? Differentiating formally in (5.10.1) we obtain

$$P'_{ij}(t) = -\alpha e^{-\alpha t}\delta_{ij} + \sum_{n=1}^{\infty}\frac{e^{-\alpha t}(\alpha t)^{n-1}}{(n-1)!}\alpha K_{ij}^{(n)} - \sum_{n=1}^{\infty}\frac{(\alpha t)^n}{n!}\alpha e^{-\alpha t}K_{ij}^{(n)},$$

and setting $t = 0$ we get

$$(5.10.2) \qquad A_{ij}^\# = \begin{cases} P'_{ij}(0) = \alpha K_{ij}, & \text{if } i \neq j, \\ P'_{ii}(0) = \alpha K_{ii} - \alpha\delta_{ii}, & \text{if } i = j. \end{cases}$$

In matrix form, this equation is

$$(5.10.2') \qquad\qquad A^\# = \alpha(K - I).$$

To verify this from first principles, let $\{X_n^\#\}$ be the Markov chain derived from $\{X_n\}$ by only observing transitions between different states. To make this precise, we define

$$n(0) = 0,$$
$$n(1) = \inf\{k > 0 : X_k \neq X_0\}$$
$$n(2) = \inf\{k > n(1) : X_k \neq X_{n(1)}\}$$
$$\vdots$$

and

$$\{X_k^\#\} = \{X_{n(k)}\}.$$

Exercise 5.8 helps us calculate the transition matrix $Q^\#$ of the Markov chain $X^\#$, and we have

$$Q_{ij}^\# = \begin{cases} 0, & \text{if } i = j \\ \frac{K_{ij}}{1-K_{ii}}, & \text{if } i \neq j. \end{cases}$$

To calculate the holding time parameter in state i, $\lambda^\#(i)$, we observe that starting in state i, the process $X^\#(t)$ leaves state i after $\sum_{j=1}^{n(1)} E_j(\alpha)$, where $n(1)$ is independent of the iid exponential random variables $\{E_j(\alpha)\}$ with parameter α. Thus

$$\lambda^\#(i) = \left(\mathbf{E}\sum_{j=1}^{n(1)} E_j(\alpha)\right)^{-1} = (\mathbf{E}(n(1))\mathbf{E}(E_1(\alpha)))^{-1},$$

so

$$\lambda^{\#}(i) = \alpha(1 - K_{ii}).$$

$\lambda^{\#}(i)$ and $K^{\#}$ are the infinitesimal parameters corresponding to $X^{\#}(t)$, and from these we easily get $A^{\#}$ and verify equation (5.10.2).

Now suppose we are given some other continuous time Markov chain $\{X(t)\}$ with generator matrix A. The only assumption about this process $X(t)$ is that the generator A satisfies the assumption that for all i in the state space, the holding times $\lambda(i) = -A_{ii} < \alpha < \infty$ for some finite α. With this assumption, we will find that we can mimic the process $X(t)$ with a Poisson paced process as discussed in the previous paragraph. To do this, define the matrix

$$K = \alpha^{-1}A + I,$$

so that

$$K_{ii} = 1 + \alpha^{-1}A_{ii} = 1 - \alpha^{-1}\lambda(i) > 0,$$

and for $i \neq j$

$$K_{ij} = A_{ij}/\alpha \geq 0.$$

Furthermore, for all i in the state space,

$$\sum_{j} K_{ij} = \alpha^{-1}\sum_{j} A_{ij} + \sum_{j} \delta_{ij} = 0 + 1 = 1.$$

Thus K is a stochastic matrix.

Let us now use this K and α to construct the continuous time process $\{X^{\#}(t), t \geq 0\}$ as described at the beginning of this section. We know the generator matrix of $X^{\#}$ is $A^{\#} = \alpha(K - I)$; from the definition of K, this is A, so $A = A^{\#}$. Thus the two processses $X(t)$ and $X^{\#}(t)$ have the same generator. Assuming they are regular, this means that both processes have the same distribution.

If a continuous time Markov chain can be constructed by placing the transitions of a discrete time Markov chain at times of Poisson jumps, then we call the continuous time process *uniformizable*. We have checked that any process with bounded holding time parameters $\{\lambda(i), i \in S\}$ is uniformizable. In particular, a finite state Markov chain is uniformizable. This provides a way for us to simulate a finite Markov chain in continuous time: Simulate a discrete time Markov chain and put the transitions at points of an independent Poisson process.

5.11. THE LINEAR BIRTH PROCESS AS A POINT PROCESS*.

In this section we treat the linear birth process as a point process and show that we may consider the linear birth process as a mixed Poisson process; i.e., we will express the linear birth process as a non-homogeneous Poisson process whose time scale has been subjected to multiplication by a non-negative random variable. This result was learned from Kendall, 1966 (see also Waugh, 1971). It is included because of its surpassing beauty.

Recall that the linear birth process has embedded jump chain

$$\{X_n, n \geq 1\} = \{1, 2, 3, \ldots\}$$

equal to the deterministically monotone Markov chain. The jump times $\{T_n\}$ have the structure

$$(5.11.1) \qquad \{T_{n+1} - T_n, n \geq 0\} \stackrel{d}{=} \left(\frac{E_1}{\lambda}, \frac{E_2}{2\lambda}, \frac{E_3}{3\lambda}, \cdots\right),$$

where $\{E_j, j \geq 1\}$ are iid, $\lambda > 0$ and $P[E_j > x] = e^{-x}$, $x > 0$. Set $T_0 = 0$. The number of births (as opposed to the population size) in $(0, t]$ is

$$(5.11.2) \qquad N(t) = \sum_{n=1}^{\infty} 1_{[T_n \leq t]},$$

so that the point process under consideration can be expressed in the notation of Chapter 4 as

$$(5.11.3) \qquad N(t) = \sum_{n=1}^{\infty} \epsilon_{T_n}((0, t]).$$

We explore how N can be related to a non-homogeneous Poisson process.

We need the following lemma, which is usually attributed to Renyi, 1953.

Lemma 5.11.1. *If $Z_1, \ldots, Z_n$ are iid, exponentially distributed with parameter λ, and if $Z_{(1)} \leq \cdots \leq Z_{(n)}$ represent the order statistics $(Z_{(1)} = \wedge_{i=1}^n Z_i, \ldots Z_{(n)} = \vee_{j=1}^n Z_j)$ then the spacings satisfy*

$$(Z_{(1)}, Z_{(2)} - Z_{(1)}, \ldots, Z_{(n)} - Z_{(n-1)}) \stackrel{d}{=} \left(\frac{Z_n}{n}, \frac{Z_{n-1}}{n-1}, \ldots, \frac{Z_1}{1}\right)$$

---

* This section contains advanced material.

so that the spacings are independent and exponentially distributed (with varying parameters).

Proof. The formal method of proof is to write down the joint density of $Z_{(1)}, ..., Z_{(n)}$ and then transform to get the joint density of the spacings. (See Feller, 1971, page 19, for example.) More heuristically, consider the following argument. At time 0, initiate the lifetimes of n machines. The lifetimes of the n machines are represented by the random variables $Z_1, ..., Z_n$. The first failure occurs at $Z_{(1)}$. After $Z_{(1)}$, the remaining $n-1$ unfailed machines have exponentially distributed lifetimes because of the forgetfulness property. The elapsed time until the next failure is distributed as the minimum of $n-1$ exponentially distributed random variables, so is exponentially distributed with parameter $\lambda(n-1)$. Continue this process until all the machines have failed. Independence of the spacings results from the forgetfulness property of the exponential density. ∎

We conclude from this lemma that, for $n \geq 1$,

$$(T_1, T_2 - T_1, ..., T_n - T_{n-1}) \stackrel{d}{=} \left(Z_{(n)} - Z_{(n-1)}, Z_{(n-1)} - Z_{(n-2)}, \ldots, Z_{(1)} \right),$$

where the spacings on the right are independent. Thus for any $n \geq 1$

$$(5.11.4) \quad (T_1, T_2, \ldots, T_n) \stackrel{d}{=} \left(Z_{(n)} - Z_{(n-1)}, Z_{(n)} - Z_{(n-2)}, \ldots, Z_{(n)} \right).$$

In particular,

$$T_n \stackrel{d}{=} Z_{(n)} = \bigvee_{i=1}^n Z_i,$$

and therefore, for $x > 0$,

$$
\begin{aligned}
P[T_n \leq x] &= P[\vee_{j=1}^n Z_j \leq x] \\
&= P\{\cap_{j=1}^n [Z_j \leq x]\} = (P\{Z_1 \leq x\})^n \\
(5.11.5) \quad &= (1 - e^{-\lambda x})^n.
\end{aligned}
$$

Thus using (5.11.2) and (5.11.5)

$$
\begin{aligned}
EN(t) &= E \sum_{n=1}^\infty P[T_n \leq t] \\
&= \frac{1 - e^{-\lambda t}}{e^{-\lambda t}} \sim e^{\lambda t},
\end{aligned}
$$

as $t \to \infty$.

Proposition 5.11.2. *There exists a unit exponential random variable W such that almost surely*

$$(a) \ \lim_{t \to \infty} \frac{N(t)}{EN(t)} = \lim_{t \to \infty} \frac{N(t)}{e^{\lambda t}} = W$$

and

$$(b) \ \lim_{n \to \infty} (\lambda T_n - \log n) = -\log W.$$

Remark. It turns out that $\{N(t)/EN(t), t \geq 0\}$ is a martingale and the martingale convergence theorem makes short work of (a). Here is a proof by classical means.

Proof. Observe first from (5.11.1) that

$$ET_n = \frac{1}{\lambda} \sum_{j=1}^{n} \frac{1}{j}$$

and

$$\text{Var}(T_n) = \frac{1}{\lambda^2} \sum_{j=1}^{n} \frac{1}{j^2}.$$

Since $\sum_{j=1}^{\infty} j^{-2} < \infty$, we obtain from the classical Kolmogorov convergence criterion (Billingsley, 1986, p. 298) that

$$\lim_{n \to \infty} (T_n - ET_n) = \lim_{n \to \infty} \sum_{j=1}^{n} ((T_j - T_{j-1}) - E(T_j - T_{j-1}))$$

exists finite almost surely. Since as $n \to \infty$ we have $ET_n - \frac{1}{\lambda} \log n$ converges to a constant (Abramowitz and Stegun, 1970, page 255), we get that the limit in (b) exists. Call this limit $-\log W$.

From the definition of $N(t)$ in (5.11.2) we have

(5.11.6) $$T_{N(t)} \leq t < T_{N(t)+1}.$$

Since the linear birth process is regular, $T_n \to \infty$, and therefore

$$P[N(t) \geq n] = P[T_n \leq t] = (1 - e^{-\lambda t})^n \to 1$$

as $t \to \infty$. Thus as $t \to \infty$, $N(t)$ converges in probability to ∞, and being monotone, $N(t)$ must thus converge almost surely to ∞. In (b) replace n by $N(t)$ to get

$$\lim_{t \to \infty} (\lambda T_{N(t)} - \log N(t)) = -\log W.$$

It is also true that if we replace n by $N(t) + 1$ in (b) we get

$$\lim_{t \to \infty} \lambda T_{N(t)+1} - \log(N(t) + 1) = -\log W,$$

and, since

$$\log(N(t) + 1) - \log N(t) = \log \frac{N(t) + 1}{N(t)} \to \log 1 = 0,$$

we get

$$\lim_{t \to \infty} \lambda T_{N(t)+1} - \log N(t) = -\log W.$$

Now write the sandwich using (5.11.6):

$$\lambda T_{N(t)} - \log N(t) \le \lambda t - \log N(t) \le \lambda T_{N(t)+1} - \log N(t),$$

and since the extreme sides of the sandwich converge to $-\log W$ we conclude

$$\lim_{t \to \infty} \lambda t - \log N(t) = -\log W.$$

Exponentiating gives (a).

It remains to prove W is unit exponential. Let us try brute force. We have, for $x > 0$,

$$P[e^{-\lambda t} N(t) > x] = P[N(t) > x e^{\lambda t}]$$

and since $N(t)$ is integer valued this is

$$= P[N(t) > [x e^{\lambda t}]],$$

where $[y]$ is the greatest integer $\le y$. From the definition of $N(t)$ in (5.11.2), the previous probability is

$$= P\left[T_{[x e^{\lambda t}]+1} \le t\right].$$

Applying (5.11.5), this is

$$= (1 - e^{-\lambda t})^{[x e^{\lambda t}]+1}$$

$$= \left\{ (1 - \frac{1}{e^{\lambda t}})^{e^{\lambda t}} \right\}^{([x e^{\lambda t}]+1)/e^{\lambda t}}.$$

Since $[x e^{\lambda t}]/e^{\lambda t} \to x$ as $t \to \infty$, we conclude

$$\lim_{t \to \infty} P[\frac{N(t)}{e^{\lambda t}} > x] = (e^{-1})^{\lim_{t \to \infty} [x e^{\lambda t}]/e^{\lambda t}}$$

$$= (e^{-1})^x = e^{-x},$$

as desired. ∎

In order to understand the dependence of N and W better, we investigate some conditional distributions.

Proposition 5.11.3.

(a) *The conditional density of* $T_1, \ldots, T_{m-1}$ *given* T_m $(m \geq 2)$ *is*

(5.11.7)
$$f(t_1, \ldots, t_{m-1}|t_m) = \begin{cases} \frac{(m-1)!\lambda^{m-1}e^{\lambda \Sigma_{i=1}^{m-1} t_i}}{(\exp\{\lambda t_m\}-1)^{m-1}} & \text{if } 0 \leq t_1 \leq \cdots \leq t_{m-1} \leq t_m, \\ 0 & \text{otherwise.} \end{cases}$$

(b) *If* g *is a bounded function on* R^{m-1}

(5.11.8) $\quad E\left(g(T_1, T_2 - T_1, \ldots, T_{m-1} - T_{m-2})|T_m, W\right)$
$$= E\left(g(T_1, \ldots, T_{m-1} - T_{m-2})|T_m\right).$$

Proof. (a) The joint density of $T_1, T_2 - T_1, \ldots, T_m - T_{m-1}$ is a product of exponential densities

$$\lambda e^{-\lambda t_1} 2\lambda e^{-2\lambda t_2} \ldots (m-1)\lambda e^{-(m-1)\lambda t_{m-1}} m\lambda e^{-m\lambda t_m}.$$

From this, by transformation of variables, we have that the joint density of $T_1, \ldots, T_m$ is

$$f(t_1, \ldots, t_m) = \lambda e^{-\lambda t_1} 2\lambda e^{-2\lambda(t_2-t_1)} \ldots m\lambda e^{-m\lambda(t_m - t_{m-1})}$$
$$= \lambda^m m! \exp\{\lambda \sum_{i=1}^{m-1} t_i\} \exp\{-\lambda m t_m\},$$

for $0 \leq t_1 \leq \cdots \leq t_m$. To get the density of T_m we can use (5.11.5) to get

$$f(t_m) = \frac{d}{dt_m} P[T_m \leq t_m] = \frac{d}{dt_m}(1 - e^{-\lambda t_m})^m$$
$$= m(1 - e^{-\lambda t_m})^{m-1}\lambda e^{-\lambda t_m},$$

for $t_m > 0$. From the last two formulas we get

$$f(t_1, \ldots, t_{m-1}|t_m) = f(t_1, \ldots, t_m)/f(t_m),$$

which is (a).

(b) From Proposition 5.11.2

$$-\log W = \lim_{n \to \infty} T_n - \lambda^{-1} \log n$$
$$= \lim_{n \to \infty} \sum_{j=1}^{n}(T_j - T_{j-1}) - \lambda^{-1} \log n$$
$$= \lim_{n \to \infty} \sum_{j=m+1}^{n} (T_j - T_{j-1}) - \lambda^{-1} \log n + T_m$$
$$= T'_m + T_m$$

for any integer m, where

$$T'_m = \lim_{n \to \infty} \sum_{j=m+1}^{n} (T_j - T_{j-1}) - \lambda^{-1} \log n$$

is independent of T_m. If $\sigma(\chi_i, i \in T)$ is the σ-algebra generated by the variables $\chi_i, i \in T$, we have

$$\sigma(T_m, W) = \sigma(T_m, -\log W)$$
$$= \sigma(T_m, T_m + T'_m)$$
$$= \sigma(T_m, T'_m)$$

so that

$$E\left(g(T_1, T_2 - T_1, \ldots, T_{m-1} - T_{m-2}) | T_m, W\right)$$
$$= E\left(g(T_1, T_2 - T_1, \ldots, T_{m-1} - T_{m-2}) | T_m, T'_m\right)$$
$$= E\left(g(T_1, T_2 - T_1, \ldots, T_{m-1} - T_{m-2}) | T_m\right),$$

since T'_m and $g(T_1, T_2 - T_1, \ldots, T_{m-1} - T_{m-2})$ are independent. ∎

We now verify that the counting function N of birth times is a non-homogeneous Poisson process conditional on W.

Theorem 5.11.4. *The linear birth point process N is a mixed Poisson process. Conditional on W, N is a non-homogeneous Poisson process with local intensity $W \lambda e^{\lambda t}$. The Laplace functional of N is ($f \geq 0$)*

(5.11.9) $$\psi_N(f) = E \exp\left\{ -\int_0^\infty (1 - e^{-f(t)}) W \lambda e^{\lambda t} dt \right\}.$$

Proof. It suffices to show that for f continuous with compact support and $f \geq 0$,

(5.11.10)
$$E\left(\exp\left\{ -\int_0^\infty f(t) N(dt) \right\} | W \right) = \exp\left\{ -\int_0^\infty (1 - e^{-f(t)}) \lambda W e^{\lambda t} dt \right\}.$$

Suppose the support of f is contained in $[0, K]$ so that $f(x) = 0$, $x > K$. Then

$$E\left(\exp\left\{ -\int_0^\infty f(t) N(dt) \right\} | W \right)$$

$$= E\left(\exp\left\{ -\sum_{n=1}^\infty f(T_n) \right\} | W \right)$$

$$= \lim_{m \to \infty} E\left(\exp\left\{ -\sum_{n=1}^{m-1} f(T_n) \right\} | W \right)$$

$$= \lim_{m \to \infty} E\left(E\left(\exp\left\{ -\sum_{n=1}^{m-1} f(T_n) \right\} | W, T_m \right) | W \right).$$

Applying (5.11.8), this is

$$= \lim_{m \to \infty} E\left(E\left(\exp\left\{-\sum_{n=1}^{m-1} f(T_n)\right\} |T_m\right) |W\right).$$

From (5.11.7),

$$E\left(\exp\left\{-\sum_{n=1}^{m-1} f(T_n)\right\} |T_m = t_m\right)$$

$$= \int \cdots \int_{0 < t_1 < \cdots < t_{m-1} < t_m}$$

$$\frac{\exp\left\{-\sum_{n=1}^{m-1} f(t_n) + \sum_{n=1}^{m-1} \lambda t_n\right\} \lambda^{m-1}(m-1)!}{(e^{\lambda t_m} - 1)^{m-1}} dt_1 \ldots dt_{m-1}.$$

Since the integrand is symmetric in $t_1, \ldots, t_{m-1}$, this is

$$= \frac{\lambda^{m-1}}{(e^{\lambda t_m} - 1)^{m-1}} \left(\int_0^{t_m} e^{-f(t) + \lambda t} dt\right)^{m-1}.$$

Because $f(x) = 0$ for $x > K$, for m large enough that $T_m > K$, we have

$$E\left(\exp\left\{-\sum_{n=1}^{m-1} f(T_n)\right\} |T_m\right)$$

$$= \frac{\lambda^{m-1}}{(e^{\lambda T_m} - 1)^{m-1}} \left(\int_0^K e^{-f(t) + \lambda t} dt + \int_K^{T_m} e^{\lambda t} dt\right)^{m-1}$$

$$= \frac{\left(\int_0^K e^{-f(t)} \lambda e^{\lambda t} dt + e^{\lambda T_m} - e^{\lambda K}\right)^{m-1}}{(e^{\lambda T_m} - 1)^{m-1}}.$$

The denominator equals

$$(e^{\lambda T_m})^{m-1} \left(1 - \frac{1}{e^{\lambda T_m}}\right)^{e^{\lambda T_m} e^{-\lambda T_m}(m-1)},$$

which, because of Proposition 5.11.2(b), is asymptotic to

$$(e^{\lambda T_m})^{m-1} e^{-W}.$$

Thus, as $m \to \infty$,

$$E\left(\exp\left\{-\sum_{n=1}^{m-1} f(T_n)\right\} | T_m\right)$$

$$\sim \left(\frac{-\int_0^K (1 - e^{-f(t)})\lambda e^{\lambda t} dt + e^{\lambda T_m} - 1}{e^{\lambda T_m}}\right)^{m-1} e^W$$

$$= \left(1 - \frac{1 + \int_0^K (1 - e^{-f(t)})\lambda e^{\lambda t} dt}{e^{\lambda T_m}}\right)^{e^{\lambda T_m} e^{-\lambda T_m}(m-1)} e^W$$

$$\to \exp\left\{-1 - \int_0^K (1 - e^{-f(t)})\lambda e^{\lambda t} dt W\right\} e^W$$

$$= \exp\{-\int_0^\infty (1 - e^{-f(x)}) W \lambda e^{\lambda t} dt,$$

since from Proposition 5.11.2(b) we have $e^{-\lambda T_m}(m - 1) \to W$. Finally, dominated convergence yields

$$E\left(\exp\left\{-\int_0^\infty f(t) N(dt)\right\} | W\right)$$

$$= \lim_{m\to\infty} E\left(E\left(\exp\left\{-\sum_{n=1}^{m-1} f(T_n)\right\} | T_m\right) | W\right)$$

$$= E\left(\exp\left\{-\int_0^\infty (1 - e^{-f(t)})\lambda e^{\lambda t} W dt\right\} | W\right)$$

$$= \exp\left\{-\int_0^\infty (1 - e^{-f(t)})\lambda e^{\lambda t} W dt\right\}. \blacksquare$$

EXERCISES

5.1. Harry and the Hollywood Mogul (continued). Recall from Problem 2.52 that one of Uncle Zeke's business acquaintances is the Hollywood mogul Sam Darling and that Harry and Sam Darling were negotiating about the possibility of doing a real down-home show entitled *Optima Street Restaurateur.*

Assume terms have been agreed upon and *Optima Street Restaurateur* goes into production. To save money, round-the-clock shooting is used. Realistic locales are essential, so the restaurant is used for all scenes. Scenery and furniture must be added, however, as the usual grungy decor in Happy Harry's does not reflect light properly. There are four different sets, labelled 1 through 4 that are brought in and out, depending on the needs of the script:

(1) neo Julia Childs, cheerful but antiseptic for quiet discussions of the art of cooking;

(2) 19th century saloon, for those action scenes where someone is shot and falls from the balcony;

(3) student modern (boards on bricks for bookcases, etc.), giving the impression of the serious artist studying his craft and planning carefully;

(4) quasi-*Cheers*, for that warm atmosphere conducive to mating and other rituals.

The changes in the sets follow a continuous time Markov chain. Holding times in each state are exponential with parameters

$$(\lambda(1),\ldots,\lambda(4)) = (2,2,3,1)$$

and changes of state are achieved according to a uniform distribution: given the system is in state i, the next state is $j \neq i$ with probability 1/3.

(a) What is the long run percentage of time each set is used?

(b) If production starts in state 4, what is the expected time until Harry can quietly discuss cooking; i.e., until set 1 is used for the first time?

(c) The mutant creepazoids are infuriated that a production company has taken over their turf. They will stage a noxious demonstration at a random time T which will halt production. The random variable T is exponentially distributed with parameter 8. This demonstration will cost the production company $200, $300, $100 or $500, if, at the time of the demonstration, the set numbered 1, 2, 3, or 4, respectively, is in use. What is the expected cost to the production company due to this demonstration? How much should be production company be willing to pay in bribes to the leaders of the mutant creepazoids to try to persuade them to call the demonstration off?

5.2. Harry's Talent Agency. After the success of amateur night, Harry decides to start a talent agency which will match amateur performers with talent seekers, i.e., people who need entertainment at weddings, Bar Mitzvah's, etc. Talent seekers arrive at Happy Harry's according to a Poisson process of rate 1, and performers arrive via an independent stream according to a Poisson process of rate 2. A performer arriving to find no talent seekers leaves immediately rather than wait around. If a performer arrives to find a talent seeker present, however, the talent seeker immediately books the gig, and performer and seeker leave together. A talent seeker arriving to find no performers waits until one arrives.

(a) Let $S(t)$ be the number of talent seekers present at time t. What sort of process is $\{S(t), t \geq 0\}$?

(b) What is the mean number of talent seekers present (computed relative to the stationary distribution)?

(c) What is the long run proportion of arriving performers who do not get gigs? (You might want to think conditionally; thin a Poisson process.)

Eventually performers get more business savvy and negotiate with talent seekers before departing. The negotiations take an exponential length of time with parameter 10. Let $\{\mathbf{X}(t) = (A(t), S(t)), t \geq 0\}$ be the bivariate process giving the number of performers and number of talent seekers present at time t.

(d) Is $\mathbf{X}(t)$ Markov? If so, give its state space and specify its generator.

5.3. A Markov Modulated Poisson Process. Suppose $\{X(t)\}$ is a Markov process with state space $S = \{1, \ldots, m\}$. As usual, the succession of states is given by the Markov chain $\{X_n\}$ with transition probabilities $Q = \{Q_{ij}\}$, and the holding time parameters are $\{\lambda(i), i = 1, \ldots, m\}$. The times of jumps are $\{T_n, n \geq 0\}$. Let $\{N(t), t \geq 0\}$ be a homogeneous Poisson process with unit rate 1, and suppose the Poisson process $\{N(t)\}$ is independent of the Markov chain $\{X(t)\}$. Construct a new point process $\{N^*(t), t \geq 0\}$ to have the property that, when the Markov chain is in state i, the process N^* behaves like a homogeneous Poisson process of rate $\alpha(i)$. Thus the rate of the new counting process changes according to the behavior of the Markov chain. Formally, we have $\alpha : S \mapsto [0, \infty)$, and

$$N^*(t) = N\left(\int_0^t \alpha(X(u))du\right),$$

so the new process N^* is the original process N with its time scale jiggled by the integral of a function of the Markov chain.

(a) Verify that

$$\{T_{n+1} - T_n, N^*(T_n, T_{n+1}], n \geq 0\}$$

are conditionally independent given $\{X_n\}$.

(b) Compute

$$P[X_{n+1} = j, T_{n+1} - T_n \leq x, N^*(T_n, T_{n+1}] = k | X_n = i]$$

in terms of $Q, \{\lambda(m), m \in S\}, \{\alpha(m), m \in S\}$.

(c) Show $N^*(0, t_1], N^*(t_1, t_2], \ldots, N^*(t_{m-1}, t_m]$ are conditionally independent given $X(0), \ldots, X(t_m)$.

(d) Consider the bivariate process $\{\xi(t) = (X(t), N^*(t)), t \geq 0\}$:

(d1) Give an argument that $\{X(t), N^*(t), t \geq 0\}$ is a Markov chain on the state space $\{1, \ldots, m\} \times \{0, 1, \ldots\}$.

(d2) Is this system regular? Why or why not?

(d3) Define

$$P_{ij}(n, t) = P_i[X(t) = j, N^*(t) = n]$$

and give a system of equations which $\{P_{ij}(n,t), i, j \in S, n \geq 0, t \geq 0\}$ satisfy.

(d4) What are the holding time parameters $\{\lambda(i, m), i \in S, m \geq 0\}$ and the transition probabilities $Q_{(i,m),(j,n)}$ for the chain $\{\xi(t)\}$? What is the generator matrix?

(d5) Does a stationary distribution exist for the process $\{\xi(t)\}$? Does a stationary distribution exist for the process $\{X(t)\}$?

(e) What is the long run arrival rate of points

$$\lim_{t \to \infty} N^*(t)/t \ ?$$

Express the limit in terms of $\{\alpha(k), k \in S\}$ and the stationary distribution of $\{X(t)\}$.

(f) Compute the Laplace functional of N^* at the function $f(t)$. (You might want to do this by conditioning on either on $\{X(t)\}$ or by conditioning on $\{X_n\}, \{T_n\}$.) Express the answer as an expectation involving $f(t), \{\alpha(k), k \in S\}, \{X(t)\}$.

(f1) From the Laplace functional, compute the Laplace transform at the point $\theta > 0$ of $N^*(a, b]$ where $0 \leq a < b$.

(f2) From the Laplace transform compute $EN^*(a, b]$. Express the answer in terms of $\{\alpha(k), k \in S\}, \{X(t)\}$.

(f3) If the Markov chain $\{X(t)\}$ is stationary (begun with the stationary distribution) does the expression for the mean number of points found in (f2) simplify? To what?

(Hint: Think conditionally.)

5.4. Harry and the Comely Young Lady, the Sequel. Recall from Exercise 2.35 that Harry had his eye on a comely young lady whom he originally spotted at a high cholesterol cooking course, and recall that eventually Harry worked up the courage to ask her for a date.

Once the ice is broken, Harry and the young lady see each other regularly. Their relationship, although close, is somewhat stormy and fluctuates between periods of amorous bliss (state 1), abrasive criticism of each other caused by spending too much time together (state 2), confusion about where the relationship is heading (state 3), and emotional exhaustion caused in part by Harry's relentless devotion to improving his business (state 4). The fluctuations follow a Markov chain with matrix

$$Q = \begin{matrix} 1 \\ 2 \\ 3 \\ 4 \end{matrix} \begin{pmatrix} 0 & .7 & .3 & 0 \\ .3 & 0 & .6 & .1 \\ .2 & .3 & 0 & .5 \\ .7 & 0 & .3 & 0 \end{pmatrix},$$

and the times spent in states 1, 2, 3, 4, respectively, are exponentially distributed with parameters $\frac{1}{3}, \frac{1}{2}, \frac{1}{4}, 1$. Let $X(t)$ be the state at time t of the relationship.

(a) What is the long run percentage of time that the couple spends in amorous bliss?

(b) Harry observes that when he is happy, his business seems to do well. The earning rates for the restaurant per day as a function of the states are \$400 when in state 1, \$200 when in state 2, \$300 when in state 3 and only \$150 when in state 4. Using these earning rates, compute the long run earning rate of the restaurant.

5.5. Phase Distributions. Consider an $m + 1$ state Markov chain $\{X(t), t \geq 0\}$ with generator matrix A, and define

$$\tau = \inf\{t > 0 : X(t) = m + 1\},$$
$$a_i = P[\tau < \infty | X(0) = i], \quad i = 1, \ldots, m.$$

Write

$$A = \begin{pmatrix} T & T^0 \\ T_1 & T_2 \end{pmatrix},$$

where T is $m \times m$ and corresponds to states $1, \ldots, m$, T^0 is $m \times 1$ and T_2 is 1×1. The distribution of τ is called a *phase distribution*.

(a) Write a system of linear equations that $\mathbf{a} = (a_1, \ldots, a_m)'$ satisfies. Express the system as a matrix equation in $T, T^0, \mathbf{a}$. (Recall the linear system you get in the discrete time case, and make adjustments for the random sojourn times.) Show that $a_i = 1$, $i = 1, \ldots, m$, iff T is nonsingular.

(b) Let $\bar{F}_i(x) = P_i[\tau > x]$, $i = 1, \ldots, m$, and set

$$\mathbf{F}(x) = (\bar{F}_1(x), \ldots, \bar{F}_m(x))'.$$

Write a system of integral equations whose unknowns are $\bar{F}_1(x), \ldots, \bar{F}_m(x)$.

(c) In (b) take Laplace transforms. Let

$$\Psi_i(\theta) = \int_0^\infty e^{-\theta x} \bar{F}_i(x) dx, \quad \theta > 0, \quad i = 1, \ldots, m.$$

Let $\Psi(\theta) = (\Psi_1(\theta), \ldots, \Psi_m(\theta))'$. Express the resulting equations as a matrix equation in $\Psi(\theta), T$, and solve for $\Psi(\theta)$ in terms of T.

(d) Invert and express $\bar{\mathbf{F}}(x)$ in terms of T. If $\alpha = (\alpha_1, \ldots, \alpha_m)'$ is an initial distribution concentrating on $\{1, \ldots, m\}$ and T is nonsingular, show

$$E_\alpha \tau = -\alpha T^{-1} \mathbf{1}.$$

(e) **Harry's Stressful Life.** Harry's stressful life causes his mental state to fluctuate among three states: 1 (depressed), 2 (hopeful), 3 (suicidal). Upon entering state 3, he must immediately be institutionalized. Sojourns in states 1 and 2 are exponentially distributed with parameters 2 and transitions of mental state occur according to the matrix

$$Q = \begin{matrix} 1 \\ 2 \\ 3 \end{matrix} \begin{pmatrix} 0 & .5 & .5 \\ .5 & 0 & .5 \\ 0 & 0 & 1 \end{pmatrix}.$$

Let $X(t)$ be Harry's mental state at time t.

(e1) Give the portion of the generator matrix corresponding to T.

(e2) If $\tau = \inf\{t \geq 0 : X(t) = 3\}$ is the time to institutionalization, find $P_2[\tau > x]$ explicitly if possible; otherwise express as a series. (This can be done independent of the foregoing theory.)

(e3) Find $E_2\tau$.

(More on phase distributions in Problem 5.50. (cf. Neuts, 1981).)

5.6. Harry Meets Sleeping Beauty. Harry dreams he is Prince Charming coming to rescue Sleeping Beauty (SB) from her slumbering imprisonment with a kiss. The situation is more complicated than in the original tale, however, as SB sleeps in one of three positions:

(1) flat on her back, in which case she looks truly radiant;
(2) fetal position, in which case she looks less than radiant;
(3) fetal position and sucking her thumb in which case she looks radiant only to an orthodontist.

SB's changes of position occur according to a Markov chain with transition matrix

$$Q = \begin{matrix} 1 \\ 2 \\ 3 \end{matrix} \begin{pmatrix} 0 & .75 & .25 \\ .25 & 0 & .75 \\ .25 & .75 & 0 \end{pmatrix}.$$

SB stays in each position for an exponential amount of time with parameter $\lambda(i)$, $1 \leq i \leq 3$, measured in hours, where

$$\lambda(1) = 1/2, \quad \lambda(2) = 1/3, \quad \lambda(3) = 1.$$

Assume for the first two questions that SB starts sleeping in the truly radiant position.

(a) What is the long run percentage of time SB looks truly radiant?

(b) If Harry arrives after an exponential length of time (parameter α), what is the probability he finds SB looking truely radiant? (Try Laplace transform and matrix techniques.)

(c) SB, being a delicate princess, gets bed sores if she stays in any one position for too long, namely if she stays in any position longer than three hours. Define for $t > 0$ and $i = 1, 2, 3$,

$$S_i(t) = P[\text{no bed sores up to time } t| \text{ SB's initial position is } i],$$

so that $S_i(t) = 1$ for $t \le 3$. Write a recursive system of equations satisfied by the functions $S_i(t), t > 0, i = 1, 2, 3$. You do not have to solve this system.

5.7. M/M/∞ Queue. Arrivals are Poisson with rate a, service times are exponential with parameter b, and there are an infinite number of servers. Let $Q(t)$ be the number in the system at time t. Give the generator matrix and show by solving $\eta' A = 0$ that the stationary and limiting distribution is Poisson.

5.8. Suppose $\{X_n\}$ is a Markov chain with transition matrix P where for some i we have $p_{ii} > 0$. Let $\{Y_n\}$ be $\{X_n\}$ observed only when the X-process changes state. That is, Y ignores transitions which are back to the same state. More precisely, we define

$$n(0) = 0,$$
$$n(1) = \inf\{k > 0 : X_k \ne X_0\}$$
$$n(2) = \inf\{k > n(1) : X_k \ne X_{n(1)}\}$$
$$\vdots$$

and define $\{Y_k\} = \{X_{n(k)}\}$. Show $\{Y_k\}$ is a Markov chain and give its transition matrix. Show $P[Y_{k+1} = i| Y_k = i] = 0$. Discuss the relevance of this result to constructing a continuous time Markov chain.

5.9. Amateur Night at Happy Harry's, the Sequel. Recall that Friday night is amateur night at Happy Harry's (Example 2.11.1). Assume in addition to the information given previously that the length of time a performer of class i is allowed to perform is limited by management and the indulgence of the crowd and is random with an exponential distribution with parameter $2i$. The amount of time it takes to quell a riot is exponentially distributed with parameter 1. After the riot is quelled, performances resume—the show must go on. Define the continuous time Markov chain $\{X(t), t \ge 0\}$ by $X(t) = i$ if a performer of class i is being endured at time t or $X(t) = 6$ if a riot is being endured.

(a) Give $\{\lambda(i)\}, Q, A$ for this chain.

(b) If the evening starts off with a class 2 performer, what is the expected waiting time until a class 1 performer is encountered?

(c) What is the long run proportion of time one spends enduring riots?

(d) Customers arrive at the door of Happy Harry's according to a Poisson process rate α. If, however, upon peeking inside, a customer notes a riot in progress, he will not enter, preferring instead the more sedate entertainment of home video. What is the structure of the process of customers who do not enter Happy Harry's?

5.10. A service center consists of two servers, each working at an exponential rate of two services per hour. If customers arrive at a Poisson rate of three per hour, then, assuming the capacity of at most three customers,

(a) What fraction of potential customers enter the system?

(b) What would the value of (a) be if there was only a single server and his rate was twice as fast (that is, rate 4)?

5.11. Show

$$P(t) = \begin{pmatrix} .6 + .4e^{-5t} & .4 - .4e^{-5t} \\ .6 - .6e^{-5t} & .4 + .6e^{-5t} \end{pmatrix}$$

is a standard transition function of a two-state Markov chain $\{X(t)\}$. Compute

$$P[X(3.4) = 1, X(3.8) = 0 | X(0) = 0].$$

5.12. Harry's Boxing Career. Harry gets into a fight with a brutish mutant creepazoid over the dress code at the restaurant. Harry does well but suffers increased dullness every time he is hit. Initially the probability he is hit in a period of length δt is $\lambda \delta t + o(\delta t)$, but when he has been hit τ times, this probability is $(\lambda + \alpha\tau)\delta t + o(t)$ where $\lambda, \alpha > 0$. Formulate this as a pure birth process, and find the probability that he suffers n hits in time $(0, t)$.

5.13. Yule Shot Noise.* Consider the linear birth process (sometimes called the *Yule process*), and represent it as a point process

$$N = \sum_{n=1}^{\infty} \epsilon_{T_n},$$

so that T_n is the time of the nth jump (cf. Section 5.11). Let $\{g_i(t), t \geq 0\}$, $i = 1, 2, \ldots$, be independent identically distributed non-negative valued stochastic processes and suppose these processes are independent of the linear birth process. Define the *Yule shot noise process*

$$Y(t) = \sum_{i=1}^{N((0,t])} g_i(t - T_i).$$

---

* This problem is done best using material from Section 5.11.

Compute the Laplace transform of $Y(t)$, and show

$$EY(t) = \lambda e^{\lambda t} \int_0^t e^{-\lambda x} Eg(x)dx,$$

where λ is the birth rate. (Hint: Since the Yule process is a non-homogeneous mixed Poisson process, a version of the order statistic property holds.)

5.14. Power starved Optima Street receives electricity through the following grid:

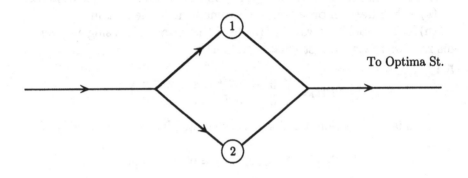

FIGURE 5.2. TRANSMISSION ROUTE

At points 1 and 2 are transformers which operate independently. Transformer i operates for an exponentially distributed amount of time with parameter a_i ($i = 1, 2$). When a breakdown occurs, repairs commence, which take an exponentially distributed amount of time with parameter $b_i, i = 1, 2$. When both transformers 1 and 2 are broken, the flow of power to Optima Street ceases and a blackout begins. What is the probability of an eventual blackout? Explain. (Hint: Don't panic. If you are doing lots of computation you are on the wrong track.) What is the distribution of the length of time the street is blacked out? What is the long run frequency of blackouts?

5.15. Markov Branching Process. If a given particle is alive at a certain time, its additional life length is a random variable which is exponentially distributed with parameter a. Upon death, it leaves k offspring with probability $p_k, k \geq 0$. As usual, particles act independently of other particles and of the history of the process. For simplicity, assume $p_1 = 0$. Let $X(t)$ be the number in the population at time t. Find the generator matrix and write out the forward and backward systems of differential equations.

Specialize to the binary splitting case where either a particle splits in two or vanishes. Find the stationary distribution of $\{X(t)\}$.

5.16. M/M/1 Queue with Batch Arrivals. As usual, services are iid random variables with exponential distribution and parameter b. Services are independent of the input. Customers arrive in batches. The batches arrive according to a Poisson process rate a. Each Poisson arrival event delivers a batch of customers, the batch size being distributed as a random variable N with

$$P[N = n] = q_n, \quad n \geq 1, \sum_{j=1}^{\infty} q_n = 1.$$

Find the generator matrix.

5.17. Harry's Scheduling Problems. In the interests of community harmony, Harry decides to attend both the Young Republicans and the Young Democrats meeting, but they meet at the same time. To be fair, when he goes to the bus terminal, if buses to both meetings are available, he flips a fair coin to decide which to choose. If only a bus to one meeting is available, he boards that bus. Otherwise he waits for the first suitable bus and boards that one. Loading times in the bus station for any bus are exponential with parameter λ_1. The time between the departure of one bus to the Young Republicans meeting and the arrival of the next bus for the Young Republican meeting is exponentially distributed with parameter λ_2. The corresponding time for Young Democrats buses is exponential with parameter λ_3. Loading times are independent of the waiting times for buses after departures, and the two types of buses are in no way dependent.

(a) What is the equilibrium (limiting) probability that Harry arrives at the terminal and finds:

(1) only a bus to the Young Republicans meeting available?
(2) only a bus to the Young Democrats meeting available?
(3) both buses available?
(4) neither bus available?

(b) What fraction of Harry's visits are to the Young Democrats meetings?

5.18. Harry's Buffet Line. Harry initiates a buffet and as an experiment organizes it as a Moscow queue: Customers queue at the cashier to pay for the buffet, then proceed to a second line where they wait to be served by a single server. Suppose customers arrive at the cashier according to a Poisson process with parameter one per minute and have exponential service times with mean $1/2$ minutes for the first queue and mean $1/3$ minutes for the second queue. Further, assume that the interarrival times and service times are all independent. Set $X(t)$ to be the length of the first queue and $Y(t)$ to be the length of the second queue.

(a) Let $\pi(i, j)$ be the distribution of $(X(t), Y(t))$ at equilibrium ($\pi(i, j) = \lim_{t\to\infty} P[X(t) = i, Y(t) = j]$). Argue heuristically (if necessary) that $\pi(i, j)$ must satisfy the forward equations

$$3\pi(i, j+1) = 2\pi(i+1, j-1) - \pi(i, j)(1+2+3) + \pi(i-1, j)$$

for $i \geq 1, j \geq 1$.

(b) What are the corresponding forward equations for the cases $i = 0, j \geq 1$ and $i = 0, j = 0$. Note for $i \geq 1$

$$3\pi(i, 1) + \pi(i-1, 0) - \pi(i, 0)(1+2) = 0.$$

(c) Verify that

$$\pi(i, j) = (\frac{1}{2})^{i+1} 2(\frac{1}{3})^{j+1}$$

is a solution to the forward equations and hence the two queue lengths are independent in equilibrium.

5.19. Weather Affects Business. During snow storms, potential patrons arrive at Harry's place according to a Poisson process at a rate of one per hour. Because his help is unable to arrive during bad weather, Harry can only serve one customer at a time, and the service time is exponentially distributed with mean 30 minutes. A prospective customer, knowing this situation, will only enter Harry's place if there are no other customers being served.

(a) Find the limiting distribution of queue length.

(b) What fraction of potential customers will be lost?

5.20. M/G/∞ Queue with Fictitious Services. Consider a single server queue with Poisson input of rate a. Service times are iid with common distribution $G(x)$. Assume $\int_0^\infty x\, dG(x) = m < \infty$, and if you like, you may assume G has a density g. If, upon completing a service, the system is empty, the server starts a fictitious service period (coffee break?) whose length is independent of input and other services and also is distributed as G. The server is considered busy during such fictitious services and cannot work on new arrivals. Let X_n be the number in the system just after the nth service (fictitious or actual) is completed. $\{X_n\}$ is a Markov chain. What is its invariant measure when $\rho = am \leq 1$? Let $Q(t)$ be the number in the system at time t. Compute

$$\lim_{t\to\infty} P[Q(t) = k | X_0 = j]$$

for all j, k. Why does this limit exist? Does this limit always equal

$$\lim_{n \to \infty} P[X_n = k | X_0 = j]?$$

5.21. More on the M/M/∞ Queue. Imagine Poisson arrivals with rate a at an infinite number of servers and that service times are iid exponentially distributed with parameter b. Let $Q(t)$ be the number of busy servers. Suppose $Q(0) = i$. Model this as a birth-death process. Set

$$P_n(t) = P[Q(t) = n].$$

Derive the forward equations

$$P_0'(t) = -aP_0(t) + bP_1(t)$$
$$P_n'(t) = -(a + nb)P_n(t) + aP_{n-1}(t) + (n+1)bP_{n+1}(t), \quad n \geq 1.$$

Set $G(s, t) = \sum_{n=0}^{\infty} P_n(t)s^n$ and show G satisfies the partial differential equation

$$\frac{\partial G}{\partial t} = (1 - s)\left[-aG + b\frac{\partial G}{\partial s}\right].$$

Solve to obtain

$$G(s, t) = \left(1 - (1 - s)e^{-bt}\right)^i \exp\{-a(1 - s)b^{-1}(1 - e^{-bt})\}.$$

Identify this distribution when $i = 0$. What happens as $t \to \infty$?

5.22. Erlang Loss Formula. Consider the M/M/s queue with Poisson input of rate a where there are s servers. Service times are iid exponential random variables with parameter b. Customers, who arrive when all s servers are busy, leave; nobody waits. Model this as a birth-death process. What are the parameters of the birth-death process? Let $Q(t)$ be the number of busy servers at time t. Find

$$p_s = \lim_{t \to \infty} P[Q(t) = s].$$

This represents the probability that an arriving customer finds the system blocked and is lost. Why is this also the long run percentage of customers who are lost?

5.23. Little's Formula. In a queueing model suppose customers arrive at times $t_1, t_2, \ldots$ and that the number of arrivals by time t is $N_A(t)$. Label the customers $c_1, c_2, \ldots$ and suppose w_j is the time in the system (queueing and service time) of c_j. Let $Q(t)$ be the number in the system at time t. Assume some limits exist:

(a) the long run average time in the system of a customer,

$$\lim_{n\to\infty} \frac{\sum_{i=1}^{n} w_i}{n} = w;$$

(b) the long run average arrival rate,

$$\lim_{t\to\infty} \frac{N_a(t)}{t} = \lambda;$$

(c) the long run average number of customers in the system,

$$\lim_{t\to\infty} \frac{\int_0^t Q(s)ds}{t} = L.$$

Prove Little's formula, $L = \lambda w$, which equates the average number in the system to the arrival rate times the average wait per customer in the system.

(Hint: Write

$$\int_0^t Q(s)ds = \sum_{j=1}^{N_A(t)} w_j \wedge (t - t_j).)$$

Consider the stable M/M/1 queue with Poisson input rate a and exponential service times parameter b. Show that L is the mean of the stationary distribution for the process $Q(t)$. Do this also for the M/M/s queue.

5.24. Virtual Waiting Times. Let $Q(t)$ be the number in the system for a stable M/M/1 queue, and let $W(t)$ be the load on the server at time t, the length of time the server serving at unit rate would have to work if the input was shut off at time t. Another description is that $W(t)$ is the waiting time of a fictitious customer named FIC who arrives at time t. Check that

$$W(t) = \begin{cases} 0, & \text{if } Q(t) = 0, \\ v_1' + v_2 + \ldots v_{Q(t)}, & \text{if } Q(t) > 0. \end{cases}$$

Here $v_2, v_3, \ldots$ are iid service times and v_1' is the residual service time of the customer in service at time t. Show that if $\rho < 1$ and $x \geq 0$

$$P[W(t) \leq x] \to P[W(\infty) \leq x] := 1 - \rho e^{-b(1-\rho)x},$$

which is a distribution of mixed type having an atom at 0 and a density away from 0 on the region $(0, \infty)$. Show also that

$$EW(\infty) = \frac{\rho}{b(1-\rho)}, \quad \text{Var}(W(\infty)) = \frac{2\rho - \rho^2}{b^2(1-\rho)^2}.$$

Check also that

$$EW(\infty) + b^{-1} = w,$$

where w is from Little's formula.

Hint: Proceed by noting that

$$P[W(t) \leq x] = P[Q(t) = 0] + \sum_{n=1}^{\infty} P[\sum_{1}^{n} v_i \leq x]P[Q(t) = n].$$

5.25. For the stable M/M/s queueing model with $Q(t)$ representing the number of customers in the system at time t, let the traffic intensity be $\rho = a/(sb)$ and show that the stationary distribution can be written in the form

$$\eta_n = \begin{cases} \frac{\eta_0 s^n \rho^n}{n!}, & \text{if } 0 \leq n \leq s \\ \eta_s \rho^{n-s}, & \text{if } n \geq s. \end{cases}$$

Write out an explicit expression for η_0, ugly though it may be. Furthermore, show that

$$L = EQ(\infty) = \sum_n n\eta_n = s\rho + \frac{\eta_s \rho}{(1-\rho)^2}.$$

For the virtual waiting time $W(t)$, note

$$W(t) = \begin{cases} 0, & \text{if } Q(t) < s, \\ D_1 + \cdots + D_{Q(t)-s+1}, & \text{if } Q(t) \geq s, \end{cases}$$

where $D_1, D_2, \ldots$ are independent and

$$D_i \overset{d}{=} \bigwedge_{j=1}^{s} E_{ij}(b) \overset{d}{=} E(sb)$$

and the $E's$ are independent exponential random variables with the parameter in the parentheses. Show that, for $x \geq 0$, $W(t)$ has a limit distribution given by

$$\lim_{t \to \infty} P[W(t) \leq x] = P[W(\infty) \leq x]$$

$$= 1 - \frac{\eta_s}{1-\rho} e^{-bs(1-\rho)x}$$

with mean and variance

$$EW(\infty) = \frac{\eta_s}{(1-\rho)^2 bs}$$

$$\text{Var}(W(\infty)) = \frac{\eta_s}{(sb)^2(1-\rho)^4}(2 - 2\rho - \eta_s).$$

5.26. For M/M/2 show that when $\rho < 1$

$$\eta_0 = \frac{1-\rho}{1+\rho}.$$

What is η_n?

5.27. For a stable M/M/1 queue, what is the long run fraction of time the server is busy? For M/M/s, if $\rho < 1$ and each server is utilized equally, what is the long run proportion that a given server, say the sth, is free? Hint: The proportion should be

$$\sum_{n=0}^{s-1} \frac{(s-n)}{s} \eta_n.$$

5.28. For an M/M/1 queue, let $N_q(t)$ be the number in the queue exclusive of the customer being served so that

$$N_q(t) = (Q(t) - 1)_+.$$

The process $\{N_q(t), t \geq 0\}$ is not Markov. Check this.

5.29. In the M/M/s queue, let $N_q(t)$ be the number of customers (if any) actually waiting. Assuming $\rho < 1$, prove
   (a) the limit distribution of $N_q(t)$ is given by $\{v_n\}$ where

$$v_0 = \eta_0 \sum_0^s \frac{1}{j!} (\frac{a}{b})^j, \quad v_n = \eta_s \rho^n$$

for $n \geq 1$;
   (b) the mean number of waiting customers, given that someone waits, is

$$E(N_q(\infty)) | N_q(\infty) > 0) = (1 - \rho)^{-1};$$

   (c) The mean waiting time, given a customer waits, is

$$E(W(\infty) | W(\infty) > 0) = (sb - a)^{-1}.$$

5.30. For M/M/1 compute
   (a) the expected number of arrivals during a service period; and
   (b) the probability that no customers arrive during a service period.

5.31. If the queue length $Q(t)$ in some queueing model can be characterized as a birth-death process with parameters λ_n and μ_n $(n \geq 0)$, for which the limit distribution $\{\eta_n\}$ exists (so that

$$\lim_{t \to \infty} P[Q(t) = n] = \eta_n$$

for any $n \geq 0$), prove that

$$\sum_n \lambda_n \eta_n = \sum_n \mu_n \eta_n.$$

This means that the expected input rate equals the expected output rate.

5.32. In M/M/s let $R(t)$ be the number of busy servers at time t. If $\rho < 1$, show that a limit distribution exists. Find it. What is the mean of the limit distribution?

5.33. **Rework–Dissatisfied Customers at Harry's.** The grill at Harry's restaurant is modelled as an M/M/1 queue with a wrinkle. Harry has hired an inexperienced college student to do the grilling; customers may be dissatisfied with the item they receive and may ask to be served again. (This paradigm corresponds to *rework* in manufacturing settings.) The college student is only capable of working on one item at a time. Suppose on receiving an order, there is probability $1 - \alpha$ (where $0 < \alpha < 1$) that the customer is dissatisfied, independent of whether that customer had been dissatisfied one or more times previously. Subsequent service times (times to get the new order), if any, are also independent and have the same exponential distribution with parameter b.

(a) Check that the length of time it takes for a customer to be satisfied with his order is exponentially distributed. What is the parameter?

(b) Supppose dissatisfied customers are served again immediately until they are satisfied. What are the conditions for the queue to be stable? What is the total amount of time a hungry customer can expect to wait between arriving and receiving a satisfactory order?

(c) Some patrons object to a dissatisfied customer being treated with priority, so a new rule is initiated requiring a dissatisfied customer to go to the end of the line. How does this affect the birth and death rates? How are answers in (b) affected?

5.34. Define the efficiency EFF of the M/M/1 system relative to the M/M/s system with the same total service rate as

$$EFF = \frac{E(\text{ time in system in M/M/s })}{E(\text{ time in system M/M/1 })}.$$

Show that as $\rho \to 0$, $EFF \to s$ while as $\rho \to 1$, $EFF \to 1$. (Prabhu, 1981.)

5.35. In M/M/1 with finite capacity c, prove the following:

(a) The stationary distribution of the queue length is

$$\eta_n = \frac{(1-\rho)\rho^n}{1-\rho^{c+1}}, \quad 0 \le n \le c,$$

and find the mean.

(b) Define the delay at time t (time to enter service) of our virtual customer FIC as

$$D(t) = \begin{cases} 0, & \text{if } Q(t) = 0 \text{ or } c \\ \sum_{i=1}^{Q(t)} v_i, & \text{if } 1 \le Q(t) \le c-1, \end{cases}$$

since if $Q(t) = 0$ or c there is no waiting; here $v_1, \ldots, v_{c-1}$ are iid with exponential distribution with parameter b. Show that the distribution function of the delay at stationarity is given by

$$P[D(\infty) \le x] = 1 - \sum_{n=1}^{c-1} \eta_n \sum_{r=0}^{n-1} \frac{e^{-bx}(bx)^r}{r!},$$

and the mean delay is

$$ED(\infty) = \frac{\rho}{b(1-\rho)} - \frac{(\rho+c)\rho^c}{b(1-\rho^{c+1})}.$$

(c) If $c = 1$, verify

$$\eta_0 = \frac{1}{1+\rho}, \quad \eta_1 = \frac{\rho}{1+\rho}.$$

5.36. More Loss Systems. In an M/M/2 queueing model with no queue allowed (arriving customers at a busy server depart without waiting), we suppose each channel is operated by one server. According to the records of the servers, during 10,000 hours of operation, 40,000 customers received service and 8,000 man-hours of service were dispensed. These figures are the total for both channels.

(a) During the 10,000 hours above, estimate the number of lost customers.

(b) Suppose lost revenue per customer is $5.00 and the cost of operating a channel (busy or not) is $4.00 per hour. Is it desirable to add another channel? (Wolf, 1989.)

5.37. Records and the Linear Birth Process. Consider a linear birth process and thin the birth time according to the mechanism: The kth birth time is deleted with probability $1 - k^{-1}$ and retained with probability k^{-1} independently of all other points. Show that the process of retained points is a homogeneous Poisson process.

This has the following interpretation: Given an iid sequence of random variables $\{\xi_n, n \geq 1\}$ from common continuous distribution $F(x)$. Mark the kth birth time with the random variable ξ_k and retain this birth time only if ξ_k is a record value of the ξ-sequence. The times of records form a homogeneous Poisson process. (Browne, 1991; Bunge and Nagaraja, 1992.)

5.38. More Loss Systems. Harry Plans Scientifically. During a particularly prosperous period, Harry plans for expansion. He buys the vacant lot next door and plans to erect a parking garage for the convenience of his customers who drive by to purchase take out food. He estimates that during rush hour cars will arrive to the restaurant according to a Poisson process at rate of 10 per minute and that the length of time cars stay in the garage has an exponential density with mean three minutes. How large should the parking garage be if there is to be only a 1% chance or less of a car being turned away because the garage is full? (This is not a problem you should work out with pencil and paper; you may wish to make a dash for the nearest computer terminal.)

5.39. Separate Lines at Harry's. Harry tries organizing the traffic flow in the restaurant. He organizes three separate counters: One is for take out food, and the other two are buffet lines for customers who eat in the restaurant. One of the buffet lines is for vegetarian food, and the other is for flesh eaters. Each counter has a server, and each counter has its own input. Assume that customers arrive needing service from only one line. The service times at all three counters have an exponential density with a mean of 15 seconds. The arrivals to the take out, vegetarian and flesh-eating counters form three Poisson processes, with mean interarrival times of 20, 18, 30 seconds, respectively.

(a) What is the average waiting time $EW(\infty)$ for each of the three counters? (Give three numbers corresponding to take out, vegetarian, flesh.)

(b) Average the three numbers obtained in (a) as a measure of the overall waiting times at the restaurant.

(c) Harry considers the possibility of a better traffic flow if the restaurant is redesigned so that each counter is capable of handling all three types of customers. Merge the three input streams into one input Poisson process so that the system can be analyzed as M/M/3. What is the average waiting time for this system?

(d) Compare the average obtained in (c) with that obtained in (b). Conclusions? What is Harry's best strategy?

5.40. A Queue with Balking—Impatient Customers During the Super Bowl. During the Super Bowl, Harry's take out customers are very impatient. Consider the take out line an M/M/1 system with input parameter a and service rate b in which customers are impatient. Upon arrival, a customer estimates his wait by n/b if there are n people in the system. The customer then joins the queue with probability $\exp\{-\alpha n/b\}$ (or leaves with probability $1 - \exp\{-\alpha n/b\}$). Assume $\alpha \geq 0$.

(a) For $\alpha > 0$, $b > 0$, under what conditions will a stationary distribution for the queue exist; what are the stability conditions?

(b) When the queue is stable, give an expression for the stationary probabilities $\{\eta_n\}$.

(c) Let $\alpha \to \infty$. To what does the stationary distribution converge as $\alpha \to \infty$? Use this limit to compute an approximation of the average number in the system when α is large.

5.41. M/M/2 Queue with Heterogeneous Servers. Consider a two-server queue with a Poisson input rate a. Suppose servers 1 and 2 have exponential service rates $b_1 > b_2$. If server 1 becomes idle, then the customer being served by server 2 switches to server 1. Give the birth-death rates, stability condition and stationary distribution for $Q(t)$.

5.42. Another Balking Model. In the system M/M/1, customers join the system with probability 1 if the server is free and with probability $p < 1$ otherwise. Find the limit distributions of $Q(t)$ and of the virtual waiting time $W(t)$.

5.43. More Impatience—Reneging. Consider an M/M/1 system with arrival rate a and service rate b where customers in the queue (but not the one in service) may get discouraged and leave without receiving service. Each customer who joins the queue will leave after a time distributed like an exponential random variable with parameter γ if the customer has not entered service before that time.

(a) Represent the number in the system as a birth-death model.

(b) What fraction of arrivals are served?

(c) Suppose an arriving customer finds one customer in the system and that the order of service is first come, first served. What is the probability this customer is served, and, given that this customer is served, what is the expected delay in queue?

(d) Compute the Laplace transform of the waiting time until service of a fictitious arrival at time t named FIC, assuming FIC cannot reneg and that at time t we have $Q(t) = n$. (Those waiting in front of FIC can reneg.)

5.44. Harry Hires Scientifically. Harry must decide which of two potential work/study students to hire as bartenders. Amos, who is willing but slow, can be hired for $C_1 = \$6.00$ per hour , while Boris, who is faster, demands a higher rate C_2 per hour. Both dispense service at exponential rates $b_1 = 20$ customers per hour and $b_2 = 30$ customers per hour. Bar customers arrive according to a Poisson process of rate 10 per hour. Harry guesses that on the average a customers time is worth five cents per minute and should be considered.

(a) Give the expected cost per hour incurred by hiring either Amos or Boris.

(b) At most, how much should Harry be willing to pay Boris?

5.45. Harry Compares One Fast Worker with Two Slow Workers. Harry is unsure whether it is better to hire a speedy energetic worker or to split the work between two slower workers. To decide which is a better strategy, he compares an M/M/1 system with an M/M/2 system in the following manner: Suppose the input for both is Poisson with parameter a, and suppose overall service rates are the same. In the M/M/s system $(s = 1, 2)$, each worker works at rate $b(s)$ so that the overall rate is $sb(s) = b'$ where b' is constant, independent of s.

(a) Check that $w^{(s)}$ from Little's formula, the expected time in the system, is

$$w^{(s)} = \frac{s}{b'} + \frac{(s\rho)^s/s!}{b'(1-\rho)^2(\sum_{n=0}^{s-1}\frac{(s\rho)^n}{n!} + \frac{(s\rho)^n}{s!(1-\rho)})}$$

where $\rho = a/b'$ is a constant independent of s. Verify that $w^{(s)}$ is minimum for $s = 1$.

(b) As before, let $N_q^{(s)}(\infty)$ represent for the M/M/s system the number waiting in the queue (not counting the customer in service) at equilibrium, with respect to the limit distribution. Verify

$$EN_q^{(s)}(\infty) = \frac{\rho^{s+1}s^s}{(1-\rho)^2 s!(\sum_{n=0}^{s-1}\frac{(s\rho)^n}{n!} + \frac{(s\rho)^s}{s!(1-\rho)})},$$

and show that $EN_q^{(2)}(\infty) < EN_q^{(1)}(\infty)$.

5.46. Consider the Markov chain $\{X(t), t \geq 0\}$ with state space $\{1, 2\}$ and generator matrix

$$A = \begin{pmatrix} -1 & 1 \\ \frac{1}{2} & -\frac{1}{2} \end{pmatrix}.$$

Suppose $\{X(t)\}$ is stationary. Is it reversible?

5.47. Consider two queues in tandem, and suppose $a/b_i < 1$, $i = 1, 2$. Compute the average number in the system L, and compute the average time in system w of a customer.

5.48. Consider three queues labelled 1, 2, 3. The input is to queue 1 and has rate a. Customers leaving queue 1 do not necessarily go to queue 2 but choose between 2 and 3 with probabilities p and $q = 1 - p$. Show that if queue 1 has Poisson arrivals and has a stationary distribution, then in equilibrium the input processes for queues 2 and 3 are independent Poisson processes. (Recall what happens when you thin a Poisson process.) Let $\mathbf{Q}(t) = (Q_1(t), Q_2(t), Q_3(t))$ be the number at each queue at time t. When does this have a stationary distribution? Show that the stationary distribution has a product form.

5.49. Consider a closed queueing network with k nodes and m customers. Suppose at node j service times are exponentially distributed with parameter $b_j(n_j)$ if there are n_j customers at the jth node.

(a) If node j operates as an M/M/s queue, what is $b_j(n_j)$?

(b) Mimic the derivation of the invariant distribution of a closed queueing network with homogeneous service rates to conclude that the stationary distribution is ($\mathbf{n} = (n_1, \ldots, n_k)$)

$$\eta_{\mathbf{n}} = \beta_m \prod_{j=1}^{k} \frac{\pi_j^{n_j}}{\prod_{i=1}^{n_j} b_j(i)},$$

where π is the stationary distribution of the routing matrix and β_m is chosen to guarantee $\sum_{\mathbf{n}} \eta_{\mathbf{n}} = 1$.

(c) Show that the number of states in the state space S is

$$\binom{m + k - 1}{k - 1}.$$

Thus, calculating the norming constant β_m appearing in the expression for the stationary probabilities can be impractical even for small values of k and m.

(d) Consider the following alternative. Define the generating functions

$$\Phi_j(z) = \sum_{n=0}^{\infty} \frac{(\pi_j z)^n}{\prod_{r=1}^{n} b_j(r)}$$

$$\beta(z) = \sum_{i=0}^{\infty} b_i^{-1} z^i.$$

Show that

$$\beta(z) = \prod_{i=1}^{k} \Phi_i(z).$$

Thus β_m can be computed by multiplying together the functions $\Phi_i(z)$, $i = 1, 2, \ldots, k$ after they have each been truncated to the first $m + 1$ terms. The number of steps required to do this is of order km^2 (Kelly, 1979).

(e) UPS in Ithaca has 35 trucks. Times between breakdowns for each truck are exponentially distributed with mean 90 days. There are three mechanics on the UPS staff who each work at an exponential rate of six per day. At equilibrium, what is the expected number of operating vehicles? (Pen and paper will not suffice to get a number. *Mathematica* is helpful, or some programming is necessary.)

5.50. More on Phase Distributions. Consider a finite Markov chain $\{X(t), t \geq 0\}$ with state space $\{1, \ldots, m+1\}$. Assume that state $m+1$ is absorbing, and let $\alpha' = (\alpha_1, \ldots, \alpha_m, 0)$ be some initial distribution. Let

$$\tau = \inf\{t > 0 : X(t) = m+1\}$$

be the absorbtion time in state $m+1$. Then

$$F(t) = P_\alpha[\tau \leq t]$$

is a phase distribution. Call the class of phase distributions (as $m+1$, α, and A vary) the *PH*-class.

(a) If $A = \{A_{i,j}, 1 \leq i, j \leq m+1\}$ is the generator matrix and $A^- = \{A_{i,j}, 1 \leq i, j \leq m\}$ and $\alpha^- = (\alpha_i, 1 \leq i \leq m)'$, show

$$F(t) = 1 - \sum_{j=1}^{m}\sum_{i=1}^{m} \alpha_i \left(e^{A^- t}\right)_{i,j} = 1 - \alpha' e^{A^- t}\mathbf{1},$$

where

$$e^{A^- t} = \sum_{k=0}^{\infty} \frac{(A^-)^k t^k}{k!}.$$

Also, the Laplace transform is

$$\hat{F}(\theta) = 1 - \sum_{j=1}^{m}\sum_{i=1}^{m} \alpha_i (I - \theta^{-1} A^-)_{i,j}^{-1} = 1 - \alpha'(I - \theta^{-1} A^-)^{-1}\mathbf{1}.$$

(Unlike Problem 5.5, try not to derive the transform equation first. What is the probability of the Markov chain wandering around without absorption until time t? $e^{A^- t}$ should govern the behavior of the process up to time τ.)

(b) A hyper-exponential density is a density of the form

$$f(x) = \sum_{i=1}^{k} \alpha_i b_i e^{-b_i x}, \quad x > 0.$$

Show such a density is also in the *PH*-class. (Do not use (a). Construct a suitable Markov chain.)

(c) Show an Erlang density is a phase density.

(d) Show mixtures of Erlangs are phase densities. A density is a mixture of Erlangs if it is of the form $\sum_{i=1}^{k} \alpha_i f_i(x)$, where f_i, $1 \le i \le k$ are Erlangs of perhaps varying number of phases. Describe the associated Markov chain for generating the phase distribution.

(e) The squared coefficient of variation of a random variable X is defined to be $\text{Var}(X)/(EX)^2$. Compute this in (b), (c), (d). Check that for hyperexponential it is at least 1 and for Erlang it is at most 1. A crude fitting method is to compute an empirical squared coefficient of variation, and if this is at least 1 we try to fit hyperexponential, and if this is at most 1 we try Erlang.

(f) Let X and Y be two independent non-negative random variables, each with a phase distribution. Show that $X + Y$ has a phase distribution. (Each of the random variables has an interpretation as the absorption time in some Markov chain. Put the Markov chains together suitably.)

5.51. At the library, requisitions for new books (monographs) and periodicals (serials) are received at random. The requisitions are first processed by Aaron, who verifies certain information and, if necessary, completes the requisition form. Aaron then sends requisitions for monographs to Brown and requisitions for serials to Craig. Brown and Craig check to see if the monograph or serial is either in the library collection or already on order. If neither is the case, then the requisition is sent to Davis, who enters the necessary data on an official university purchase order form. The cost of the monograph or serial is determined, and the appropriate account is charged.

Aaron receives requisitions according to a Poisson process at a rate of eight per hour. Approximately 70% of the requisitions are for monographs. For both monographs and serials, 60% of the requisitions are returned to the person who initiated the request with the notification that the monograph or serial is either on hand or on order. Service time at each stage of the acquisition process is exponentially distributed with a mean of 6 minutes for Aaron, 7.5 minutes for Brown, 15 minutes for Craig and 15 minutes for Davis.

(a) Find the expected total number of requisitions being processed at any time in the library (at equilibrium).

(b) Find the expected processing time for a requisition for a monograph. (This is the expected elapsed time in equilibrium between when a requisition for a monograph arrives until either the order is completed or the initiator of the requisition is notified the item is already on order or in the collection. Prabhu, 1981.)

5.52. A Closed Cyclic Network with Feedback. Consider a closed network consisting of k nodes and m customers. Customers cycle through the nodes in increasing order, meaning that $p_{j,j+1} = 1$, $j < k$. Upon leaving the last node k, a migration to node i takes place with probability p_i ($\sum_{i=1}^{k} p_i = 1$) so that $p_{ki} = p_i$. Assume the service rate at node j is b_j. Find the throughput rates $\{r_j\}$ and the stationary probability distribution. Give the condition that node j should be a bottleneck. If $b_1 = \cdots = b_k = b$, where is the bottleneck?

5.53. Suppose that a stream of customers arriving at a queue forms a Poisson process of rate a and that there are two servers who possibly differ in efficiency so that a customer's service time at server i is exponentially distributed with mean b_i^{-1}, for $i = 1, 2$. To ensure that equilibrium is possible, suppose $b_1 + b_2 > a$. If a customer arrives to find both servers free, he is equally likely to be allocated to either server. The queue can be represented as a Markov process where the state variable is the number in the system; the state space should be taken to be $\{0, 1a, 1b, 2, 3, \ldots\}$, where $1a$ represents one person in the system at server 1 and $1b$ represents one person in the system at server 2. What is the stationary distribution? Is the process reversible? If so, describe the departure process. (Warning: This is not a birth–death process. Compare this problem with 5.41; the difference is that, in this problem, a customer being served by the slow server cannot switch to the faster server.)

5.54. An arbitrary stochastic process $\{X(t), -\infty < t < \infty\}$ is called time reversible if for any τ we have

$$\{X(t), -\infty < t < \infty\} \overset{d}{=} \{X(\tau - t), -\infty < t < \infty\}.$$

Show such a process is stationary. If $\{X(t)\}$ is reversible, then so is $\{f(X(t))\}$ even though the latter process may not be Markov even if X is Markov (Kelly, 1979).

5.55. Use reversibility to verify that a stationary M/M/∞ has a Poisson output.

5.56. Consider a Markov chain $\{X(t), t \geq 0\}$ on the state space with generator matrix

$$A_{jk} = \begin{cases} a^j, & \text{if } k = j + 1, \\ b, & \text{if } k = 0, \\ 0, & \text{otherwise.} \end{cases}$$

Is the process regular for $a > 1$? Find the stationary distribution when $a > 1, b > 0$ (Kelly, 1979).

5.57. If $X_1(t)$ and $X_2(t)$ are two independent reversible Markov chains, then $(X_1(t), X_2(t))$ is a reversible Markov chain (Kelly, 1979).

5.58. Consider the following generalizations of Proposition 5.9.4.

(a) Suppose there are r queues in series, with $r > 2$, and the ith queue has one server working at rate b_i. The servers operate independently of each other.

(b) Suppose there are r queues in series, and the ith queue has s_i servers with each serving at exponential rate b_i. All the servers serve independently of each other.

(c) Suppose there are r queues in series, and the ith queue has one server serving at a state dependent rate of $b_i(n)$ when $Q_i(t) = n$.

In each case show that the stationary distribution has a product form.

5.59. Consider two Markovian queues in series as in Section 5.9. Compute the average number L in the system and the average wait in the system.

5.60. Consider two tandem Markovian queues as in Section 5.9. Let $W_i(t)$ for $i = 1, 2$ be the virtual waiting time in queue i; that is, $W_i(t)$ is the time that a customer would have to wait for service at the ith counter if he arrived at time t. Check that for a fixed t, $W_1(t)$ and $W_2(t)$ are independent if η is the initial distribution.

5.61. Consider two tandem Markovian queues as in Section 5.9. Compute the generator matrix of $\mathbf{Q}(\cdot) = (Q_1(\cdot), Q_2(\cdot))$, and use this to verify that the stationary distribution is a product distribution.

5.62. System Busy Period for M/M/s. Consider an M/M/s queue and define
$$T = \inf\{t \geq 0 : Q(t) < s\},$$
where, as usual, $Q(t)$ is the number in the system at time t. We are interested in T when the initial condition is $Q(0) = i$, where $i \geq s$. In this case, T has the interpretation of the first time some server is free and is called the system busy period. You should review Problem 3.26.

Show

(a) $\qquad P[T < \infty | Q(0) = s] = 1$ iff $\rho = \dfrac{a}{bs} \leq 1$.

In the case $\rho < 1$ show

(b) $\qquad E(T | Q(0) = s) = \dfrac{1}{sb(1 - \rho)},$

$$\mathrm{Var}(T | Q(0) = s) = \frac{1 + \rho}{(sb)^2(1 - \rho)^3}.$$

(c) $\qquad E(T | Q(0) = s) = \infty$ iff $\rho = 1$.

(Hint: Let $\{X(t)\}$ be a birth-death process with birth parameters a and $\mu_k = sb$ for $k \geq 1$. Argue that the distribution of T is the distribution of the first passage time of X from s to $s - 1$. Note for our problem the values of the birth-death parameters for $k \geq s$ are irrelevant. The problem is equivalent to finding the distribution of the first time X moves down by 1 and hence is equivalent to finding the distribution of the first time an M/M/1 queue moves from state 1 to state 0. This is the busy period for M/M/1. Apply Problem 3.26.

5.63. Queue Discipline. Consider the M/M/s queue under two queue disciplines. Let $\{Q_f)\}$ be the queue length process when the order of service is first come, first served, and let $\{Q_l(t)\}$ be the queue length when the order of service is last come, first served. Show

$$\{Q_f(t)\} \stackrel{d}{=} \{Q_l(t)\},$$

by showing both are birth-death processes with the same parameters. The queue length process ignores the differences between the queue disciplines, but the waiting times will be different. Let $W_l(t)$ be the virtual waiting time; i.e., the waiting time for service of a fictitious customer named FIC who arrives at time t assuming the queue discipline is last come, first served. Argue that

$$W_l(t) = \begin{cases} 0, & \text{if } Q_l(t) < s, \\ T^*, & \text{if } Q_l(t) \geq s, \end{cases}$$

where T^* has the distribution of the system busy period of an M/M/s queue with initial condition that there are s customers present at time 0. Show

$$\lim_{t \to \infty} P[W_l(t) \leq x] = \begin{cases} P[T^* \leq x], & \text{if } \rho \geq 1, \\ 1 - P[T^* > x]\frac{\eta_s}{1-\rho}, & \text{if } \rho < 1, \end{cases}$$

where the distribution of T^* is as discussed in Problem 5.62 and η_s is given as in Problem 5.25.

5.64. Harry's Bar. The ladies' room in Harry's bar has one toilet and, in accordance with local ordinances, also possesses a couch where two ladies may sit and wait their turn. Others who arrive when the couch and toilet are fully occupied must stand and wait their turn. Ladies spend an average of five minutes in the toilet, the actual time being exponentially distributed. Between 6 PM and midnight, ladies come to use the facility according to a Poisson process with a rate of one every 15 minutes.

(a) What percentage of time is the toilet busy?

(b) What is the expected time that any particular lady will have to spend standing up?

The men's room at the bar contains one toilet and no couch. Men arrive at the men's room according to a Poisson process with average time of 10 minutes between one arrival and the next. Each user of the toilet takes an exponential amount of time with mean three minutes. Arriving men who find the toilet occupied wait; the queue spills out into the hall if necessary.

(c) What is the probability that an arriving man will have to wait more than 10 minutes to use the toilet?

(d) What is the probability that an arriving man will spend more than 10 minutes away from the bar?

(e) Harry will install a second toilet in the men's room if he is convinced the expected time away from the bar is at least three minutes. What would the arrival rate to the men's room have to be to justify the extra toilet? (Cf. Problem 5.24.)

5.65. A gas station has room for seven cars including the ones at the pumps. The installation of a pump costs \$50 per week, and the average profit on a customer is 40 cents. Customers arrive in a Poisson process at a rate of four per minute, and the service times have exponential density with mean 1 minute. Find the number of pumps which will maximize the expected net profit (Prabhu, 1981).

5.66. Suppose $\{X(t), t \geq 0\}$ is a Markov chain with discrete state space S. Let $C \subset S$ be a closed set (with respect to the embedded discrete time Markov chain $\{X_n\}$). Suppose $\eta' = (\eta_i, i \in S)$ is a stationary distribution for $\{X(t)\}$. Show

$$\eta'_C = (\frac{\eta_i}{\sum_{k \in C} \eta_k}, i \in C)$$

is also a stationary distribution for $\{X(t)\}$.

5.67. Consider a two node closed queueing network with routing matrix

$$P = \begin{pmatrix} 0 & 1 \\ 1 & 0 \end{pmatrix}$$

and N people and with service rates at the two nodes being b_1, b_2. Show the stationary distribution can be written as

$$\eta(n, N - n) = \frac{a^n}{\sum_{j=0}^{n} a^j},$$

where $a = b_2/b_1$.

5.68. Kelly's Lemma. Let $\{X(t)\}$ be a Markov chain with discrete state space S and infinitesimal generator matrix A. Let η be a probability distribution on S, and let $\hat{A}$ be an infinitesimal generator matrix of some Markov chain with state space S. If

$$\eta_i A_{ij} = \eta_j \hat{A}_{ji}, \quad i, j \in S,$$

then prove

(a) η is a stationary distribution for $\{X(t)\}$ and

(b) $\hat{A}$ is the generator matrix of the reversed time process (Kelly, 1979).

(c) Suppose now that we have a two node queue in series with feedback. The service rate in node i is b_i for $i = 1, 2$, and customers arrive at node 1 according to a Poisson process of rate a. Upon service completion in node 1, customers proceed to node 2. After service completion at node 2, there is probability p $(0 < p < 1)$ that rework will be necessary, in which case the customer needing rework rejoins the queue at node 1. Let $Q_i(t)$ be the number at node i at time t for $i = 1, 2$. Then $\mathbf{Q}(t) = (Q_1(t), Q_2(t))$ is Markov. Guess what the structure of the model obtained by reversing the time should be and use Kelly's Lemma to prove that the stationary distribution is of product form

$$\eta(n, m) = (1 - \rho_1)\rho_1^n (1 - \rho_2)\rho_2^m.$$

What are the values of ρ_i, $i = 1, 2$?

(d) **Multiclass M/M/1 Queue.** Consider a first come , first served single server queue with the added wrinkle that customers are of different types $1, 2, \ldots, T$. Assume each type customer arrives via an independent Poisson process with rate a_i but that service rates are independent of customer type and equal to b. Let the state space be of the form

$$\{(n, c_1, \ldots, c_n) : n \geq 0, c_i \in \{1, \ldots, T\}\},$$

where $c_1, \ldots, c_n$ represent the types of customers present in the order of their arrival. Using Kelly's Lemma or by thinning a Poisson process of rate $a := a_1 + \cdots + a_T$, show the stationary distribution is of the form

$$\eta(n, c_1, \ldots, c_n) = \rho^n (1 - \rho) p_{c_1} \cdots p_{c_n},$$

where $\rho = a/b$ and $p_i = a_i/a$. Of course, assume $\rho < 1$. (Walrand, 1989.)

5.69. Erlang Systems, M/E$_k$/1 Queue. Consider a single server queue with Poisson input of rate a. Assume service requires k stages, and that each stage of service requires a random amount of time with exponential density with mean b^{-1}. Stages are independent of each other and of the input. Let $Q(t)$ be the number of customers in the system at time t, and let $R(t)$ be the number of residual stages a customer must pass through before exiting the queue. If at time t it happens that the queue is empty, set $R(t) = 0$. (Note that whenever a customer is traversing the stages, no customer can enter service.) For example, if at time t the customer is still in the first of the k stages, we have $R(t) = k$.

(a) Argue that the traffic intensity should be ka/b.

(b) Argue that $\{Q(t)\}$ is not Markov.

(c) Argue that $\{(Q(t), R(t))\}$ is Markov with state space $S = \{0, 1, \ldots\} \times \{0, 1, \ldots, k\}$.

(d) Check that the following are permitted transitions of this system: For $i > 0$ and $m \geq 2$,

$$(i, m) \mapsto (i, m - 1) \text{ (phase completion before arrival)}$$
$$(i, m) \mapsto (i + 1, m) \text{ (arrival before phase completion)}$$

and, for $i > 0$,

$$(0, 0) \mapsto (1, k) \text{ (arrival to empty system)}$$
$$(1, 1) \mapsto (0, 0) \text{ (phase completion and customer leaves)}$$
$$(i, 1) \mapsto (i - 1, k) \text{ (departure before arrival)}.$$

Compute the generator of this Markov process. Write out the equations determining the stationary distribution.

(e) Let $Q_1(t)$ be the total number of stages of work in the system; i.e.,

$$Q_1(t) = \begin{cases} 0, & \text{if } Q(t) = 0, \\ k(Q(t) - 1) + R(t), & \text{if } Q(t) > 0 \end{cases}.$$

Check that $\{Q_1(t)\}$ is Markov. Give the generator. Write out the equations determining the stationary distribution.

The equations in (d) and/or (e) can be solved using generating functions. The solution requires skill and patience (cf. Prabhu, 1981; Wolff, 1989). One obtains

$$EQ(\infty) = L = \rho + \frac{\rho^2}{2(1 - \rho)}(1 + \frac{1}{k}),$$

where ρ is the traffic intensity assumed to be less than 1. For the mean of the limit distribution of the virtual waiting time, under the same conditions we have

$$EW(\infty) = \frac{k + 1}{2ka} \frac{\rho^2}{1 - \rho}.$$

Note that both L and $EW(\infty)$ decrease as k increases (assuming ρ is held fixed). The square coefficient of variation (Problem 5.50) decreases with k. The more regular the system, the better the performance.

(f) The Optima Street Computer Company carefully checks each 486 class computer shipped to a customer. Machines arrive at the inspector according to a Poisson process rate of five per hour. Inspection consists of 10 separate tests, each taking one minute on average. The tests must be done sequentially in an order specified by company policy. Actual times for each test were found to be roughly exponentially distributed.

(a) Find the average waiting time a machine experiences.

(b) Find the average number of sets with the inspector.

(c) Suppose there were only two tests that had to be done. The company has the choice of hiring two inspectors with each inspector specializing in one test or hiring one inspector to do both the tests. Assume the single inspector would work at rate $2b$ on each inspection and that each of the single inspectors would work at rate b. Compare the two schemes from the point of view of the average waiting time a machine experiences. Assume machines arrive according to a Poisson process rate a. (Distinguish between single server queues in series, i.e., tandom queues, and single server queues with Erlang service times.)

5.70. The Arrival Theorem. Review or do problems 2.26, 2.53 and 2.55. Consider a continuous time Markov chain $\{X(t), t \geq 0\}$ with state space S, an embedded jump process $\{X_n\}$ with transition probability matrix Q. Assume the continuous time process jumps at $\{T_n\}$ and, as usual, the generator matrix is A. Suppose the stationary distribution η exists. Suppose further that

$$\beta = \sum_i \eta_i \lambda(i) < \infty.$$

Check that this condition implies that $\{X_n\}$ has a stationary distribution π.

Suppose $J \subset S^2 \setminus \{(s,s) : s \in S\}$. Think of the elements of J as highlighted state transitions. Such movements as occur in J can be interpreted as traffic due to customers moving about a queueing system. For example, if $J = \{(n, n-1) : n \geq 1\}$, we could interpret J as transitions corresponding to departures.

Define $\lambda = \sum_{(i,j) \in J} \eta_i A_{ij}$ and assume $\lambda > 0$. Also define for $n \geq 0$

$$s_1 = \inf\{n : (X_n, X_{n+1}) \in J\}$$
$$s_{n+1} = \inf\{k > s_n : (X_k, X_{k+1}) \in J\}.$$

Let

$$S_n = T_{s_n+1}$$

be the time of the nth jump within J, and let the counting function be $K(t) = \sum_n 1_{[S_n \leq t]}$.

(a) Show $\{(X(t), K(t), t \geq 0\}$ is Markov. What is its generator? Express $EK(t)$ in terms of A and $\{P[X(u) = j], j \in S, u \leq t\}$.

(b) (Harder) If for every t we have $X(t)$ and $K(t)$ independent, then $\{K(t), t \geq 0\}$ is Poisson. What is the mean function? (Check that K has independent increments.)

(c) $\{X(T_{s_n+1}), n \geq 1\}$ is a Markov chain with stationary distribution

$$\pi_j^+ = \lambda^{-1} \sum_i \eta_i A_{ij} 1_{[(i,j) \in J]}.$$

(Recall that $\{Z_n = (X_n, X_{n+1}\}$ is Markov and so is $\{Z_{s_n}\}$. With the stationary distribution as an initial distribution, $\{Z_{s_n}\}$ is stationary, and hence so are the marginal processes. It remains to show that the second marginal process is Markov. But $\{X_{s_n}\}$ is $\{X_n\}$ sampled at successive iterates of a stopping time and hence is a Markov chain (cf. Problem 2.55).)

(d) By time reversal, show that $\{X(T_{s_n}), n \geq 1\}$ is a Markov chain with stationary distribution

$$\pi_i^- = \lambda^{-1} \sum_j \eta_i A_{ij} 1_{[(i,j) \in J]}.$$

(e) (Harder) For every $t \geq 0$ we have $\{K(t+s) - K(t), s \geq 0\}$ independent of $X(t)$ with respect to P_η iff $\pi_j^- = \eta_j$ for all $j \in S$. In this case K is Poisson with rate λ.

(f) (Harder) By time reversal show that for every $t \geq 0$ we have $\{K(s), s \leq t\}$ and $X(t)$ independent with respect to P_η iff $\pi_j^+ = \eta_j$ for all $j \in S$. In this case K is Poisson with rate λ. (Walrand, 1988, page 75; Melamed, 1979).

5.71. Open Networks with State Dependent Service Rates. Consider an open network with k nodes such that if n_j customers are present at node j, then the service time is exponential with parameter $b_j(n_j)$. Verify that the invariant measure is of the form

$$\eta_{\mathbf{n}} = \prod_{j=1}^{k} \frac{r_j^{n_j}}{\prod_{i=1}^{n_j} b_j(i)}.$$

Give a condition that a stationary probability distribution exists.

5.72. The Batch Markovian Arrival Process. (a) A Poisson process of buses arrives at Harry's restaurant according to a Poisson process of rate a. Each bus contains k hungry passengers with probability p_k. Let $N(t)$ be the total number of arrivals by time t. Show $\{N(t), t \geq 0\}$ is Markov with generator of the form

$$A = \begin{pmatrix} d_0 & d_1 & d_2 & d_3 & \cdots \\ 0 & d_0 & d_1 & d_2 & \cdots \\ 0 & 0 & d_0 & d_1 & \cdots \\ \vdots & & & \ddots & \end{pmatrix}.$$

Give an expression for $\{d_i\}$.

(b) Consider the following generalization of (a). We have a two dimensional Markov chain $\{(N(t), J(t)), t \geq 0\}$ on the state space $S = \{(i, j) :$

$i \geq 0, 1 \leq j \leq m$}. Think of $N(t)$ as a counting variable, and think of $J(t)$ as an environmental variable which helps to determine the distribution of $N(t)$. The infinitesimal parameters are specified as follows: For the holding time parameters in a state (l, i) we have

$$\lambda(l, i) = \lambda(i),$$

and the embedded discrete time Markov chain has transition probabilities of the form

$$Q_{(l,i),(l+j,k)} = p_i(j, k),$$

which we interpret as the probability starting from (l, i) that the environment changes from i to k and that a batch of size j arrives.

(b1) Check that the infinitesimal generator A is

$$A_{(l,i),(l,i)} = -\lambda(l, i) = \lambda(i)$$
$$A_{(l,i),(l+j,k)} = \lambda(i)p_i(j, k)$$

and that, if we order S as

$$S = \{(0, 1), (0, 2), \ldots, (0, m), (1, 1), \ldots, (1, m), \ldots$$

this matrix can be written as

$$A = \begin{pmatrix} D_0 & D_1 & D_2 & \cdots \\ & D_0 & D_1 & D_2 & \cdots \\ & & D_0 & D_1 & \cdots \\ \vdots & & & \ddots \end{pmatrix},$$

where D_k, $k \geq 0$ are $m \times m$ matrices, D_0 has negative diagonal elements and nonnegative off-diagonal elements and $D_k, k \geq 1$, are nonnegative. Specify the matrices $D_k, k \geq 0$.

(b2) Check that $\{J(t), t \geq 0\}$ is Markov with infinitesimal parameters $\lambda(i)$ and $Q_{ik} = \sum_{j=0}^{\infty} p_i(j, k)$ and, therefore, for $k \neq i$,

$$A_{ik} = \sum_{j=0}^{\infty} \lambda(i)p_i(j, k) = \sum_{j=0}^{\infty} D_j(i, k).$$

Assume $\{J(t)\}$ is irreducible, so that it is a Markov chain with generator matrix $A = \sum_{j=0}^{\infty} D_j$.

(b3) Let $\eta_j, 1 \leq j \leq m$, be the stationary distribution of J so that

$$\eta' D = 0.$$

Show the long run arrival rate for the arrival process $N(t)$ is

$$\eta' \sum_{k=1}^{\infty} k D_k \mathbf{1},$$

where $\mathbf{1}$ is a column vector of 1's.

(b4) Define

$$P_{ij}(n,t) = P[N(t) = n, J(t) = j | N(0) = 0, J(0) = i].$$

From the Chapman-Kolmogorov equations show that the matrices $P(n,t)$ satisfy

$$P'(n,t) = \sum_{j=0}^{n} P(j,t) D_{n-j}, \quad n \geq 1, t \geq 0,$$

$$P(0,0) = I.$$

Set $(0 \leq s \leq 1)$

$$P^*(s,t) = \sum_{n=0}^{\infty} P(n,t) s^n$$

so that

$$\frac{d}{dt} P^*(s,t) = P^*(s,t) D(s), \quad P^*(s,0) = I,$$

and therefore

$$P^*(s,t) = e^{D(s)t}.$$

How can one compute $EN(t)$?

(b5) Discuss: D_0 governs transitions that correspond to no arrivals, and for $j \geq 1$, we have D_j governing transitions that correspond to arrivals of batches of size j.

(b6) If $D_0 = -\alpha, D_1 = \alpha$, check that N is Poisson with rate α.

(b7) If $D_j = p_j D_1$ for $j \geq 1$ where $\{p_j, j \geq 1\}$ is a probability distribution, discuss the structure of N.

(b8) Discuss why the Markov modulated Poisson process (Problem 5.3) is an example of a batch Markovian arrival process (Lucantoni, 1991; Neuts, 1989.)

5.73. Terminal Feedback. Consider an open network where $a_1 = a, a_2 = \cdots = a_k = 0$. Customers enter the system only at node 1. Routing is as in Problem 5.52 in that $p_{i,i+1} = 1$, $i = 1, \ldots, k-1$ and feedback from k occurs in that $p_{kj} = p_j$, $j = 1, \ldots, k$. Suppose $q = 1 - \sum_{j=1}^{k} p_j > 0$, so from node k there is a possibility of exiting from the system. Assume the service rate at node j is b_j. Find the throughput rates and the condition that a stationary distribution exists. Give the stationary distribution when it exists. Specialize to the case $b_1 = \cdots = b_k$.

5.74. Reversing an Open Network. Consider a stationary open Markovian network $\{Q(t), -\infty < t < \infty\}$ with k nodes, input rates $a_1, \ldots, a_k$, service rates $b_1, \ldots, b_k$, throughput rates $r_1, \ldots, r_k$ and routing matrix $(p_{ij}, 1 \le i, j \le k)$. Let $\hat{Q}(t) = Q(-t)$ be the reversed process.

(a) Compute $\hat{A}$ the generator of $\hat{Q}$.

(b) Conclude $\hat{Q}$ is a Markov chain model of an open queue with parameters

$$\hat{a}_j = q_j r_j, \quad \hat{b}_j = b_j, \quad \hat{p}_{ij} = p_{ji} \frac{r_j}{r_i}.$$

What is the stationary distribution of $\hat{Q}$?

(c) Conclude that the departure processes which count exits from the different nodes of Q are independent Poisson processes. From node i, departures from the system are Poisson with rate $r_i q_i$.

5.75. Apportioning Effort in an Open Network. Consider an open network with k nodes. Suppose $a_1 = \cdots = a_k = 1$, $b_1, \ldots, b_k$ are the service rates and $r_1, \ldots, r_k$ are the throughput rates. Suppose the stationary distribution exists. Let $Q(\infty) = (Q_1(\infty), \ldots, Q_k(\infty))$ be a vector with the stationary distribution. Set $L_j = EQ_j(\infty)$.

(a) Compute L_j for $j = 1, \ldots, k$ and $\sum_{j=1}^{k} L_j$.

(b) Suppose we can control $b_1, \ldots, b_k$ subject to $\sum_{j=1}^{k} b_j = b$. How do we assign service rates $b_1, \ldots, b_k$ so as to minimize $\sum_{j=1}^{k} L_j$, subject to the constraint that $\sum_{j=1}^{k} b_j = b$? (Hint: Use the Lagrange method.) (Kelly, 1979.)

5.76. Let $\eta = (\eta_0, \eta_1, \ldots)$ be a probability distribution. Show there is a birth-death process with η as the stationary distribution.

5.77. Harry Visits the Customs Office. Harry is sent a present from a relative living abroad and must go to the customs office to claim it. He takes his statistically inclined niece along and together they become fascinated by the traffic patterns in the office. There are three windows. Harry's niece notices that external arrivals at the windows follow independent Poisson processes with rates 5, 10, 15. Service times at the three windows seem to be exponentially distributed with rates 10, 50, 100. A customer completing his business at window 1 seems equally likely to migrate to window 2, window 3 or leave the office. A customer departing from window 2 always goes to window 3. A customer from window 3 is equally likely to migrate to window 2 or leave the system.

(a) Compute the stationary distribution for this system.

(b) Find the average number in the system in equilibrium (with respect to the stationary distribution) and the average time spent in the system by a customer.

5.78. Consider a closed network with two nodes and two customers. Suppose the routing matrix is

$$P = \begin{pmatrix} .5 & .5 \\ .5 & .5 \end{pmatrix}.$$

Let the service rates be b_1, b_2. Find the stationary distribution and the average number at node i, $i = 1, 2$.

5.79. Consider an infinite server queue. Service times at each server are iid exponentially distributed random variables with mean b^{-1}. Customers arrive in pairs, the arrival times constituting a Poisson process with mean rate a. Let $Q(t)$ be the number of customers in the system at time t. Assuming that the service times and the interarrival times are independent, argue that Q is a Markov chain and write down the generator of $\{Q(t)\}$. If $Q(0) = 0$, find the expected time at which the last of the first pair of arrivals leaves the system.

5.80. Particles arrive at the surface of a bacterium according to a Poisson process with mean rate λ. Each arriving particle is either of type 2 with probability β $(0 < \beta < 1)$ or of type 1 with probability $1 - \beta$. The state $X(t)$ of the bacterium at time t is a Markov chain with state space $S = \{0, 1, 2\}$ and initial value $X(0) = 0$. If the bacterium is in state 0, then an arriving type 2 particle takes it to state 2 where it stays forever, and an arriving type 1 particle takes it to state 1, where it remains for an exponentially distributed time with mean μ^{-1} before returning to state 0. If the bacterium is in states 1 or 2 then arriving particles have no effect.

Show that

$$P[X(t) = 2 | X(0) = 0] = 1 - \frac{\beta\lambda}{s_2 - s_1}\left((1 + \frac{\mu}{s_1})e^{s_1 t} - (1 + \frac{\mu}{s_2})e^{s_2 t} \right),$$

where s_1, s_2 are the roots of the equation

$$s^2 + s(\lambda + \mu) + \beta\lambda\mu = 0.$$

5.81. The Poisson Process is Markov.

(a) Treat a Poisson process $N(t), t \geq 0$, as a Markov process and write down the generator.

(b) Now pretend you know only the generator. Solve the forward equations to determine

$$P[N(t) = n | N(0) = 0], \quad n \geq 0.$$

5.82. Happy Hour. Harry's happy hour customers arrive according to a Poisson process of rate α. Each customer is either a yuppie or a computer nerd with respective probabilities p and q (where $p, q > 0$, $p + q = 1$) independent of past arrivals. Let $X(t)$ be the type of the last arrival before time t. Show $\{X(t)\}$ is a Markov chain, and give the transition probability matrix $P(t)$.

5.83. The Order Statistic Property for a Linear Birth Process. (a) Consider the linear birth process of Section 5.11. Prove a version of the order statistic property: Given n births in $(0, t]$, what is the joint conditional distribution of the times of the n births?

(b) Generalize Problem 4.53 to the non-homogeneous case: Let $N = \sum_k \epsilon_{T_k}$ be a point process on $(0, \infty)$ with points $\{T_k\}$ satisfying $0 < T_1 < T_2 \ldots$ and set $N(t) = N((0, t])$. Suppose $m(t) = EN(t) < \infty$. Also suppose that for each $t > 0$ there is a distribution function $F_t(x)$ such that the conditional distribution of $(T_1, \ldots, T_n)$ given $N(t) = n$ is the same as the distribution of the order statistics from a sample of size n from the distribution function $F_t(x)$. Show that

$$F_t(x) = \frac{m(x)}{m(t)}, \quad 0 < x \le t.$$

Then show that there exists a homogeneous Poisson process N^* on $(0, \infty)$ and an independent non-negative random variable W, both defined on the same probability space as N, such that

$$N(\cdot) = N^*(Wm(\cdot)).$$

(c) Combine (a) and (b) to provide an alternate proof of Kendall's result in Section 5.11. (Feigin, 1979; Neuts and Resnick, 1971; Waugh, 1970; Keiding, 1974.)

CHAPTER 6

Brownian Motion

THE BROWNIAN motion process, sometimes called the Wiener process, was originally posed by the English botanist Robert Brown as a model for the motion of a small particle immersed in a liquid and thus subject to molecular collisions. Brownian motion assumes a central role in the modern theory of stochastic processes and in the modern large sample theory of statistics. It is basic to descriptions of financial markets, the construction of a large class of Markov processes called *diffusions*, approximations to many queueing models and the calculation of asymptotic distributions in large sample statistical estimation problems.

6.1. INTRODUCTION.

We will need some notation before the basic definitions. Let $C[0, \infty)$ denote the collection of real valued continuous functions defined on the domain $[0, \infty)$. We denote the normal distribution function with mean μ and variance σ^2 by $N(\mu, \sigma^2, x)$, and similarly, the normal density will be denoted by $n(\mu, \sigma^2, x)$. We write $N(x) = N(0, 1, x)$ and $n(x) = N'(x) = n(0, 1, x)$.

Definition. The stochastic process $B = \{B(t), t \geq 0\}$ is *standard Brownian motion* if the following holds:

    (1) B has independent increments.
    (2) For $0 \leq s < t$,

$$B(t) - B(s) \sim N(0, t - s),$$

        meaning the increment $B(t) - B(s)$ is normally distributed with mean 0 and variance equal to the length of the increment separating s and t.
    (3) With probability 1, paths of B are continuous; that is,

$$P[B \in C[0, \infty)] = 1.$$

    (4) $B(0) = 0.$

Property (4) is merely a normalization and is a convenience rather than a basic requirement. If a process $\{W(t), t \geq 0\}$ satisfied the first three postulates but not the last, then the process $\{W(t) - W(0), t \geq 0\}$ would be standard Brownian motion. Property (3) follows from (1) and (2) in the sense that given a process W satisfying (1) and (2), there always exists a version satisfying (3) (see Doob, 1953) so properties (1) and (2) are the most basic. Property (1) says incremental behavior in one interval does not affect incremental behavior in a disjoint interval. This is a rather strong assumption. Likewise, assuming Brownian increments are normally distributed in property (2) is also a rather special assumption.

If you have nagging doubts about whether there really exists a process satisfying properties (1)–(4), these should be laid to rest by reading Section 6.3, which shows how to construct Brownian motion starting from a supply of iid $N(0,1)$ random variables.

Brownian motion can be thought of as a continuous time approximation of a random walk where the size of the steps is scaled to become smaller (necessary if an approximation is to have continuous paths) and the rate at which steps are taken is speeded up. Let $\{X_n, n \geq 0\}$ be iid with $EX_n = 0$ and $\mathrm{Var}(X_n) = 1$. Define the random walk by $S_0 = 0$, and, for $n \geq 1$, let $S_n = X_1 + \cdots + X_n$. Define the continuous time process

$$(6.1.1) \qquad B_n(t) = \frac{S_{[nt]}}{\sqrt{n}}, \quad t \geq 0,$$

where $[t]$ is the integer part of t, i.e., the greatest integer less than or equal to t. Then we have that

$$(6.1.2) \qquad B_n \Rightarrow B,$$

where "$\Rightarrow$" denotes convergence in distribution. For the time being, we interpret (6.1.2) as meaning convergence of the finite dimensional distributions, that is, for any k and $0 \leq t_1 < \cdots < t_k$ and real numbers $x_1, \ldots, x_k$, we have

$$\lim_{n \to \infty} P[B_n(t_i) \leq x_i, i = 1, \ldots, k] = P[B(t_i) \leq x_i, i = 1, \ldots, k].$$

To check (6.1.2), we rely on the central limit theorem. First note that for any $t > 0$ we have

$$\frac{S_{[nt]}}{\sqrt{n}} = \frac{S_{[nt]}}{\sqrt{[nt]}} \sqrt{\frac{[nt]}{n}}$$

and since

$$\frac{S_n}{\sqrt{n}} \Rightarrow N(0,1),$$

by the central limit theorem, where $N(0,1)$ is a normal random variable, we have that

$$\frac{S_{[nt]}}{\sqrt{n}} \Rightarrow tN(0,1) \overset{d}{=} B(t).$$

Also, for $t > s$, we have

$$B_n(t) - B_n(s) = \frac{\sum_{j=[ns]+1}^{[nt]} X_j}{\sqrt{n}}$$

$$\overset{d}{=} \frac{S_{[nt]-[ns]}}{\sqrt{n}}$$

$$\Rightarrow (t-s)N(0,1) \overset{d}{=} B(t) - B(s).$$

Since the variables

$$(B_n(t_1), B_n(t_2) - B_n(t_1), \ldots, B_n(t_k) - B_n(t_{k-1}))$$

are independent due to being composed of sums of X's from disjoint blocks, we have joint convergence:

$$P[B_n(t_1) \le x_1, B_n(t_2) - B_n(t_1) \le x_2, \ldots, B_n(t_k) - B_n(t_{k-1}) \le x_k]$$
$$= P[B_n(t_1) \le x_1]P[B_n(t_2) - B_n(t_1) \le x_2] \ldots P[B_n(t_k) - B_n(t_{k-1}) \le x_k]$$
$$\to P[B(t_1) \le x_1]P[B(t_2) - B(t_1) \le x_2] \ldots P[B(t_k) - B(t_{k-1}) \le x_k].$$

Therefore,

$$(B_n(t_1), B_n(t_2) - B_n(t_1), \ldots, B_n(t_k) - B_n(t_{k-1}))$$
$$\Rightarrow (B(t_1), B(t_2) - B(t_1), \ldots, B(t_k) - B(t_{k-1}))$$

in R^k, and applying the map

$$(b_1, b_2, \ldots, b_k) \mapsto (b_1, b_1 + b_2, \ldots, b_1 + b_2 + \cdots + b_k)$$

from $R^k \mapsto R^k$ yields the k-dimensional convergence

$$(B_n(t_1), B_n(t_2), \ldots, B_n(t_k)) \Rightarrow (B(t_1), B(t_2), \ldots, B_n(t_k)).$$

The invariance principle. The convergence in (6.1.2) is, in fact, stronger than convergence of the finite dimensional distributions. We now explain this stronger form of convergence which is called an *invariance principle* or *functional central limit theorem*. Suppose we make $C[0, \infty)$ into a metric space in such a way that convergence in the metric space

$C[0, \infty)$ is the same as local uniform convergence. The usual metric is defined for $f, g \in C[0, \infty)$ by

$$\rho(f, g) = \sum_{n=1}^{\infty} \frac{1 \wedge \sup_{0 \le t \le n} |f(t) - g(t)|}{2^n}.$$

Thus for $f_n \in C[0, \infty)$, $n \ge 0$, $\rho(f_n, f_0) \to 0$ if and only if for any k

$$\lim_{n \to \infty} \sup_{0 \le t \le k} |f_n(t) - f_0(t)| = 0.$$

The process $\{B_n(t), t \ge 0\}$ defined in (6.1.1) is not continuous, so we modify it so that the paths of the modification are in $C[0, \infty)$. Define

$$B_n^{(c)}(t) = \frac{S_{[nt]}}{\sqrt{n}} + (nt - [nt]) \frac{X_{[nt]+1}}{\sqrt{n}}, \quad t \ge 0.$$

This is a continuous function which is obtained by placing dots in the $[0, \infty) \times R$ plane at points $\{(\frac{i}{n}, \frac{S_i}{\sqrt{n}}), j \ge 0\}$ and then connecting the dots. See Figure 6.1.

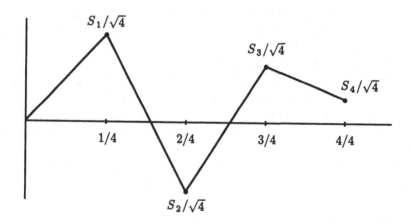

FIGURE 6.1.

The strengthening of (6.1.2) is that for any real valued function

$$T : C[0, \infty) \mapsto R,$$

which is continuous from its metric space domain $C[0, \infty)$ into its range R, we have

$$T(B_n^{(c)}) \Rightarrow T(B).$$

For instance, if $T(f) = \sup_{0 \leq t \leq 1} f(t)$ then, as $n \to \infty$,

$$T(B_n^{(c)}) = \sup_{0 \leq t \leq 1} \frac{S_{[nt]}}{\sqrt{n}}$$

$$= \frac{\sup_{0 \leq j \leq n} S_j}{\sqrt{n}}$$

$$\Rightarrow \sup_{0 \leq t \leq 1} B(t) =: M(1).$$

The distribution of $M(t)$ will be found in Section 6.5, and thus we obtain

$$\lim_{n \to \infty} P[\frac{\sup_{0 \leq j \leq n} S_j}{\sqrt{n}} \leq x] = P[M(1) \leq x].$$

What makes an invariance principle so powerful and flexible is something called the *continuous mapping theorem*: If $\psi : C[0, \infty) \to \mathcal{X}$ is any map from $C[0, \infty)$ into a nice (complete and separable) metric space $\mathcal{X}$ and ψ satisfies

(6.1.3) $\qquad P[B \in \{f \in C[0, \infty) : \psi \text{ is continuous at } f\}] = 1,$

then

$$\psi(B_n^{(c)}) \Rightarrow \psi(B)$$

as random elements of the space $\mathcal{X}$. If ψ is continuous (i.e., continuous at all $f \in C[0, \infty)$), then (6.1.3) is automatically satisfied. However, condition (6.1.3) says that ψ does not have to be continuous everywhere, only on Brownian paths constituting a full set.

This helps explain the importance of Brownian motion in large sample statistics. An estimator which is a function of partial sums of a random sample will frequently converge in distribution to some function of Brownian motion. An example of this use for Brownian motion is given in Section 6.10.

Another reason for the importance of Brownian motion is the role it plays in constructing diffusions. Briefly, imagine a continuous path process $\{X(t), t \geq 0\}$ which, given that it is at a state x of its state space, adds an increment which is $N(\mu(x), \sigma^2(x))$ distributed. A way to think about such a process is through stochastic differentials:

(6.1.4) $\qquad dX(t) = \mu(X(t)) \, dt + \sigma(X(t)) \, dB(t),$

so that the change in the process at time t results from a drift of $\mu(X(t))$ and a Brownian increment with variance $\sigma^2(X(t))$. The quantities $\mu(x)$ and $\sigma^2(x)$ are called the infinitesimal mean and variance respectively. Existence and properties of solutions of the equations (6.1.4) is the subject of the theory of *stochastic differential equations* and such processes transform according to rules called the *Ito calculus*.

6.2. PRELIMINARIES.

Before undertaking the study of the properties of Brownian motion, we pause to survey some results needed for later work.

6.2.1. BOREL-CANTELLI LEMMA. For events $\{A_n\}$, define

$$[\,A_n \text{ i.o. }\,] = [\sum_n 1_{A_n} = \infty]$$
$$= \cap_{m=1}^{\infty} \cup_{n \geq m} A_n$$
$$= \{\omega : \omega \in A_n \text{ for infinitely many } n\}$$

to be the event that infinitely many of the events A_n occur.

Borel-Cantelli Lemma 6.2.1. *If* $\sum_n PA_n < \infty$, *we have*

$$P[\,A_n \text{ i.o. }\,] = 0.$$

If the events $\{A_n\}$ *are independent, the converse is true as well.*

Proof. If $\sum_n PA_n < \infty$ then $\infty > \sum_n PA_n = E\sum_n 1_{A_n}$, and thus $P[\sum_n 1_{A_n} < \infty] = 1$ which is the same as $P[A_n \text{ i.o. }] = 0$.

Conversely, if the events $\{A_n\}$ are independent and $\sum_n PA_n = \infty$, then, for $0 < s < 1$,

$$Es^{\sum_n 1_{A_n}} = \prod_n Es^{1_{A_n}} = \prod_n(1 - (1-s)PA_n)$$

by independence. From Lemma 2.9.1, this is zero since

$$\sum_n(1 - (1 - (1-s)PA_n)) = (1-s)\sum_n PA_n = \infty.$$

Therefore, $P[\sum_n 1_{A_n} = \infty] = 1$, as required. ∎

Mill's ratio for the normal distribution tail. Calculations involving Brownian motion invariably involve the normal density and distribution and facility with handling tail probabilities is essential. The following result saves much effort.

Lemma 6.2.2. Mill's Ratio. *As* $x \to \infty$,

(6.2.1) $1 - N(x) \sim n(x)/x$

and

(6.2.2) $\frac{x}{1+x^2}e^{-x^2/2} < \int_x^{\infty} e^{-u^2/2}du < \frac{1}{x}e^{-x^2/2}.$

Proof. We have

$$\frac{1}{x^2} \int_x^\infty e^{-u^2/2} duv > \int_x^\infty \frac{1}{u^2} e^{-u^2/2} du.$$

An integration by parts gives $(d(-\frac{1}{u}) = \frac{du}{u^2})$ the right side equal to

$$= \frac{1}{x} e^{-x^2/2} - \int_x^\infty e^{-u^2/2} du.$$

Thus

$$\left(1 + \frac{1}{x^2}\right) \int_x^\infty e^{-u^2/2} du \geq \frac{1}{x} e^{-x^2/2},$$

which is equivalent to the left hand inequality in (6.2.2). Furthermore,

$$\frac{1}{x} e^{-x^2/2} = \int_x^\infty 1 e^{-u^2/2} du + \int_x^\infty \frac{1}{u^2} e^{-u^2/2} du$$

$$= \int_x^\infty \left(1 + \frac{1}{u^2}\right) e^{-u^2/2} du$$

$$\geq \int_x^\infty e^{-u^2/2} du,$$

which is the right hand inequality in (6.2.2). ∎

The Cauchy and normal densities. Suppose N_1, N_2 are iid normal random variables with mean 0 and variance 1. Then the ratio $N_1/|N_2|$ has a Cauchy density

(6.2.3) $$f(x) = \frac{1}{\pi(1 + x^2)}, \quad -\infty < x < \infty,$$

with distribution function

(6.2.4) $$F(x) = \frac{1}{2} + \pi^{-1} \arctan x, \quad -\infty < x < \infty.$$

To see this, observe that for $x > 0$

$$P[\frac{N_1}{|N_2|} > x] = \int_{v>0} P[N_1 > vx] P[|N_2| \in dv].$$

Differentiating with respect to x, the density is

$$f(x) = \int_0^\infty vn(vx)P[|N_2| \in dv]$$

$$= \int_0^\infty vn(vx)(n(v) + n(-v))dv$$

$$= 2 \int_0^\infty vn(vx)n(v)dv$$

$$= \frac{2}{2\pi} \int_0^\infty ve^{-\frac{1}{2}(v^2x^2+v^2)}dv.$$

Making the change of variable $y = \frac{v^2}{2}(1+x^2)$ yields

$$= \frac{1}{\pi(1+x^2)} \int_0^\infty e^{-y}dy$$

$$= \frac{1}{\pi(1+x^2)}$$

as desired. The formula for the distribution function can be verified by differentiating (6.2.4) to get (6.2.3).

6.3. CONSTRUCTION OF BROWNIAN MOTION*.

There are numerous ways to construct Brownian motion. One way uses as primitive building blocks a supply of iid $N(0,1)$ random variables.

Theorem 6.3.1. *Brownian motion exists and may be constructed from a sequence of iid $N(0,1)$ random variables.*

The construction depends heavily on the following elementary lemma, which uses the fact that normal random variables are independent if and only if they are uncorrelated.

Lemma 6.3.2. *Suppose we have two random variables $X(s), X(t)$ defined on the same probability space such that $X(t) - X(s)$ has a $N(0, t-s)$ distribution. Then there exists a random variable $X(\frac{t+s}{2})$ defined on the same space such that*

$$X(\frac{t+s}{2}) - X(s) \stackrel{d}{=} X(t) - X(\frac{t+s}{2}),$$

---

* This section contains advanced material which may be skipped by beginning readers, who are advised to read the statement of Theorem 6.3.1 and then proceed to Section 6.4.

with $X(\frac{t+s}{2}) - X(s)$ and $X(t) - X(\frac{t+s}{2})$ independent and each having a $N(0, \frac{t-s}{2})$ distribution.

Proof of Lemma 6.3.2. Define $U := X(t) - X(s)$. Suppose $V \sim N(0, t-s)$ is independent of U, and define $X(\frac{t+s}{2})$ by

$$X(t) - X(\frac{t+s}{2}) = \frac{U+V}{2},$$

$$X(\frac{t+s}{2}) - X(s) = \frac{U-V}{2},$$

so that

$$X(t) - X(s) = U$$

(6.3.1) $$X(t) + X(s) - 2X(\frac{t+s}{2}) = V.$$

Thus $X(t) - X(\frac{t+s}{2}) \overset{d}{=} X(\frac{t+s}{2}) - X(s)$, and both have a $N(0, \frac{t-s}{2})$ distribution.

To check independence, merely note

$$E(U+V)(U-V) = EU^2 + 0 - 0 - EV^2 = 0,$$

since U and V are independent and identically distributed. ∎

For later use, observe from (6.3.1) that

(6.3.2) $$\left| X(\frac{t+s}{2}) - \frac{X(t) + X(s)}{2} \right| = \frac{1}{2}|V|, \quad V \sim N(0, t-s).$$

Construction of Brownian motion. Now we show how to construct Brownian motion on $[0, 1]$. The extension of the construction to $[0, \infty)$ is easy.

Find a probability space in which there is a supply of independent normal random variables:

$$V(1), V(\tfrac{1}{2}), V(\tfrac{1}{4}), V(\tfrac{3}{4}) \dots,$$

that is,

$$\{V(\frac{k}{2^n}), k = 1, 2, \dots, 2^n, n \geq 1\}.$$

It will turn out that we need

$$V(\frac{i}{2^{n+1}}) \sim N(0, \frac{1}{2^n}).$$

Define $X(0) = 0, X(1) = V(1)$, and use $V(\frac{1}{2})$ to construct $X(\frac{1}{2})$ using the procedure given in Lemma 6.3.2, so that $X(\frac{1}{2}) - X(0)$ and $X(1) - X(\frac{1}{2})$ are iid $N(0, \frac{1}{2})$ random variables.

Now we use induction: Suppose $\{X(\frac{k}{2^n}), 0 \le k \le 2^n\}$ is defined using $\{V(\frac{k}{2^n}), 0 \le k \le 2^n\}$ where $\{X(\frac{k}{2^n}) - X(\frac{k-1}{2^n}), 1 \le k \le 2^n\}$ are iid with $N(0, \frac{1}{2^n})$ distributions. For each $k \le 2^n$, construct $X(\frac{2k+1}{2^{n+1}})$ using $V(\frac{2k+1}{2^{n+1}})$ in such a way that

$$X(\frac{2k+1}{2^{n+1}}) - X(\frac{k}{2^n}) \overset{d}{=} X(\frac{k+1}{2^n}) - X(\frac{2k+1}{2^{n+1}}) \sim N(0, \frac{1}{2^{n+1}}),$$

and the sequence of random variables $\{X(\frac{k}{2^{n+1}}), 0 \le k \le 2^{n+1}\}$ has independent increments.

For each $n = 1, 2, \ldots$, define the process $\{B^{(n)}(t), t \ge 0\}$ by

$$B^{(n)}(t, \omega) = X(t, \omega), \quad \text{for } t \in \{\frac{k}{2^n}, 0 \le k \le 2^n\}$$

and make $B(n)$ linear in each interval $[\frac{k}{2^n}, \frac{k+1}{2^n}]$. Also define

$$\Delta^{(n)}(\omega) = \max_{0 \le t \le 1} \left| B^{(n+1)}(t, \omega) - B^{(n)}(t, \omega) \right|$$

$$= \max_{0 \le k \le 2^n - 1} \max_{\frac{k}{2^n} \le t \le \frac{k+1}{2^n}} \left| B^{(n+1)}(t, \omega) - B^{(n)}(t, \omega) \right|.$$

In Figure 6.2, the solid line represents $B^{(n)}(t)$ on the interval $[\frac{k}{2^n}, \frac{k+1}{2^n}]$, and the dotted line is $B^{(n+1)}(t)$ on $[\frac{k}{2^n}, \frac{k+1}{2^n}]$.

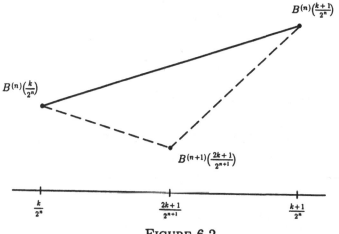

FIGURE 6.2.

From the figure or geometry,

$$\max_{\frac{k}{2^n} \leq t \leq \frac{k+1}{2^n}} |B^{(n+1)}(t) - B^{(n)}(t)| = \left| \frac{B^{(n)}(\frac{k+1}{2^n}) + B^{(n)}(\frac{k}{2^n})}{2} - B^{(n+1)}(\frac{2k+1}{2^{n+1}}) \right|$$

$$= \left| \frac{X(\frac{k+1}{2^n}) + X(\frac{k}{2^n})}{2} - X(\frac{2k+1}{2^{n+1}}) \right|,$$

which from (6.3.2), is

$$= \frac{1}{2}|V(\frac{2k+1}{2^{n+1}})|.$$

Therefore,

$$\Delta^{(n)}(\omega) = \frac{1}{2} \max_{0 \leq k \leq 2^n - 1} \left| V(\frac{2k+1}{2^{n+1}}) \right|,$$

where $V(\frac{2k+1}{2^{n+1}})$ is $N(0, \frac{1}{2^n})$. For the following calculations we let $N(\mu, \sigma^2)$ stand both for the normal distribution and for a random variable with that distribution. We have for $x \geq 1$ that

$$P[\Delta^{(n)} > \frac{x/2}{\sqrt{2^n}}] = P[\frac{1}{2} \max_{k \leq 2^n - 1} \left| V(\frac{2k+1}{2^{n+1}}) \right| > \frac{1}{2} \frac{x}{\sqrt{2^n}}]$$

$$\leq 2^n P[\left| N(0, \frac{1}{2^n}) \right| > \frac{x}{\sqrt{2^n}}]$$

$$\leq 2^n 2 P[N(0, \frac{1}{2^n}) > \frac{x}{\sqrt{2^n}}]$$

$$= 2^{n+1} P[\frac{N(0, \frac{1}{2^n})}{\sqrt{\frac{1}{2^n}}} > x]$$

$$= 2^{n+1} P[N(0, 1) > x]$$

$$\leq 2^{n+1} \frac{n(x)}{x} = \frac{2^{n+1}}{\sqrt{2\pi}} \frac{e^{-x^2/2}}{x}.$$

Let $x = 2\sqrt{n}$, and from the previous calculation, with $c > 0$, we get

$$P[\Delta^{(n)} > \frac{\sqrt{n}}{\sqrt{2^n}}] \leq c 2^n e^{-(2\sqrt{n})^2/2} = c 2^n e^{-2n} = c \left(\frac{2}{e^2} \right)^n,$$

and, therefore,

$$\sum_n P[\Delta^{(n)} > \frac{\sqrt{n}}{\sqrt{2^n}}] \leq \sum_n c(\frac{2}{e^2})^n < \infty.$$

The Borel-Cantelli Lemma 6.2.1 gives

$$P[\Delta^{(n)} > \frac{\sqrt{n}}{\sqrt{2^n}} \text{ i.o.}] = 0,$$

so there is a last time (depending on ω) that $\Delta^{(n)} > \frac{\sqrt{n}}{\sqrt{2^n}}$. Eventually, for all large n, we have $\Delta^{(n)} \le \sqrt{\frac{n}{2^n}}$, and therefore (with probability 1)

$$\sum_n \Delta^{(n)} < \infty.$$

The sequence of continuous functions

$$\{B^{(n)}, n \ge 1\} = \{\{B^{(n)}(t), 0 \le t \le 1\}, n \ge 1\}$$

is Cauchy in the metric space $C[0,1]$, the space of real valued, continuous functions on $[0, 1]$, since for $n < m$

$$\sup_{0 \le t \le 1} |B^{(n)}(t) - B^{(m)}(t)| \le \Delta^{(n)} + \cdots + \Delta^{(m-1)} \to 0$$

as $n, m \to \infty$. Therefore, since $C[0, 1]$ is a complete metric space, we have $\lim_{n \to \infty} B^{(n)}$ exists in C[0,1], with probability 1. Define the process

$$B = \begin{cases} \lim_{n \to \infty} B^{(n)}, & \text{if } \lim_{n \to \infty} B^{(n)} \text{ exists} \\ = 0, & \text{otherwise (on a set of probability 0).} \end{cases}$$

This gives a Brownian motion process on $[0, 1]$. Note the following:

(1) B is continuous since it is the uniform limit of $B^{(n)}$ which is continuous. (Recall uniform limits of continuous functions are continuous.)

(2) $B(t) - B(s)$ has a $N(0, t - s)$ distribution since

$$B(t) - B(s) = \lim_{n \to \infty} B^{(n)} \left(\frac{[t2^n]}{2^n} \right) - B^{(n)} \left(\frac{[s2^n]}{2^n} \right)$$

and $B^{(n)} \left(\frac{[t2^n]}{2^n} \right) - B^{(n)} \left(\frac{[s2^n]}{2^n} \right)$ has a $N \left(0, \frac{[t2^n]-[s2^n]}{2^n} \right)$ distribution which converges, as $n \to \infty$, to a $N(0, t - s)$ distribution.

(3) B has independent increments, since this is true for $B^{(n)}$, and the independent increment property is preserved by taking limits. For instance, using the characteristic function transform for $0 \le u < v < s < t$,

$$Ee^{i(\theta_1(B(t)-B(s))+\theta_2(B(v)-B(u))}$$

$$= \lim_{n \to \infty} Ee^{i(\theta_1(B^{(n)}(t)-B^{(n)}(s))+\theta_2(B^{(n)}(v)-B^{(n)}(u)))}$$

$$= \lim_{n \to \infty} Ee^{i\theta_1(B^{(n)}(t)-B^{(n)}(s))} Ee^{i\theta_2(B^{(n)}(v)-B^{(n)}(u))}$$

(since $B^{(n)}$ has independent increments)

$$= Ee^{i\theta_1(B(t)-B(s))}Ee^{i\theta_2(B(v)-B(u))}.$$

This completes the construction. ∎

6.4. SIMPLE PROPERTIES OF STANDARD BROWNIAN MOTION.

In this section we discuss some fairly simple properties of standard Brownian motion $B(\cdot)$, which will increase our familiarity with the process and be useful later. We will frequently write

$$\{A(t), t \geq 0\} \stackrel{d}{=} \{B(t), t \geq 0\}$$

to mean that the finite dimensional distributions of the processes A and B are the same.

   1. The Markov Property. Brownian motion is a Markov process with stationary transition probabilities

$$
\begin{aligned}
p_t(x, A) : &= P[B(t+s) \in A | B(s) = x] \\
&= P[B(t+s) - B(s) \in A - x | B(s) = x] \\
&= \int_{A-x} n(0, t, v)dv \\
&= \int_A n(x, t, v)dv.
\end{aligned}
$$

To check this we merely need to write

$$B(t+s) = B(s) + (B(t+s) - B(s)),$$

which represents the new state $B(t+s)$ as a sum of the old state $B(s)$ and an independent normal random variable $B(t+s) - B(s)$. The Markov property and the calculation of the transition probabilities follow from this.

   2. Differential Property. For any $s > 0$,

$$\{B_s(t) = B(t+s) - B(s), t \geq 0\}$$

is a standard Brownian motion process independent of $\{B(u), u \leq s\}$. (This will be generalized later so that s can be replaced by special random times.) To check this, note that $B_s(t)$ is continuous in t and has independent increments and the increments are normally distributed with the correct variance.

3. Scaling Property. For any $c > 0$ we have

$$\{\sqrt{c}B(\frac{t}{c}), t \geq 0\} \overset{d}{=} \{B(t), t \geq 0\},$$

and the process

$$\{\sqrt{c}B(\frac{t}{c}), t \geq 0\}$$

is standard Brownian motion.

Check this by noting that the process $\{\sqrt{c}B(\frac{t}{c}), t \geq 0\}$ has continuous paths, stationary independent increments, normal marginal distributions and the correct variance.

4. Symmetry. The negative of a Brownian motion is still a Brownian motion:

$$-B(\cdot) \overset{d}{=} B(\cdot).$$

Again, we just note that the process $-B$ has continuous paths, stationary independent increments, normal marginal distributions and the correct variance.

5. Gaussian Process. A standard Brownian motion B is characterized as a continuous path, zero mean Gaussian process (that is, the finite dimensional joint distributions are multivariate normal) with $B(0) = 0$ and covariance function

$$\text{Cov}(B(s), B(t)) = EB(s)B(t) = s \wedge t$$

for $s \geq 0, t \geq 0$. To check the covariance calculation we use the independent increment property: For $s < t$

$$\begin{aligned} EB(s)B(t) &= EB(s)(B(s) + B(t) - B(s)) \\ &= EB^2(s) + EB(s)(B(t) - B(s)) \\ &= s + 0 = s, \end{aligned}$$

so $EB(s)B(t) = s \wedge t$ as asserted.

6. Time Reversal. We have

$$\{tB(1/t), t > 0\} \overset{d}{=} \{B(t), t > 0\}.$$

If we define

$$B^{(1)}(t) = \begin{cases} tB(\frac{1}{t}), & \text{if } t > 0, \\ 0, & \text{if } t = 0, \end{cases}$$

then

$$B^{(1)}(\cdot) \overset{d}{=} B(\cdot),$$

and $B^{(1)}$ is a standard Brownian motion.

Since $B(\cdot)$ is a Gaussian process, $B^{(1)}$ is as well. Also $B^{(1)}$ has zero means and is continuous (we must check continuity at zero). Finally, if $0 < s < t$,

$$EB^{(1)}(s)B^{(1)}(t) = EsB(\frac{1}{s})tB(\frac{1}{t})$$

$$= st(\frac{1}{s}) \wedge (\frac{1}{t}) = st\frac{1}{t}$$

$$= s = EB(s)B(t).$$

Therefore, the finite dimensional distributions of $B^{(1)}$ and B are the same multivariate normal distributions, and $B^{(1)} \stackrel{d}{=} B$.

It remains to verify that $B^{(1)}$ is continuous at 0; i.e.,

$$\lim_{t\to 0} B^{(1)}(t) = \lim_{t\to 0} tB(\frac{1}{t}) = \lim_{s\to\infty} \frac{B(s)}{s} = 0$$

almost surely. To verify $\lim_{s\to\infty} B(s)/s = 0$, note $B(n)/n \to 0$ by the strong law of large numbers. For $s \in [n, n+1]$ we have

$$|\frac{B(s)}{s} - \frac{B(n)}{n}| \le |\frac{B(s)}{s} - \frac{B(n)}{s}| + |\frac{B(n)}{s} - \frac{B(n)}{n}|$$

$$\le |B(n)|\, |\frac{1}{s} - \frac{1}{n}| + \frac{1}{n} \sup_{n\le s\le n+1} |B(s) - B(n)|$$

$$\le \frac{|B(n)|}{n^2} + \frac{Z_n}{n},$$

where

$$Z_n = \sup_{0\le s\le 1} |B(s+n) - B(n)| \stackrel{d}{=} \vee_{0\le s\le 1}|B(s)|,$$

and $\{Z_n, n \ge 0\}$ are iid by the stationary independent increment property of B. By the strong law of large numbers,

$$\frac{B(n)}{n^2} = \frac{\sum_{j=1}^{n}(B(j) - B(j-1))}{n^2} \to 0.$$

To show $Z_n/n \to 0$ it suffices to show for any $\epsilon > 0$ that

$$P[|Z_n/n| > \epsilon \text{ i.o. }] = 0,$$

which, by the Borel-Cantelli Lemma 6.2.1, will be true if

$$\sum_n P[|Z_n| > \epsilon n] = \sum_n P[|Z_1| > \epsilon n] < \infty.$$

Since $E|Z_1| = \int_0^\infty P[|Z_1| > x]dx$, it is readily checked that the latter sum converges if $E|Z_1| < \infty$. This will become apparent after the discussion of the distribution of the maximum of a Brownian motion over a finite interval.

This completes the discussion of simple properties.

6.5. The Reflection Principle and the Distribution of the Maximum.

There is an easy reflection argument yielding the distribution of the maximum $M(t) = \bigvee_{s=0}^{t} B(s)$ of Brownian motion up to time t. The argument rests on the following fact.

Theorem 6.5.1. Reflection. *For $a > 0$ let*

$$T_a = \inf\{t : B(t) = a\}$$

be the first time the Brownian motion hits level a. Define

$$B^*(t) = \begin{cases} B(t), & \text{if } t \le T_a \\ 2a - B(t), & \text{if } t > T_a. \end{cases}$$

Then B^* is a Brownian motion.

To get B^*, we wait until B reaches height a and then reflect the part of the path past the hitting time of a about the horizontal line at height a. This is illustrated in Figure 6.3.

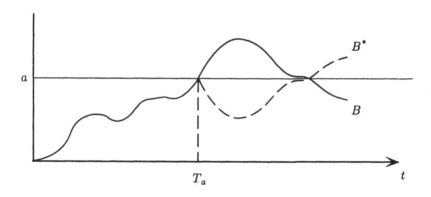

Figure 6.3.

Note that $T_a < \infty$, since $\{B(n)\}$ is a mean 0 random walk and hence

$$\limsup_{n \to \infty} B(n) = +\infty, \liminf_{n \to \infty} B(n) = -\infty.$$

This fact about random walks is proven in Chapter 7.

The reflection principle is proven rigorously in Section 6.6, but it is easy to see why it must be true. Suppose that $t > T_a$. Run the Brownian motion up to time T_a. The differential property (2) of Section 6.4 suggests that the process $\{B(T_a + s) - B(T_a) = B(T_a + s) - a, s \geq 0\}$ is a standard Brownian motion independent of the process B up until the time level a has been attained. (But this requires a proof—see Section 6.6). By the symmetry property (4), $\{-B(T_a + s) + a, s \geq 0\}$ should also be a Brownian motion. Setting $s + T_a = t > T_a$, this means that if

$$a + (B(t) - B(T_a)) = B(t)$$

is a Brownian motion, then

$$a + (B(T_a) - B(t)) = 2a - B(t) = B^*(t)$$

should also be a Brownian motion.

The reflection principle readily yields the distribution of the maximum. Recall the notation $M(t) = \vee_{0 \leq s \leq t} B(s)$.

Proposition 6.5.2. *For standard Brownian motion and $a > 0$, $y \geq 0$,*

$$P[B(t) \leq a - y, M(t) \geq a] = P[B(t) > a + y].$$

Proof. We have

$$[M(t) \geq a] = [T_a \leq t]$$

and

$$T_a = T_a^* := \inf\{t : B^*(t) = a\}.$$

Therefore,

$$P[B(t) \leq a - y, M(t) \geq a] = P[B(t) \leq a - y, T_a \leq t].$$

Since $B \overset{d}{=} B^*$ and $T_a = T_a^*$, this equals

$$\begin{aligned}
&= P[B^*(t) \leq a - y, T_a \leq t] \\
&= P[2a - B(t) \leq a - y, T_a \leq t] \\
&= P[B(t) \geq a + y, T_a \leq t] \\
&= P[B(t) \geq a + y]
\end{aligned}$$

since, for $y \geq 0$,

$$[B(t) \geq a + y] \subset [T_a \leq t]. \quad \blacksquare$$

Corollary 6.5.3. For $a \geq 0$,

(6.5.1) $P[M(t) \geq a] = 2P[B(t) \geq a] = P[|B(t)| \geq a].$

Proof. We have

$$P[M(t) \geq a] = P[M(t) \geq a, B(t) \leq a] + P[M(t) \geq a, B(t) > a].$$

Applying Proposition 6.5.2 with $y = 0$, we get

$$= P[B(t) > a] + P[M(t) \geq a, B(t) > a]$$
$$= 2P[B(t) > a]. \blacksquare$$

It is now easy to compute the densities of the random variables $M(t)$ and T_a explicitly.

Corollary 6.5.4. For $a \geq 0$, the density of $M(t)$ is

(6.5.2) $$f_{M(t)}(a) = \sqrt{\frac{2}{\pi t}} e^{-a^2/2t} I_{[0,\infty)}(a),$$

and the distribution function is

(6.5.3) $$P[M_t \leq a] = \int_0^a \sqrt{\frac{2}{\pi t}} e^{-s^2/2t} ds.$$

The distribution function of T_a is

$$P[T_a \leq x] = 2(1 - N(a/\sqrt{x})), \quad x \geq 0,$$

and its density is

(6.5.4) $$f_{T_a}(x) = \frac{a}{\sqrt{2\pi}} e^{-\frac{a^2}{2x}} x^{-\frac{3}{2}}, \quad x > 0.$$

The Laplace transform of T_a is

(6.5.5) $$Ee^{-\lambda T_a} = e^{-\sqrt{2\lambda}a}, \quad \lambda > 0.$$

Proof. Recall (6.5.1), so that

$$P[T_a \leq t] = P[M(t) > a]$$
$$= 2P[B(t) > a]$$
$$= 2P\left[\frac{B(t)}{\sqrt{t}} > \frac{a}{\sqrt{t}}\right]$$
$$= 2(1 - N(a/\sqrt{t})).$$

Then the density of $M(t)$ is obtained by differentiating with respect to a,

$$f_{M(t)}(a) = -\frac{d}{da}2P[B(t) > a] = 2n(0, t, a),$$

and the density of T_a is obtained by differentiating with respect to t.

To obtain the Laplace transform of T_a we must evaluate the integral

$$\int_0^\infty e^{-\lambda t} \frac{a}{\sqrt{2\pi}} t^{-\frac{3}{2}} e^{-\frac{a^2}{2t}} dt.$$

If you are good with integrals you can try to evaluate this one; alternatively, one can use a table of Laplace transfroms to find the Laplace transform of the density of T_a. (Abramowitz and Stegun, 1972, page 1026, Formula 29.3.82 gives the answer.) ∎

We can also obtain the joint density of $(M(t), B(t))$.

Corollary 6.5.5. For $\xi > 0$ and $x \le \xi$, we have

$$P[M(t) \in d\xi, B(t) \in dx]$$

$$= \sqrt{\frac{2}{\pi}} \left(\frac{2\xi - x}{t^{3/2}} \right) e^{-\frac{(2\xi-x)^2}{2t}} d\xi dx \cdot 1_{[0,\infty)}(\xi) 1_{(-\infty,\xi]}(x).$$

Proof. Recall from Proposition 6.5.2

$$P[B(t) \le a - y, M(t) \ge a] = P[B(t) \ge a + y].$$

Let $a = \xi, \xi - y = x$ to get

$$P[B(t) \le x, M(t) \ge \xi] = P[B(t) \ge 2\xi - x].$$

The required density is obtained by taking $\frac{-\partial}{\partial \xi} \frac{\partial}{\partial x}$ of the right side to obtain

$$\frac{-\partial}{\partial x} \frac{\partial}{\partial \xi} P[B(t > 2\xi - x] = \frac{\partial}{\partial x} 2n(0, t, 2\xi - x)$$

$$= \frac{\partial}{\partial x} \frac{2}{\sqrt{2\pi t}} e^{-\frac{(2\xi-x)^2}{2t}}$$

$$= \frac{2}{\sqrt{2\pi t}} e^{-\frac{(2\xi-x)^2}{2t}} 2\frac{(2\xi - x)}{2t}$$

$$= \frac{2}{\sqrt{2\pi t}} \frac{2\xi - x}{t} e^{-\frac{(2\xi-x)^2}{2t}}$$

$$= \sqrt{\frac{2}{\pi}} \left(\frac{2\xi - x)}{t^{3/2}} \right) e^{-\frac{(2\xi-x)^2}{2t}}. \quad ∎$$

Corollary 6.5.6. *Let* $\{X_n, n \geq 1\}$ *be iid random variables with* $EX_1 = 0$ *and* $\text{Var}(X_1) = \sigma^2$. *Then*

$$P[\frac{\vee_{j=0}^n S_j}{\sigma\sqrt{n}} \leq x] \to P[M(1) \leq x]$$
$$= P[\|B(1)\| \leq x]$$
$$= \frac{2}{\sqrt{2\pi}} \int_0^x e^{-u^2/2} du.$$

Proof. The invariance principle discussed in Section 6.1 says

$$B_n^{(c)} \Rightarrow B;$$

recall that $B_n^{(c)}$ is the continuous path process obtained by putting dots at the points $\{(\frac{i}{n}, \frac{S_i}{\sigma\sqrt{n}}), j \geq 0\}$ in the plane and then connecting them. For $f \in C[0, \infty)$, $T : C[0, \infty) \mapsto R$ defined by

$$Tf = \sup_{0 \leq s \leq 1} f(s)$$

is continuous and hence

$$TB_n^{(c)} = \bigvee_{j=0}^n \frac{S_j}{\sigma\sqrt{n}} \Rightarrow M(1)$$

in R. The distribution of $M(1)$ is given in Corollary 6.5.3. ∎

There are some fairly easy extensions of these techniques that yield the probability that B hits level 0 in the time interval (t_0, t_1) $(0 < t_0 < t_1)$.

Proposition 6.5.7. *For standard Brownian motion* $\{B(t), t \geq 0\}$, *the probability of a zero in* (t_0, t_1) *is*

$$P\{\bigcup_{t_0 < t < t_1} [B(t) = 0]\} = \frac{2}{\pi} \arccos \sqrt{\frac{t_0}{t_1}}.$$

Proof. We begin by observing that, for $a > 0$,

$$P[\bigwedge_{0 \leq u \leq t} B(u) \leq 0 | B(0) = a] = 1 - P[\bigwedge_{0 \leq u \leq t} B(u) \geq 0 | B(0) = a].$$

Since $B \overset{d}{=} -B$, this is

$$= 1 - P[- \bigwedge_{0 \leq u \leq t} -B(u) \leq 0| - B(0) = a]$$

$$= 1 - P[\bigvee_{0 \leq u \leq t} B(u) \leq 0|B(0) = -a]$$

$$= P[\bigvee_{0 \leq u \leq t} B(u) \geq 0|B(0) = -a]$$

$$= P[\bigvee_{0 \leq u \leq t} B(u) \geq a|B(0) = 0].$$

Applying Corollary 6.5.3 yields

$$= 2P[B(t) > a].$$

Thus if we define for $a \in R$

$$g(a) = P\{ \bigcup_{0 < t < t_1 - t_0} [B(t) = 0]|B(0) = a\},$$

we have for $a > 0$

$$g(a) = P[\bigwedge_{0 \leq u \leq t_1 - t_0} B(u) \leq 0|B(0) = a]$$
$$= 2P[B(t_1 - t_0) > a].$$

Therefore, for any $a \in R$, one readily checks that

$$g(a) = 2P[B(t_1 - t_0) > |a|].$$

Thus the probability that B has a zero in (t_0, t_1) can now be computed:

$$P\{ \bigcup_{t_0 < t < t_1} [B(t) = 0]\} = \int_{-\infty}^{\infty} P\{ \bigcup_{t_0 < t < t_1} [B(t) = 0]|B(t_0) = a\}P[B(t_0) \in da],$$

and using the Markov or stationary, independent increment property of Brownian motion yields

$$= \int_{-\infty}^{\infty} P\{ \bigcup_{0 < t < t_1 - t_0} [B(t) = 0]|B(0) = a\}P[B(t_0) \in da]$$

$$= \int_{-\infty}^{\infty} g(a)P[B(t_0) \in da]$$

$$= \int_{-\infty}^{\infty} 2P[B(t_1 - t_0) > |a|]n(0, t_0, a)da,$$

which, by the stationary independent increment property of Brownian motion

$$= 2P[B(t_1) - B(t_0) > |B(t_0)|].$$

Let N_1, N_2 be two iid $N(0,1)$ random variables. Since $B(t) \stackrel{d}{=} \sqrt{t}N_1$, we have

$$= 2P[\sqrt{t_1 - t_0}N_1 > \sqrt{t_0}|N_2|]$$

$$= 2P[\frac{N_1}{|N_2|} > \sqrt{\frac{t_0}{t_1 - t_0}}].$$

Recalling from (6.2.3) and (6.2.4) that $N_1/|N_2|$ has a Cauchy distribution, we obtain

$$= 2\left(1 - (\frac{1}{2} + \frac{1}{\pi}\arctan\sqrt{\frac{t_0}{t_1 - t_0}})\right)$$

$$= 1 - \frac{2}{\pi}\arctan\sqrt{\frac{t_0}{t_1 - t_0}}$$

$$= \frac{2}{\pi}\left(\frac{\pi}{2} - \arctan\sqrt{\frac{t_0}{t_1 - t_0}}\right)$$

$$= \frac{2}{\pi}\arccos\sqrt{\frac{t_0}{t_1}}.$$

The last step is justified by trigonometry. See Figure 6.4. ■

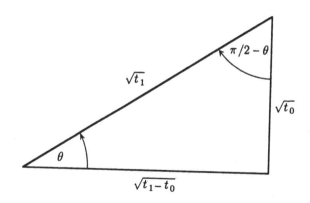

FIGURE 6.4.

6.6. THE STRONG INDEPENDENT INCREMENT PROPERTY AND REFLECTION*.

If you had misgivings about the rigor of the derivation of the reflection principle in the previous section, this section should make you happier. Suppose $\{B(t), t \geq 0\}$ is a Brownian motion. We frequently have need of the differential property (2) of Section 6.4 when the differences are taken relative to a random time. Provided this random time is a *stopping time* the differential property will continue to hold. Recall that a non-negative random variable T is a stopping time with respect to the process B if for every $t \geq 0$ the event $[T \leq t]$ is determined by the process up to time t. This means that the information necessary to determine whether stopping occurred at or before time t is contained in the path $B(s), 0 \leq s \leq t$.

For us, the most important examples of stopping times are the hitting times defined, for $a \neq 0$, by

$$T_a = \inf\{t > 0 : B(t) = a\}.$$

The next result may be compared to Proposition 1.8.2 about splitting iid sequences.

Proposition 6.6.1. Strong Independent Increments. *Suppose* $\{B(t), t \geq 0\}$ *is a Brownian motion and* T *is a stopping time of* B. *Suppose both are defined on the probability space* $(\Omega, \mathcal{A}, P)$.

(a) *If* T *is a stopping time which is finite almost surely, then*

(6.6.1) $$\{\tilde{B}(t), t \geq 0\} := \{B(T + t) - B(T), t \geq 0\}$$

is a Brownian motion process which is independent of events determined by B *up to time* T.

(b) *If* T *is a stopping time such that* $P[T = \infty] > 0$, *then on the trace probability space*

$$(\Omega^\#, \mathcal{A}^\#, P^\#) := (\Omega \cap [T < \infty], \{B \cap [T < \infty] : B \in \mathcal{A}\}, P\{\cdot | [T < \infty]\})$$

$\tilde{B}$ *defined in (6.6.1) is a Brownian motion independent of events in* $\mathcal{A}^\#$ *determined by* B *up to time* T.

Remark. For those who know σ-fields, the more precise wording of these concepts is as follows: Suppose $\{\mathcal{B}_t, t \geq 0\}$ is an increasing family of σ-fields

---

*A beginning student should concentrate on understanding the statement of Proposition 6.6.1(a).

such that for every $t \geq 0$ we have that $B(t)$ is measurable with respect to $\mathcal{B}_t$. Then T is a stopping time, if for $t \geq 0$, we have

$$[T \leq t] \in \mathcal{B}_t.$$

Events determined by B up to time T constitute the σ-field $\mathcal{B}_T$ which contains all events Λ such that for every $t \geq 0$

$$\Lambda \cap [T \leq t] \in \mathcal{B}_t.$$

(One readily checks that $\mathcal{B}_T$ is indeed a σ-field.) The strong independent increments property then says that if T is a stopping time which is finite almost surely, then $B(T+\cdot) - B(T)$ is Brownian motion and is independent of $\mathcal{B}_T$.

Proof. Let $E^{\#}$ denote expectation with respect to $P^{\#}$. It suffices to show for $f : R^k \to R$ bounded and continuous, $0 < t_1 < \cdots < t_k$, and an event A determined by the Brownian motion up to time T, that

$$E^{\#}\left(f(\tilde{B}(t_1), \ldots, \tilde{B}(t_k)) 1_A \right) = E^{\#} f(\tilde{B}(t_1), \ldots, \tilde{B}(t_k)) P^{\#}(A)$$

(6.6.2)
$$= E f(B(t_1), \ldots, B(t_k)) P^{\#}(A).$$

(The reason this is sufficient is that we can imagine f as a continuous approximation to the indicator function of a k-dimensional set in the state space of B.)

Suppose first that T restricted to $\Omega^{\#}$ has a countable range $\tau_1, \tau_2, \ldots$. Then the left side of (6.6.2) yields

$$\sum_{\alpha} E^{\#} f(\tilde{B}(t_1), \ldots, \tilde{B}(t_k)) 1_{A[T=\tau_\alpha]}$$

$$= \sum_{\alpha} \frac{E f(B(t_1 + \tau_\alpha) - B(\tau_\alpha), \ldots, B(t_k + \tau_\alpha) - B(\tau_\alpha)) 1_{A \cap [T=\tau_\alpha]}}{P[T < \infty]},$$

which, because Brownian motion has stationary independent increments and $A \cap [T = \tau_\alpha]$ is determined by the process up to time τ_α, is the same as

$$= \sum_{\alpha} E f(B(t_1 + \tau_\alpha) - B(\tau_\alpha), \ldots, B(t_k + \tau_\alpha) - B(\tau_\alpha)) \frac{P(A \cap [T = \tau_\alpha])}{P[T < \infty]}$$

$$= \sum_{\alpha} E f(B(t_1), \ldots, B(t_k)) \frac{P(A \cap [T = \tau_\alpha])}{P[T < \infty]}$$

$$= E f(B(t_1), \ldots, B(t_k)) P^{\#}(A).$$

Thus we get

$$E^{\#}\left(f(\tilde{B}(t_1),\ldots,\tilde{B}(t_k))1_A\right) = Ef(B(t_1),\ldots,B(t_k))P^{\#}(A),$$

and setting $A = [T < \infty]$ shows

$$E^{\#}f(\tilde{B}(t_1),\ldots,\tilde{B}(t_k)) = Ef(B(t_1),\ldots,B(t_k)),$$

so

$$E^{\#}\left(f(\tilde{B}(t_1),\ldots,\tilde{B}(t_k))1_A\right) = E^{\#}f(\tilde{B}(t_1),\ldots,\tilde{B}(t_k))P^{\#}(A)$$
$$= Ef(B(t_1),\ldots,B(t_k))P^{\#}(A),$$

as required.

If T does not have a countable range, we approximate T on $\Omega^{\#}$ from above by a countably valued variable T_n. Define

$$T_n = \sum_{j=1}^{\infty} \frac{j}{2^n} 1_{[T \in [\frac{j-1}{2^n}, \frac{j}{2^n})]}$$

so that $T_n \downarrow T$ as $n \to \infty$. Approximating T from above lets T_n be a stopping time, since if $t \in [\frac{j-1}{2^n}, \frac{j}{2^n})$ then

$$[T_n \le t] = [T \le \frac{j-1}{2^n}],$$

which is determined by the process up to time $\frac{j-1}{2^n}$ and hence by the process up to time t. This verifies that T_n is a stopping time. Also, any event A determined by the process up to time T will be contained in the collection of events determined by the process up to time T_n and (still supposing $t \in [\frac{j-1}{2^n}, \frac{j}{2^n})$)

$$A \cap [T_n \le t] = A \cap [T \le \frac{j-1}{2^n}].$$

Thus we have

$$E^{\#}\left(f^{(}\tilde{B}(t_1),\ldots,\tilde{B}(t_k))1_A\right)$$
$$= \lim_{n\to\infty} E^{\#}\left(f(B(t_1 + T_n) - B(T_n),\ldots,B(t_k + T_n) - B(T_n))1_A\right)$$

(since both B and f are continuous, f is bounded, and $T_n \downarrow T$)

$$= \lim_{n \to \infty} E^{\#} \{ f(B(t_1 + T_n) - B(T_n)), \ldots, B(t_k + T_n) - B(T_n)) \, P^{\#}(A)$$

$$= \lim_{n \to \infty} E f(B(t_1), \ldots, B(t_k)) P^{\#}(A)$$

(by the first part of this proof)

$$= E^{\#} f(B(t_1), \ldots, B(t_k)) P^{\#}(A)$$

$$= E f(B(t_1), \ldots, B(t_k)) P^{\#}(A). \ \blacksquare$$

Now we give a more rigorous proof of the reflection principle using Proposition 6.6.1. Since T_a is finite almost surely (random walk theory in Chapter 7 assures us that $\limsup_{n \to \infty} B(n) \to \infty$ almost surely) part (a) of Proposition 6.6.1 is applicable. For convenience, we review the statement of the reflection principle.

Theorem 6.5.1. Reflection. *For $a > 0$ let*

$$T_a = \inf\{t : B(t) = a\}$$

be the first time the Brownian motion hits level a. Define

$$B^*(t) = \begin{cases} B(t), & \text{if } t \leq T_a \\ 2a - B(t), & \text{if } t > T_a. \end{cases}$$

Then B^* is a Brownian motion.

Proof. (Freedman (1971)) Let f and g be continuous functions on $[0, \infty)$ with $g(0) = 0$. Suppose $t_0 \geq 0$. Define $C[0, \infty)$ to be the continuous functions on $[0, \infty)$, and let C_0 be the subset of $C[0, \infty)$ consisting of functions g with $g(0) = 0$. Define $\psi : C[0, \infty) \times [0, \infty) \times C_0 \mapsto C[0, \infty)$ by

$$\psi(f, t_0, g) = \begin{cases} f(t), & \text{if } t \leq t_0 \\ f(t_0) + g(t - t_0), & \text{if } t \geq t_0. \end{cases}$$

Now let

$$f(t, \omega) = B(t \wedge T_a(\omega)) = \begin{cases} B(t, \omega), & \text{if } t \leq T_a(\omega), \\ a, & \text{if } t \geq T_a(\omega), \end{cases}$$

and define

$$g(t, \omega) = B(T_a(\omega) + t) - B(T_a(\omega)).$$

Then the random variables T_a and f are determined by the Brownian motion up to time T_a and are independent of g or $-g$. Note by the strong independent increments property that

$$g \overset{d}{=} -g \overset{d}{=} B.$$

Therefore, $(f, T_a, g) \stackrel{d}{=} (f, T_a, -g)$ as random elements of $C[0, \infty) \times [0, \infty) \times C_0$, and by applying the ψ map, we obtain

(6.6.3) $$\psi(f, T_a, g) \stackrel{d}{=} \psi(f, T_a, -g).$$

On the left side of (6.6.2) we have

$$\psi(f, T_a, g) = \begin{cases} f(t), & \text{if } t \le T_a \\ f(T_a) + g(t - T_a), & \text{if } t > T_a, \end{cases}$$
$$= \begin{cases} B(t), & \text{if } t \le T_a \\ B(T_a) + B(T_a + t - T_a) - B(T_a), & \text{if } t > T_a \end{cases}$$
$$= B(t),$$

and on the right side of (6.6.2) we find

$$\psi(f, T_a, -g) = \begin{cases} f(t), & \text{if } t \le T_a \\ f(t) - g(t - T_a), & \text{if } t > T_a \end{cases}$$
$$= \begin{cases} B(t), & \text{if } t \le T_a \\ a - (B(T_a + t - T_a) - B(T_a)), & \text{if } t > T_a \end{cases}$$
$$= B^*(t). \quad \blacksquare$$

6.7. ESCAPE FROM A STRIP.

As in Section 6.5, let T_x be the first time Brownian motion hits the point $x \ne 0$, and suppose $b < 0 < a$. Starting from position 0, the first time standard Brownian motion escapes from the strip $[b, a]$ is

$$T_{ab} := T_a \wedge T_b.$$

We will determine the Laplace transform of T_{ab} and its mean and the probabilities
$$P[B(T_{ab}) = a], \quad P[B(T_{ab}) = b]$$

of exiting the strip at the upper or lower boundaries.

Comparable calculations for diffusions determine basic characteristics of the diffusion. There are many ways to perform the calculations, martingale methods being among the shortest. We rely on elementary arguments using Laplace transforms and the strong independent increments property.

We begin by computing the Laplace transform of T_{ab}. From (6.5.5) we know the Laplace transform of T_a, so

$$\begin{aligned}
Ee^{-\lambda T_a} = e^{-\sqrt{2\lambda}a} &= Ee^{-\lambda T_a}1_{[T_a<T_b]} + Ee^{-\lambda T_a}1_{[T_a\geq T_b]} \\
&= Ee^{-\lambda T_{ab}}1_{[T_a<T_b]} + Ee^{-\lambda T_a}1_{[T_a\geq T_b]} \\
&= I + II.
\end{aligned}$$

Concentrate on the second summand II. We have that the set

$$[T_a \geq T_b] = [B(u) \neq a, 0 \leq u \leq T_b]$$

depends on the Brownian path up to time T_b and on this set

$$\begin{aligned}
T_a &= T_b + \inf\{s > 0 : B(T_b + s) = a\} \\
&= T_b + \inf\{s > 0 : B(T_b + s) - B(T_b) = a - b\} \\
&= T_b + \tilde{T}_{a+|b|},
\end{aligned}$$

where $\tilde{T}_{a+|b|}$ refers to the first passage time of the Brownian motion $\{\tilde{B}(s) = B(T_b + s) - B(T_b), s \geq 0\}$. By the strong independent increments property,

$$\inf\{s > 0 : B(T_b + s) - B(T_b) = a - b\} = \tilde{T}_{a+|b|} \overset{d}{=} T_{a+|b|}$$

and is independent of the path up to time T_b. Thus, Proposition 6.6.1 allows us to rewrite II as

$$\begin{aligned}
II &= Ee^{-\lambda T_{ab}}1_{[T_a\geq T_b]}Ee^{-\lambda T_{a+|b|}} \\
&= Ee^{-\lambda T_{ab}}1_{[T_a\geq T_b]}e^{-\sqrt{2\lambda}(a+|b|)}.
\end{aligned}$$

Putting I and II together, we conclude

$$e^{-\sqrt{2\lambda}|a|} = Ee^{-\lambda T_{ab}}1_{[T_a<T_b]} + e^{-\sqrt{2\lambda}(|a|+|b|)}Ee^{-\lambda T_{ab}}1_{[T_a\geq T_b]}.$$

Likewise, interchanging the roles of a and b, we get

$$e^{-\sqrt{2\lambda}|b|} = Ee^{-\lambda T_{ab}}1_{[T_b<T_a]} + e^{-\sqrt{2\lambda}(|a|+|b|)}Ee^{-\lambda T_{ab}}1_{[T_b\geq T_a]}.$$

We now have two equations in two unknowns. Set

$$e_1 = e^{-\sqrt{2\lambda}|a|}, \quad e_2 = e^{-\sqrt{2\lambda}|b|}$$

and

$$x_1 = Ee^{-\lambda T_{ab}} 1_{[T_a < T_b]}, \quad x_2 = Ee^{-\lambda T_{ab}} 1_{[T_b < T_a]},$$

and the two equations become

$$e_1 = x_1 + e_1 e_2 x_2$$
$$e_2 = e_1 e_2 x_1 + x_2.$$

Solving the equations gives

$$x_1 = \frac{e_2 - e_2^{-1}}{e_1 e_2 - e_1^{-1} e_2^{-1}},$$

so we find

$$(6.7.1) \qquad Ee^{-\lambda T_{ab}} 1_{[T_a < T_b]} = \frac{e^{\sqrt{2\lambda}|b|} - e^{-\sqrt{2\lambda}|b|}}{e^{\sqrt{2\lambda}(|a|+|b|)} - e^{-\sqrt{2\lambda}(|a|+|b|)}}$$

$$(6.7.2) \qquad\qquad\qquad = \frac{\sinh(\sqrt{2\lambda}|b|)}{\sinh(\sqrt{2\lambda}(|a| + |b|))},$$

and

$$(6.7.3) \qquad Ee^{-\lambda T_{ab}} 1_{[T_b < T_a]} = \frac{e^{\sqrt{2\lambda}|a|} - e^{-\sqrt{2\lambda}|a|}}{e^{\sqrt{2\lambda}(|a|+|b|)} - e^{-\sqrt{2\lambda}(|a|+|b|)}}$$

$$(6.7.4) \qquad\qquad\qquad = \frac{\sinh(\sqrt{2\lambda}|a|)}{\sinh(\sqrt{2\lambda}(|a| + |b|))}.$$

We may sum these two expressions to get the Laplace transform of T_{ab}.

Observe that if we let $\lambda \to 0$ we obtain

$$Ee^{-\lambda T_{ab}} 1_{[T_a < T_b]} \to P[T_a < T_b].$$

In order to evaluate this limit, note that for any $c > 0$

$$\lim_{\lambda \downarrow 0} \frac{e^{\lambda c} - e^{-\lambda c}}{\lambda} = 2c$$

by L'Hôpital's rule or by a Taylor expansion. Thus we find by taking limits in (6.7.1) that

$$P[B(T_{ab}) = a] = P[T_a < T_b] = \frac{|b|}{|a| + |b|}.$$

We summarize our findings.

Theorem 6.7.1. *Let $b < 0 < a$. For standard Brownian motion we have the following facts about how the process escapes from a strip $[b, a]$:*

$$\phi(\lambda) := Ee^{-\lambda T_{ab}} = \frac{\sinh(\sqrt{2\lambda}|a|) + \sinh(\sqrt{2\lambda}|b|)}{\sinh(\sqrt{2\lambda}(|b| + |a|))}$$

and

$$P[B(T_{ab}) = a] = \frac{|b|}{|a| + |b|}, \quad P[B(T_{ab}) = b] = \frac{|a|}{|a| + |b|}$$

and

$$ET_{ab} = |a||b|.$$

Verifying the last fact about the expected value is elementary but a bit tedious. We must evaluate $-\phi'(\lambda)$ at $\lambda = 0$ to get ET_{ab}. A bit of work can be saved by noting that

$$-\frac{\frac{d}{d\lambda}\phi(\frac{\lambda^2}{2})}{\lambda}\bigg|_{\lambda=0} = ET_{ab}.$$

The advantage of this is that

$$Ee^{-\frac{\lambda^2}{2}T_{ab}} = \frac{e^{\lambda|b|} - e^{-\lambda|b|}}{e^{\lambda(|a|+|b|)} + e^{-\lambda(|a|+|b|)}}$$

is a simpler expression to differentiate. If you differentiate and use L'Hôpital's rule three times you get the correct answer.

6.8. Brownian Motion with Drift.

Let B be a standard Brownian motion. For $\mu \in R$ define

$$B_\mu(t) = B(t) + \mu t, \quad t \geq 0.$$

The process B_μ is called *Brownian motion with drift μ.*

Suppose the drift is negative; that is, $\mu < 0$. Since almost surely

$$\lim_{t \to \infty} \frac{B_\mu(t)}{t} = \lim_{t \to \infty} \frac{B(t) + \mu t}{t} = \mu < 0$$

(see Section 6.4, property 6), we have $\lim_{t \to \infty} B_\mu(t) = -\infty$, and thus

$$M_\mu(\infty) := \bigvee_{s=0}^{\infty} B_\mu(s) < \infty$$

almost surely. Here we compute the distribution of $M_\mu(\infty)$ and show that it is exponential. Section 6.9 discusses an interesting application of this result to heavy traffic approximations in queueing theory.

The approach is to use the strong independent increments property of B to derive a functional equation for the tail of the distribution of $M_\mu(\infty)$. Analogous to the hitting times of Sections 6.5 and 6.7, we define for $x \neq 0$

$$\tau_x = \inf\{t > 0 : B_\mu(t) = x\} = \inf\{t > 0 : B(t) + \mu t = x\},$$

which is the first hitting time of x by B_μ. Note that τ_x is a stopping time of the standard Brownian motion B. We continue to use the notation

$$T_x = \inf\{s > 0 : B(s) = x\}.$$

For $x_1 > 0, x_2 > 0$, we have

$$\begin{aligned} P[M_\mu(\infty) &\geq x_1 + x_2] \\ &= P[B_\mu(t) = x_1 + x_2, \text{ for some } t > 0]. \end{aligned}$$

Since B_μ cannot hit $x_1 + x_2$ without first hitting x_1, this is

$$\begin{aligned} &= P[B(t) + \mu t = x_1 + x_2 \text{ for some } t > 0, \ \tau_{x_1} < \infty] \\ &= P[B(s + \tau_{x_1}) + \mu(s + \tau_{x_1}) = x_1 + x_2 \text{ for some } s > 0, \ \tau_{x_1} < \infty]. \end{aligned}$$

Since $B(\tau_{x_1}) + \mu\tau_{x_1} = x_1$, we get

$$= P[B(s + \tau_{x_1}) - B(\tau_{x_1}) + \mu s = x_2 \text{ for some } s > 0, \ \tau_{x_1} < \infty].$$

Now apply the strong independent increments property of Proposition 6.6.1(b). Since the post-τ_{x_1} process is a standard Brownian motion independent of the process up to time τ_{x_1} we get the above equal to

$$\begin{aligned} &= P^{\#}[B(s + \tau_{x_1}) - B(\tau_{x_1}) + \mu s = x_2 \text{ for some } s > 0] P[\tau_{x_1} < \infty] \\ &= P[M_\mu(\infty) \geq x_2] P[M_\mu(\infty) \geq x_1]. \end{aligned}$$

Therefore, $P[M_\mu(\infty) \geq x]$ satisfies the functional equation

$$P[M_\mu(\infty) \geq x_1 + x_2] = P[M_\mu(\infty) \geq x_1] P[M_\mu(\infty) \geq x_2].$$

The only solutions to this functional equation are the exponential functions; hence for some $c = c(\mu) > 0$ we must have

$$(6.8.1) \qquad P[M_\mu(\infty) \geq x] = e^{-cx}, \quad x > 0.$$

To determine the unknown constant c, for $x > 0$ we write

$$P[M_\mu(\infty) \geq x] = P[B(s) + \mu s = x \text{ for some } s > 0]$$

and since B_μ crosses x after B because of the negative drift, this is

$$= P[B(s + T_x) + \mu(s + T_x) = x \text{ for some } s > 0].$$

We now use $B(T_x) = x$ to get

$$= P[B(s + T_x) - B(T_x) + \mu(s + T_x) = 0 \text{ for some } s > 0].$$

Now let us recall the fact that $\{\tilde{B}(s) = B(s + T_x) - B(T_x), s \geq 0\}$ is a Brownian motion independent of the path of B up to time T_x and hence $\tilde{B}$ is independent of T_x. Conditioning on T_x shows that the previous expression can be rewritten as

$$= \int_0^\infty P[B(s) + \mu(s + y) = 0 \text{ for some } s > 0] P[T_x \in dy]$$

$$= \int_0^\infty P[B(s) + \mu s = |\mu|y \text{ for some } s > 0] P[T_x \in dy]$$

$$= \int_0^\infty P[M_\mu(\infty) \geq |\mu|y] P[T_x \in dy],$$

and (6.8.1) gives

$$= \int_0^\infty e^{-c|\mu|y} P[T_x \in dy]$$

$$= E e^{-c|\mu|T_x},$$

which we recognize as the Laplace transform of T_x at $c|\mu|$. From (6.5.5) this is equal to

(6.8.2) $$= e^{-\sqrt{2c|\mu|}x}.$$

Comparing (6.8.1) and (6.8.2), we get the equality

$$e^{-\sqrt{2c|\mu|}x} = e^{-cx}, \quad x > 0,$$

and thus

$$\sqrt{2c|\mu|} = c,$$

which implies

$$c = 2|\mu|.$$

We summarize these findings.

Proposition 6.8.1. For Brownian motion B_μ with negative drift $\mu < 0$, the supremum $M_\mu(\infty) := \bigvee_{s=0}^\infty B_\mu(s)$ is finite and has exponential distribution:

$$P[M_\mu(\infty) > x] = e^{-2|\mu|x}, \quad x \geq 0.$$

6.9. HEAVY TRAFFIC APPROXIMATIONS IN QUEUEING THEORY.

Brownian motion with drift aids in obtaining approximations for the moments and distribution of the equilibrium waiting time in a fairly general queueing model. Related material is discussed in Sections 3.12.3 and 7.8 where regenerative and random walk methods are used. Here we emphasize the reliance on functional central limit theorems and Brownian motion to obtain our results.

In what follows we again describe the G/G/1 queueing model. Recall that the symbols G/G/1 stand for a queueing model with general input (arrivals occur according to a renewal process), general service times (service times of successive customers are iid) and one server. Customer number 0 arrives at time 0. Let σ_{n+1} be the interarrival time between the nth and the $(n+1)$st arriving customer, and we assume $\sigma_n, n \geq 1$ are iid with common distribution $A(x) = P[\sigma_1 \leq x]$. Set $a^{-1} = E\sigma_1 < \infty$ and customers arrive at rate a. Let t_k be the time of arrival of customer $k, k \geq 0$, so that $t_0 = 0$, $t_k = \sigma_1 + \cdots + \sigma_k, k \geq 1$. The renewal counting function ν is

$$\nu(t) = \text{number of arrivals in } [0, t]$$

$$= \sum_{k=0}^{\infty} 1_{[t_k \leq t]}.$$

To describe the service mechanism, let τ_n be the service time of the nth arriving customer, and suppose $\{\tau_n, n \geq 0\}$ is iid with common distribution $B(x) = P[\tau_0 \leq x]$. Set $b^{-1} = E\tau_0 < \infty$ for customers to be served at rate b. Define the traffic intensity ρ by

$$(6.9.1) \qquad \rho = a/b = E\tau_0/E\sigma_1 = (E\sigma_1)^{-1}/(E\tau_0)^{-1},$$

and ρ is the ratio of the arrival rate to the service rate. If $\rho < 1$, then on the average, the server is able to cope with his load. Assume $\{\tau_n\}$ and $\{\sigma_n\}$ are independent. (Sometimes it suffices that $\{(\tau_n, \sigma_{n+1}), n \geq 0\}$ be iid.)

We assume that there is one server and that he serves customers on a first come, first served basis. A basic process of interest is W_n, the waiting time of the nth customer until his service commences. This is the elapsed time between the arrival of the nth customer and the beginning of his service period. A basic recursion for W_n is

$$(6.9.2) \qquad W_0 = 0, \quad W_{n+1} = (W_n + \tau_n - \sigma_{n+1})^+, \quad n \geq 0,$$

where $x^+ = x$ if $x \geq 0,$, and $x^+ = 0$ if $x < 0$. The validity of (6.9.2) is checked at the beginning of Section 3.12.3. The reader is referred to that discussion and to Figure 3.12.16.

For $n \geq 0$, define

$$(6.9.3) \qquad\qquad X_{n+1} = \tau_n - \sigma_{n+1},$$

so that $\{X_n, n \geq 1\}$ is iid. With this notation,

$$W_{n+1} = (W_n + X_{n+1})^+.$$

It is evident that $\{W_n\}$ is a random walk with a boundary at 0, that is, a sum of iid random variables, that is prevented from going negative. (See also Chapter 7.) Denote the random walk by $\{S_n, n \geq 0\}$, where $S_n = X_1 + \cdots + X_n$. Note that if $\mu = EX_1$ then $\mu = a^{-1}(\rho - 1)$, and

$$\mu < 0 \text{ if and only if } \rho < 1,$$
$$\mu = 0 \text{ if and only if } \rho = 1,$$

and

$$\mu > 0 \text{ if and only if } \rho > 1.$$

Proposition 6.9.1. *For the waiting time W_n of the $G/G/1$ queueing model, we have*

$$(6.9.4) \qquad W_n = \max\{0, X_n, X_n + X_{n-1}, \ldots, \sum_{i=2}^{n} X_i, S_n\}$$

$$\stackrel{d}{=} \bigvee_{j=0}^{n} S_j.$$

Proof. The proof of (6.9.4) proceeds by induction. The equality (6.9.4) is trivially true for $n = 0$ and $n = 1$. Assume that it holds for n. Then

$$W_{n+1} = (W_n + X_{n+1})^+$$

$$= (\max\{0, X_n, X_n + X_{n-1}, \ldots, \sum_{i=1}^{n} X_i\} + X_{n+1})^+$$

(by the induction hypothesis)

$$= (\max\{X_{n+1}, X_{n+1} + X_n, \ldots, \sum_{i=1}^{n+1} X_i\})^+$$

$$= \max\{0, X_{n+1}, X_{n+1} + X_n, \ldots, \sum_{i=1}^{n+1} X_i\}.$$

Therefore, if (6.9.4) holds for n, it also holds for $n+1$.

To prove the equality in distribution $W_n \overset{d}{=} \vee_{j=0}^n S_j$, we observe that

$$(X_1, \ldots, X_n) \overset{d}{=} (X_n, \ldots, X_1),$$

since both vectors consist of iid random variables. Therefore,

$$W_n = \max\{0, X_n, X_n + X_{n-1}, \ldots, \sum_{i=2}^n X_i, S_n\}$$

$$\overset{d}{=} \max\{0, X_1, X_1 + X_2, \ldots, \sum_{i=1}^{n-1} X_i, S_n\}$$

$$= \bigvee_{j=0}^n S_j. \quad \blacksquare$$

This simple result allows us to calculate the asymptotic distribution of $\{W_n\}$.

Proposition 6.9.2. *For the waiting time W_n of a G/G/1 queueing model, the following are true.*

(a) *If $\rho = 1$ and $\sigma := \mathrm{Var}(X_1) < \infty$, then for $x > 0$ as $n \to \infty$ we have*

$$P[\frac{W_n}{\sigma \sqrt{n}} \le x] = P[\frac{\vee_{j=0}^n S_j}{\sigma \sqrt{n}} \le x]$$

$$\to P[M(1) \le x] = P[|N| \le x],$$

where $M(1) = \vee_{0 \le s \le 1} B(s)$ is the maximum of a standard Brownian motion on $[0, 1]$ and N is a $N(0, 1)$ random variable.

(b) *If $\rho < 1$, then $W_\infty := \vee_{j=1}^\infty S_j < \infty$ and*

$$P[W_n \le x] \to P[W_\infty \le x].$$

(c) *If $\rho > 1$, then as $n \to \infty$*

$$\frac{W_n}{n} \to a(\rho - 1) = \mu = EX_1 \quad \text{almost surely.}$$

Proof. For (a) and (b), the critical fact is that $W_n \overset{d}{=} \vee_{j=0}^n S_j$. Then (a) follows from the invariance principle for sums of iid random variables with mean 0 and finite variance converging to Brownian motion. (This was discussed in Section 6.1.) That $M(1) \overset{d}{=} |N|$ is easily checked, since we know the explicit distribution of $M(1)$; cf. Exercise 6.4. For (b), note that since $\rho < 1$, we have $\mu < 0$, so by the strong law of large numbers, $S_n \to -\infty$. Thus $W_\infty < \infty$, and

$$P[W_n \leq x] = P[\bigvee_{j=0}^{n} S_j \leq x] \to P[\bigvee_{j=0}^{\infty} S_j \leq x].$$

For (c), observe that $\rho > 1$ means $\mu > 0$, so $S_n \to \infty$. Since

$$W_n = \max\{0, X_n, X_n + X_{n-1}, \ldots, \sum_{i=2}^{n} X_i, S_n\} \geq S_n \to \infty,$$

there is a last time that $W_n = 0$. Therefore, define

$$\eta = \sup\{n : W_n = 0\},$$

then $P[\eta < \infty] = 1$. For $k \geq 1$ we must have $W_{\eta+k} > 0$, so

$$W_{\eta+1} = W_\eta + X_{\eta+1} = X_{\eta+1}$$
$$W_{\eta+2} = W_{\eta+1} + X_{\eta+2} = X_{\eta+1} + X_{\eta+2}$$
$$W_{\eta+3} = X_{\eta+1} + X_{\eta+2} + X_{\eta+3}$$

$$\vdots \quad \vdots$$

$$W_{\eta+k} = S_{\eta+k} - S_\eta.$$

We have

$$\lim_{n\to\infty} \frac{W_n}{n} = \lim_{k\to\infty} \frac{W_{\eta+k}}{\eta + k}$$
$$= \lim_{k\to\infty} (\frac{S_{\eta+k}}{\eta + k} - \frac{S_\eta}{\eta + k})$$
$$= EX_1 = \mu. \quad \blacksquare$$

When evaluating the usefulness of Proposition 6.9.2, one should keep in mind that the most important case is when $\rho < 1$, since this corresponds to a queue where the server is able to cope with the work input and not fall hopelessly behind. However, it is just in the $\rho < 1$ case that the previous result is not particularly helpful, since we will see in Chapter 7 that it is difficult to compute the distribution of the maximum of a random walk with negative drift, W_∞; thus for the case $\rho < 1$ Proposition 6.9.2 does not provide a particularly helpful approximation. Keeping in mind that ρ can be interpreted as the long run percentage of time that the server is occupied, we see that economic constraints will force us to consider models with $\rho < 1$ (stability) but with ρ near 1 where server utilization is high. The case where $\rho < 1$ but ρ is close to 1 is the *heavy traffic* case. If an approximation to the distribution of W_∞ can be developed as $\rho \uparrow 1$, we will call it a *heavy traffic approximation*. We will consider a sequence of G/G/1 queueing models indexed by k, and the traffic intensity for the kth model will be ρ_k. We will assume $\rho_k < 1$ and $\rho_k \uparrow 1$ as $k \to \infty$.

Since we know waiting times in the G/G/1 model are related to the maxima of random walks, we begin by considering a sequence of random walks indexed by k. For each k, let $\{X_n(\mu_k, \sigma_k^2), n \geq 1\}$ be iid random variables with $EX_1(\mu_k, \sigma_k^2) = \mu_k$ and $\text{Var}(X_1(\mu_k, \sigma_k^2)) = \sigma_k^2$. Let the associated random walk be

$$S_0^{(k)} = 0, \quad S_n^{(k)} = \sum_{i=1}^{n} X_i(\mu_k, \sigma_k^2), \ n \geq 1.$$

We need an invariance principle for such random walks. An outline of the proof will be given later in this section.

Proposition 6.9.3. *Suppose* $\mu_k < 0$ *and*

$$\mu_k \uparrow 0, \sigma_k^2 \to 1$$

as $k \to \infty$. *In addition, suppose that*

$$\{X_1(\mu_k, \sigma_k^2)^2, k \geq 1\}$$

is uniformly integrable, that is,

$$\lim_{b \to \infty} \sup_{k \geq 1} EX_1^2(\mu_k, \sigma_k^2) 1_{[|X_1(\mu_k, \sigma_k^2)| > b]} = 0.$$

Define for $t \geq 0$

$$X_k(t) := |\mu_k| \sum_{i=1}^{[t/\mu_k^2]} \left(X_i(\mu_k, \sigma_k^2) - \mu_k \right).$$

Then an invariance principle holds, and the sequence of processes $\{X_k(\cdot)\}$ *converges to a standard Brownian motion: As* $k \to \infty$

$$X_k(t) \Rightarrow B(t).$$

If this is your first encounter with the notion of *uniform integrability*, think of it as a condition which requires the tails of the distributions of

$X_1^2(\mu_k, \sigma_k^2), k \geq 1$ to be small uniformly over k. It is a standard condition which relates convergence of moments of a sequence of random variables to convergence of the sequence of random variables.

Now this invariance principle is stretched into a new one. For the kth random walk write

$$\begin{aligned}|\mu_k| S_{[t/\mu_k^2]}^{(k)} &= |\mu_k| \left(S_{[t/\mu_k^2]}^{(k)} - [t/\mu_k^2]\mu_k \right) - \mu_k^2 [t/\mu_k^2] \\ &= X_k(t) - \mu_k^2 [t/\mu_k^2].\end{aligned}$$

From Proposition 6.9.3 we get an invariance principle:

$$|\mu_k| S_{[t/\mu_k^2]}^{(k)} \Rightarrow B(t) - t =: B_{-1}(t),$$

where the limit is Brownian motion with drift -1. The magic of the invariance principle allows us to apply to the previous convergence the functional that takes a function into its maximum on $[0, T]$. Therefore, for any $T > 0$, as $k \to \infty$ we have

(6.9.5) $$|\mu_k| \bigvee_{t=0}^{T} S_{[t/\mu_k^2]}^{(k)} \Rightarrow \bigvee_{t=0}^{T} B_{-1}(t).$$

We are interested in an approximation for the distribution of $\bigvee_{j=0}^{\infty} S_j^{(k)}$ and so we write

$$\begin{aligned}Y^{(k)} &= |\mu_k| \bigvee_{j=0}^{\infty} S_j^{(k)} \\ &= \left(|\mu_k| \bigvee_{t=0}^{T} S_{[t/\mu_k^2]}^{(k)} \right) \vee \left(|\mu_k| \bigvee_{t=T}^{\infty} S_{[t/\mu_k^2]}^{(k)} \right) \\ &=: Y^{(k)}(\leq T) \vee Y^{(k)}(> T).\end{aligned}$$

The second piece turns out to be asymptotically negligible for large T and large k which is the content of the next proposition. Comments on the proof of Proposition 6.9.4 are reserved for the end of this section.

Proposition 6.9.4. *Suppose the assumptions of Proposition 6.9.3 hold. Then*

$$\lim_{T \to \infty} \limsup_{k \to \infty} P[|\mu_k| \bigvee_{j > T|\mu_k|^{-2}} S_j^{(k)} > 0] = 0.$$

Therefore for any $T > 0$,

$$Y^{(k)} = Y^{(k)}(\leq T) \vee Y^{(k)}(> T),$$

and we know from (6.9.5) that the first piece on the right converges in distribution as $k \to \infty$:

$$Y^{(k)}(\leq T) \Rightarrow \bigvee_{t=0}^{T} B_{-1}(t).$$

As $T \to \infty$, it is clear that

$$\bigvee_{t=0}^{T} B_{-1}(t) \to \bigvee_{t=0}^{\infty} B_{-1}(t).$$

Also, for any $\epsilon > 0$, from Proposition 6.9.4 we have

$$\lim_{T \to \infty} \limsup_{k \to \infty} P[|Y^{(k)} - Y^{(k)}(\leq T)| > \epsilon]$$

$$\leq \lim_{T \to \infty} \limsup_{k \to \infty} P[Y^{(k)}(> T) > 0] = 0.$$

This says the distributions of $Y^{(k)}$ and $Y^{(k)}(\leq T)$ must be close for large T and large k. A standard technique for proving convergence in distribution (Billingsley, 1986, Theorem 25.5, page 342) then allows the conclusion that

$$Y^{(k)} \Rightarrow \bigvee_{0 \leq t < \infty} B_{-1}(t).$$

If you believe Propositions 6.9.3 and 6.9.4, we have the following result.

Proposition 6.9.5. *For a sequence of random walks with negative drift satisfying the assumptions of Proposition 6.9.3,*

$$\lim_{k \to \infty} P[Y^{(k)} \leq x] = \lim_{k \to \infty} P[|\mu_k| \bigvee_{j=0}^{\infty} S_j^{(k)} \leq x]$$

$$= P[\bigvee_{0 \leq t < \infty} B_{-1}(t) \leq x]$$

$$= 1 - e^{-2x}$$

for $x > 0$.

Recall the distribution of $\bigvee_{0 \leq t < \infty} B_{-1}(t)$ was shown to be exponential in Proposition 6.8.1.

Before grappling with the proofs of Propositions 6.9.3 and 6.9.4, consider again a sequence of G/G/1 queueing models. For the kth model suppose the following: the interarrival distribution is $A_k(x)$ with mean

a_k^{-1}; the service time distribution is $B_k(x)$ with mean b_k^{-1}; and the traffic intensity is $\rho_k = a_k/b_k < 1$ so that $\mu_k = b_k^{-1} - a_k^{-1} < 0$. For each fixed k, Proposition 6.9.2 assures us that the limiting waiting time distribution exists:

$$\lim_{n\to\infty} P[W_n^{(k)} \le x] = P[W_\infty^{(k)} \le x].$$

Suppose $\mu_k \uparrow 0$ or equivalently $\rho_k \uparrow 1$ and also the variances converge

$$\sigma_k^2 = \int_0^\infty (x - a_k^{-1})^2 dA_k(x) + \int_0^\infty (x - b_k^{-1})^2 dB_k(x) \to 1.$$

Further, assume the uniform integrability conditions

$$\lim_{b\to\infty} \sup_{k\ge 1} \int_b^\infty x^2 dA_k(x) = 0$$

and

$$\lim_{b\to\infty} \sup_{k\ge 1} \int_b^\infty x^2 dB_k(x) = 0.$$

Then

(6.9.6) $$\lim_{k\to\infty} P[|\mu_k|W_\infty^{(k)} \le x] = 1 - e^{-2x}.$$

The assumption that $\sigma_k^2 \to 1$ is stronger than needed is used here to simplify matters. Further discussion of this and related matters is presented nicely in Asmussen (1987). It is sufficient to suppose that $\{\sigma_k^2, k \ge 1\}$ is a sequence which is bounded away from 0. (As we will show, the main ingredient in the proof of Proposition 6.9.3 is the verification of a condition called Lindeberg's condition, which is easily seen to hold under the weaker condition that the σ's are bounded away from 0.) Under this weaker condition, Proposition 6.9.3 can be modified to state that the following invariance principle holds:

$$\frac{|\mu_k|}{\sigma_k^2} \sum_{i=1}^{[t\sigma_k^2/\mu_k^2]} (X_i(\mu_k, \sigma_k^2) - \mu_k) \Rightarrow B(t).$$

Under the weaker condition, (6.9.6) can be modified to read

(6.9.6') $$\lim_{k\to\infty} P[\frac{|\mu_k|}{\sigma_k^2} W_\infty^{(k)} \le x] = 1 - e^{-2x}.$$

One can also justify that the means converge in (6.9.6') and so we get

$$\frac{|\mu_k|E(W_\infty^{(k)})}{\sigma_k^2} \to \frac{1}{2},$$

1/2 being the mean of the limiting exponential distribution. This suggests that the approximation for the limiting waiting time distribution in a stable G/G/1 queue with traffic intensity less than but near 1 should be

$$P[W_\infty > y] = P[|\mu|\frac{W_\infty}{\sigma^2} > \frac{|\mu|y}{\sigma^2}]$$
$$\approx e^{-2|EX_1|y/\mathrm{Var}(X_1)}.$$

Now we discuss the justifications for Proposition 6.9.3 and Proposition 6.9.4. A beginning reader may wish to omit this discussion or proceed only with the first paragraph of the proof of Proposition 6.9.3.

Proof of Proposition 6.9.3. We first show $X_k(1) \Rightarrow B(1)$, which means we need to show that $X_k(1)$ is asymptotically normally distributed. The standard method of proving the central limit theorem for triangular arrays where each variable in the kth row is iid is to verify the Lindeberg condition (Feller, 1971). Thus we must show as $k \to \infty$

$$(6.9.7) \quad \sum_{i=1}^{[1/\mu_k^2]} E\left(|\mu_k|(X_i(\mu_k,\sigma_k^2) - \mu_k)\right)^2 1_{[|X_i(\mu_k,\sigma_k^2)-\mu_k|>\epsilon/|\mu_k|]} \to 0.$$

Because the random variables are iid for different i, we have the left side equal to

$$= [1/\mu_k^2]\mu_k^2 E|X_1(\mu_k,\sigma_k^2) - \mu_k|^2 1_{[|X_1(\mu_k,\sigma_k^2)-\mu_k|>\epsilon/|\mu_k|]}$$
$$\leq 2E\left(X_1(\mu_k,\sigma_k^2)^2 + \mu_k^2\right) 1_{[|X_1|\mu_k,\sigma_k^2)|>\epsilon/(2|\mu_k|)]}$$
$$\leq 2EX_1(\mu_k,\sigma_k^2)^2 1_{[|X_1(\mu_k,\sigma_k^2)|>\epsilon/(2|\mu_k|)]} + \mu_k^2$$
$$\leq 2\sup_{j\geq 1} EX_1(\mu_j,\sigma_j^2)^2 1_{[|X_1(\mu_j,\sigma_j^2)|>\epsilon/(2|\mu_k|)]} + o(1)$$
$$\to 0$$

as $k \to \infty$ since $\{X_1(\mu_j,\sigma_j^2), j \geq 1\}$ is uniformly integrable.

This verifies the Lindeberg condition (6.9.7) and shows $X_k(1) \Rightarrow B(1)$. Now we must extend this to show that finite dimensional distributions converge; for any $m \geq 1$ and $0 \leq t_1 < \cdots < t_m$, as $k \to \infty$ we have

$$(X_k(t_1),\ldots,X_k(t_m)) \Rightarrow (B(t_1),\ldots,B(t_m))$$

in R^m. This can be accomplished exactly as in Section 6.1. The final step in the proof is to show a property called tightness (Billingsley, 1968), and this can be accomplished in a manner almost identical to the proof of tightness in Donsker's theorem (Billingsley, 1968, page 68; see also the discussion of Prohorov's theorem in Billingsley, 1968, page 77). ∎

An elegant martingale proof of Proposition 6.9.4 is contained in Asmussen, 1989. Here is a plausibility argument based on the invariance principle that should be convincing. We need to show

$$\lim_{T \to \infty} \limsup_{k \to \infty} P[| \bigvee_{j > T|\mu_k|^{-2}} S_j^{(k)} > 0] = 0.$$

For continuous functions $f(t), t \geq 0$, such that $\lim_{t \to \infty} f(t) = -\infty$, define a functional $\chi(f)$ by

$$\chi(f) = \sup\{s : f(s) > 0\}.$$

For the kth random walk define

$$\beta_k = \sup\{m : S_m^{(k)} > 0\}$$

to be the last time the kth random walk is positive. The assertion we are trying to prove is rephrased as

$$\lim_{T \to \infty} \lim_{k \to \infty} P[\beta_k > \frac{T}{\mu_k^2}] = 0.$$

We know that the following invariance principle holds:

$$|\mu_k| S_{[t/\mu_k^2]}^{(k)} \Rightarrow B_{-1}(t)$$

as $k \to \infty$. We hope it is true that if we apply χ then convergence still holds in R; namely

$$\chi(|\mu_k| S_{[t/\mu_k^2]}^{(k)}) \Rightarrow \chi(B_{-1}(t)).$$

Since

$$\chi(|\mu_k| S_{[t/\mu_k^2]}^{(k)}) = \mu_k^2 \beta_k,$$

then, as $k \to \infty$,

$$P[\beta_k > \frac{T}{\mu_k^2}] = P[\mu_k^2 \beta_k > T] \to P[\chi(B_{-1}) > T].$$

Since the probability on the right tends to zero as $T \to \infty$ (recall that $B_{-1}(t) \to -\infty$ as $t \to \infty$), the desired result should follow.

6.10. THE BROWNIAN BRIDGE AND THE KOLMOGOROV–SMIRNOV STATISTIC.

Consider the problem of either estimating a population distribution or testing the null hypothesis that a random sample came from a particular distribution against the alternative that the null hypothesis is false. Suppose $X_1, \ldots, X_n$ is a random sample from the distribution function $F(x)$; that is, in probabilists terms, $X_1, \ldots, X_n$ are iid with common distribution F.

In order to estimate $F(x)$ we may use the *empirical distribution function* $\hat{F}_n(x)$ defined by

$$\hat{F}_n(x) = \frac{1}{n} \sum_{i=1}^{n} \epsilon_{X_i}((-\infty, x]) = \frac{1}{n} \sum_{i=1}^{n} 1_{[X_i \leq x]}.$$

Thus $\hat{F}_n(x)$ is the fraction of the sample which is less than or equal to x.

Here are some elementary properties of the empirical distribution function:

(1) For fixed x, $n\hat{F}_n(x)$ is a binomial random variable with mean $nF(x)$ and variance $nF(x)(1 - F(x))$. Thus

$$E\hat{F}_n(x) = F(x), \quad \mathrm{Var}(\hat{F}_n(x)) = \frac{1}{n}F(x)(1 - F(x)).$$

(2) By the strong law of large numbers, we get for each fixed x

$$\hat{F}_n(x) = \frac{1}{n} \sum_{i=1}^{n} 1_{[X_i \leq x]} \to E1_{[X_1 \leq x]} = F(x)$$

almost surely as $n \to \infty$. If F were continuous, then by Dini's Theorem, the convergence would automatically be uniform; see for instance Resnick, 1987, page 3. However, there is a result called the *Glivenko–Cantelli Theorem* (Chung, 1974, page 133) which guarantees uniform convergence even if F is not continuous:

$$\sup_{x \in R} |\hat{F}_n(x) - F(x)| \to 0$$

almost surely as $n \to \infty$. Thus the empirical distribution function closely approximates the true distribution for large sample sizes.

(3) For fixed x, the central limit theorem applied to the sums of iid random variables $\sum_{i=1}^{n} 1_{[X_i \leq x]}$ implies

$$\frac{\sum_{i=1}^{n}(1_{[X_i \leq x]} - F(x))}{\sqrt{nF(x)(1 - F(x))}} \Rightarrow N(0, 1)$$

as $n \to \infty$ (remember that "$\Rightarrow$" denotes convergence in distribution), and therefore

$$\sqrt{n}(\hat{F}_n(x) - F(x)) \Rightarrow N(0, F(x)(1 - F(x))).$$

We may elaborate this statement to get a multivariate version. Note first that if $x_1 < x_2$ then

$$\begin{aligned}
\text{Cov}(1_{[X_i \leq x_1]}, 1_{[X_i \leq x_2]}) &= E 1_{[X_i \leq x_1]} 1_{[X_i \leq x_2]} - E 1_{[X_i \leq x_1]} E 1_{[X_i \leq x_2]} \\
&= F(x_1 \wedge x_2) - F(x_1)F(x_2) \\
&= F(x_1)(1 - F(x_2)),
\end{aligned}$$

so by the multivariate central limit theorem (see, for example, Billingsley, 1986, page 398),

$$\frac{1}{\sqrt{n}} \sum_{i=1}^{n} \left(\begin{pmatrix} 1_{[X_i \leq x_1]} \\ 1_{[X_i \leq x_2]} \end{pmatrix} - \begin{pmatrix} F(x_1) \\ F(x_2) \end{pmatrix} \right)$$

$$= \sqrt{n} \left(\begin{pmatrix} \hat{F}_n(x_1) \\ \hat{F}_n(x_2) \end{pmatrix} - \begin{pmatrix} F(x_1) \\ F(x_2) \end{pmatrix} \right)$$

$$\Rightarrow N \left(0, \begin{pmatrix} F(x_1)(1 - F(x_1)) & F(x_1)(1 - F(x_2)) \\ F(x_1)(1 - F(x_2)) & F(x_2)(1 - F(x_2)) \end{pmatrix} \right)$$

in R^2. Suppose $G(x), x \in R$ is a zero mean Gaussian process with covariance function

(6.10.1) $$\text{Cov}(G(x_1), G(x_2)) = F(x_1)(1 - F(x_2)), \quad x_1 < x_2.$$

Then an elaboration of the previous argument shows that in the sense of convergence of finite dimensional distributions

(6.10.2) $$\sqrt{n}(\hat{F}_n(x) - F(x)) \Rightarrow G(x).$$

The Brownian Bridge. Let $\{B(t), t \geq 0\}$ be a standard Brownian motion and define a new process $\{B^{(0)}(t), 0 \leq t \leq 1\}$ by

$$B^{(0)}(t) = B(t) - tB(1), \quad 0 \leq t \leq 1.$$

Then $B^{(0)}$ is a Gaussian process and $B^{(0)}(0) = B^{(0)}(1) = 0$ and $EB^{(0)}(t) = 0$. Furthermore if $0 \leq t_1 < t_2 \leq 1$ then

$$(6.10.3) \qquad \text{Cov}(B^{(0)}(t_1), B^{(0)}(t_2)) = t_1(1 - t_2).$$

To check (6.10.3) is easy since we know the covariance function of B is $\text{Cov}(B(s), B(t)) = s \wedge t$. Thus, if $F(x) = U(x) := x$ for $0 \leq x \leq 1$, that is, if F is the uniform distribution function, then G in (6.10.2) is the *Brownian bridge*. However, more importantly, we may represent the limit G in (6.10.2) as a function of a Brownian bridge. If $\{B^{(0)}(t), 0 \leq t \leq 1\}$ is a Brownian bridge, define a new process $G^{\#}$ by

$$G^{\#}(x) = B^{(0)}(F(x)), \quad x \in R.$$

Then using (6.10.1) and (6.10.3), we see that

$$G^{\#}(x) \overset{d}{=} G(x)$$

in the sense of equality of finite dimensional distributions, since both G and $G^{\#}$ are Gaussian processes with the same covariance function given in (6.10.1). Thus (6.10.2) can be rewritten

$$(6.10.4) \qquad \sqrt{n}(\hat{F}_n(x) - F(x)) \Rightarrow B^{(0)}(F(x)).$$

This hints at the relationship between the Brownian bridge and the empirical distribution function.

The Kolmogorov–Smirnov Statistic. Suppose we have a random sample of size n and we wish to test the simple hypothesis H_0 that the sample comes from the distribution function $F(x)$ against the composite alternative hypothesis that H_0 is false. We compute

$$\hat{F}_n(x) = \frac{1}{n} \sum_{i=1}^{n} 1_{[X_i \leq x]}$$

as an approximation to the true population distribution. We reject the null hypothesis if the discrepancies of $\hat{F}_n(x)$ from $F(x)$ are large for some x. One way to measure these discrepancies is to compute the *Kolmogorov–Smirnov statistic* D_n given by

$$D_n = \sup_{x \in R} |\hat{F}_n(x) - F(x)|.$$

We hope to use D_n as a test statistic and reject when D_n is large. The use of this statistic is feasible because, for any continuous distribution F, the distribution of D_n is the same as if $F = U$, the uniform distribution. Thus, within the class of continuous distributions, D_n is a distribution free statistic. We state this formally.

Proposition 6.10.1. *Suppose $X_1, \ldots, X_n$ are iid with common continuous distribution $F(x)$ and $U_1, \ldots, U_n$ are iid uniformly distributed random variables with common distribution $U(x) = x$, $0 \leq x \leq 1$, and empirical distribution function*

$$\hat{U}_n(x) = \frac{1}{n} \sum_{i=1}^{n} 1_{[U_i \leq x]}.$$

Then

$$D_n := \sup_{x \in R} |\hat{F}_n(x) - F(x)| \overset{d}{=} \sup_{x \in R} |\hat{U}_n(x) - U(x)|.$$

Proof. Some basic facts that we need, which follow from the continuity of F, are as follows:

(1) The inverse function of F

$$F^{\leftarrow}(y) := \inf\{u : F(u) \geq y\}$$

is strictly increasing for $y \in (0, 1)$, and

(6.10.5) $$F^{\leftarrow}(y) \leq t \text{ iff } y \leq F(t).$$

The last equivalence follows because F is a right continuous distribution function and therefore the set $\{u : F(u) \geq y\}$ is closed.

(2) We have

(6.10.6) $$F(F^{\leftarrow}(t)) = t.$$

(3) If X is a random variable with distribution $F(x)$ then $F(X)$ is a random variable with uniform distribution $U(x)$. To check this, note that, for $0 \leq x \leq 1$, (6.10.5) yields

$$P[F(X) \geq x] = P[X \geq F^{\leftarrow}(x)]$$
$$= 1 - F(F^{\leftarrow}(x))$$

which, by (6.10.6), is

$$= 1 - x = P[U_1 \geq x].$$

To understand the idea of the proof of Proposition 6.10.1, suppose not only that F is continuous, but also that F is strictly increasing on R. Then

$F^{\leftarrow}$ is continuous and strictly increasing with domain $(0,1)$ and range R. Thus

$$D_n = \sup_{x \in R} |\hat{F}_n(x) - F(x)|$$

$$= \sup_{u \in (0,1)} |\hat{F}_n(F^{\leftarrow}(u)) - F(F^{\leftarrow}(u))|$$

$$= \sup_{u \in (0,1)} |\hat{F}_n(F^{\leftarrow}(u)) - u|.$$

Now

$$\hat{F}_n(F^{\leftarrow}(u)) = \frac{1}{n} \sum_{i=1}^{n} 1_{[X_i \leq F^{\leftarrow}(u)]}$$

$$= \frac{1}{n} \sum_{i=1}^{n} 1_{[F(X_i) \leq u]}$$

$$\stackrel{d}{=} \frac{1}{n} \sum_{i=1}^{n} 1_{[U_i \leq u]} = \hat{U}_n(u),$$

and the desired result follows.

If you are happy understanding the proof under the extra assumption that F is strictly increasing, skip the rest of this discussion. We now proceed with the proof assuming only the continuity of F.

Let S be the support of F; that is, the smallest closed set carrying all the probability in F. So $F(S) = 1$ and S^c is open in R. Any open set can be represented as a union of open intervals, so we write $S^c = \cup_n I_n$, where $\{I_n = (a_n, b_n)\}$ are open intervals with $F(I_n) = 0$. This means $F(a_n) = F(b_n)$, and since

$$P[X_j \in I_n] = F(I_n) = 0, \quad j = 1, \dots, n,$$

we have

$$\hat{F}_n(I_n) = \frac{1}{n} \sum_{i=1}^{n} \epsilon_{X_i}(I_n) = 0,$$

from which $\hat{F}_n(b_n) = \hat{F}_n(a_n)$. Therefore, neither F nor $\hat{F}_n$ change on the intervals I_n, and thus

$$D_n = \sup_{x \in R \setminus \cup_n I_n} |\hat{F}_n(x) - F(x)|$$

$$= \sup_{x \in S} |\hat{F}_n(x) - F(x)|.$$

The proof now proceeds as in the simpler case discussed above. As u sweeps through $(0,1)$, $F^{\leftarrow}(u)$ sweeps through S and therefore

$$\sup_{x \in S} |\hat{F}_n(x) - F(x)| = \sup_{u \in (0,1)} |\hat{F}_n(F^{\leftarrow}(u)) - F(F^{\leftarrow}(u))|. \quad \blacksquare$$

Because the Kolmogorov–Smirnov statistic has the same distribution for all continuous distributions F, the test is fairly easy to apply. Critical values k_α satisfying

$$P[D_n \geq k_\alpha] = \alpha$$

are given for various values of α and moderate sized values of n, say, $n \leq 80$, in many texts. See, for example, Bickel and Doksum (1977) and for additional information, Birnbaum (1952). In practice, D_n can be computed as follows: Let $X_{(1)} < \cdots < X_{(n)}$ be the order statistics of the sample $X_1, \ldots, X_n$. Then, since $\hat{F}_n(X_{(i)}) = i/n$ and $\hat{F}_n(x)$ does not change value except at an order statistic, we get

$$(6.10.7) \qquad D_n = \max_{1 \leq i \leq n} |\frac{i}{n} - F(X_{(i)})| \vee |F(X_{(i)}) - \frac{i-1}{n}|.$$

Asymptotic Theory. For small or moderate values of n, say $n \leq 80$, we know how to use the Kolmogorov–Smirnov statistic. For larger values of n, we rely on the fact that D_n has a limiting distribution when properly normalized. As $n \to \infty$, we will show

$$\sqrt{n} D_n \Rightarrow D,$$

where

$$(6.10.8) \qquad P[D > d] = 2 \sum_{k=1}^{\infty} (-1)^{k-1} e^{-2k^2 d^2}.$$

Critical points can now be evaluated numerically for D_n by using the limit distribution.

To understand why $\sqrt{n} D_n$ has a limit distribution, review (6.10.2). The convergence in (6.10.2) was stated for the sense of convergence of finite dimensional distributions. If this convergence could be strengthened to an invariance principle, then we could apply the supremum functional to the convergence to obtain

$$\sqrt{n} D_n = \sqrt{n} \sup_x |\hat{F}_n(x) - F(x)| \Rightarrow \sup_x |G(x)|.$$

Since, for continuous distributions the distribution of D_n is the same as if F is the uniform distribution, we would then obtain

$$(6.10.9) \qquad \sqrt{n} D_n = \sqrt{n} \sup_x |\hat{F}_n(x) - F(x)| \Rightarrow \sup_{0 \leq x \leq 1} |B^{(0)}(x)|.$$

In fact, the invariance principle strength version of (6.10.2) does exist when formulated properly. See, for example, Billingsley, 1968 or Pollard, 1984. However, it is relatively easy to prove (6.10.9) directly.

Proposition 6.10.2. *If $\{X_n, n \geq 1\}$ are iid with a common continuous distribution function F, then*

$$(6.10.9) \qquad \sqrt{n}D_n = \sqrt{n}\sup_x |\hat{F}_n(x) - F(x)| \Rightarrow \sup_{0 \leq x \leq 1} |B^{(0)}(x)| =: D.$$

Proof. Since D_n has the same distribution as if $F = U$, the uniform distribution function, we may as well suppose that $\{X_n\}$ is replaced by an iid sequence $\{U_n\}$ of random variables uniformly distributed on $(0, 1)$. We will choose our uniform random variables in a particular way. Recall from Lemma 4.5.1(b) that if $\{E_n\}$ are iid unit exponential random variables, and if $\{\Gamma_n = E_1 + \cdots + E_n, n \geq 1\}$ is the renewal Poisson process generated by the E's, then, in R^n

$$(U_{(1)}, \ldots, U_{(n)}) \stackrel{d}{=} \left(\frac{\Gamma_1}{\Gamma_{n+1}}, \ldots, \frac{\Gamma_n}{\Gamma_{n+1}} \right),$$

where $(U_{(1)}, \ldots, U_{(n)})$ are the order statistics from a sample of size n from $U(x)$. Thus

$$\sqrt{n}D_n \stackrel{d}{=} \sqrt{n}\sup_x |\hat{U}_n(x) - U(x)|,$$

which, from (6.10.7), is

$$= \sqrt{n}\bigvee_{i=1}^n |U_{(i)} - \frac{i}{n}| \vee |U_{(i)} - \frac{i-1}{n}|$$

$$= \sqrt{n}\bigvee_{i=1}^n |\frac{i}{n} - U_{(i)}| + O(\frac{1}{\sqrt{n}})$$

$$\stackrel{d}{=} \sqrt{n}\bigvee_{i=1}^n |\frac{i}{n} - \frac{\Gamma_i}{\Gamma_{n+1}}| + O(\frac{1}{\sqrt{n}})$$

$$= \frac{n}{\Gamma_{n+1}}\bigvee_{i=1}^n \left| \frac{\Gamma_i - i}{\sqrt{n}} - \frac{i}{n}\frac{\Gamma_{n+1} - n}{\sqrt{n}} \right| + O(\frac{1}{\sqrt{n}})$$

$$= \frac{n}{\Gamma_{n+1}}\bigvee_{i=1}^n \left| \frac{\Gamma_i - i}{\sqrt{n}} - \frac{i}{n}\frac{\Gamma_n - n}{\sqrt{n}} \right| + o_p(1),$$

where $o_p(1) \Rightarrow 0$ as $n \to \infty$. Note $o_p(1)$ is of the order of

$$\frac{\Gamma_{n+1} - \Gamma_n}{\sqrt{n}} = \frac{E_{n+1}}{\sqrt{n}},$$

which converges to zero in distribution.

Now, as described in Section 6.1, let $B_n^{(c)}(t)$ be the continuous modification of the process

$$\frac{\Gamma_{[nt]} - [nt]}{\sqrt{n}}, \quad 0 \le t \le 1,$$

so that the paths of $B_n^{(c)}(t)$ are continuous. The invariance principle holds and $B_n^{(c)}(t) \Rightarrow B(t)$. The mapping

$$f(\cdot) \to \sup_{0 \le t \le 1} |f(t) - tf(1)|$$

is continuous from $C[0, \infty)$ into R. Thus, since $\frac{n}{\Gamma_{n+1}} \to 1$ almost surely by the strong law of large numbers, we get

$$\sqrt{n} D_n \overset{d}{=} \frac{n}{\Gamma_{n+1}} \bigvee_{i=1}^{n} |B_n^{(c)}(\frac{i}{n}) - \frac{i}{n} B_n^{(c)}(1)| + o_p(1)$$

$$= \frac{n}{\Gamma_{n+1}} \sup_{0 \le t \le 1} |B_n^{(c)}(t) - t B_n^{(c)}(1)| + o_p(1)$$

$$\Rightarrow \sup_{0 \le t \le 1} |B(t) - t B(1)|$$

$$= \sup_{0 \le t \le 1} |B^{(0)}(t)|,$$

as desired. ∎

We now know the asymptotic distribution of D_n, the Kolmogorov-Smirnov statistic, is determined by D, the supremum of the absolute value of the Brownian bridge process, and we must figure out how to compute the distribution of D. Since we know a fair amount about Brownian motion, a good approach may be to try to relate $B^{(0)}$ to B. Since we know that $B^{(0)}(1) = 0$, we may try to tie down Brownian motion at time 1 by conditioning on $B(1)$ being in a neighborhood of 0. We hope this produces an approximation to $B^{(0)}$. The next proposition shows that this works.

Proposition 6.10.3. *Suppose, as usual, that B is a standard Brownian motion. Let $\{B^{(\epsilon)}(t), 0 \le t \le 1\}$ be a continuous path process whose finite dimensional distributions satisfy*

$$P[B^{(\epsilon)}(t_1) \le x_1, \ldots, B^{(\epsilon)}(t_k) \le x_k]$$
$$= P[B(t_1) \le x_1, \ldots, B(t_k) \le x_k | 0 \le B(1) \le \epsilon].$$

for any k, $0 \le t_1 \le t_2 \le \cdots \le t_k \le 1$. Then, as $\epsilon \to 0$, we have

(6.10.10) $$B^{(\epsilon)}(\cdot) \Rightarrow B^{(0)}(\cdot),$$

in the sense of convergence of finite dimensional distributions. In fact, (6.10.10) is also an invariance principle, which implies

$$(6.10.11) \qquad \bigvee_{s=0}^{1} |B^{(\epsilon)}(s)| \Rightarrow \bigvee_{s=0}^{1} |B^{(0)}(s)|.$$

This says that we may think of a Brownian bridge as a Brownian motion conditioned to be zero at time 1. Sometimes a Brownian bridge is called a tied-down Brownian motion.

The proof of this result is made much easier by the following simple lemma.

Lemma 6.10.4. *If B is standard Brownian motion and $B^{(0)}(t) = B(t) - tB(1)$ is the corresponding Brownian bridge, then $B(1)$ is independent of the Brownian bridge process $\{B^{(0)}(t), 0 \leq t \leq 1\}$.*

Proof of Lemma 6.10.4. It suffices to show for fixed t that $B^{(0)}(t)$ and $B(1)$ are independent. Since $(B^{(0)}(t), B(1))$ are jointly normally distributed, it suffices to show they are uncorrelated. This is easy to show, however, since

$$EB^{(0)}(t)B(1) = E\left((B(t) - tB(1))B(1)\right) = EB(t)B(1) - tE(B(1)^2),$$

and, from the formula for the covariance function of Brownian motion (property (5) of Section 6.4), this becomes

$$= t - t1 = 0,$$

as desired. ∎

Proof of Proposition 6.10.3. We only prove finite dimensional convergence in (6.10.10). For a proof of the invariance principle, consult, for example, Billingsley, 1968.

Once we write the calculation for $k = 2$, the procedure for general k will be clear. For $0 \leq t_1 \leq t_2 \leq 1$, consider the densities

$$P[B^{(\epsilon)}(t_1) \in dx_1, B^{(\epsilon)}(t_2) \in dx_2]$$
$$= P[B(t_1) \in dx_1, B(t_2) \in dx_2 | 0 \leq B(1) \leq \epsilon]$$
$$= \int_0^\epsilon P[B(t_1) \in dx_1, B(t_2) \in dx_2 | B(1) = w] \frac{P[B(1) \in dw]}{P[0 \leq B(1) \leq \epsilon]}$$
$$= \int_0^\epsilon P[B(t_1) - t_1 w \in dx_1 - t_1 w, B(t_2) - t_2 w \in dx_2 - t_2 w | B(1) = w]$$
$$\qquad\qquad \frac{n(w)dw}{N((0, \epsilon])}$$
$$= \int_0^\epsilon P[B^{(0)}(t_1) \in dx_1 - t_1 w, B^{(0)}(t_2) \in dx_2 - t_2 w | B(1) = w] \frac{n(w)dw}{N((0, \epsilon])}$$

and, because of the independence proven in Lemma 6.10.4, the conditioning on the previous line disappears to yield

$$= \int_0^\epsilon P[B^{(0)}(t_1) \in dx_1 - t_1 w, B^{(0)}(t_2) \in dx_2 - t_2 w] \frac{n(w)dw}{N((0,\epsilon])}.$$

Divide numerator and denominator by ϵ, and let $\epsilon \to 0$. By the mean value theorem for integrals the previous expression converges to

$$\to P[B^{(0)}(t_1) \in dx_1, B^{(0)}(t_2) \in dx_2]n(0)/n(0)$$
$$= P[B^{(0)}(t_1) \in dx_1, B^{(0)}(t_2) \in dx_2]. \quad \blacksquare$$

This gives us a way to calculate the limit distribution of the Kolmogorov–Smirnov statistic.

Corollary 6.10.5. *We have*

$$P[D > d] = P[\bigvee_{s=0}^1 |B^{(0)}(s)| > d]$$

$$= \lim_{\epsilon \to 0} P[\bigvee_{s=0}^1 |B^{(\epsilon)}(s)| > d]$$

$$= \lim_{\epsilon \to 0} P[\bigvee_{s=0}^1 |B(s)| > d | 0 \le B(1) \le \epsilon].$$

This reduces the problem of computing the distribution of D to a Brownian motion calculation. The result that we need is contained in the next proposition, whose proof using reflection and the strong independent increments property of Brownian motion is deferred until the end of the section.

Proposition 6.10.6. *For a standard Brownian motion B, we have*

$$P[a_1 < \bigwedge_{s=0}^1 B(s) < \bigvee_{s=0}^1 B(s) < a_2, B(1) \in [c,d]]$$

$$= \sum_{k=-\infty}^\infty N(c + 2k(a_2 - a_1), d + 2k(a_2 - a_1))$$

$$- \sum_{k=-\infty}^\infty N(2a_2 - d + 2k(a_2 - a_1), 2a_2 - c + 2k(a_2 - a_1))$$

for $a_1 < 0 < a_2$ and $[c,d] \subset [a_1, a_2]$.

We use this result to derive the formula for the distribution of D. We will check (6.10.8) in the equivalent form

(6.10.12) $$P[D \le v] = 1 + 2 \sum_{k=1}^\infty (-1)^k e^{-2k^2 v^2}, \quad v > 0.$$

If we apply Proposition 6.10.6 with $a_1 = -v, a_2 = v, c = 0$ and $d = \epsilon$, we get

$P[D \leq v]$

$$= \lim_{\epsilon \to 0} P[-v < \bigwedge_{s=0}^{1} B(s) \leq \bigvee_{s=0}^{1} B(s) < v, B(1) \in [0, \epsilon]]/P[0 \leq B(1) \leq \epsilon]$$

$$= \lim_{\epsilon \to 0} \frac{\sum_{k=-\infty}^{\infty} N((4kv, \epsilon + 4kv]) - \sum_{k=-\infty}^{\infty} N((2v - \epsilon + 4kv, 2v + 4kv])}{N((0, \epsilon])}.$$

Since one can check that the series converge uniformly in ϵ, the limit may be taken under the summations to yield

$$= \left(\sum_{k=-\infty}^{\infty} n(4kv) - \sum_{k=-\infty}^{\infty} n((4k+2)v) \right) / n(0)$$

$$= 1 + \left(\sum_{k=1}^{\infty} 2n(4kv) - \sum_{k=0}^{\infty} 2n((4k+2)v) \right) / n(0)$$

$$= 1 + 2 \left(\sum_{k=1}^{\infty} e^{-8k^2 v^2} - \sum_{k=0}^{\infty} e^{-\frac{1}{2}(4k+2)^2 v^2} \right).$$

We can check that the expression on the previous line is the same as the series in (6.10.12).

Our last task in this section is to prove Proposition 6.10.6. The proof is an elaborate application of the reflection principle.

Proof of Proposition 6.10.6. We set $I = [c, d]$. For $a, b \in R$, we write

$$a + bI = \{a + bx : x \in I\}.$$

Then we have

$$P[a_1 < \bigwedge_{s=0}^{1} B(s) < \bigvee_{s=0}^{1} B(s) < a_2, B(1) \in I]$$

$$= P[B(1) \in I] - P\{[B(1) \in I] \cap [a_1 < \bigwedge_{s=0}^{1} B(s) < \bigvee_{s=0}^{1} B(s) < a_2]^c\}$$

(6.10.13)
$$= I - II.$$

To evaluate II, let

$$j^* = \begin{cases} 1, & \text{if } T_{a_1} \wedge T_{a_2} = T_{a_1}, \\ 2, & \text{if } T_{a_1} \wedge T_{a_2} = T_{a_2}, \end{cases}$$

and define

$$\tau_1 = T_{a_1} \wedge T_{a_2} = T_{a_{j^*}},$$
$$\tau_2 = \tau_1 + \inf\{s : B(\tau_1 + s) = a_{j^*+1 \bmod 2}\},$$
$$\tau_3 = \tau_2 + \inf\{s : B(\tau_2 + s) = a_{j^*+2 \bmod 2}\},$$

$$\vdots \qquad \vdots$$

etc. If $j^* = 1$, B hits a_1 before a_2, and at time τ_2, hits a_2, and at time τ_3, hits a_1, and $\ldots$. Let

$$N = \sum_{k=1}^{\infty} \epsilon_{T_k}([0,1]) = \sum_{k=1}^{\infty} 1[\tau_k \leq 1]$$

be the number of alternating hits of the boundary of the interval (a_1, a_2) before time 1. The relevance of N is that

$$II = P[N \geq 1, B(1) \in I].$$

Because B is continuous with probability 1, we have $N < \infty$ almost surely. Thus $P[N \geq n] \to 0$ as $n \to \infty$, and therefore

$$II = \lim_{n \to \infty} \sum_{k=1}^{n} (P[N \geq k, B(1) \in I] - P[N \geq k+1, B(1) \in I])$$

$$= \sum_{k=1}^{\infty} (P[N \geq k, B(1) \in I] - P[N \geq k+1, B(1) \in I])$$

$$= \sum_{k=1}^{\infty} (-1)^{k+1} (P[N \geq k, B(1) \in I] + P[N \geq k+1, B(1) \in I])$$

$$= \sum_{k=1}^{\infty} (-1)^{k+1} (P[N \geq k, j^* = 1, B(1) \in I]$$

$$+ P[N \geq k+1, j^* = 2, B(1) \in I])$$

$$+ \sum_{k=1}^{\infty} (-1)^{k+1} (P[N \geq k, j^* = 2, B(1) \in I]$$

(6.10.14) $$+ P[N \geq k+1, j^* = 1, B(1) \in I]).$$

Now

$$P[N \geq k, j^* = 1, B(1) \in I] + P[N \geq k+1, j^* = 2, B(1) \in I]$$
(6.10.15) $$= P\{([N \geq k, j^* = 1] \cup [N \geq k+1, j^* = 2]) \cap [B(1) \in I]\}.$$

This is the probability that $B(1) \in I$, that B hits a_1 before time 1, then hits a_2 before 1, then a_1 before 1, ... alternating at least $k-1$ times after the initial hit of a_1. To evaluate (6.10.15), we define

$$S_1 = T_{a_1},$$
$$S_2 = S_1 + \inf\{s : B(S_1 + s) = a_2\},$$
$$S_3 = S_2 + \inf\{s : B(S_2 + s) = a_1\},$$

$$\vdots \qquad \vdots$$

etc. Set $\Delta = a_2 - a_1$, and define

$$B_0(t) = B(t)$$

$$B_1(t) = \begin{cases} B_0(t), & \text{if } t < S_1, \\ 2B_0(S_1) - B_0(t), & \text{if } t \geq S_1, \end{cases}$$

$$B_2(t) = \begin{cases} B_1(t), & \text{if } t < S_2, \\ 2B_1(S_2) - B_0(t), & \text{if } t \geq S_2, \end{cases}$$

$$\vdots \qquad \vdots$$

etc. Then we may write

$$S_1 = \inf\{s : B_0(s) = a_1\}$$
$$S_2 = \inf\{s : B_1(s) = a_1 - \Delta\}$$
$$S_3 = \inf\{s : B_2(s) = a_1 - 2\Delta\},$$

and, in general,

$$S_k = \inf\{s : B_{k-1}(s) = a_1 - (k-1)\Delta\}.$$

See Figure 6.5 for a caricature of a path. The solid line in this figure is B, the dotted line is B_1, and the little circles represent B_2.

So B_k is B_{k-1} reflected about the horizontal line at height $a_1 - (k-1)\Delta$, and

(6.10.16) $$B_{k-1}(S_k) = a_1 - (k-1)\Delta.$$

By the strong independent increments property, each process B_k is a Brownian motion. Also, if $S_5 \leq 1$, we have

$B(1) \in I$

    iff $B_1(1) = 2B_0(S_1) - B_0(1) \in 2a_1 - I =: I_1,$

    iff $B_2(1) = 2B_1(S_2) - B_1(1) \in 2(a_1 - \Delta) - I_1 = I - 2\Delta =: I_2,$

    iff $B_3(1) = 2B_2(S_3) - B_2(1) \in 2(a_1 - 2\Delta) - I_2 = 2a_1 - 2\Delta - I =: I_3$

    iff $B_4(1) = 2B_3(S_4) - B_3(1) \in 2(a_1 - 3\Delta) - I_3 = I - 4\Delta =: I_4,$

    iff $B_2(1) = 2B_4(S_5) - B_4(1) \in 2(a_1 - 4\Delta) - I_4 = 2a_1 - 4\Delta - I =: I_5.$

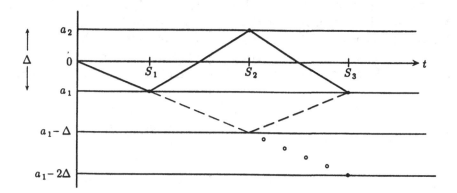

FIGURE 6.5

In general, if $S_k \leq 1$, we observe that

$$B(1) \in I \text{ iff } B_k(1) \in I_k,$$

where

$$I_k = \begin{cases} I - k\Delta, & \text{if } k \text{ is odd,} \\ 2a_1 - (k-1)\Delta - I, & \text{if } k \text{ is even.} \end{cases}$$

The crucial thing about this construction is that

$$[N \geq k, j^* = 1] \cup [N \geq k+1, j^* = 2] = [S_k \leq 1]$$

and

$$[S_k \leq 1, B(1) \in I] = [B_k(1) \in I_k],$$

and, therefore for the expression in (6.10.15),

$$P\{([N \geq k, j^* = 1] \cup [N \geq k+1, j^* = 2]) \cap [B(1) \in I]$$
$$= P[S_k \leq 1, B(1) \in I] = P[B_k(1) \in I_k]$$
$$\text{(6.10.17)} \qquad = P[B(1) \in I_k]$$

since $B \overset{d}{=} B_k$ by the strong independent increments property.

   For the second infinite sum in (6.14) we have similarly that a typical term,

$$P[N \geq k, j^* = 2, B(1) \in I] + P[N \geq k+1, j^* = 1, B(1) \in I],$$

is the probability that $B(1) \in I$, B hits a_2 and then makes alternating visits to the boundary of (a_1, a_2) at least $k-1$ more times before time 1.

Since $B \stackrel{d}{=} -B$, this is also the probability that $-B(1) \in I$, $-B$ hits a_2 and then makes alternating visits to the boundary of (a_1, a_2) at least $k-1$ more times before time 1, which is the same as the probability that $B(1) \in -I$, B hits $-a_2$ and then makes alternating visits to the boundary of $(-a_2, -a_1)$ at least $k-1$ more times before time 1. By examining (6.10.17), we obtain

$$P[N \geq k, j^* = 2, B(1) \in I] + P[N \geq k+1, j^* = 1, B(1) \in I] = P[B(1) \in I'_k]$$

where

$$I'_k = \begin{cases} -I - j\Delta, & \text{if } k \text{ is even,} \\ -2a_2 - (k-1)\Delta + I, & \text{if } k \text{ is odd.} \end{cases}$$

The expression for I'_k is obtained from the definition of I_k by replacing a_1 with $-a_2$ and replacing a_2 with $-a_1$. Thus (6.10.13) becomes

$$P[a_1 < \bigwedge_{s=0}^{1} B(s) < \bigvee_{s=0}^{1} B(s) < a_2, B(1) \in I]$$

$$= P[B(1) \in I] - \sum_{k=1}^{\infty} (-1)^{k+1} \left(P[B(1) \in I_k] + P[B(1) \in I'_k] \right)$$

$$= P[B(1) \in I] + \sum_{k=1}^{\infty} (-1)^{k} \left(P[B(1) \in I_k] + P[B(1) \in I'_k] \right)$$

$$= P[B(1) \in I] + \sum_{k=1}^{\infty} \left(P[B(1) \in I_{2k}] + P[B(1) \in I'_{2k}] \right)$$

$$\quad - \sum_{k=0}^{\infty} \left(P[B(1) \in I_{2k+1}] + P[B(1) \in I'_{2k+1}] \right)$$

$$= P[B(1) \in I] + \sum_{k=1}^{\infty} \left(P[B(1) \in I - 2k\Delta] + P[B(1) \in I + 2k\Delta] \right)$$

$$\quad - \sum_{k=0}^{\infty} \left(P[B(1) \in 2a_1 - 2k\Delta - I] + P[B(1) \in 2a_2 + 2k\Delta - I] \right)$$

$$= \sum_{k=-\infty}^{\infty} P[B(1) \in I + 2k\Delta]$$

$$-\sum_{k=0}^{\infty}(P[B(1) \in -2(a_2 - a_1) + 2a_2 - 2k\Delta - I]$$

$$+P[B(1) \in 2a_2 + 2k\Delta - I])$$

$$= \sum_{k=-\infty}^{\infty} P[B(1) \in I + 2k\Delta]$$

$$-\sum_{k=0}^{\infty}(P[B(1) \in 2a_2 - 2(k+1)\Delta - I] + P[B(1) \in 2a_2 + 2k\Delta - I])$$

$$= \sum_{k=-\infty}^{\infty} P[B(1) \in I + 2k\Delta] - \sum_{k=-\infty}^{\infty} P[B(1) \in 2a_2 + 2k\Delta - I],$$

which is the desired result. ∎

6.11. PATH PROPERTIES*.

The paths of Brownian motion possess many intriguing and bizarre properties; they are continuous but badly behaved. We begin by showing that the paths are nowhere differentiable.

It is convenient to assume the probability space is complete, which means that all subsets of events of probability 0 are events. This is no loss of generality.

Theorem 6.11.1. *Almost all paths of Brownian motion are nowhere differentiable, since*

$$P[\forall t \geq 0 : \limsup_{h \to 0} |\frac{B(t+h) - B(t)}{h}| = +\infty] = 1.$$

Proof. It suffices to show that for any M we have $PA^{(M)} = 0$ where

$$A^{(M)} = \{\omega : \text{There exists some } t \in [0,1] \text{ such that}$$

$$\limsup_{h \to 0} |\frac{B(t+h) - B(t)}{h}| \leq M\}.$$

If $\omega \in A^{(M)}$, there exists a t and n_0 such that $n \geq n_0$ implies

$$|B(s) - B(t)| \leq 2M|t - s|$$

---

* This section is more technically demanding and can be skipped by a beginning student; or the student can just examine the statements of the results without proofs.

if $|s - t| \le 2/n$. Define

$A_n = \{\omega :$ There exists $t \in [0,1]$ such that

$$|B(t,\omega) - B(s,\omega)| \le 2M|t - s| \text{ if } |t - s| \le \frac{2}{n}\}$$

Then we know $A^{(M)} \subset \cup_n A_n$ and that $\{A_n\}$ is monotone: For any n, we have $A_n \subset A_{n+1}$. Suppose $\omega \in A_n$ and t has the property

$$|B(s,\omega) - B(t,\omega)| \le 2M|t - s|$$

if $|t - s| \le 2n$. Let $k = \sup\{j : \frac{j}{n} \le t\}$. Define

$$Y_k = \max\{|B(\frac{k+2}{n}) - B(\frac{k+1}{n})|, |B(\frac{k+1}{n}) - B(\frac{k}{n})|, |B(\frac{k}{n}) - B(\frac{k-1}{n})|\}.$$

We claim $Y_k(\omega) \le 6M/n$ if $\omega \in A_n$. To verify this claim, we observe that

$$(1) \quad |B(\frac{k+2}{n}) - B(\frac{k+1}{n})| \le |B(\frac{k+2}{n}) - B(t)| + |B(t) - B(\frac{k+1}{n})|,$$

and, since $\omega \in A_n$, we have

$$\le 2M|\frac{k+2}{n} - t| + 2M|t - \frac{k+1}{n}|$$

$$\le 2M\frac{2}{n} + 2M\frac{1}{n} = \frac{6M}{n}.$$

$$(2) \qquad |B(\frac{k+1}{n}) - B(\frac{k}{n})| \le |B(\frac{k+1}{n}) - B(t)| + |B(t) - B(\frac{k}{n})|$$

which, by the fact that $\omega \in A_n$, yields the upper bound

$$\le \frac{2M}{n} + \frac{2M}{n} \le \frac{6M}{n}.$$

(3)   The third case follows similarly; the verification of the assertion is concluded.

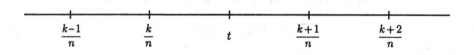

FIGURE 6.6.

We now know that

$$A_n \subset \bigcup_{k=1}^{n-2} \{\omega : Y_k(\omega) \le \frac{6M}{n}\} =: B_n.$$

Therefore

$$PA^{(M)} \le P\{\bigcup_n A_n\} = \lim_{n \to \infty} PA_n \le \lim_{n \to \infty} PB_n.$$

Thus it suffices to show that $PB_n \to 0$. We have that

$$PB_n = P\{\bigcup_{k=1}^{n-2} [Y_k \le \frac{6M}{n}]\}$$

$$\le \sum_{k=1}^{n-2} P[Y_k \le \frac{6M}{n}]$$

and, using the stationary independent increments property of Brownian motion, we get the bound

$$\le nP[\{|B(\frac{3}{n}) - B(\frac{2}{n})| \vee |B(\frac{2}{n}) - B(\frac{1}{n})| \vee |B(\frac{1}{n})|\} \le \frac{6M}{n}],$$

which, from independent increments, is

$$= n(P[|B(\frac{1}{n})| \le \frac{6M}{n}])^3$$

$$= nP[\frac{|B(\frac{1}{n})|}{\frac{1}{\sqrt{n}}} \le \frac{6M}{\sqrt{n}}]^3$$

$$= nP[|N(0,1)| \le \frac{6M}{\sqrt{n}}]^3$$

$$= n(\int_{-6M/\sqrt{n}}^{6M/\sqrt{n}} n(x)dx)^3.$$

Because the normal density $n(x)$ is non-zero at 0, this is asymptotically equivalent to

$$\sim n(\frac{2}{\sqrt{2\pi}} \frac{6M}{\sqrt{n}})^3$$

$$= K\frac{n}{n^{3/2}} = K\frac{1}{n^{1/2}} \to 0$$

as $n \to \infty$ where $K > 0$ is a positive constant. This proves the desired result. ∎

A real valued function f with domain $[0,1]$ has *bounded variation* if it is rectifiable; that is, the graph has finite length. This means that

$$\sup \sum_{i=1}^{n} |f(t_i) - f(t_{i-1})| < \infty,$$

where the sup is taken over all finite partitions of $[0,1]$ of the form $0 \leq t_0 < t_1 < \cdots < t_n = 1$ and where we allow n and $t_1, \ldots, t_n$ to vary. A well known result from analysis is that a function which has bounded variation is almost everywhere differentiable. Since we know from the previous result that almost no Brownian path is differentiable anywhere, we conclude that Brownian paths do not have bounded variation.

Corollary 6.11.2. *Almost no path of Brownian motion has bounded variation.*

Therefore, almost no path is rectifiable.

6.12. QUADRATIC VARIATION.

We know Brownian paths are badly behaved continuous functions. They are neither differentiable nor rectifiable. In trying to calculate the length of a path $B(t), 0 \leq t \leq 1$, we choose division points for $[0,1]$ which we call

$$\Pi = \{0 = t_0 < t_1 < \cdots < t_n = 1\}.$$

An important characteristic of the division points Π is how far apart any two successive points can be, so we define

$$\Delta(\Pi) = \max_{1 \leq k \leq n} |t_k - t_{k-1}|.$$

Any two successive division points are at most $\Delta(\Pi)$ units apart. To calculate the length of the path, we compute

$$\sum_{k=1}^{n} |B(t_k, \omega) - B(t_{k-1}, \omega)|$$

and let $\Delta(\Pi)$ go to zero. The resulting limit is almost surely infinite. If $\Delta(\Pi)$ is small, each individual difference $|B(t_k, \omega) - B(t_{k-1}, \omega)|$ is small by continuity, but not small enough to make the sums converge. If we make the individual differences smaller by, say, squaring each, then there is a chance the limit will be finite.

Thus we define

$$Q(\Pi, \omega) := \sum_{k=1}^{n}{}'|B(t_k, \omega) - B(t_{k-1}, \omega)|^2$$

and call $Q(\Pi, \omega)$ the *quadratic variation* of $\{B(t, \omega), 0 \le t \le 1\}$ over Π. If the limit exists as $\Delta \to 0$, we call the limit the *quadratic variation* of B on $[0, 1]$. If instead of $[0, 1]$ we had taken $[0, t]$ for $t > 0$, then the quadratic variation of B on $[0, t]$ would be a function of t and hence a stochastic process which is called the quadratic variation process. In the theory of stochastic integration such processes are important as increasing processes which can be subtracted from submartingales to give martingales. Quadratic variation processes are also used as random time changes which turn processes constructed from Brownian motions using stochastic integration into new Brownian motions.

Theorem 6.12.1. *If*

$$\Pi^{(i)} \subset \Pi^{(i+1)}$$

and

$$\Delta(\Pi^{(i)}) \to 0$$

fast enough, then almost surely

$$Q(\Pi^{(i)}) \to 1,$$

which is the length of the interval we have decomposed. (If, instead of decomposing $[0, 1]$, we decompose $[0, t]$, then the limit would be t.)

Fancier versions of this result exist; the present statement and proof, modelled after the treatment in Breiman, 1968, are given because of their simplicity. See, for example, McKean, 1969, page 28; Doob, 1953; and Karatzas and Shreve, 1988.

Before giving the proof, we need to recall the following simple fact.

Lemma 6.12.2. *Let N be a $N(0, 1)$ random variable. Then the moment generating function is*

$$Ee^{\alpha N} = e^{\alpha^2/2}, \quad -\infty < \alpha < \infty,$$

and

$$E(N^4) = 3.$$

Proof of the lemma. We have

$$
\begin{aligned}
Ee^{\alpha N} &= \int_{-\infty}^{\infty} e^{\alpha x} \frac{e^{-x^2/2}}{\sqrt{2\pi}} dx \\
&= \int_{-\infty}^{\infty} \frac{e^{-\frac{1}{2}(x^2 - 2\dot{\alpha}x)}}{\sqrt{2\pi}} dx \\
&= e^{\alpha^2/2} \int_{-\infty}^{\infty} \frac{e^{-\frac{1}{2}(x^2 - 2\alpha x + \alpha^2)}}{\sqrt{2\pi}} dx \\
&= e^{\alpha^2/2} \int_{-\infty}^{\infty} \frac{e^{-\frac{1}{2}(x-\alpha)^2}}{\sqrt{2\pi}} dx \\
&= e^{\alpha^2/2}
\end{aligned}
$$

since the integral is the integral of a normal density and hence equals 1. Thus, on the one hand,

$$
e^{\alpha^2/2} = \sum_{k=0}^{\infty} \frac{E(N^k)\alpha^k}{k!}
$$

and, on the other,

$$
e^{\alpha^2/2} = \sum_{j=0}^{\infty} \frac{(\frac{1}{2}\alpha^2)^j}{j!} = \sum_{j=0}^{\infty} \frac{2^{-j}\alpha^{2j}}{j!}.
$$

Equating powers of α^4 yields the result. ∎

Proof Theorem. We begin by observing that if

$$
\Pi^{(i)} = \{0 = t_0 < t_1 < \ldots < t_n = 1\}
$$

then

$$
\begin{aligned}
Q(\Pi^{(i)}) - 1 &= \sum_{i=1}^{n} \{[B(t_i) - B(t_{i-1})]^2 - (t_i - t_{i-1})\} \\
&= \sum_{i=1}^{n} \theta_i,
\end{aligned}
$$

where $\{\theta_i := [B(t_i) - B(t_{i-1})]^2 - (t_i - t_{i-1})\}$ forms a sequence of independent, zero mean random variables. Because the variance of a sum of independent random variables is the sum of the variances, we have

$$E\left(Q(\Pi^{(i)}) - 1\right)^2 = E\left(\sum_{i=1}^{n} \theta_i\right)^2 = \sum_{i=1}^{n} E\theta_i^2$$

$$= \sum_{i=1}^{n} E\left((B(t_i) - B(t_{i-1}))^4\right.$$
$$\left. -2(B(t_i) - B(t_{i-1}))^2(t_i - t_{i-1}) + (t_i - t_{i-1})^2\right).$$

Applying Lemma 6.12.2 yields

$$= \sum_{i=1}^{n} \left(3(t_i - t_{i-1})^2 - 2(t_i - t_{i-1})^2 + (t_i - t_{i-1})^2\right)$$

$$= \sum_{i=1}^{n} 2(t_i - t_{i-1})^2$$

$$\leq 2\Delta(\Pi^{(i)}) \sum_{i=1}^{n}(t_i - t_{i-1})$$

$$= 2\Delta(\Pi^{(i)})1.$$

For typographical ease, set $\Delta^{(i)} = \Delta(\Pi^{(i)})$. If, for instance,

$$\Delta^{(i)} = \frac{\epsilon_i}{i^2},$$

where $\epsilon_i \to 0$, then

$$P[|Q(\Pi^{(i)}) - 1| > i\sqrt{2\Delta(\Pi^{(i)})}] = P[|Q(\Pi^{(i)}) - 1|^2 > 2\epsilon_i],$$

which, by Chebychev's inequality, is bounded by

$$\leq \frac{E(Q(\Pi^{(i)}) - 1)^2}{2\epsilon_i}$$

$$\leq 2\frac{\Delta(\Pi^{(i)})}{2\epsilon_i} = \frac{2\epsilon_i}{2\epsilon_i i^2} = i^{-2}.$$

Therefore,

$$\sum_i P[|Q(\Pi^{(i)}) - 1| > i\sqrt{2\Delta(\Pi^{(i)})}] < \infty.$$

From the Borel-Cantelli lemma, we conclude

$$P\{[|Q(\Pi^{(i)}) - 1| > i\sqrt{2\Delta(\Pi^{(i)})}] \text{ i.o. }\} = 0,$$

and for all large i (depending on the sample point ω),

$$|Q(\Pi^{(i)}) - 1| \leq \sqrt{2\epsilon_i} \to 0,$$

as required. ∎

6.13. KHINTCHINE'S LAW OF THE ITERATED LOGARITHM FOR BROWNIAN MOTION*.

The law of the iterated logarithm is a very precise statement about how Brownian motion oscillates in a neighborhood of the origin. Let

$$h(t) = \sqrt{2t \log \log t^{-1}}.$$

Then we have the following result.

Theorem 6.13.1. *For standard Brownian motion B*

$$P[\limsup_{t \downarrow 0} \frac{B(t)}{h(t)} = 1] = 1$$

and

$$P[\liminf_{t \downarrow 0} \frac{B(t)}{h(t)} = -1] = 1.$$

Remark 6.13.1. The second statement follows from the first. Since $B(\cdot) \overset{d}{=} -B(\cdot)$, the first result yields

$$\limsup_{t \downarrow 0} \frac{-B(t)}{h(t)} = 1$$

almost surely; that is

$$-\limsup_{t \downarrow 0} -\frac{B(t)}{h(t)} = \liminf_{t \downarrow 0} \frac{B(t)}{h(t)} = -1$$

almost surely.

Remark 6.13.2. Khintchine's law also gives us a law of the iterated logarithm near infinity:

$$P[\limsup_{t \to \infty} \frac{B(t)}{\sqrt{2t \log \log t}} = 1, \liminf_{t \to \infty} \frac{B(t)}{\sqrt{2t \log \log t}} = -1] = 1.$$

This follows from the time reversal property (6) in Section 6.4: We have

$$B^{(1)}(t) =: \begin{cases} tB(1/t), & \text{if } t > 0 \\ 0, & \text{if } t = 0, \end{cases}$$

---

* Beginners should study the statement of Theorem 6.13.1 as well as the remarks that follow, but skip the proofs.

is a Brownian motion, so almost surely

$$1 = \limsup_{t \to 0} \frac{B^{(1)}(t)}{h(t)}$$

$$= \limsup_{t \to 0} \frac{tB(1/t)}{\sqrt{2t \log \log t^{-1}}}.$$

Let $t = \frac{1}{s}$ so $s \to \infty$, and we find

$$\limsup_{s \to \infty} \frac{\frac{B(s)}{s}}{\sqrt{2\frac{1}{s} \log \log s}} = \limsup_{s \to \infty} \frac{B(s)}{\sqrt{2s \log \log s}} = 1$$

almost surely.

Remark 6.13.3. An important conclusion from the law of the iterated logarithm is that almost all paths $B(t, \omega)$ of Brownian motion pass through 0 infinitely often in every neighborhood of zero. Thus, for any $\epsilon > 0, \delta > 0$, $B(t, \omega)$ is infinitely often in the region $\{(t, y) : 0 \le t \le \delta, (1 - \epsilon)h(t) < y < (1 + \epsilon)h(t)\}$ and also infinitely often in the region $\{(t, y) : 0 \le t \le \delta, -(1 + \epsilon)h(t) < y < -(1 - \epsilon)h(t)\}$. Therefore, the path oscillates near 0 jumping from being positive to negative infinitely many times. See Figure 6.7.

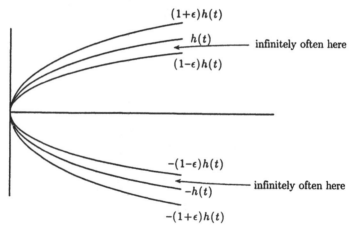

FIGURE 6.7.

The proof is rather technical. The first part is easier than the second part, which requires some patience. We need the following lemma.

Lemma 6.13.2. *Suppose B is standard Brownian motion. For $\alpha > 0$, $\beta > 0$ and any $t > 0$, we have*

$$P[\bigvee_{s \le t} (B(s) - \frac{\alpha s}{2}) > \beta] \le e^{-\alpha \beta}.$$

Proof of Lemma 6.13.2. Let $\mu = -\alpha/2 < 0$. As in Section 6, let $B_\mu(t) = B(t) + \mu t = B(t) - \frac{1}{2}\alpha t$ be Brownian motion with drift. Then

$$P[\bigvee_{s \le t} (B(s) - \frac{\alpha s}{2}) > \beta] = P[\bigvee_{s \le t} (B(s) + \mu s) > \beta]$$

$$= P[\bigvee_{s \le t} B_\mu(s) > \beta]$$

$$\le P[\bigvee_{0 \le s < \infty} B_\mu(s) > \beta].$$

Applying Proposition 6.8.1, this is

$$= e^{-2|\mu|\beta} = e^{-\alpha\beta},$$

as required. ∎

Proof of Theorem 6.13.1. **Part 1.** We show

$$\limsup_{t \downarrow 0} \frac{B(t)}{h(t)} \le 1.$$

For $0 < \theta < 1, 0 < \delta < 1$, let

$$t_n = \theta^{n-1}$$

$$\alpha_n = \frac{(1+\delta)}{\theta^n} h(\theta_n)$$

$$\beta_n = \frac{1}{2} h(\theta^n).$$

Thus

$$\alpha_n \beta_n = \frac{(1+\delta)}{\theta^n} \frac{h^2(\theta^n)}{2}$$

$$= \frac{(1+\delta)2\theta^n \log\log\theta^{-n}}{2\theta^n}$$

$$= (1+\delta) \log\log\theta^{-n}$$

$$= \log\left(\log\theta^{-n}\right)^{1+\delta}$$

and

$$e^{-\alpha_n\beta_n} = e^{-\log\left(\log\theta^{-n}\right)^{1+\delta}}$$

$$= \frac{1}{(\log\theta^{-n})^{1+\delta}} = \left(\frac{1}{n\log\theta^{-1}}\right)^{1+\delta}$$

$$= \frac{K}{n^{1+\delta}}$$

for a constant $K > 0$. Then, applying Lemma 6.13.2, we have

$$P[\bigvee_{s \leq t_n} [B(s) - \frac{\alpha_n s}{2}] > \beta_n] \leq e^{-\alpha_n \beta_n} = \frac{K}{n^{1+\delta}},$$

so

$$\sum_n P[\bigvee_{s \leq t_n} [B(s) - \frac{\alpha_n s}{2}] > \beta_n] < \infty.$$

By the Borel-Cantelli lemma, ultimately, for all sufficiently large (depending on ω) n,

$$\bigvee_{s \leq t_n} [B(s) - \frac{\alpha_n s}{2}] \leq \beta_n.$$

For such n we have

$$\bigvee_{s \leq t_n} B(s) - \frac{\alpha_n t_n}{2} \leq \beta_n.$$

Paraphrasing the previous statement, we have

$$\bigvee_{s \leq t_n} B(s) \leq \beta_n + \frac{\alpha_n t_n}{2}.$$

For such n, if $t \in [\theta^n, \theta^{n-1}) = [t_{n+1}, t_n)$, we have

$$B(t) \leq \bigvee_{s \leq t_n} B(s) \leq \beta_n + \frac{\alpha_n t_n}{2}$$

$$= \frac{h(\theta^n)}{2} + \frac{1+\delta}{2\theta^n} h(\theta^n)\theta^{n-1}$$

$$= h(\theta^n)[\frac{1}{2} + \frac{1+\delta}{2\theta}].$$

Using the fact that $h(t)$ is non-decreasing in a neighborhood of 0, we get the bound

$$\leq h(t)[\frac{1}{2} + \frac{1+\delta}{2\theta}].$$

We may conclude that for any $0 < \theta < 1$ and $0 < \delta < 1$

$$\limsup_{t \downarrow 0} \frac{B(t)}{h(t)} \leq [\frac{1}{2} + \frac{1+\delta}{2\theta}].$$

Observe that the left side is independent of θ and δ, so in right side we let $\delta \downarrow 0$, and $\theta \uparrow 1$. This gives the desired assertion for part 1.

Part 2. We show

$$\limsup_{t \downarrow 0} \frac{B(t)}{h(t)} \geq 1.$$

Define independent events

$$A_n := [B(\theta^n) - B(\theta^{n+1}) \geq (1 - \sqrt{\theta})h(\theta^n)].$$

Thus

$$PA_n = P\left[\frac{B(\theta^n) - B(\theta^{n+1})}{\sqrt{\theta^n(1-\theta)}} > (1 - \sqrt{\theta})\sqrt{\frac{2\theta^n \log\log\theta^{-n}}{\theta^n(1-\theta)}}\right].$$

The random variable

$$\frac{B(\theta^n) - B(\theta^{n+1})}{\sqrt{\theta^n(1-\theta)}}$$

is $N(0,1)$. Call the right side of the inequality above x_n. Since $\theta^{-n} \to \infty$, we have $x_n \to \infty$. Applying Mill's ratio we have

$$PA_n = P[N(0,1) > x_n] \sim \frac{n(x_n)}{x_n}$$

$$= \frac{\frac{1}{\sqrt{2\pi}}e^{-\frac{1}{2}[\frac{1-\sqrt{\theta}}{\sqrt{1-\theta}}\sqrt{2\log\log\theta^{-n}}]^2}}{\frac{1-\sqrt{\theta}}{\sqrt{1-\theta}}\sqrt{\log n}}$$

$$= \frac{K\exp\{-\frac{1}{2}\left[\frac{1-\sqrt{\theta}}{\sqrt{1-\theta}}\right]^2(2\log\log\theta^{-n})\}}{\sqrt{\log n}}.$$

Since

$$\frac{1-\sqrt{\theta}}{\sqrt{1-\theta}} < 1,$$

we obtain the lower bound

$$\geq \frac{K}{\sqrt{\log n}}e^{-\log\log\theta^{-n}}$$

$$= \frac{K'}{n\sqrt{\log n}}.$$

We conclude $\sum_n PA_n = \infty$, and by the Borel–Cantelli lemma (Section 6.2), we get that $P\{A_n \text{ i.o }\} = 1$. Thus, for infinitely many n, we have

$$B(\theta^n) - B(\theta^{n+1}) > (1 - \sqrt{\theta})h(\theta^n)$$

or, equivalently,

(6.13.1) $B(\theta^n) > B(\theta^{n+1}) + (1 - \sqrt{\theta})h(\theta^n).$

From part 1 of this proof, we have that $B(t) \leq 2h(t)$ for small t. Since $-B(t)$ is also Brownian motion, we have that $-B(t) \leq 2h(t)$ for t sufficiently small; i.e., $B(t) \geq -2h(t)$. For n sufficiently large, $B(\theta^{n+1}) \geq -2h(\theta^{n+1})$. From (6.13.1), for infinitely many n,

$$B(\theta^n) > (1 - \sqrt{\theta})h(\theta^n) - 2h(\theta^{n+1}).$$

Thus for infinitely many sufficiently large n, we have

$$\frac{B(\theta^n)}{h(\theta^n)} > 1 - \sqrt{\theta} - \frac{2h(\theta^{n+1})}{h(\theta^n)}$$

$$= 1 - \sqrt{\theta} - 2\sqrt{\frac{2\theta^{n+1} \log\log\theta^{-(n+1)}}{2\theta^n \log\log\theta^{-n}}}$$

$$= 1 - \sqrt{\theta} - 2\sqrt{\theta(1 + o(1))}$$

$$> 1 - \sqrt{\theta} - 3\sqrt{\theta} = 1 - 4\sqrt{\theta}.$$

Our conclusion is that

$$\limsup_{t\downarrow 0} \frac{B(t)}{h(t)} \geq 1 - 4\sqrt{\theta}.$$

Note that the left side is independent of θ so that on the right side we may let $\theta \downarrow 0$ to obtain the desired result. ∎

EXERCISES

6.1. Derive the joint density of $(M(t), M(t) - B(t))$ for standard Brownian motion.

6.2. Prove $M(t) - B(t) \overset{d}{=} |B(t)|$. Prove that $\{|B(t)|, t \geq 0\}$ is Markov, and give its transition density.

6.3. For $b < 0 < a$ and $x \in (b, a)$, compute $P[T_{ab} = b | B(0) = x]$ and $E(T_{ab} | B(0) = x)$.

6.4. Compute the conditional density of $B(s)$ given that $B(t_1) = A, B(t_2) = B$ where $0 < t_1 < s < t_2$. Specialize to the case that $A = 0 = B$.

6.5. Check $ET_a = \infty$, for standard Brownian motion, $a \neq 0$.

6.6. Check that for standard Brownian motion, $\{T_a, a \geq 0\}$ is a stable Lévy motion: Show it has stationary independent increments.

6.7. Compare the covariance function of Brownian motion

$$\text{Cov}(B(t), B(s)), \quad 0 \leq s \leq t$$

with that of a homogeneous Poisson process $\{N(t), t \geq 0\}$ of rate λ. Are they ever equal?

6.8. For standard Brownian motion, derive

$$P[B(t) > y, \bigwedge_{0 \leq u \leq t} B(u) > 0 | B(0) = x].$$

(Use reflection.)

6.9. Let B be standard Brownian motion, and define *geometric Brownian motion* Y by

$$Y(t) = e^{B(t)}.$$

Determine the diffusion coefficients:

$$a(y) := \lim_{h \downarrow 0} \frac{E\left(Y(t+h) - Y(t))^2 | Y(t) = y\right)}{h},$$

and

$$b(y) := \lim_{h \downarrow 0} \frac{E\left(Y(t+h) - Y(t)) | Y(t) = y\right)}{h}$$

for $y > 0$.

6.10. For standard Brownian motion, compute

$$P[\bigvee_{u_1 < t < u_2} B(t) > x]$$

for $0 \leq u_1 < u_2$.

6.11. Let $B^{(0)}(t), 0 \leq t \leq 1\}$ be a Brownian bridge; that is, a continuous path Gaussian process with covariance function

$$\text{Cov}(B^{(0)}(s), B^{(0)}(t)) = s(1 - t), \quad 0 \leq s \leq t \leq 1.$$

Define $B(t)$ by

$$B(t) := (1 + t)B^{(0)}(\frac{t}{1+t}),$$

and show B is a Brownian motion.

6.12. For standard Brownian motion, verify

$$P[\bigvee_{0 \le t \le 1} |B(t)| < b] = \sum_{k=-\infty}^{\infty} (-1)^k P[(2k-1)b < B(1) < (2k+1)b].$$

6.13. If $\{X_1, \ldots, X_n\}$ are iid with common continuous distribution $F(x)$, show that the statistics

$$D_n^+ = \sup_x(\hat{F}_n(x) - F(x))$$

and

$$D_n^- = \inf_x(\hat{F}_n(x) - F(x))$$

are distribution free, and show they have limit distributions when multiplied by $\sqrt{n}$. Identify the limit distributions in terms of the Brownian bridge.

6.14. For $i = 1, 2$, consider the maps $T_i : C([0,1]) \mapsto R$ defined for $f \in C([0,1])$ by

$$T_1(f) = \sup_{0 \le t \le 1} |f(t)|,$$
$$T_2(f) = \sup_{0 \le t \le 1} |f(t) - tf(1)|.$$

Show that both T_1 and T_2 are continuous.

6.15. Let B be standard Brownian motion.
   (a) Prove $S(\omega) := \{t \in [0,1] : B(t, \omega) = 0\}$ has Lebesgue measure zero for almost all ω. (Hint: Compute the expectation of

$$\int_0^1 1_0(B(t))dt.$$

   (b) Prove that with probability 1, $S(\omega)$ is a closed *perfect* set. (A perfect set S is a compact set dense in itself; that is, for each point $x \in S$, every neighborhood of x contains points of S other than x. There can be no isolated points.) Show that $S(\omega)$ is therefore non-countable with probability 1.
   (c) For any $a \in R$, show $\{t \in [0,] : B(t, \omega) = a\}$ is, with probability 1, either empty or a closed perfect set of Lebesgue measure zero.

6.16. Let $\{Z_n, n \geq 0\}$ be iid $N(0,1)$ random variables. Show

$$B(t) = \frac{t}{\sqrt{\pi}} Z_0 + \sqrt{\frac{2}{\pi}} \sum_{m=1}^{\infty} \frac{\sin mt}{m} Z_m$$

is a standard Brownian motion.

6.17. Let B be a standard Brownian motion. Find expressions for the following quantities:

(1) $\quad P[\bigvee_{s=0}^{1} B(s) \leq x, u < B(1) < v]$;

(2) $\quad P[\bigvee_{s=0}^{1} B(s) \leq x, \bigwedge_{s=0}^{1} B(s) \geq y], \quad y < 0 < x$;

(3) $\quad P[\bigvee_{s=0}^{1} |B(s)| \leq x]$.

      6.18. Let $\{B^{(0)}(t), 0 \leq t \leq 1\}$ be a Brownian bridge. Find expressions for the following quantities:

(1) $\quad P[a_1 < \bigwedge_{s=0}^{1} B^{(0)}(s) \leq \bigvee_{s=0}^{1} B^{(0)}(s) \leq a_2], \quad a_1 < 0 < a_2$;

(2) $\quad P[\bigvee_{s=0}^{1} |B^{(0)}(s)| < x], \quad x \geq 0$;

(3) $\quad P[\bigvee_{s=0}^{1} B^{(0)}(s) < x], \quad x \geq 0$.

6.19. Let $\{B^{(-\epsilon)}(t), 0 \leq t \leq 1\}$ be a continuous path process whose finite dimensional distributions are given by

$$P[B^{(-\epsilon)}(t_i) \leq x_i, i = 1, \ldots, k] = P[B(t_i) \leq x_i, i = 1, \ldots, k | -\epsilon < B(1) \leq 0]$$

for $x_i \in R$ and $0 \leq t_i \leq 1$, $i = 1, \ldots, k$. Prove, as $\epsilon \to 0$,

$$B^{(-\epsilon)}(\cdot) \Rightarrow B^{(0)}(\cdot)$$

in the sense of convergence of finite dimensional distributions.

6.20. Let $\{B(t), t \geq 0\}$ be a continuous path process which has stationary, independent increments, $EB(t) = 0$, and $\text{Var}(B(t)) = t$. In addition, suppose the finite dimensional distributions of B are the same as the scaled process

$$\{c^{-1}B(c^2t), t \geq 0\}$$

for any $c > 0$. Show B is a Brownian motion.

6.21. Let B be standard Brownian motion. Define

$$Y_n = \sum_{i=1}^{2n} |B(\frac{i}{2^n}) - B(\frac{i-1}{2^n})|.$$

Show

$$EY_n = 2^{n/2}E(|B(1)|), \quad \text{Var}(Y_n) = \text{Var}(|B(1)|)$$

and that

$$\sum_n P[Y_n < n] < \infty.$$

Conclude that B does not have paths which have bounded variation on $[0, 1]$ (Billingsley, 1986) .

6.22. For standard Brownian motion, show that the probability of no zero in $(t, 1)$ is

$$\frac{2}{\pi} \arcsin \sqrt{t},$$

and hence that the position of the last zero preceding 1 is distributed over $(0, 1)$ with density

$$\pi^{-1}(t(1-t))^{-1/2}, \quad 0 \leq t \leq 1.$$

(Hint: Review Proposition 6.5.7.)

6.23. For standard Brownian motion:

(a) calculate the distribution of the position of the first zero following time 1;

(b) calculate the joint distribution of the last zero before time 1 and the first zero after time 1;

(c) show

$$P[B \text{ has no zeros in } (xt, t)] = \frac{2}{\pi} \arcsin \sqrt{x}.$$

6.24. Compute

$$P[T_1 < T_{-1} < T_2].$$

6.25. You own one share of a stock whose price is approximated by a Brownian motion. Suppose you purchased the stock at a price $b+c$, $c > 0$, and the present price is b. You will sell the stock either when it reaches price $b+c$ or when t time units have passed, which ever comes first. What is the probability that you do not recover your purchase price (Ross, 1985).

6.26. Suppose the present value of a stock is y and that its price varies according to a geometric Brownian motion (see problem 6.9) $Y(t) = \exp\{B(t)\}$. One has the option of purchasing at a future time T one unit of stock at a fixed price K. Compute the expected worth of owning the option; that is, compute

$$E((Y(T) - K) \vee 0).$$

6.27. Ornstein–Uhlenbeck process. Let $\{B(t), t \geq 0\}$ be standard Brownian motion. Define the *Ornstein–Uhlenbeck process* by

$$I(t) = e^{-at/2} B(e^{at}), \quad t \geq 0, \alpha > 0.$$

Verify that the process I is stationary.

6.28. Let B be a standard Brownian motion.
    (a) Show

$$P[\bigvee_{s=0}^{t} B(s) > a \mid \bigvee_{s=0}^{t} B(s) = B(t)] = e^{-a^2/2t}, \quad a > 0.$$

(Hint: Find the conditional density given $Y(t) = \bigvee_{s=0}^{t} B(s) - B(t) = 0$.
    (b) Prove

$$P[B(1) \leq x \mid \bigwedge_{u=0}^{1} B(u) \geq 0] = 1 - e^{-x^2/2}.$$

(Hint: $B'(t) := B(1) - B(1-t), 0 \leq t \leq 1$ is Brownian motion on the index set $[0, 1]$.)

6.29. A Lindley Process in Continuous Time. For $\mu \in R$, define Brownian motion with drift, as usual, by

$$B_\mu(t) = B(t) + \mu t.$$

Now define the Lindley process $W(t)$ by

$$W(t) = (W(0) + B_\mu(t)) \vee \bigvee_{s=0}^{t} (B_\mu(t) - B_\mu(s)).$$

Verify that $\{W(t)\}$ is Markov.
    Now set

$$M_\mu(t) = \bigvee_{s=0}^{t} B_\mu(s),$$

and show that for each t

$$W(t) \stackrel{d}{=} (W(0) + B_\mu(t)) \vee M_\mu(t).$$

If $\mu < 0$, verify that $M_\mu(\infty) < \infty$ almost surely and $W(t) \Rightarrow M_\mu(\infty)$.
Verify that

$$\tau := \inf\{t \geq 0 : W(0) + B_\mu(t) \leq 0\} = \inf\{t \geq 0 : W(t) = 0\},$$

and then $W(t) = W(0) + B_\mu(t)$ for $t < \tau$.
    If $\mu = 0$, show $\{W(t)\}$ is a Brownian motion with reflection at zero which can be represented as $\{|W(0) + B(t)|\}$.

6.30. Let B_1, B_2 be two independent standard Brownian motions, and let

$$T_y^{(2)} = \inf\{t \geq 0 : B_2(t) = y\}, \quad y > 0.$$

Calculate the distribution of $B_1(T_y^{(2)})$.

6.31. Brownian meander. Let B be a standard Brownian motion and define *Brownian meander* $\{B^+(t), 0 \leq t \leq 1\}$ to be the continuous path process whose finite dimensional distributions are obtained in the following way: For any k and $0 \leq t_1 < t_2 < \cdots < t_k \leq 1$,

$$P[B^+(t_i) \leq x_i, i = 1, \ldots, k]$$

$$= \lim_{\epsilon \to 0} P[B(t_i) \leq x_i, i = 1, \ldots, k | \bigwedge_{s=0}^{1} B(s) > -\epsilon].$$

(a) Give the finite dimensional distributions of B^+. Verify

$$P[B^+(1) \leq x] = 1 - e^{-x^2/2}, \quad x > 0.$$

(b) Verify

$$P[\bigvee_{s=0}^{1} B^+(s) \leq x, B^+(1) \leq y]$$

$$= \sum_{k=-\infty}^{\infty} [e^{-(2ks)^2/2} - e^{-(2kx+y)^2/2}], \quad 0 < y << x.$$

(c) Compute $P[\bigvee_{s=0}^{1} B^+(s) \leq x]$. (Set $y = x$ in (b).)
(d) Compute

$$E\left(\bigvee_{s=0}^{1} B^+(s)\right) = \sqrt{2\pi} \log 2$$

(Durrett and Iglehart, 1977).

6.32. For standard Brownian motion, check that for $\mu > 0, \nu > 0$

$$P[B(t) \leq \mu t + \nu, \text{ for all } t \geq 0 | B(0) = w] = 1 - e^{-2\mu(\nu - w)},$$

for $w \leq \nu$.

CHAPTER 7

The General Random Walk*

WHAT IS this mathematical model called a random walk? Let $\{X_n, n \geq 1\}$ be iid real valued random variables. Define $S_0 = 0$, and for $n \geq 1$ define $S_n = X_1 + \ldots + X_n$. Then $\{S_n, n \geq 0\}$ is a random walk. The random variables $\{X_i\}$ are called the *steps* of the random walk, and the distribution $F(x) = P[X_1 \leq x]$ is called the *step distribution*.

This chapter contains advanced material. Part of the treatment is somewhat non-standard and follows ideas learned from P. Greenwood. (See Greenwood, 1976; Greenwood and Shaked, 1977; Greenwood and Pitman, 1979.) Conscientious readers are urged to broaden their view of this subject by consulting the excellent standard treatments presented in, for example, the books of Feller (1971), Breiman (1968), and Chung (1974).

Basic topics in the study of the the random walk are

(1) Recurrence: How often does the process hit neighborhoods of points? We do not treat this subject. Consult the references just given.

(2) Global properties of $\{S_n, n \geq 0\}$ studied by means of *ladder variables*.

(3) Wiener-Hopf factorizations, which lead to the joint distribution of the ladder height and ladder time in terms of a double transform.

(4) Behavior of $\{\vee_{j=0}^n S_j, \wedge_{j=0}^n S_j, n \geq 0\}$.

(5) Connections with queueing theory and storage models.

The random walk is a basic model underlying many other models. It is simple to describe but offers many challenging problems. Many of the problem solving methods used for the random walk are common to more complicated models. Standard models in queueing theory, storage theory and time series analysis contain embedded random walks. Also, the random walk is the discrete time prototype of the fundamental continuous time process Brownian motion studied in Chapter 6.

We begin by reviewing some material on stopping times before launching into a discussion of the global properties of the random walk.

---

* This chapter contains advanced material not suitable for beginning readers.

7.1. STOPPING TIMES.

We have encountered the notion of stopping time several times in previous chapters with varying degrees of formality. It is now time to review our knowledge of this important technical concept, so please reread carefully the material contained in Sections 1.8, 1.8.1, and 1.8.2.

For random walks, prominent stopping times are the first strict ascending ladder epoch

$$N = \inf\{n \geq 1 : S_n > 0\}$$

and the first descending ladder epoch

$$\bar{N} = \inf\{n \geq 1 : S_n \leq 0\}.$$

It is frequently the case that N or $\bar{N}$ is infinite with positive probability so that in reviewing the important result Proposition 1.8.2 on splitting an iid sequence at a stopping time, keep in mind that the case where the stopping time is infinite with positive probability is crucial and not something of only pathological interest.

Iterates of stopping times: Let $E = \cup_{n=0}^{\infty} R^n$ be the set of all finite sequences of reals where R^0 is $\emptyset$. Define the natural σ-algebra $\mathcal{E}$ by

$$\mathcal{E} = \{G \subset E : G \cap R^n \in \mathcal{B}(R^n) \text{ for } n \geq 1\},$$

where $\mathcal{B}(R^n)$ are the Borel subsets of R^n.

Set $\mathcal{F}_n = \sigma(X_1, \ldots, X_n)$ and $\mathcal{F}'_n = \sigma(X_{n+1}, X_{n+2}, \ldots)$. Now suppose that α is a finite stopping time for $\{\mathcal{F}_n, n \geq 1\}$. For this entire chapter, $\{X_n, n \geq 1\}$ is iid. For each n there exists $B_n \in \mathcal{B}(R^n)$ such that

$$[\alpha = n] = [(X_1, \ldots, X_n) \in B_n].$$

Define $\alpha(0) = 0, \alpha(1) = \alpha$; $\alpha(2)$ is defined by

$$[\alpha(2) = n] = [(X_{\alpha(1)+j}, 1 \leq j \leq n) \in B_n]$$

for $n \geq 1$, so $\alpha(2)$ is obtained by stopping $(X_{\alpha(1)+n}, n \geq 1)$ in the same way we stopped $\{X_n, n \geq 1\}$. Proceeding inductively, define $\alpha(k+1)$ by

$$[\alpha(k+1) = n] = [(X_{\alpha(k)+j}, 1 \leq j \leq n) \in B_n]$$

for all n. The next result shows that $\beta(0) = 0, \beta(k) = \alpha(1)+\cdots+\alpha(k)$, $k \geq 1$ is a renewal process. (Cf. Chung, 1974, page 261.)

Proposition 7.1.1. *Let α be a finite stopping time for the iid sequence $\{X_n, n \geq 1\}$. Then*

$$\{(\alpha(k), X_{\beta(k-1)+1}, \ldots, X_{\beta(k)}), k \geq 1\}$$

are iid random elements of $\{1, 2, \ldots\} \times E$.

Proof. For any $A \in \mathcal{E}$,

$$[\alpha(1) = n] \cap [(X_1, \ldots, X_{\alpha(1)}) \in A] = [\alpha(1) = n] \cap [(X_1, \ldots, X_n) \in A] \in \mathcal{F}_n$$

so the random element $(\alpha(1), X_1, \ldots, X_{\alpha(1)})$ is $\mathcal{F}_{\alpha(1)}$ measurable. Similarly,

$$[\alpha(2) = n] \cap [(X_{\alpha(1)+1}, \ldots, X_{\alpha(1)+\alpha(2)}) \in A]$$
$$= [(X_{\alpha(1)+1}, \ldots, X_{\alpha(1)+n} \in B_n] \cap [(X_{\alpha(1)+1}, \ldots, X_{\alpha(1)+n}) \in A] \in \mathcal{F}'_{\alpha(1)},$$

so $(\alpha(2), X_{\alpha(1)+1}, \ldots, X_{\alpha(1)+\alpha(2)}) \in \mathcal{F}'_{\alpha(1)}$. From Proposition 1.8.2 we know that

$$(\alpha(1), X_1, \ldots, X_{\alpha(1)}) \text{ and } (\alpha(2), X_{\alpha(1)+1}, \ldots, X_{\alpha(1)+\alpha(2)})$$

are iid random elements of E. Proceed by induction for the general statement. ∎

In the case $P[\alpha = \infty] > 0$ we get from Proposition 1.8.2 that

$$\{(\alpha(k), X_{\beta(k-1)+1}, \ldots, X_{\beta(k)}), k \leq n\}$$

is iid on the trace probability space

$$\Omega \cap_{k=1}^n [\alpha(k) < \infty], \mathcal{F} \cap_{k=1}^n [\alpha(k) < \infty], \tilde{P} = \frac{P\{\cdot \cap_{k=1}^n [\alpha(k) < \infty]\}}{P\{\cap_{k=1}^n [\alpha(k) < \infty]\}}.$$

In this case, $\{\beta(k), k \geq 0\}$ is a terminating renewal process.

Special cases of outstanding interest are when $\{\beta(k), k \geq 0\} = \{N(k), k \geq 0\}$, the ascending ladder indices, or $\{\beta(k)\} = \{\bar{N}(k), k \geq 0\}$, the descending ladder indices. The previous results tell us

$$\{X_{\beta(k-1)} + \cdots + X_{\beta(k)} = S_{\beta(k)} - S_{\beta(k-1)}, k \geq 1\}$$

is iid (possibly with respect to appropriate trace spaces if $P[\beta(1) = \infty] > 0$), so $\{S_{\beta(k)}, k \geq 0\}$ is a sum of iid variables. In the case of N and $\bar{N}$, we may thus conclude that the following are all renewal processes:

$$\{N(k), k \geq 0\}, \{\bar{N}(k), k \geq 0\}, \{S_{N(k)}, k \geq 0\}, \{S_{\bar{N}(k)}, k \geq 0\}.$$

$\{N_k\}$ are the *ascending ladder epochs*, and $\{S_{N(k)}\}$ are the *ascending ladder heights*. Similarly, $\{\bar{N}(k)\}$ are the descending ladder epochs, and $\{S_{\bar{N}(k)}\}$ are the *descending ladder heights*. Note that $\{N(k)\}$ are the epochs when the random walk achieves record heights strictly larger than previous heights. The values of these record heights are the ascending ladder heights.

7.2. Global Properties.

Here we investigate relations between $N, \limsup_{n\to\infty} S_n, M_\infty := \vee_{j=0}^\infty S_j$ and classify random walks according to whether or not ladder renewal processes are terminating.

Recall

$$N = \inf\{n > 0 : S_n > 0\}, \quad \bar{N} = \inf\{n > 0 : S_n \leq 0\}$$

and set $M_n = \vee_{j=0}^n S_j$, $0 \leq n \leq \infty$.

Proposition 7.2.1. *The following are equivalent:*
  (a) $P[N < \infty] = 1$,
  (b) $P[\limsup_{n\to\infty} S_n = \infty] = 1$,
  (c) $P[M_\infty = \infty] = 1$.
Also, the following are equivalent:
  (a') $P[N = \infty] > 0$,
  (b') $P[\limsup_{n\to\infty} S_n = \infty] = 0$,
  (c') $P[M_\infty = \infty] = 0$.

Remarks. The first set of equivalences describes the situation where both $\{N(k)\}$ and $\{S_{N(k)}\}$ are non-terminating renewal processes. Similar statements obviously hold for $\bar{N}, \liminf_{n\to\infty} S_n$ and $\wedge_{j=0}^\infty S_j$ and describe when $\{\bar{N}(k)\}$ and $\{-S_{N(k)}\}$ are non-terminating.

Proof. (a) $\rightarrow$ (b): If $N < \infty$ a.s. then $S_N > 0$ is a.s. well-defined and $ES_N > 0$, since if $ES_N = 0$ we have $S_N = 0$ a.s., which contradicts the definition of N. Apply (7.1.1) to get that

$$(S_{N(k)} - S_{N(k-1)}, k \geq 1)$$

are iid proper random variables. By the strong law of large numbers, we have

$$k^{-1} S_{N(k)} = \sum_{j=1}^k (S_{N(j)} - S_{N(j-1)})/k \to ES_N > 0$$

as $k \to \infty$, and hence $S_{N(k)} \to \infty$ a.s. This entails $\limsup_{n\to\infty} S_n = \infty$, so (b) holds.

  (b) $\rightarrow$ (c): This should be clear since $\limsup_{n\to\infty} S_n \leq M_\infty$.
  (c) $\rightarrow$ (a): This should be clear since on the set $[M_\infty = \infty]$ we have $N < \infty$.

  (a') $\rightarrow$ (b') : If (a') holds, then (a) fails and therefore (b) fails. This means $P[\limsup_{n\to\infty} S_n = \infty] < 1$. However, $\limsup_{n\to\infty} S_n$ is a permutable random variable (interchanging X_i and X_j leaves the value of $\limsup_{n\to\infty} S_n$ unchanged), and so, by the Hewitt-Savage 0-1 Law (Billingsley, 1986, page 304; Chung, 1974, page 255; Feller, 1971, page 124) we have $\limsup_{n\to\infty} S_n$ is a.s. constant. Therefore, if

$$P[\limsup_{n\to\infty} S_n = \infty] < 1,$$

we must have

$$P[\limsup_{n\to\infty} S_n = \infty] = 0,$$

and consequently (b') holds.

$(b') \to (c')$: If $\limsup_{n\to\infty} S_n(\omega) < \infty$ for a.a. ω, we have a finite $K(\omega)$ with

$$\limsup S_n(\omega) \le K(\omega).$$

Given some $\epsilon > 0$, there exists $n_0(\omega)$ such that $n \ge n_0(\omega)$ implies

$$S_n(\omega) \le K(\omega) + \epsilon.$$

Hence, $M_\infty(\omega) < \infty$.

$(c') \to (a')$: If (c') holds, then (c) does not, and hence (a) fails. If (a) fails, (a') holds. ∎

The next result says a random walk either drifts to an infinity or oscillates between infinities.

Proposition 7.2.2. *Suppose $P[X_1 = 0] < 1$. Then one of the following holds:*

(i) $S_n \to +\infty$ a.s.,

(ii) $S_n \to -\infty$ a.s.,

or

(iii) $-\infty = \liminf_{n\to\infty} S_n < \limsup_{n\to\infty} S_n = +\infty$ a.s.

Proof. Set $S_\infty = \limsup_{n\to\infty} S_n$. Then S_∞ is a permutable random variable and hence a.s. constant, i.e., there exists a constant c (possibly $\pm\infty$) and $S_\infty = c$ a.s. Now

$$S_\infty = X_1 + \limsup_{n\to\infty} \sum_{i=2}^{n} X_i.$$

Since $\{X_n, n \ge 1\} \overset{d}{=} \{X_n, n \ge 2\}$, we have $S_\infty \overset{d}{=} \limsup_{n\to\infty} \sum_{i=2}^{n} X_i$ and thus almost surely

$$c = X_1 + c.$$

Since $P[X_1 \ne 0] > 0$, we see that $c = \pm\infty$. So either

(A) $\limsup_{n\to\infty} S_n = +\infty$ or (B) $\limsup_{n\to\infty} S_n = -\infty$, and the latter possibility means $\lim_{n\to\infty} S_n = -\infty$. A similar argument with $\liminf$ gives

(A') $\liminf_{n\to\infty} S_n = +\infty$, so that $\lim_{n\to\infty} S_n = +\infty$, or alternatively (B') $\liminf_{n\to\infty} S_n = -\infty$. Combine the possibilities $A+A' = (i)$, $A+B' = (iii)$, $B + A' = \emptyset$, $B + B' = (ii)$. ∎

When the mean exists and equals zero, we get the oscillation case.

Proposition 7.2.3. *If $E|X_1| < \infty$ and $P[X_1 = 0] < 1$, then*
(i) *if $EX_1 > 0$, $S_n \to \infty$ a.s.,*
(ii) *if $EX_1 < 0$, $S_n \to -\infty$ a.s.,*
(iii) *if $EX_1 = 0$, we have*

$$-\infty = \liminf_{n\to\infty} S_n < \limsup_{n\to\infty} S_n = +\infty.$$

Proof. Statements (i) and (ii) follow straightforwardly from the strong law of large numbers, so we concentrate on (iii).

Suppose, for the purposes of obtaining a contradiction, that

$$P[N = \infty] = q > 0.$$

From Proposition 7.2.1(c), we have $M_\infty = \vee_{j\geq 0} S_j < \infty$ a.s. Let $\nu = \inf\{j \geq 0 : S_j = M_\infty\}$ with $\inf \emptyset = \infty$ if necessary. We have

$$1 \geq \sum_{n=0}^{\infty} P[\nu = n]$$

$$= \sum_{n=0}^{\infty} P[S_j < S_n, 0 \leq j < n, S_k \leq S_n, k > n]$$

(interpret $n = 0$ sensibly)

$$= \sum_{n=0}^{\infty} P\{[\sum_{i=j+1}^{n} X_i > 0, 0 \leq j < n] \cap [\sum_{i=n+1}^{k} X_i \leq 0, k > n]\};$$

by independence, this is

$$= \sum_{n=0}^{\infty} P[\sum_{i=j+1}^{n} X_i > 0, 0 \leq j < n] P[\sum_{i=n+1}^{k} X_i \leq 0, k > n]$$

$$= \sum_{n=0}^{\infty} P[X_n > 0, X_n + X_{n-1} > 0, \ldots, \sum_{i=1}^{n} X_i > 0] P[S_j \leq 0, j \geq 0]$$

$$= \sum_{n=0}^{\infty} P[X_1 > 0, X_1 + X_2 > 0, \ldots, S_n > 0] P[N = \infty]$$

(since $(X_n, X_{n-1}, \ldots, X_1) \overset{d}{=} (X_1, \ldots, X_n)$)

$$= \sum_{n=0}^{\infty} P[\bar{N} > n]q = E\bar{N}q.$$

Because $q > 0$, we conclude $E\bar{N} < \infty$. Wald's identity (Proposition 1.8.1) implies

$$ES_{\bar{N}} = E\bar{N}EX_1 = E\bar{N}0 = 0.$$

Since $S_{\bar{N}} \leq 0$ a.s. we know $S_{\bar{N}} = 0$ a.s., which violates the definition of $\bar{N}$ because

$$P[S_{\bar{N}} < 0] \geq P[X_1 < 0] > 0 \text{ if } EX_1 = 0.$$

We get a contradiction, which means $q = 0$. Similarly, $P[\bar{N} = \infty] = 0$. Finish by applying Proposition 7.2.1(b) and its lim inf counterpart. ∎

7.3. PRELUDE TO WIENER-HOPF THEORY: PROBABILISTIC INTERPRETATIONS OF TRANSFORMS.

The Wiener-Hopf factorization concerns itself with joint transforms such as

$$Ee^{i\zeta S_N}q^N, \quad 0 \leq q \leq 1, \zeta \in R.$$

Our treatment of such transforms is based on probabilistic rather than analytic arguments. To better understand this procedure, it is beneficial to discuss some historical attempts to interpret transforms as probabilities. This interpretation is frequently called the method of collective marks and is usually attributed to David Van Dantzig (1948). See also Neuts, 1973, p. 137, and Runnenburg, 1958, 1965.

Here is one interpretation of generating functions: Let $\{p_k, k \geq 0\}$ be a probability sequence so that $p_k \geq 0$ and $\sum p_k = 1$. Imagine k objects are chosen with probability p_k. Independent of the choice mechanism is a marking mechanism: An object is marked with probability $1 - q$ and left unmarked with probability q, $0 \leq q \leq 1$. Then

$$P[\text{ no object is marked }] = \sum_{k=0}^{\infty} P[k \text{ objects chosen, none of } k \text{ is marked }]$$

$$= \sum_{k=0}^{\infty} p_k q^k = P(q),$$

which, of course, is the generating function corresponding to $\{p_k\}$.

Here is a second interpretation using geometric random variables: Let Y be a non-negative integer valued random variable with $P[Y = k] =$

p_k, $p_k \geq 0$, $\sum_{k=0}^{\infty} p_k = 1$. Suppose the random variable T is independent of Y and T has a geometric distribution, and suppose

$$P[T \geq k] = q^k, k \geq 0.$$

Then

$$P[Y \leq T] = \sum_{k=0}^{\infty} P[Y = k]P[k \leq T]$$

$$= \sum_{0}^{\infty} p_k q^k = P(q).$$

Similar interpretations are possible for Laplace transforms. For example, suppose T is the waiting time for a catastrophe which has an exponentially distributed waiting time with parameter λ. (Why this subject traditionally uses such apocalyptic terminology is not clear.) Let Y be the waiting time until a picnic; Y has distribution F and is independent of the catastrophe time T. Then

$$P[\text{ picnic takes place before the catastrophe }]$$

$$= P[Y \leq T] = \int_0^{\infty} P[t \leq T]F(dt)$$

$$= \int_0^{\infty} e^{-\lambda t} F(dt) = \hat{F}(\lambda).$$

We concentrate on the integer-valued case. If $P[T \geq k] = q^k$, then a useful property for T is the discrete forgetfulness property:

(7.3.1) $$P[T \geq n + k|T \geq n] = P[T \geq k].$$

This holds since

$$P[T \geq n + k|T \geq n] = \frac{P[T \geq n+k, T \geq n]}{P[T \geq n]}$$

$$= P[T \geq n + k]/P[T \geq n] = q^{n+k}/q^n$$

$$= q^k = P[T \geq k].$$

There is a random analogue: If Y and Z are independent, non-negative and integer valued and each is independent of T, then

(7.3.2) $$P[T \geq Y + Z|T \geq Y] = P[T \geq Z].$$

This is readily checked once one observes

$$P[T \geq Y + Z] = \sum_0^\infty q^k P[Y + Z = k] = Eq^{Y+Z}$$

$$= Eq^Y Eq^Z = P[T \geq Y]P[T \geq Z];$$

the generating function $P_{Y+Z}(q)$ of $Y + Z$ satisfies

$$P_{Y+Z}(q) = P_Y(q)P_Z(q).$$

For (7.3.2), we now have

$$P[T \geq Y + Z | T \geq Y] = P[T \geq Y + Z]/P[T \geq Y]$$

$$= P[T \geq Y]P[T \geq Z]/P[T \geq Y] = P[T \geq Z].$$

As a final warm-up, consider again the material of Section 1.6 on the Bernoulli random walk. We have $P[X_1 = 1] = a = 1 - P[X_1 = -1]$. Let $N = \inf\{n \geq 1 : S_n = 1\}$, and compute $P(q) = Eq^N$. The classical method of solution discussed in Section 1.6 is to write a difference equation for $P[N = n]$, multiply through by q^n, and sum over n to get an equation in the unknown $P(q)$. The methods discussed above allow us to bypass the step of writing a difference equation and go directly to an equation in which the unknown is the generating function. This is done as follows.

Let T be geometrically distributed, $P[T \geq n] = q^n$, and independent of the random walk. Then

$$P(q) = P[N \leq T] = P[N \leq T, X_1 = -1] + P[N \leq T, X_1 = +1].$$

On the set $[X_1 = +1]$ we have $N = 1$, so

$$P[N \leq T, X_1 = +1] = P[1 \leq T, X_1 = 1] = qa.$$

If $X_1 = -1$, then the random walk must start at -1, move eventually to 0 in $N_{-1,0}$ steps, and then from 0 to 1 in $N_{0,1}$ steps. From Sections 1.6, 1.8.2 and Proposition 7.1.1 we confidently assert $N_{-1,0}, N_{0,1}$ are independent and

$$N \overset{d}{=} N_{-1,0} \overset{d}{=} N_{0,1},$$

so

$$P[N \leq T, X_1 = -1] = P[1 + N_{-1,0} + N_{0,1} \leq T](1 - a)$$

$$= P[T \geq 1]P[T \geq N_{-1,0}]P[T \geq N_{0,1}](1 - a)$$

$$= qP^2(q)(1 - a).$$

Therefore,

$$P(q) = qa + q(1-a)P^2(q),$$

and solving the quadratic for $P(q)$ yields

$$P(q) = \frac{1 \pm \sqrt{1 - 4q^2 a(1-a)}}{2q(1-a)}, \quad 0 \le q \le 1.$$

The solution with the plus sign is ruled out, since it is unbounded as $q \downarrow 0$ in violation of the behavior of generating functions.

In general random walk theory, basic results describe the joint distribution of (N, S_N) as well as the joint distribution of $(\bar{N}, S_{\bar{N}})$. In deriving these distributions we will be led to consider the measures

$$(7.3.3) \qquad\qquad H_q(\cdot) = P[S_N \in \cdot, N \le T]$$

and

$$(7.3.4) \qquad\qquad \bar{H}_q(\cdot) = P[S_{\bar{N}} \in \cdot, \bar{N} \le T],$$

where T is independent of $\{S_n\}$, $P[T \ge n] = q^n$. The relevance of H_q and $\bar{H}_q$ comes from the fact that, for $\zeta \in R$,

$$\int_{(0,\infty)} e^{i\zeta x} H_q(dx) = \sum_{n=1}^{\infty} \int_{(0,\infty)} e^{i\zeta x} P[S_n \in dx, n \le T, N = n]$$

$$= \sum_{n=1}^{\infty} \int_{(0,\infty)} e^{i\zeta x} P[S_n \in dx, N = n] q^n$$

$$(7.3.5) \qquad\qquad = E q^N e^{i\zeta S_N},$$

which is the joint transform of (N, S_N). Similarly,

$$(7.3.6) \qquad\qquad \int_{(-\infty,0]} e^{i\zeta x} \bar{H}_q(dx) = E q^{\bar{N}} e^{i\zeta S_{\bar{N}}}.$$

Note H_q concentrates on $(0, \infty)$ and $\bar{H}_q$ concentrates on $(-\infty, 0]$. The Wiener-Hopf factorization informs us that if F is the step distribution of the random walk and δ is the measure concentrating mass one at the origin then

$$(\delta - qF) = (\delta - H_q) * (\delta - \bar{H}_q).$$

Techniques exist which allow us to solve for each factor. We pursue these issues in the next two sections.

7.4. DUAL PAIRS OF STOPPING TIMES.

Duality is the basic property leading to a Wiener-Hopf factorization.

Henceforth it will be convenient to have a concrete probability space, so suppose $\Omega = R^\infty, \mathcal{F} = \mathcal{B}(R^\infty)$. Suppose P is a product measure on $(R^\infty, \mathcal{B}(R^\infty))$ such that each marginal measure is $F(\cdot)$, the step distribution of the random walk. When $\omega = (x_1, x_2, \dots)$, we define the coordinate maps by

$$X_n(\omega) = X_n((x_1, x_2, \dots)) = x_n, \quad n \geq 1.$$

Under P, $\{X_n, n \geq 1\}$ is iid. If α is a stopping time for $\mathcal{F}_n = \sigma(X_1, \dots, X_n)$, $n \geq 1$, then the iterates of α can be reinterpreted as follows: $\alpha_0 = 0, \alpha_1 = \alpha$ and $\alpha_2(\omega) = \alpha_2(x_1, x_2, \dots) = \alpha(x_{\alpha+1}, x_{\alpha+2}, \dots)$, etc. These α_k's form an iid sequence. Let $M_\alpha(\omega) = \{\sum_{i=0}^n \alpha_i(\omega), n \geq 0\}$, i.e., $M_\alpha(\omega)$ is the (random) set formed by the renewal sequence generated by iterates of α. Note M_α is a.s. a finite set if $P[\alpha = \infty] > 0$.

As a final preparation, define for each $n \geq 1$ the reversal maps $r_n : R^\infty \to R^\infty$ by

$$r_n(x_1, x_2, \dots, x_n, x_{n+1}, x_{n+2}, \dots) = (x_n, x_{n-1}, \dots, x_1, x_{n+1}, x_{n+2}, \dots).$$

Define r_0 as the identity from $R^\infty \to R^\infty$, and note $r_n = r_n^{-1}$ and $r_n \circ r_n$ is the identity. Further, $P = P \circ r_n^{-1}$, which merely says that in R^∞

$$(X_1, X_2, \dots) \stackrel{d}{=} (X_n, X_{n-1}, \dots, X_1, X_{n+1}, X_{n+2}, \dots).$$

Definition. Suppose τ and η are stopping times for $\{\mathcal{F}_n, n \geq 1\}$. Then τ is dual to η if for every n we have

(7.4.1) $$\{\omega : n \in M_\tau(\omega)\} = \{\omega : n < \eta \circ r_n(\omega)\}.$$

Thus n is an iterate of τ iff when we reverse time from n, looking backward, we see no iterates of η. The most important example of this concept is $N, \bar{N}$ as is shown in the next proposition.

Proposition 7.4.1. N is dual to $\bar{N}$.

Proof. Try drawing a picture. The analytic approach is as follows: We have that $n \in M_N(\omega)$ iff $n = N_k(\omega)$ for some $k \geq 0$, iff $S_n(\omega) > S_j(\omega)$, for $j = 0, 1, \dots, n-1$, iff $S_n(\omega) - S_j(\omega) > 0$, $j = 0, \dots, n-1$, iff $\sum_{i=j+1}^n X_i(\omega) > 0$, $j = 0, \dots, n-1$, iff $S_{n-j}(r_n\omega) > 0$, $j = 0, \dots, n-1$, iff $S_k(r_n\omega) > 0$, $k = 1, \dots, n$, iff $n < \bar{N} \circ r_n\omega$. ∎

Re-examine Proposition 7.2.3 to see how duality has already been used.

We will need the following characterization of duality due to Greenwood and Shaked (1978) with an assist from J. Pitman. For a stopping time τ, define for $n \geq 0$

$$L(\tau, n)(\omega) = \max\{i \leq n : i \in M_\tau(\omega)\},$$

so that $L(\tau, n)$ is the last renewal generated by iterates of τ at or before index n.

Proposition 7.4.2. *Let τ and η be stopping times for $\{\mathcal{F}_n\}$. Then τ is dual to η iff for all n*

$$(7.4.2) \qquad n - L(\tau, n) = L(\eta, n) \circ r_n.$$

(Note that $L(\eta, n) \circ r_n$ is the random variable with value at ω equal to $L(\eta, n)(r_n \omega)$.)

If we rephrase (7.4.2) by substituting $r_n \omega$ in both sides, we get the equivalent statement

$$n - L(\tau, n) \circ r_n = L(\eta, n),$$

which is illustrated in Figure 7.1.

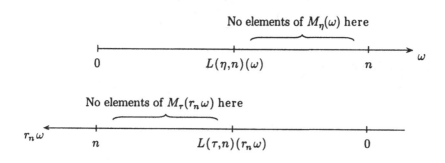

FIGURE 7.1

Before the proof of Proposition 7.4.2, we give some corollaries.

Corollary 7.4.3. Reflexivity. *τ is dual to η iff η is dual to τ.*

Proof of Corollary 7.4.3. We have τ is dual to η iff $n - L(\tau, n) = L(\eta, n) \circ r_n$ for all n, iff $n - L(\tau, n) \circ r_n = L(\eta, n) \circ r_n \circ r_n$ (recall that $r_n \circ r_n$ is the identity), iff $n - L(\eta, n) = L(\tau, n) \circ r_n$ for all n. Applying Proposition 7.4.2, the above holds true iff η is dual to τ. ∎

Henceforth we talk of dual pairs.

Corollary 7.4.4. *For $n \geq 0$,*

$$L(\eta, n) \stackrel{d}{=} n - L(\tau, n).$$

Proof. Since r_n preserves measure (that is, $P = P \circ r_n^{-1}$), we have

$$n - L(\tau, n) = L(\eta, n) \circ r_n$$

(from Proposition 7.4.2)

$$\stackrel{d}{=} L(\eta, n). \blacksquare$$

Corollary 7.4.5. *For $n \geq 0$,*

$$\sum_{i=1}^{L(\tau,n)} X_i \stackrel{d}{=} \sum_{i=L(\eta,n)+1}^{n} X_i.$$

Proof. Since r_n preserves measure,

$$\sum_{i=1}^{L(\tau,n)} X_i \stackrel{d}{=} \sum_{i=1}^{L(\tau,n) \circ r_n} X_i \circ r_n,$$

which, by Proposition 7.4.2, is

$$= \sum_{i=1}^{n-L(\eta,n)} X_i \circ r_n.$$

Since

$$(X_1(r_n), X_2(r_n), \ldots, X_n(r_n), X_{n+1}(r_n), \ldots)$$
$$= (X_n, X_{n-1}, \ldots, X_1, X_{n+1}, \ldots),$$

we know for $i = 1, \ldots, n$, that $X_i \circ r_n = X_{n-i+1}$. Therefore,

$$\sum_{i=1}^{n-L(\eta,n)} X_i \circ r_n = \sum_{i=1}^{n-L(\eta,n)} X_{n-i+1},$$

and changing dummy indices, say $j = n - i + 1$, gives the above equal to $\sum_{j=L(\eta,n)+1}^{n} X_j$, as required. $\blacksquare$

Proof of Proposition 7.4.2. We first show (7.4.2) implies (7.4.1). We have

$$n \in \mathcal{M}_\tau(\omega) \text{ iff } n - L(\tau, n)(\omega) = 0.$$

If we apply (7.4.2), this last statement holds iff $L(\eta, n)(r_n\omega) = 0$ which is equivalent to $\eta(r_n\omega) > n$, as desired.

For the converse, we now prove characterization 7.4.2 based on the definition of duality. It is convenient to define the shift operators $\theta_k : R^\infty \to R^\infty$ for $k \geq 1$ by

$$\theta_k(x_1, x_2, \dots) = (x_{k+1}, x_{k+2}, \dots).$$

Now suppose $L(\tau, n)(\omega) = v$, $n - v = m$. We must show

$$L(\eta, n)(r_n\omega) = m.$$

Suppose for now $m < n$ as a separate (but simpler) argument for $m = n$ (i.e., $v = 0$) is necessary. If $\omega = (x_1, x_2, \dots) \in R^\infty$, observe

$$\theta_m r_n\omega = \theta_m(x_n, x_{n-1}, \dots, x_1, x_{n+1}, x_{n+2}, \dots)$$
$$= (x_{n-m}, x_{n-m-1}, \dots, x_1, x_{n+1}, x_{n+2}, \dots)$$

and also

$$r_{n-m}\omega = (x_{n-m}, x_{n-m-1}, \dots, x_1, x_{n-m+1}, x_{n-m+2}, \dots).$$

We therefore conclude

(7.4.3)   the first $n - m$ components of $\theta_m r_n\omega$ and $r_{n-m}\omega$ are equal.

Since $v = L(\tau, n)(\omega)$, we have $v \in \mathcal{M}_\tau(\omega)$, and hence, from the definition of duality

$$n - m = v < \eta \circ r_v\omega = \eta \circ r_{n-m}\omega.$$

Since η is a stopping time, and since the first v terms of the sequence $r_{n-m}\omega$ match the first v terms of $\theta_m r_n\omega$, we have

(7.4.4) $n - m < \eta \circ \theta_m \circ r_n\omega.$

If (to be proved) it is also the case that

(7.4.5) $m \in \mathcal{M}_\eta(r_n\omega),$

then on the one hand m is a renewal epoch generated by iterates of η for the realization $r_n\omega$ and on the other hand (7.4.4) says there are no other renewal epochs between $n - m$ and n for this realization. Hence m is the last renewal epoch before n, and

$$m = L(\eta, n)(r_n\omega),$$

as was to be shown.

To prove (7.4.5) it is enough to prove

(7.4.6) $$m = L(\eta, m)(r_n\omega).$$

Set $k = L(\eta, m)(r_n\omega)$ so that $k \leq m$. We wish to show $k = m$. For the purpose of obtaining a contradiction, suppose $k < m$. As before,

$$\theta_k r_n\omega = (x_{n-k}, x_{n-k-1}, \ldots, x_1, x_{n+1}, x_{n+2}, \ldots)$$

and

$$r_{m-k}\theta_v\omega = r_{m-k}(x_{v+1}, x_{v+2}, \ldots, x_{v+m-k}, x_{v+m-k+1}, \ldots)$$
$$= (x_{v+m-k}, x_{v+m-k-1}, \ldots, x_{v+2}, x_{v+1}, x_{v+m-k+1}, \ldots),$$

so because $v + m = n$, we have that

(7.4.7)   the first $m - k$ components of $\theta_k r_n\omega$ and $r_{m-k}\theta_v\omega$ are equal.

Since $k = L(\eta, m)(r_n\omega)$ there are no η iterates on the $r_n\omega$ path between k and m and thus

$$\eta(\theta_k r_n\omega) > m - k.$$

Using (7.4.7) gives

$$\eta(r_{m-k}\theta_v\omega) > m - k,$$

which implies, since τ is dual to η, that

(7.4.8) $$m - k \in M_\tau(\theta_v\omega).$$

The original assumption, however, was that $L(\tau, n)(\omega) = v > 0$, and thus $(v, n] \cap M_\tau(\omega) = \emptyset$. If $k < m$, however, then $0 < m - k \leq m \leq n$, and a reinterpretation of (7.4.8) gives

$$(v, n] \cap M_\tau(\omega) \neq \emptyset,$$

a contradiction. Hence $k = m$, and (7.4.6) holds.

If $m = n$, the proof is handled by reading the foregoing starting from (7.4.6). ∎

7.5. WIENER-HOPF DECOMPOSITIONS.

As a first step in deriving the distributions of (N, S_N) and $(\bar{N}, S_{\bar{N}})$, we factor the step distribution F into a convolution product. The only property of $(N, \bar{N})$ that is used is duality, so we proceed with an arbitrary dual pair of stopping times τ and η. Pick $q \in [0, 1]$, and let T be a non-negative integer valued random variable independent of $\{S_n\}$ with a geometric distribution: $P[T \geq n] = q^n$. Set

$$H_{\tau,q}(\cdot) = P[S_\tau \in \cdot, \tau \leq T], \quad H_{\eta,q}(\cdot) = P[S_\eta \in \cdot, \eta \leq T].$$

We denote by δ the probability measure which concentrates all mass at 0.

Theorem 7.5.1. *We have the following factorization of F, the step distribution:*

$$(7.5.1) \qquad \delta - qF = (\delta - H_{\tau,q}) * (\delta - H_{\eta,q}).$$

The decomposition in (7.5.1) is known as the *Wiener-Hopf factorization*. Eventually we will apply a separation technique to the factorization in (7.5.1) which allows us to solve for each factor on the right side.

The proof proceeds via a series of fairly simple lemmas. The idea is to write

$$S_T = \sum_{i=1}^{L(\tau,T)} X_i + \sum_{i=L(\tau,T)+1}^{T} X_i.$$

The two sums on the right must be shown to be independent, and they lead to the convolution factors in (7.5.1).

Let us choose a convenient probability space on which to do our analysis. Let $(\Omega', \mathcal{F}') = (R^\infty, \mathcal{B}(R^\infty))$, and set P' to be product measure which makes the coordinate random variables $\{X'_n\}$ iid with common distribution F. Let $(\Omega'', \mathcal{F}'', P'')$ be any probability space supporting a geometrically distributed random variable T'', i.e., $P''[T'' \geq n] = q^n$. Define as usual

$$(\Omega, \mathcal{F}, P) = (\Omega' \times \Omega'', \mathcal{F}' \times \mathcal{F}'', P' \times P'')$$

and

$$X_n(\omega', \omega'') = X'_n(\omega'), \quad n \geq 1,$$
$$T(\omega', \omega'') = T''(\omega'').$$

Define $r_n : \Omega \to \Omega$, $\theta_k : \Omega \to \Omega, n \geq 1, k \geq 1$, in the following manner. For $(\omega' = (x_1, x_2, \dots))$

$$r_n(\omega', \omega'') = ((x_n, x_{n-1}, \dots, x_1, x_{n+1}, \dots), \omega'')$$
$$\theta_k(\omega', \omega'') = ((x_{k+1}, x_{k+2}, \dots), \omega''),$$

so that shift and reversal operators have no effect on T.

We begin the proof of the Wiener-Hopf decomposition (7.5.1) with the following lemma.

Lemma 7.5.2. *We have*

$$\sum_{i=L(\tau,T)+1}^{T} X_i \stackrel{d}{=} \sum_{i=1}^{L(\eta,T)} X_i.$$

Proof. Corollary 7.4.5 yields, for $n \geq 0$,

$$\sum_{i=1}^{L(\eta,n)} X_i \stackrel{d}{=} \sum_{i=L(\tau,n)+1}^{n} X_i.$$

Hence, for $x \in R$,

$$P[\sum_{i=L(\tau,T)+1}^{T} X_i \leq x] = \sum_{n=0}^{\infty} P[\sum_{i=L(\tau,n)+1}^{n} X_i \leq x]P[T=n]$$

$$= \sum_{n=0}^{\infty} P[\sum_{i=1}^{L(\eta,n)} X_i \leq x]P[T=n]$$

$$= P[\sum_{i=1}^{L(\eta,T)} X_i \leq x]. \quad \blacksquare$$

Lemma 7.5.3. *Let $\{\tau(i), i \geq 0\}$ be the renewal process generated by τ ($\tau(0) = 0, \tau(1) = \tau, \tau(2) = \tau + \tau \circ \theta_\tau$, etc.). As before, set*

$$H_{\tau,q}(x) = P[S_\tau \leq x, \tau \leq T].$$

(a) *For $x_i \in R, i = 1, \ldots, k$,*

$$P[S_{\tau(i)} - S_{\tau(i-1)} \leq x_i, i = 1, \ldots, k; \tau_k \leq T] = \prod_{i=1}^{k} H_{\tau,q}(x_i).$$

b) *For $k \geq 0$*

$$P[S_{\tau(k)} \leq x, \tau(k) \leq T] = H_{\tau,q}^{k*}(x).$$

Proof. (a) We have

$$P[S_{\tau(i)} - S_{\tau(i-1)} \le x_i, i = 1, \ldots, k; \tau(k) \le T]$$

$$= \sum_{n=0}^{\infty} P[\tau(k) - \tau(k-1) = n, S_{\tau(i)} - S_{\tau(i-1)} \le x_i,$$

$$i = 1, \ldots, k, \tau(k-1) + n \le T]$$

$$= \sum_{n=0}^{\infty} P\left\{ \cap_{i=1}^{k-1}[S_{\tau(i)} - S_{\tau(i-1)} \le x_i, \tau(k-1) + n \le T] \right.$$

$$\left. \cap [\tau(k) - \tau(k-1) = n, S_{\tau(k)} - S_{\tau(k-1)} \le x_k] \right\}.$$

The first square-bracketed event belongs to $\mathcal{F}_{\tau(k-1)} \vee \sigma(T)$, and the second belongs to $\mathcal{F}'_{\tau(k-1)}$. Hence by the facts about splitting an iid sequence at a stopping time, these two events are independent. Thus the above equals

$$\sum_{n=0}^{\infty} P[S_{\tau(i)} - S_{\tau(i-1)} \le x_i, i = 1, \ldots, k-1, \tau(k-1) + n \le T]$$

$$\cdot P[\tau = n, S_\tau \le x_k]$$

(we have applied a slight variation of Proposition 7.1.1)

$$= \sum_{n=0}^{\infty} P[\tau = n, S_\tau \le x_k] E(1_{[S_{\tau(i)} - S_{\tau(i-1)} \le x_i, i=1,\ldots,k-1]}$$

$$P[\tau(k-1) + n \le T | \mathcal{F}_{\tau(k-1)}])$$

$$= \sum_{n=0}^{\infty} P[\tau = n, S_\tau \le x_k] q^n E(1_{[S_{\tau(i)} - S_{\tau(i-1)} \le x_i, i=1,\ldots,k-1]} q^{\tau(k-1)})$$

$$= \sum_{n=0}^{\infty} P[\tau = n, n \le T, S_\tau \le x_k] E(1_{\{\cap_{i=1}^{k-1}[S_{\tau(i)} - S_{\tau(i-1)} \le x_i]\}} q^{\tau(k-1)})$$

$$= \sum_{n=0}^{\infty} P[\tau = n, \tau \le T, S_\tau \le x_k] E(1_{\{\cap_{i=1}^{k-1}[S_{\tau(i)} - S_{\tau(i-1)} \le x_i]\}} q^{\tau(k-1)})$$

$$= P[\tau \le T, S_\tau \le x_k] E(1_{\{\cap_{i=1}^{k-1}[S_{\tau(i)} - S_{\tau(i-1)} \le x_i]\}} q^{\tau(k-1)})$$

$$= H_{\tau,q}(x_k) E(E(1_{\{\cap_{i=1}^{k-1}[S_{\tau(i)} - S_{\tau(i-1)} \le x_i]\}} q^{\tau(k-1)} | \mathcal{F}_{\tau(k-1)})))$$

$$= H_{\tau,q}(x_k) E P \left\{ \cap_{i=1}^{k-1}[S_{\tau(i)} - S_{\tau(i-1)} \le x_i] \cap \tau(k-1) \le T | \mathcal{F}_{\tau(k-1)} \right\}$$

$$= H_{\tau,q}(x_k) P[S_{\tau(i)} - S_{\tau(i-1)} \le x_i, i = 1, \ldots, k-1, \tau(k-1) \le T].$$

The proof is completed by induction.

(b) We may rewrite (a) as

$$P[\cap_{i=1}^{k}[S_{\tau(i)} - S_{\tau(i-1)} \le x_i]|\tau_k \le T]$$

$$= \prod_{i=1}^{k} H_{\tau,q}(x_i)/P[\tau_k \le T]$$

$$= \prod_{i=1}^{k} H_{\tau,q}(x_i)/P[\sum_{i=1}^{k}(\tau(i) - \tau(i-1)) \le T]$$

and from (7.3.2) this is

$$= \prod_{i=1}^{k} H_{\tau,q}(x_i)/\prod_{i=1}^{k} P[\tau(i) - \tau(i-1) \le T]$$

$$= \prod_{i=1}^{k}(H_{\tau,q}(x_i)/P[\tau \le T]).$$

Therefore

$$P[S_{\tau(k)} \le x|\tau_k \le T] = (H_{\tau,q}(\cdot)/P[\tau \le T])^{k*}(x)$$

and

$$P[S_{\tau(k)} \le x, \tau_k \le T] = H_{\tau,q}^{k*}(x). \blacksquare$$

Lemma 7.5.4. The distribution of $\sum_{i=1}^{L(\tau,T)} X_i$ is

$$P[\sum_{i=1}^{L(\tau,T)} X_i \le x] = \sum_{k=0}^{\infty} H_{\tau,q}^{k*}(x)(1 - P[T \ge \tau]).$$

Proof. We have

$$P[\sum_{i=1}^{L(\tau,T)} X_i \le x]$$

$$= \sum_{k=0}^{\infty} P[\sum_{i=1}^{\tau(k)} X_i \le x, \tau(k) \le T < \tau(k+1)]$$

$$= \sum_{k=0}^{\infty} \left(P[\sum_{i=1}^{\tau(k)} X_i \le x, \tau(k) \le T] - P[\sum_{i=1}^{\tau(k)} X_i \le x, \tau(k+1) \le T] \right)$$

Applying Lemma 7.5.3, the above is

$$= \sum_{k=0}^{\infty} \left(H_{\tau,q}^{k*}(x) - P[S_{\tau(k)} \le x, S_{\tau(k+1)} - S_{\tau(k)} < \infty, \tau(k+1) \le T] \right)$$

$$= \sum_{k=0}^{\infty} \left(H_{\tau,q}^{k*}(x) - H_{\tau,q}^{k*}(x) H_{\tau,q}(\infty) \right)$$

$$= \sum_{k=0}^{\infty} H_{\tau,q}^{k*}(x)(1 - P[\tau \le T]). \blacksquare$$

Lemma 7.5.5. The random variables $\sum_{i=1}^{L(\tau,T)} X_i$ and $\sum_{i=L(\tau,T)+1}^{T} X_i$ are independent.

The proof of this important lemma rests on the following result (see for example, Freedman, 1971, p. 137) and is a special case of the *découpage de Lévy* (e.g., Resnick, 1987, pages 196ff).

Lemma 7.5.6. Suppose $(\Omega, \mathcal{F}, P)$ is a probability space and $(E, \mathcal{E})$ is a measurable space. Let $X_1, X_2, \ldots$ be iid random elements from $(\Omega, \mathcal{F})$ into $(E, \mathcal{E})$. Suppose $V \in \mathcal{E}$ is fixed and

$$P[X_1 \in V^c] = \theta \in (0,1).$$

Define

$$L = \inf\{n \ge 1 : X_n \in V^c\}.$$

Now suppose $T, Z, Y_1, Y_2, \ldots$ are independent and defined on $(\Omega, \mathcal{F})$ satisfying

(a) $P[T = n] = \theta(1 - \theta)^{n-1}$, $n \ge 1$,
(b) $P[Z \in \Lambda] = P[X_1 \in \Lambda | X_1 \in V^c]$ for $\Lambda \in \mathcal{E}$,
(c) $P[Y_i \in \Lambda] = P[X_1 \in \Lambda | X_1 \in V]$, $\Lambda \in \mathcal{E}$, $i \ge 1$.

Then

$$(X_1, X_2, \ldots, X_{L-1}, X_L, L) \overset{d}{=} (Y_1, Y_2, \ldots, Y_{T-1}, Z, T)$$

on $\cup_{k=0}^{\infty}(E^k \times \{0, 1, \ldots\})$.

Proof. For $n \geq 1$ and $A_i \in \mathcal{E} \cap V$, $i = 1, \ldots, n-1$, $B \in \mathcal{E} \cap V^c$, we have

$$P[L = n, X_i \in A_i, i = 1, \ldots, n-1, X_n \in B]$$
$$= P[X_i \in A_i, i = 1, \ldots, n-1, X_n \in B]$$
$$= \prod_{i=1}^{n-1} P[X_i \in A_i] P[X_n \in B]$$
$$= \prod_{i=1}^{n-1} \left(P[X_i \in A_i | X_i \in V](1-\theta) \right) P[X_n \in B | X_n \in V^c]\theta$$
$$= (1-\theta)^{n-1}\theta \prod_{i=1}^{n-1} P[Y_i \in A_i] P[Z \in B]$$
$$= P[T = n] \prod_{i=1}^{n-1} P[Y_i \in A_i] P[Z \in B]$$
$$= P[T = n, Y_i \in A_i, i = 1, \ldots, n-1, Z \in B]. \quad \blacksquare$$

Proof of Lemma 7.5.5. Let $\{\delta_j, j \geq 1\}$ be iid Bernoulli random variables with $P[\delta_1 = 1] = p$, $P[\delta_1 = 0] = q$, and suppose further that $\{\delta_j\}$ is independent of $\{S_n\}$. Define

$$T = \sup\{n \geq 0 : \sum_{i=1}^{n} \delta_i = 0\}$$

(interpret $\sum_{i=1}^{0} = 0$) so that

$$P[T \geq n] = P[\sum_{i=1}^{n} \delta_i = 0] = P[\delta_1 = 0 = \cdots = \delta_n] = q^n.$$

For $i \geq 1$, set

$$W_i = (X_{\tau(i-1)+1}, \ldots, X_{\tau(i)}; \delta_{\tau(i-1)+1}, \ldots, \delta_{\tau(i)})$$

so that $\{W_i, i \geq 1\}$ are iid random elements of

$$\bigcup_{n=0}^{\infty} (R^n \times \{0,1\}^n).$$

Suppose

$$L = \inf\{i \geq 1 : \sum_{j=\tau(i-1)+1}^{\tau(i)} \delta_j \neq 0\}$$

and, by Lemma 7.5.6,

$$W_1, W_2, \ldots, W_{L-1}, W_L, L$$

are independent. However,

$$\sum_{i=1}^{L(\tau,T)} X_i = \sum_{k=0}^{\infty} S_{\tau(k)} 1_{[\tau(k) \le T < \tau(k+1)]}$$

$$= \sum_{k=0}^{\infty} S_{\tau(k)} 1_{[\Sigma_{i=1}^{\tau(k)} \delta_i = 0, \Sigma_{i=\tau(k)+1}^{\tau(k+1)} \delta_i \ne 0]}$$

$$= \sum_{k=0}^{\infty} S_{\tau(k)} 1_{[L=k+1]} = S_{\tau(L-1)} \in \sigma(W_1, \ldots, W_{L-1})$$

and

$$\sum_{i=L(\tau,T)+1}^{T} X_i = \sum_{k=0}^{\infty} (S_T - S_{\tau(k)}) 1_{[\tau(k) \le T < \tau(k+1)]}$$

$$= \sum_{k=0}^{\infty} (S_T - S_{\tau(k)}) 1_{[L=k+1]}$$

$$= S_T - S_{\tau(L-1)} \in \sigma(W_L);$$

hence independence follows. ∎

We now derive the **Wiener-Hopf factorization**. Since

$$S_T = \sum_{i=1}^{L(\tau,T)} X_i + \sum_{i=L(\tau,T)+1}^{T} X_i$$

is a sum of two independent terms (Lemma 7.5.5), we have

$$P[S_T \le x] = P[\sum_{i=1}^{L(\tau,T)} X_i \le \cdot] * P[\sum_{i=L(\tau,T)+1}^{T} X_i \le \cdot](x),$$

which, from Lemma 7.5.2, is

$$= P[\sum_{i=1}^{L(\tau,T)} X_i \le \cdot] * P[\sum_{i=1}^{L(\eta,T)} X_i \le \cdot](x).$$

Note that $P[S_T \le x] = \sum_{n=0}^{\infty} P[S_n \le x] P[T = n] = \sum_0^{\infty} q^n p F^{n*}(x)$, and applying Lemma 7.5.4 yields

(7.5.2)
$$\sum_0^{\infty} pq^n F^{n*}(x) = \sum_0^{\infty} H_{\tau,q}^{n*}(\cdot)(1 - P[\tau \le T]) * \sum_0^{\infty} H_{\eta,q}^{n*}(\cdot)(1 - P[\eta \le T])(x).$$

We now invert (7.5.2) to get the Wiener-Hopf factorization using the following observation: Set $F^{0*} = \delta$, the probability measure which concentrates all mass at 0, and we have

$$\sum_0^{\infty} q^n F^{n*} = \delta + \sum_1^{\infty} q^n F^{n*} = \delta + \sum_0^{\infty} q^n F^{n*} * qF,$$

and therefore

(7.5.3)
$$\sum_0^{\infty} q^n F^{n*} * (\delta - qF) = \delta,$$

a useful inversion formula. To use this formula on the right side of (7.5.2), we need to re-express the convolution factors. Write

$$C_\tau(x) = P[S_\tau \le x | \tau \le T], \quad q_\tau = P[\tau \le T] = 1 - p_\tau$$
$$C_\eta(x) = P[S_\eta \le x | \eta \le T], \quad q_\eta = P[\eta \le T] = 1 - p_\eta.$$

Then

$$\sum_{n=0}^{\infty} H_{\tau,q}^{n*}(x)(1 - P[\tau \le T]) = \sum_{n=0}^{\infty} p_\tau q_\tau^n C_\tau^{n*}(x),$$

and likewise for the second factor, so (7.5.2) becomes

(7.5.4)
$$\sum_{n=0}^{\infty} pq^n F^{n*} = \sum_{n=0}^{\infty} p_\tau q_\tau^n C_\tau^{n*} * \sum_{n=0}^{\infty} p_\eta q_\eta^n C_\eta^{n*}.$$

Convolve (7.5.4) successively with $\delta - qF, \delta - q_\tau C_\tau$ and $\delta - q_\eta C_\eta$, and use the inversion formula (7.5.3) to obtain

$$p\delta = \sum_{n=0}^{\infty} p_\tau q_\tau^n C_\tau^{n*} * \sum_{n=0}^{\infty} p_\eta q_\eta^n C_\eta^{n*} * (\delta - qF),$$

$$p(\delta - q_\tau C_\tau) = p_\tau \sum_{n=0}^{\infty} p_\eta q_\eta^n C_\eta^{n*} * (\delta - qF),$$

$$p(\delta - q_\tau C_\tau) * (\delta - q_\eta C_\eta) = p_\tau p_\eta (\delta - qF).$$

Since $q_\tau C_\tau = H_{\tau,q}, q_\eta C_\eta = H_{\eta,q}$ we have

$$p(\delta - H_{\tau,q}) * (\delta - H_{\eta,q}) = p_\tau p_\eta (\delta - qF).$$

It remains for us to show

$$p = p_\tau p_\eta.$$

Notice, however, that

$$\begin{aligned}
p_\tau p_\eta &= (1 - P[\tau \le T])(1 - P[\eta \le T]) \\
&= (1 - Eq^\tau)(1 - Eq^\eta),
\end{aligned}$$

which, using duality (verify this as Exercise 7.3) equals p. The derivation is complete.

7.6. CONSEQUENCES OF THE WIENER-HOPF FACTORIZATION.

We now specialize to the case of $(N, \bar{N})$ and solve for the factors in the Wiener-Hopf equation. We first state Baxter's equations, which determine the joint distributions of (N, S_N) and $(\bar{N}, S_{\bar{N}})$.

Theorem 7.6.1. *For $0 \le q \le 1, \zeta \in R$:*

$$1 - Eq^N \exp\{i\zeta S_N\} = \exp\{-\sum_{n=1}^{\infty} \frac{q^n}{n} \int_{(0,\infty)} e^{i\zeta x} F^{n*}(dx)\}$$

$$1 - Eq^{\bar{N}} \exp\{i\zeta S_{\bar{N}}\} = \exp\{-\sum_{n=1}^{\infty} \frac{q^n}{n} \int_{(-\infty,0]} e^{i\zeta x} F^{n*}(dx)\}.$$

The Baxter equations of Theorem 7.6.1 (as well as Spitzer's formula presented in Section 7.7) are crowning intellectual achievements of classical random walk theory. The joint distribution of (N, S_N) is determined by a mixed transform

$$Eq^N \exp\{i\zeta S_N\}.$$

In principle, this mixed transform can be inverted to get joint probabilities associated with (N, S_N). Similar remarks apply to $(\bar{N}, S_{\bar{N}})$. These formulas are rather complex and depend on knowledge of F^{n*} for all $n \ge 0$. In this and later sections we will explore the extent to which these complex formulas can be made to yield useful information about random walks and associated queueing and storage models.

Before the proof we need the following preliminary result (cf. Breiman, 1968, page 197; Feller, 1970.) In order not to disturb the flow of logic, the proof of Lemma 7.6.2 is presented at the end of this section. The lemma will be applied in equation (7.6.2).

Lemma 7.6.2. *Suppose ν_1, ν_2 are two measures on R such that*

$$\int (|x| \wedge 1)\nu_i(dx) < \infty, \quad i = 1, 2.$$

If, for all $\zeta \in R$,

$$\int_R (e^{i\zeta x} - 1)\nu_1(dx) = \int_R (e^{i\zeta x} - 1)\nu_2(dx),$$

then $\nu_1 = \nu_2$.

Proof of Baxter's equations. First observe that, for any distribution function F, and $0 < q < 1, p = 1 - q$, we have

$$(7.6.1) \qquad \log p/(1 - q\hat{F}(\zeta)) = \int_R (e^{i\zeta x} - 1)(\sum_{n=1}^{\infty} \frac{q^n}{n} F^{n*}(dx)),$$

where $\hat{F}(\zeta) = \int e^{i\zeta x} F(dx)$. To check this, note

$$\log p/(1 - q\hat{F}(\zeta)) = \log(1 - q) - \log(1 - q\hat{F}(\zeta))$$

$$= -\sum_{n=1}^{\infty} n^{-1}q^n + \sum_{n=1}^{\infty} n^{-1}q^n \hat{F}^n(\zeta)$$

$$= \sum_{n=1}^{\infty} n^{-1}q^n \int_R (-1)F^{n*}(dx) + \sum_{n=1}^{\infty} n^{-1}q^n \int_R e^{i\zeta x} F^{n*}(dx)$$

$$= \int_R (e^{i\zeta x} - 1)(\sum_{n=1}^{\infty} n^{-1}q^n F^{n*}(dx)).$$

Set $\nu(dx) = \sum_{n=1}^{\infty} n^{-1}q^n F^{n*}(dx)$ and observe

$$\int_R (|x| \wedge 1)\nu(dx) < \infty.$$

This is readily verified since

$$\int_{\{|x| \leq 1\}} |x|\nu(dx) \leq \sum_{n=1}^{\infty} n^{-1}q^n F^{n*}[-1, 1]$$

$$= p^{-1} \sum_{n=1}^{\infty} n^{-1}pq^n P(S_n \in [-1, 1]]$$

$$= p^{-1} \sum_{n=1}^{\infty} n^{-1}P[S_n \in [-1, 1], T = n]$$

$$\leq p^{-1} \sum_{n=1}^{\infty} P[S_T \in [-1, 1], T = n]$$

$$\leq p^{-1}P\{S_T \in [-1, 1]\} < \infty.$$

Similarly, we may show

$$\int_{[|x|>1]} \nu(dx) < \infty.$$

Now set

$$H_q(x) = P[S_N \le x, T \le N], \; q_+ = P[T \le N], \; p_+ = 1 - q_+,$$
$$\bar{H}_q(x) = P[S_{\bar{N}} \le x, T \le \bar{N}], \; q_- = P[T \le \bar{N}], \; p_- = 1 - q_-,$$
$$C_+(x) = P[S_N \le x | T \le N] = H_q(x)/q_+$$
$$C_-(x) = P[S_{\bar{N}} \le x | T \le \bar{N}] = \bar{H}_q(x)/q_-.$$

Note the crucial fact that C_+ concentrates on $(0, \infty)$ and C_- concentrates on $(-\infty, 0]$. Recall from duality (see the argument at the end of Section 7.5 and Exercise 7.3)

$$p = p_+ p_-.$$

The Wiener-Hopf equation is

$$\delta - qF = (\delta - H_q) * (\delta - \bar{H}_q).$$

Transforming, we get

$$1 - q\hat{F}(\zeta) = (1 - \hat{H}_q(\zeta))(1 - \hat{\bar{H}}_q(\zeta)),$$

so that

$$\frac{p}{1 - q\hat{F}(\zeta)} = \left(\frac{p_+}{1 - q_+\hat{C}_+(\zeta)}\right)\left(\frac{p_-}{1 - q_-\hat{\bar{H}}_-(\zeta)}\right).$$

Take logarithms and apply (7.6.1) to obtain

$$\int_R (e^{i\zeta x} - 1) \sum_{n=1}^{\infty} n^{-1} q^n F^{n*}(dx)$$

$$= \int_{(0,\infty)} (e^{i\zeta x} - 1) \sum_{n=1}^{\infty} n^{-1} q_+^n C_+^{n*}(dx)$$

$$+ \int_{(-\infty,0]} (e^{i\zeta x} - 1) \sum_{n=1}^{\infty} n^{-1} q_-^n C_-^{n*}(dx).$$

Now the importance of Lemma 7.6.2 is apparent, since it permits the following conclusions:

(7.6.2) $$\sum_{n=1}^{\infty} n^{-1} q^n F^{n*}(dx) = \sum_{n=1}^{\infty} n^{-1} q_+^n C_+^{n*}(dx), \quad x > 0,$$

$$(7.6.3) \qquad \sum_{n=1}^{\infty} n^{-1} q^n F^{n*}(dx) = \sum_{n=1}^{\infty} n^{-1} q_-^n C_-^{n*}(dx), \quad x \le 0.$$

If we evaluate (7.6.2) on the set $(0, \infty)$, we get

$$\sum_{n=1}^{\infty} n^{-1} q^n F^{n*}(0, \infty) = \sum_{n=1}^{\infty} n^{-1} q_+^n C_+^{n*}(0, \infty)$$

$$(7.6.4) \qquad\qquad = \sum_{n=1}^{\infty} n^{-1} q_+^n = -\log(1 - q_+) = -\log p_+.$$

Finally, we have

$$\log(p_+ / 1 - q_+ \hat{C}_+(\zeta)) = \int_{(0,\infty)} (e^{i\zeta x} - 1) \sum_{n=1}^{\infty} n^{-1} q_+^n C_+^{n*}(dx)$$

(from (7.6.1))

$$= \int_{(0,\infty)} (e^{i\zeta x} - 1) \sum_{n=1}^{\infty} n^{-1} q^n F^{n*}(dx)$$

$$= \int_{(0,\infty)} e^{i\zeta x} \sum_{n=1}^{\infty} n^{-1} q^n F^{n*}(dx)$$

$$- \sum_{n=1}^{\infty} n^{-1} q^n F^{n*}(0, \infty)$$

$$= \int_{(0,\infty)} e^{i\zeta x} \sum_{n=1}^{\infty} n^{-1} q^n F^{n*}(dx) + \log p_+$$

(from (7.6.4)) so that

$$-\log(1 - q_+ \hat{C}_+(\zeta)) = \int_{(0,\infty)} e^{i\zeta x} \sum_{n=1}^{\infty} n^{-1} q^n F^{n*}(dx).$$

Remembering that $q_+ \hat{C}_+(\zeta) = \hat{H}_q(\zeta) = E q^N e^{i\zeta S_N}$, the desired result follows. ∎

We now specialize these results and consider some facts about the moments of N and $\bar{N}$.

The Wiener-Hopf equation can be written as

$$1 - q\hat{F}(\zeta) = (1 - E e^{i\zeta S_N} q^N)(1 - E e^{i\zeta S_{\bar{N}}} q^{\bar{N}}),$$

and setting $\zeta = 0$ yields

$$1 - q = (1 - Eq^N)(1 - Eq^{\bar{N}}).$$

Therefore,

$$(1 - Eq^N)/(1 - q) = (1 - Eq^{\bar{N}})^{-1}.$$

Let $q \uparrow 1$, and we have

$$EN = \lim_{q\uparrow 1}(1 - Eq^N)/(1 - q) = (1 - P[\bar{N} < \infty])^{-1} = (P[\bar{N} = \infty])^{-1},$$

and

$$EN = (P[\bar{N} = \infty])^{-1}, \quad E\bar{N} = (P[N = \infty])^{-1},$$

so $EN < \infty$ iff $\bar{N}$ is defective. Furthermore, if N and $\bar{N}$ are both proper (which holds iff $-\infty = \liminf_{n\to\infty} S_n < \limsup_{n\to\infty} S_n = +\infty$), then $EN = E\bar{N} = \infty$.

In Baxter's equation, set $\zeta = 0$ to obtain

$$(7.6.5) \qquad Eq^N = 1 - \exp\left\{-\sum_{n=1}^{\infty}\frac{1}{n}q^n P[S_n > 0]\right\}.$$

Proposition 7.6.3. *We have*
(a) $P[N < \infty] = 1$ iff $\sum_{n=1}^{\infty}\frac{1}{n}P[S_n > 0] = \infty$.
(b) *If* $P[N < \infty] = 1$, *then*

$$EN = \exp\left\{\sum_{n=1}^{\infty}\frac{1}{n}P[S_n \le 0]\right\}.$$

Remark. $\sum_{n=1}^{\infty}n^{-1}P[S_n > 0]$ is a useful measure of the strength with which the random walk pushes to $+\infty$.

Proof. (a) Let $q \uparrow 1$ in (7.6.5) to obtain

$$P[N < \infty] = 1 - \exp\left\{-\sum_{n=1}^{\infty}\frac{1}{n}P[S_n > 0]\right\}.$$

(b) We have

$$\frac{1 - Eq^N}{1 - q} = \frac{\exp\{-\sum_{n=1}^{\infty}n^{-1}q^n P[S_n > 0]\}}{\exp\{-\sum_{n=1}^{\infty}n^{-1}q^n\}}$$

(from (7.6.5))

$$= \exp\left\{\sum_{n=1}^{\infty}\frac{1}{n}q^n(1 - P[S_n > 0])\right\}$$

$$= \exp\left\{\sum_{n=1}^{\infty}\frac{1}{n}q^n P[S_n \leq 0]\right\}.$$

Now let $q \uparrow 1$ and apply the monotone convergence theorem. ∎

When moments of the step distribution exist we have the following criteria:

Proposition 7.6.4. *Suppose EX_1 exists.*
  (a) *If $EX_1 = 0$, then both N and $\bar{N}$ are proper and $EN = E\bar{N} = \infty$.*
  (b) *If $\infty \geq EX_1 > 0$, then N is proper, $EN < \infty$ and $\bar{N}$ is defective.*
  (c) *If $-\infty \leq EX_1 < 0$, then $\bar{N}$ is proper, $E\bar{N} < \infty$ and N is defective.*

Proof. (a) Suppose for the purposes of getting a contradiction that $\bar{N}$ is defective. Then $EN < \infty$, and, by the strong law of large numbers,

$$0 = \lim_{n\to\infty} n^{-1}S_n = \lim_{k\to\infty} k^{-1}S_{N(k)}/(k^{-1}N(k))$$
$$= ES_N/EN,$$

so $ES_N = 0$. Since $S_N > 0$, we have the desired contradiction. Thus, $\bar{N}$ is proper, and similarly, so is N. Hence $EN = E\bar{N} = \infty$.

(b) The law of large numbers implies $S_n \to \infty$, so $\mathcal{M}_{\bar{N}}$, the set of descending ladder epochs, is finite; there is a last descending ladder epoch. Hence $\bar{N}$ is defective. Therefore,

$$EN = 1/P[\bar{N} = \infty] < \infty. \quad \blacksquare$$

The last order of business in this section is the deferred proof of Lemma 7.6.2.

Proof of Lemma 7.6.2. Define $\psi_i(\zeta) = \int_R(e^{i\zeta x} - 1)\nu_i(dx)$, $i = 1, 2$, so that

$$\psi_i(\zeta) - \int_0^1 \frac{1}{2}(\psi_i(\zeta + h) + \psi_i(\zeta - h))dh$$

$$= \psi_i(\zeta) - [\int_0^1 \frac{1}{2}\int_R \{(e^{i(\zeta+h)x} - 1) + (e^{i(\zeta-h)x} - 1)\}\nu_i(dx)dh]$$

$$= \psi_i(\zeta) - [\int_0^1 \int_R e^{i\zeta x}\{\frac{e^{ihx} + e^{-ihx}}{2} - 1\}\nu_i(dx)dh]$$

$$= \psi_i(\zeta) - [\int_R e^{i\zeta x}\int_0^1 (\cos hx - 1)dh\nu_i(dx)]$$

(from Fubini's Theorem)

$$= \psi_i(\zeta) - \int_R e^{i\zeta x}(x^{-1}\sin x - 1)\nu_i(dx).$$

Therefore, if $\psi_i = \psi_2$, we have

$$\int_R e^{i\zeta x}(1 - x^{-1}\sin x)\nu_1(dx) = \int_R e^{i\zeta x}(1 - x^{-1}\sin x)\nu_2(dx),$$

and hence, by the uniqueness theorem for characteristic functions,

$$(1 - x^{-1}\sin x)\nu_1(dx) = (1 - x^{-1}\sin x)\nu_2(dx).$$

Since

$$0 < \inf_R(1 - x^{-1}\sin x) < \sup_R(1 - x^{-1}\sin x) < \infty,$$

we get

$$\nu_1 = \nu_2. \ \blacksquare$$

7.7. THE MAXIMUM OF A RANDOM WALK.

We begin by giving the celebrated *Spitzer's formula*.

Theorem 7.7.1. *We have for $\zeta \in R, 0 < q < 1$,*

$$\sum_{n=0}^{\infty} q^n E \exp\{i\zeta \bigvee_{j=0}^{n} S_j\} = \exp\{\sum_{n=1}^{\infty} \frac{q^n}{n} E e^{i\zeta S_n^+}\}.$$

Spitzer's formula is another pinnacle in classical random walk theory. The left side of Spitzer's formula is the generating function of the sequence $\{E \exp\{i\zeta \bigvee_{j=0}^{n} S_j\}, n \geq 0\}$, and knowing the generating function means we know the sequence. Since for any n we know the characteristic function $E \exp\{i\zeta \bigvee_{j=0}^{n} S_j\}$, in principle we know the distribution of $\bigvee_{j=0}^{n} S_j$ for any n. The practical implications of this formula, however, remain to be explored, since the right side of Spitzer's formula is quite complicated. Although it involves only the step distribution F, the formula requires us to calculate all the powers of F.

Proof. Since $\bigvee_{j=0}^{T} S_j$ is the last ascending ladder height before T, we have

$$\bigvee_{j=0}^{T} S_j = S_{L(N,T)}.$$

The distribution of $S_{L(N,T)}$ was computed in Lemma 7.5.4, and hence

$$P[\bigvee_{j=0}^{T} S_j \le x] = P[S_{L(N,T)} \le x]$$

$$= \sum_{k=0}^{\infty} H_q^{k*}(x)(1 - P[N \le T]).$$

Setting $q_+ = P[N \le T]$, $H_q = q_+ C_+$ gives for the above

$$\sum_{k=0}^{\infty} p_+ q_+^k C_+^{k*}(x).$$

Therefore, taking transforms,

$$E \exp\{i\zeta \bigvee_{j=0}^{T} S_j\}$$

$$= \sum_{n=0}^{\infty} pq^n E \exp\{i\zeta \bigvee_{j=0}^{n} S_j\}$$

$$= \sum_{n=0}^{\infty} p_+ q_+^k \hat{C}_+^k(\zeta)$$

$$= p_+/(1 - q_+ \hat{C}_+(\zeta))$$

$$= \exp\left\{\int_{(0,\infty)} (e^{i\zeta x} - 1) \sum_{n=1}^{\infty} \frac{q_+^n}{n} C_+^{n*}(dx)\right\}.$$

Applying (7.6.1), we have

(7.7.1)

$$= \exp\left\{\int_{(0,\infty)} (e^{i\zeta x} - 1) \sum_{n=1}^{\infty} \frac{q^n}{n} F^{n*}(dx)\right\}.$$

Now, applying (7.6.2) yields

$$= \exp\left\{\int_{(0,\infty)} e^{i\zeta x} \sum_{n=1}^{\infty} \frac{q^n}{n} F^{n*}(dx) - \sum_{n=1}^{\infty} \frac{q^n}{n} P[S_n > 0]\right\}$$

$$= \exp\left\{\sum_{n=1}^{\infty} \frac{q^n}{n} \int_{[S_n>0]} e^{i\zeta S_n} dP - \sum_{n=1}^{\infty} \frac{q^n}{n} + \sum_{n=1}^{\infty} \frac{q^n}{n} P[S_n \le 0]\right\}$$

$$= \exp\left\{\log(1-q) + \sum_{n=1}^{\infty} \frac{q^n}{n} \int e^{i\zeta S_n^+} dP\right\}$$

(7.7.2)

$$= p\exp\left\{\sum_{n=1}^{\infty} \frac{q^n}{n} E e^{i\zeta S_n^+}\right\}.$$

The result follows. ∎

We now specify the distribution of $M_\infty = \vee_{j=0}^{\infty} S_j$. Set $M_n = \vee_{j=0}^{n} S_j$ so that $0 \le M_n \uparrow M_\infty$ as $n \to \infty$.

Theorem 7.7.2. *We have*

$$M_\infty < \infty \text{ almost surely iff } \sum_{n=1}^{\infty} \frac{1}{n} P[S_n > 0] < \infty,$$

in which case

$$E e^{i\zeta M_\infty} = \exp\left\{\sum_{n=1}^{\infty} \frac{1}{n}(E e^{i\zeta S_n^+} - 1)\right\}.$$

Proof. The first statement is immediate: From the global result in Proposition 7.2.1, we have $M_\infty < \infty$ a.s. iff N is defective. From Proposition 7.6.3 we have N defective iff $\sum n^{-1} P[S_n > 0] < \infty$. Suppose now $M_\infty < \infty$. For the rest of the proof we need to have the dependence of T on q explicit, so we write $T(q)$. Note for any n

$$P[T(q) \ge n] = q^n \to 1$$

as $q \uparrow 1$, so we conclude

$$T(q) \xrightarrow{P} \infty$$

as $q \uparrow 1$. Therefore, $M_{T(q)} \xrightarrow{P} M_\infty$ as $q \to 1$, since for any $\epsilon > 0$

$$P[|M_{T(q)} - M_\infty| > \epsilon] = P[|M_{T(q)} - M_\infty| > \epsilon, T(q) \ge n]$$
$$+ P[|M_{T(q)} - M_\infty| > \epsilon, T(q) < n]$$
$$\le P[|M_n - M_\infty| > \epsilon] + P[T(q) < n].$$

Therefore

$$\limsup_{q\uparrow 1} P[|M_{T(q)} - M_\infty| > \epsilon] \le P[|M_n - M_\infty| > \epsilon].$$

This holds for all n, so upon letting $n \to \infty$ we get zero for the right side. Finally, since $M_{T(q)} \xrightarrow{P} M_\infty$, we have $Ee^{i\zeta M_{T(q)}} \to Ee^{i\zeta M_\infty}$. Since $M_{T(q)} = S_{L(N,T(q))}$, (7.7.1) yields

$$Ee^{i\zeta M_{T(q)}} = \exp\left\{ \int_{(0,\infty)} (e^{i\zeta x} - 1) \sum_{n=1}^{\infty} \frac{q^n}{n} F^{n*}(dx) \right\}$$

$$= \exp\left\{ \sum_{n=1}^{\infty} \frac{q^n}{n} \left(\int_{[S_n > 0]} e^{i\zeta S_n} dP - P[S_n > 0] \right) \right\}$$

$$= \exp\left\{ \sum_{n=1}^{\infty} \frac{q^n}{n} \left(\int_{[S_n > 0]} e^{i\zeta S_n} dP - 1 + P[S_n \le 0] \right) \right\}$$

$$= \exp\left\{ \sum_{n=1}^{\infty} \frac{q^n}{n} \left(Ee^{i\zeta S_n^+} - 1 \right) \right\}.$$

Because $E\exp\{i\zeta M_{T(q)}\} \to E\exp\{i\zeta M_\infty\}$, it remains to show

$$(7.7.3) \qquad |\sum_{n=1}^{\infty} \frac{q^n}{n}(Ee^{i\zeta S_n^+} - 1) - \sum_{n=1}^{\infty} \frac{1}{n}(Ee^{i\zeta S_n^+} - 1)| \to 0$$

as $q \uparrow 1$. Recall that we assume $\sum n^{-1} P[S_n > 0] < \infty$. The difference in (7.7.3) is bounded by

$$(7.7.4) \qquad \sum_{n=1}^{\infty} \frac{1}{n}(1 - q^n)|E\exp\{i\zeta S_n^+\} - 1|,$$

and since

$$|E\exp\{i\zeta S_n^+\} - 1| = |\int_{S_n > 0} e^{i\zeta S_n} + P[S_n \le 0] - 1] \le 2P[S_n > 0],$$

we have (7.7.4) bounded by

$$2\sum_{n=1}^{\infty} \frac{1}{n}(1 - q^n)P[S_n > 0].$$

This converges to zero as $q \uparrow 1$ by dominated convergence. ∎

7.8. RANDOM WALKS AND THE G/G/1 QUEUE.

Recall the set-up for the queueing model known as G/G/1. This is a model for a single server queueing system where arrivals occur according to a renewal process and customers are served on a first come, first served basis. The service times of successive customers are iid random variables and independent of the input to the service facility. Let interarrivals be represented by the iid sequence $\{\sigma_n, n \geq 0\}\}$ where σ_{n+1} represents the interarrival time between the nth and $(n+1)$st customer. Suppose customers arrive at rate a and that

$$P[\sigma_1 \leq x] = A(x), \quad E\sigma_1 = a^{-1} \in (0, \infty).$$

The service time of the nth customer is $\tau_n, n \geq 0$, where $\{\tau_n, n \geq 0\}$ is iid. The service rate is b, and

$$P[\tau_0 \leq x] = B(x), \quad E\tau_0 = b^{-1}.$$

Recall $\rho = a/b$. Assume $\{\tau_n\}$ and $\{\sigma_n\}$ are independent. Define

$$X_{n+1} = \tau_n - \sigma_{n+1}, \quad n \geq 0,$$

so that $S_0 = 0, S_n = X_1 + \cdots + X_n$ is a random walk. Now define the waiting time process $\{W_n, n \geq 0\}$ by $W_0 = 0$ and

$$W_{n+1} = (W_n + X_{n+1})^+, \quad n \geq 0,$$

so that W_{n+1} represents the time that customer $n + 1$ has to wait before entering service. See the related discussion in Sections 6.9 and 3.12.3, and also review Figure 3.16.

   Here are the basic facts about the waiting time process. (See also Propositions 6.9.1 and 6.9.2.)

Proposition 7.8.1. Let $M_n = \vee_{j=0}^{n} S_j$.
   (i) For each fixed $n \geq 0$,

$$M_n \overset{d}{=} W_n.$$

   (ii) Suppose the ascending ladder process is defective, so that by Theorem 7.7.2 we get $M_\infty < \infty$. (Recall that this is true iff $P[N = \infty] > 0$, iff $\sum n^{-1} P[S_n > 0] < \infty$.) Then in R we get W_n converging in distribution to M_∞:

$$W_n \Rightarrow W_\infty \overset{d}{=} M_\infty.$$

Reminder: A sufficient condition for the ascending ladder process to be defective is that

$$0 > EX_1 = E\tau_0 - E\sigma_1 = b^{-1} - a^{-1},$$

i.e., $\rho = a/b < 1$.

Proof. Since $\{M_n\}$ is non-decreasing, $M_n \uparrow M_\infty$, and thus (ii) follows directly from (i).

For the proof of (i) observe

$$W_n = S_n \text{ for } 0 \le n < \bar{N}(1),$$
$$W_{\bar{N}(1)} = 0 \text{ since } S_{\bar{N}(1)} \le 0,$$
$$W_{\bar{N}(1)+1} = (W_{\bar{N}(1)} + X_{\bar{N}(1)+1})^+ = (X_{\bar{N}(1)})^+,$$
$$W_{\bar{N}(1)+j} = X_{\bar{N}(1)+1} + \cdots + X_{\bar{N}(1)+j} \text{ for } \bar{N}(1) + j \le \bar{N}(2),$$
$$W_{N(2)} = 0.$$

In general, we have

$$W_{\bar{N}(k)+j} = X_{\bar{N}(k)} + \cdots + X_{\bar{N}(k)+j} = S_{\bar{N}(k)+j} - S_{\bar{N}(k)} \quad \text{if } \bar{N}(k)+j < \bar{N}(k+1),$$

so we conclude

$$W_n = S_n - S_{L(\bar{N},n)}.$$

Note that $S_{L(\bar{N},n)} = \wedge_{j=0}^n S_j$, so

$$W_n = S_n - \bigwedge_0^n S_j = \bigvee_{j=0}^n (S_n - S_j)$$

$$= \max\{S_n, \sum_{i=2}^n X_i, \sum_{i=3}^n X_i, \ldots, X_n + X_{n-1}, X_n, 0\}$$

$$\stackrel{d}{=} \max\{S_n, \sum_{n=1}^{n-1} X_i, \sum_{n=1}^{n-2} X_i, \ldots, X_1 + X_2, X_1, 0\}$$

$$= \vee_{j=0}^n S_j,$$

since $\{X_1, \ldots, X_n\} \stackrel{d}{=} \{X_n, X_{n-1}, \ldots, X_1\}$. ∎

With queueing applications in mind, we now turn to consideration of random walks whose step distributions have exponential tails. Consider a random walk $\{S_n\}$ with step distribution F subject to the Wiener-Hopf factorization

$$\delta - qF = (\delta - H_q) * (\delta - \bar{H}_q).$$

Recall

$$H_q(x) = P[S_N \leq x, N \leq T]$$

concentrates on $(0, \infty)$ and

$$\bar{H}_q(x) = P[S_{\bar{N}} \leq x, \bar{N} \leq T]$$

concentrates on $(-\infty, 0]$. Since

$$(\delta - \bar{H}_q) * \sum_{k=0}^{\infty} \bar{H}_q^{k*} = \delta,$$

we know from Wiener-Hopf that

$$\delta - H_q = (\delta - qF) * \sum_{k=0}^{\infty} \bar{H}_q^{k*}$$

(7.8.1)
$$= \sum_{k=0}^{\infty} \bar{H}_q^{k*} - qF * \sum_{k=0}^{\infty} \bar{H}_q^{k*}.$$

If $A \in \mathcal{B}((0, \infty))$, then since δ concentrates on $\{0\}$ and $\bar{H}_q$ concentrates on $(-\infty, 0]$ (7.8.1) yields

$$-H_q(A) = 0 - q \int_{(-\infty,0]} F(A - y) \sum_{k=0}^{\infty} \bar{H}_q^{k*}(dy).$$

Let $q \uparrow 1$, so that $T(q) \xrightarrow{P} \infty$, and set

$$H(A) = P[S_N \in A, N < \infty]$$
$$\bar{H}(\cdot) = P[S_{\bar{N}} \in \cdot, \bar{N} < \infty]$$
$$\bar{G}(dy) = \sum_{k=0}^{\infty} \bar{H}^{k*}(dy).$$

We obtain the following result.

Proposition 7.8.2. *For $A \in \mathcal{B}((0, \infty))$, we have*

$$H(A) = P[S_N \in A, N < \infty] = \int_{(-\infty,0]} F(A - y)\bar{G}(dy),$$

where

$$\bar{G}(dy) = \sum_{k=0}^{\infty} \bar{H}^{k*}(dy) = \sum_{k=0}^{\infty} P[S_{\bar{N}(k)} \in dy, \bar{N}(k) < \infty]$$

$$= \sum_{k=0}^{\infty} P[S_{\bar{N}} \in dy, \bar{N} < \infty]^{k*}.$$

(by means of Lemma 7.5.3(b) wih $q = 1$). Likewise, for $A \in \mathcal{B}((-\infty, 0])$,

$$\bar{H}(A) = P[S_{\bar{N}} \in A, \bar{N} < \infty] = \int_{[0,\infty)} F(A - y)G(dy),$$

where

$$G(dy) = \sum_{j=0}^{\infty} P[S_{N(j)} \in dy, N(j) < \infty]$$

$$= \sum_{j=0}^{\infty} P[S_N \in dy, N < \infty]^{j*}.$$

We next review the relationship between the maximum of a random walk and the ladder heights.

Proposition 7.8.3. *Suppose $P[N = \infty] > 0$ (which is true iff*

$$\sum n^{-1} P[S_n > 0] < \infty),$$

so that $M_\infty < \infty$ a.s. Then for $x \geq 0$

$$P[M_\infty \leq x] = P[N = \infty] \sum_{k=0}^{\infty} P[S_{N(k)} \leq x, N(k) < \infty]$$

$$= P[N = \infty] \sum_{k=0}^{\infty} P[S_N \leq x, N < \infty]^{k*}$$

$$= P[N = \infty]G(x).$$

Proof. We have that

$$P[M_\infty \leq x]$$

$$= \sum_{k=0}^{\infty} P[S_{N(k)} \leq x, N(k) < \infty, N(k+1) = \infty]$$

$$= \sum_{k=0}^{\infty} P[S_{N(k)} \leq x, N(k) < \infty, N(k+1) - N(k) = \infty]$$

$$= \sum_{k=0}^{\infty} P[S_{N(k)} \leq x, N(k+1) - N(k) = \infty | N(k) < \infty]P[N(k) < \infty].$$

On the trace probability space

$$(\Omega \cap [N(k) < \infty], \mathcal{F} \cap [N(k) < \infty], P(\cdot | N(k) < \infty)),$$

the events $[S_{N(k)} \le x]$ and $[N(k+1) - N(k) = \infty]$ are independent (review Proposition 1.8.2 and Proposition 7.1.1), and the above becomes

$$\sum_{k=0}^{\infty} P[S_{N(k)} \le x | N(k) < \infty] P[N(k) < \infty] P[N = \infty]$$

$$= \sum_{k=0}^{\infty} P[S_{N(k)} \le x, N(k) < \infty] P[N = \infty].$$

We finish with an application of Lemma 7.5.3(b). ■

7.8.1. EXPONENTIAL RIGHT TAIL.

Suppose the right tail of F is of the form

$$1 - F(x) = \xi e^{-bx}, \quad x \ge 0,$$

and $\xi \in (0,1)$, $b > 0$. Before considering the distribution of M_{∞}, we examine the form of G. For $x > 0$, write

$$P[N = n, S_N > x] = P[\vee_{j=0}^{n-1} S_j \le 0, S_n > x].$$

Recalling $\mathcal{F}_{n-1} = \sigma(S_0, S_1, \ldots, S_{n-1})$, this is

$$= E P[\vee_{j=0}^{n-1} S_j \le 0, S_n > x | \mathcal{F}_{n-1}]$$
$$= E 1_{[\vee_{j=0}^{n-1} S_j \le 0]} P[X_n > x - S_{n-1} | \mathcal{F}_{n-1}],$$

and, because X_n is independent of $\mathcal{F}_{n-1}$ and $S_{n-1} \le 0$, we get the above equal to

$$E 1_{[\vee_{j=0}^{n-1} S_j \le 0]} \xi e^{-b(x - S_{n-1})} = e^{-bx} E 1_{[\vee_0^{n-1} S_j \le 0]} \xi e^{bS_{n-1}}$$

$$= e^{-bx} E P[\vee_0^{n-1} S_j \le 0, X_n > -S_{n-1} | \mathcal{F}_{n-1}]$$
$$= e^{-bx} P[\vee_0^{n-1} S_j \le 0 < S_n]$$
$$= e^{-bx} P[N = n].$$

Therefore we may record the two equalities:

(7.8.1.1)
$$P[N = n, S_N > x] = P[N = n] e^{-bx}$$
$$P[S_N > x, N < \infty] = e^{-bx} P[N < \infty],$$

where the second is obtained by summing over n in the first equality. Put $p = P[N < \infty]$, and we may compute

$$G(dy) = \sum_{k=0}^{\infty} P[S_N \in dy, N < \infty]^{k*}$$

as follows: From (7.8.1.1) with $x > 0$ we get

$$P[S_N \in dx, n < \infty] = be^{-bx}p,$$

and, therefore, for $k \geq 1$,

$$P[S_N \in dx, N < \infty]^{k*} = p^k b \frac{(bx)^{k-1}e^{-bx}}{(k-1)!}.$$

Summing over k for the case $p < 1$ gives

$$\sum_{k=1}^{\infty} P[S_N \in dx, N < \infty]^{k*} = \sum_{k=1}^{\infty} p^k b(bx)^{k-1} e^{-bx}/(k-1)!$$
$$= pbe^{pbx - bx}$$
$$= pbe^{-b(1-p)x};$$

hence, for $x > 0$

$$G(x) = 1 + \int_0^x pbe^{-b(1-p)y} dy$$
$$= 1 + \frac{p}{1-p}(1 - e^{-b(1-p)x})$$
$$= 1 + \frac{p}{1-p} - \frac{p}{1-p}e^{-b(1-p)x},$$

so that

$$(7.8.1.2) \qquad G(x) = \frac{1}{1-p} - \frac{p}{1-p}e^{-b(1-p)x}.$$

We now continue under the assumption $\mu = EX_1 < 0$, which is the case of most relevance to queueing theory.

Proposition 7.8.4. *(Billingsley, 1986, p. 329ff.) Suppose $EX_1 = \mu < 0$ and, for $x \geq 0$,*

$$P[X_1 > x] = \xi e^{-bx}, \quad \xi > 0, \ b > 0.$$

Then

(i) $p = P[N < \infty] < 1$,

(ii) $P[M_\infty > x] = pe^{-b(1-p)x}$, $x \geq 0$, and

(iii) p can be determined as follows: $b(1-p)$ is the unique root of

$$f(s) = \int_{-\infty}^{\infty} e^{sx} F(dx) = 1$$

in the range $0 < s < b$.

Proof. (i) This is obvious since $\mu < 0$ implies $S_n \to -\infty$, and therefore $P[N = \infty] > 0$.

(ii) From Proposition 7.8.3, we have

$$P[M_\infty \le x] = (1 - p)G(x);$$

applying (7.8.1.2) yields

$$P[M_\infty \le x] = (1 - p)\left(\frac{1}{1-p} - \frac{p}{1-p}e^{-b(1-p)x}\right)$$
$$= 1 - pe^{-b(1-p)x}.$$

(iii) Recall from Proposition 7.8.2 that, for $x \le 0$,

$$P[S_{\bar{N}} \le x, \bar{N} < \infty] = \int_{[0,\infty)} F(x - y)G(dy).$$

When $\mu < 0$ we have $\bar{N}, S_{\bar{N}}$ both proper and $S_{\bar{N}} \le 0$. Setting $x = 0$ we have

$$P[S_{\bar{N}} \le 0, \bar{N} < \infty] = 1 = \int_{[0,\infty)} F(-y)G(dy).$$

Use the form of G given by (7.8.1.1) and recall that G has an atom of size 1 at zero to obtain

$$1 = F(0) + \int_{(0,\infty)} bpe^{-b(1-p)y}F(-y)dy$$

$$= F(0) + \int_{(-\infty,0)} bpe^{b(1-p)y}F(y)dy$$

$$= F(0) + \int_{y\in(-\infty,0)} bpe^{b(1-p)y}\int_{u\in(-\infty,y)} F(du)dy$$

$$= F(0) + \int_{u\in(-\infty,0)}\left(\int_{y\in(u,0)} bpe^{b(1-p)y}dy\right)F(du)$$

$$= F(0) + \int_{(-\infty,0)} \frac{p}{1-p}(1 - e^{b(1-p)u})F(du)$$

$$= F(0) + \frac{p}{1-p}F(0) - \frac{p}{1-p}\int_{(-\infty,0)} e^{b(1-p)u}F(dy).$$

since $1 - F(0) = \xi$, some algebra gives

(7.8.1.3) $$\int_{(-\infty,0)} e^{b(1-p)y}F(dy) = 1 - p^{-1}\xi.$$

We may now show that $b(1-p)$ is the unique root of $f(s) = 1$ in $(0, b)$. Observe that for $s \in (0, b)$

$$f(s) = \int_{-\infty}^{0} e^{sx} F(dx) + \int_{0}^{\infty} e^{sx} \xi b e^{-bx} \, dx$$

$$= \int_{-\infty}^{0} e^{sx} F(dx) + \xi b/(b - s).$$

The first term exists for all $s > 0$, and the second term holds for $s \in (0, b)$. Since $p > 0$,

$$b(1 - p) < b.$$

f exists at $b(1 - p)$, and by evaluating f we get

$$f(b(1 - p)) = \int_{(-\infty, 0)} e^{b(1-p)y} F(dy) + \xi b/(b - b(1 - p)).$$

Using (7.8.1.3), this is

$$1 - p^{-1} \xi + \xi p^{-1} = 1.$$

Now that we know $b(1 - p)$ is a root in $(0, b)$ of the equation $f(s) = 1$, we prove it is the unique root. Note that $f(0) = f(b(1 - p)) = 1$ and f is continuous on $[0, b)$. Furthermore, f is strictly convex, since

$$f''(s) = \int_{-\infty}^{\infty} x^2 e^{sx} F(dx) > 0$$

on $(0, b)$. The graph of f is pictured in Figure 7.2.

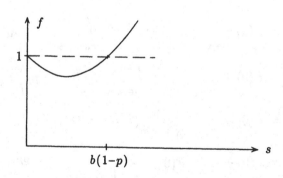

FIGURE 7.2. GRAPH OF f

There can be no other roots of $f(s) = 1$ on $(0, b)$. ∎

7.8.2. Applications to the G/M/1 Queueing Model.

This is a queueing model with renewal input and exponentially distributed service times, so the interarrival distribution $A(x)$ is an arbitrary distribution on $[0, \infty)$ with mean a^{-1} and Laplace transform

$$\hat{A}(\lambda) = \int_0^\infty e^{-\lambda x} A(dx).$$

The service time distribution B is

$$B(x) = 1 - e^{-bx}, \quad x > 0, \ b > 0,$$

and we assume $\rho = a/b < 1$. We have

$$X_{n+1} = \tau_n - \sigma_{n+1},$$

and thus, for $x \geq 0$,

$$
\begin{aligned}
1 - F(x) &= P[\tau_0 - \sigma_1 > x] \\
&= \int_0^\infty P[\tau_0 > y + x] A(dy) \\
&= \int_0^\infty e^{-b(y+x)} A(dy) \\
&= e^{-bx} \hat{A}(b)
\end{aligned}
$$

and

$$\xi = 1 - F(0) = \hat{A}(b).$$

To find $p = P[N < \infty]$, we compute

$$
\begin{aligned}
f(s) &= Ee^{sX_1} = Ee^{s(\tau_0 - \sigma_1)} \\
&= Ee^{s\tau_0} Ee^{-s\sigma_1} \\
&= \left(\int_0^\infty e^{sx} b e^{-bx} dx \right) \hat{A}(s),
\end{aligned}
$$

and, assuming $s \in (0, b)$, we get

$$= \frac{b}{b - s} \hat{A}(s).$$

The equation

$$f(s) = 1, 0 < s < b$$

becomes

$$\hat{A}(s) = \frac{b-s}{b}.$$

Let s^* be the unique solution in $(0, b)$ guaranteed by Proposition 7.8.4. Then $b(1-p) = s^*$ and

$$P[N = \infty] = 1 - p = b^{-1}s^*.$$

By applying Proposition 7.8.3 and equation (7.8.1.2), we get for $x \geq 0$

$$P[M_\infty \leq x] = P[W_\infty \leq x] = P[N = \infty]G(x)$$
$$= (1-p)\left((1-p)^{-1} - (1-p)^{-1}pe^{-b(1-p)x}\right)$$
$$= 1 - pe^{-b(1-p)x}$$
(7.8.2.1) $$= 1 - (1 - b^{-1}s^*)e^{-s^*x}.$$

If we further specialize to the M/M/1 queue and suppose

$$A(x) = 1 - e^{-ax}, \quad x > 0,$$

then

$$\hat{A}(s) = a/(a+s), \quad s > 0,$$

so that the equation

$$f(s) = 1, \quad 0 < s < b,$$

becomes

$$\hat{A}(s) = (b - s)/b = a/(a+s)$$

or, equivalently,

$$s^2 - (b - a)s = 0.$$

The roots are $s = 0$ and $b - a > 0$ (since $\rho = a/b < 1$), so $s^* = b - a$. Thus $b(1-p) = s^* = b - a$, and

$$P[N = \infty] = 1 - p = 1 - a/b = 1 - \rho.$$

We find for $x \geq 0$ that (7.8.2.1) reduces to

$$P[M_\infty \leq x] = P[W_\infty \leq x]$$
$$= 1 - (1 - (1 - b^{-1}a))e^{-(b-a)x}$$
(7.8.2.2) $$= 1 - pe^{-(b-a)x}.$$

Note $P[W_\infty = 0] = 1 - \rho$.

Before leaving exponential right tails, let us examine what happens if $\mu = EX_1 \geq 0$. We continue to suppose $1 - F(x) = \xi e^{-bx}$, $x \geq 0$, $\xi > 0, b > 0$. If $\mu \geq 0$, $p = P[N < \infty] = 1$, S_N is proper and $P[M_\infty = \infty] = 1$. From (7.8.1.1),

$$P[S_N > x] = e^{-bx}, \quad x \geq 0,$$

so that S_N has an exponential distribution. Thus, since we know the renewal function for a Poisson process, we have

$$G(x) = \sum_0^\infty P[S_N \leq x]^{k*} = 1 + bx, \quad x \geq 0.$$

From Proposition 7.8.2, if $A \in \mathcal{B}((-\infty, 0])$ we get

$$P[S_{\bar{N}} \in A, \bar{N} < \infty] = \int_{[0,\infty)} G(dy)F(A - y)$$

so for $x < 0$

$$P[S_{\bar{N}} \leq x, \bar{N} < \infty] = F(x) + \int_{(0,\infty)} F(x - y)b\,dy$$

(7.8.2.3)
$$= F(x) + b\int_{-\infty}^x F(y)dy.$$

Assuming F is continuous at 0 (which, from the previous equation, implies $P[S_{\bar{N}} \leq x, \bar{N} < \infty]$ is continuous at 0), letting $x \uparrow 0$ yields

$$P[S_{\bar{N}} \leq 0, \bar{N} < \infty] = P[\bar{N} < \infty]$$

$$= F(0) + b\int_{-\infty}^0 F(y)dy$$

$$= 1 - \xi + b\int_{-\infty}^0 \int_{-\infty}^y F(du)dy$$

$$= 1 - \xi + b\int_{-\infty}^0 \int_u^0 dy F(du)$$

$$= 1 - \xi + b\int_{-\infty}^0 -uF(du)$$

$$= 1 - \xi + bEX_1^-.$$

Observe

$$EX_1^- = EX_1^+ - EX_1 = \int_0^\infty \xi e^{-bx}dx - \mu$$

$$= \xi/b - \mu.$$

Thus

$$P[\bar{N} < \infty] = 1 - \xi + b(\xi b^{-1} - \mu)$$
$$= 1 - b\mu,$$

and $P[\bar{N} = \infty] = 1/EN = b\mu$, which can also be obtained from the Wald identity and the fact that S_N is exponentially distributed.

7.8.3. EXPONENTIAL LEFT TAIL.

If we assume F is continuous we can reduce the case of F having exponential left tail to the case just considered. We merely reflect about zero using the map $x \to -x$.

Proposition 7.8.5. *Consider a random walk with continuous step distribution F with $\mu = EX_1 < 0$ and an exponential left tail; i.e.,*

$$F(x) = P[X_1 \le x] = \xi e^{ax}, \quad x < 0, \ a > 0, \ \xi \in (0,1).$$

Then

$$p = P[N < \infty] = 1 + a\mu,$$

and, for $y > 0$,

$$P[N < \infty, S_N \le y] = F(0, y] + a \int_0^y (1 - F(u))du.$$

Proof. Consider the reflected random walk $\{S_n^{(r)}, n \ge 0\} = \{-S_n, n \ge 0\}$. With obvious notation we have

$$1 - F^{(r)}(x) = P[X_1^{(r)} > x] = P[-X_1 > x]$$
$$= P[X_1 < -x] = P[X_1 \le -x]$$

(since F is continuous)

$$= F(-x).$$

Therefore, for $y \ge 0$,

$$1 - F^{(r)}(y) = F(-y) = \xi e^{-ay},$$

and $F^{(r)}$ has exponential right tail as studied in Section 7.8.1. Note that

$$EX_1^{(r)} = -EX_1 = -\mu > 0,$$

and applying (7.8.2.3) gives, for $x < 0$,

$$(7.8.3.1) \qquad P[S^{(r)}_{\bar{N}^{(r)}} \leq x, \bar{N}^{(r)} < \infty] = F^{(r)}(x) + a \int_{-\infty}^{x} F^{(r)}(y)dy.$$

It remains to see what this formula means in terms of the original random walk $\{S_n\}$. We have for $y > 0$

$$P[S^{(r)}_{\bar{N}^{(r)}} \leq -y, \bar{N}^{(r)} < \infty] = \sum_{n=1}^{\infty} P[\bigwedge_{k<n}(-S_k) > 0, -S_n \leq -y]$$

$$= \sum_{n=1}^{\infty} P[-\bigwedge_{k<n}(-S_k) < 0, S_n \geq y],$$

and, because $-\bigwedge_{k<n}(-S_k) = \bigvee_{k<n}S_k$ and F is continuous, this is equal to

$$= \sum_{n=1}^{\infty} P[\bigvee_{k<n} S_k \leq 0, S_n > y]$$

$$= \sum_{n=1}^{\infty} P[N = n, S_N > y] = P[S_N > y, N < \infty].$$

Therefore, for $y > 0$,

$$P[S_N > y, N < \infty] = P[S^{(r)}_{\bar{N}^{(r)}} \leq -y, \bar{N}^{(r)} < \infty],$$

and applying (7.8.3.1) yields

$$F^{(r)}(-y) + a \int_{-\infty}^{-y} F^{(r)}(u)du = 1 - F(y) + a \int_{-\infty}^{-y}(1 - F(-u))du$$

$$= 1 - F(y) + a \int_{y}^{\infty}(1 - F(s))ds.$$

Thus

$$p = P[N < \infty] = P[N < \infty, S_N > 0]$$

$$= 1 - F(0) + a \int_{0}^{\infty}(1 - F(s))ds$$

$$= 1 - \xi + aEX_1^+$$

$$= 1 - \xi + a(EX_1^+ - EX_1^-) + aEX_1^-$$

$$= 1 - \xi + a\mu + a\xi a^{-1}$$

$$= 1 - \xi + \xi + a\mu$$

$$= 1 + a\mu$$

since $EX_1^+ = \int_0^\infty (1 - F(s))ds$ and $EX_1^- = \int_{-\infty}^0 \xi e^{ax} dx = \xi a^{-1}$.
Finally, for $y > 0$,

$$P[N < \infty] = P[S_N > y, N < \infty] + P[S_N \leq y, N < \infty].$$

Therefore,

$$\begin{aligned}
P[S_N \leq y, N < \infty] &= P[N < \infty] - P[S_N > y, N < \infty] \\
&= 1 + a\mu - \left(1 - F(y) + a \int_y^\infty (1 - F(s))ds\right) \\
&= F(y) + a\mu - \left(aEX^+ - a \int_0^x (1 - F(s))ds\right) \\
&= F(y) + a\mu - a(EX_1^+ - EX_1^- + EX_1^-) \\
&\quad + a \int_0^y (1 - F(s))ds \\
&= F(y) + a\mu - a\mu - a\xi a^{-1} + a \int_0^y (1 - F(s))ds \\
&= F(y) - F(0) + a \int_0^y (1 - F(s))ds,
\end{aligned}$$

since $F(0) = \xi = a\xi a^{-1}$. ∎

We now derive the distribution of the maximum M_∞ of $\{S_n\}$. Recall Proposition 7.8.3,

$$P[M_\infty \leq x] = P[N = \infty] \sum_{k=0}^\infty P[S_N \leq x, N < \infty]^{k*}, \quad x > 0,$$

so that for $\lambda > 0$ the Laplace transform of M_∞ is

$$\begin{aligned}
Ee^{-\lambda M_\infty} &= (1 - p) \sum_{k=0}^\infty \left(\int_0^\infty e^{-\lambda x} P[N < \infty, S_N \in dx]\right)^k \\
&= \frac{1 - p}{1 - \int_0^\infty e^{-\lambda x} P[N < \infty, S_N \in dx]}.
\end{aligned}$$

To evaluate this, we set

$$\hat{F}_+(\lambda) = \int_0^\infty e^{-\lambda x} F(dx)$$

and use the formula for $P[N < \infty, S_N \leq x]$ given in Proposition 7.8.5 to get

$$\int_0^\infty e^{-\lambda x} P[N < \infty, S_N \in dx] = \hat{F}_+(\lambda) + a \int_0^\infty e^{-\lambda x}(1 - F(x))dx$$

$$= \hat{F}_+(\lambda) + a \int_{x=0}^\infty \int_{u=x}^\infty F(du)e^{-\lambda x}dx$$

$$= \hat{F}_+(\lambda) + a \int_{u=0}^\infty \left(\int_{x=0}^u e^{-\lambda x}dx \right) F(du)$$

$$= \hat{F}_+(\lambda) + a \int_0^\infty (1 - e^{-\lambda u})\lambda^{-1}F(du)$$

$$= \hat{F}_+(\lambda) + a\lambda^{-1} \left(1 - F(0) - \hat{F}_+(\lambda) \right)$$

$$= \hat{F}_+(\lambda)(1 - a\lambda^{-1}) + a\lambda^{-1}(1 - \xi).$$

Using this and the formula for p given in Proposition 7.8.5, we obtain

$$(7.8.3.2) \qquad Ee^{-\lambda M_\infty} = \frac{-a\mu}{1 - a\lambda^{-1}(1 - \xi) - (1 - a\lambda^{-1})\hat{F}_+(\lambda)},$$

which is a version of the Pollaczek-Khintchine formula.

7.8.4. The M/G/1 queue.

Consider a queue with arrivals according to a Poisson process. Then $\{\sigma_{n+1}, n \geq 0\}$ are iid, $P[\sigma_1 > x] = e^{-ax}$, $x > 0$, $\{\tau_n, n \geq 0\}$ are iid and, finally, $P[\tau_0 \leq x] = B(x)$. Assume $\rho = a/b = E\tau_0/E\sigma_1 < 1$, so $\mu = EX_1 < 0$. Observe that for $x \leq 0$

$$F(x) = P[X_1 \leq x] = P[\tau_0 - \sigma_1 \leq x]$$

$$= P[\tau_0 - x \leq \sigma_1]$$

$$= \int_0^\infty P[y - x \leq \sigma_1]B(dy)$$

$$= \int_0^\infty e^{-a(y-x)}B(dy)$$

$$= e^{ax} \int_0^\infty e^{-ay}B(dy)$$

$$= e^{-ax}\hat{B}(a),$$

so that F has exponential left tail and

$$\xi = F(0) = \hat{B}(a).$$

To evaluate (7.8.3.2) we need $\hat{F}_+(\lambda)$. For this we have

$$\hat{F}_+(\lambda) = \int_0^\infty e^{-\lambda x} F(dx) = Ee^{-\lambda X_1} 1_{[X_1 \geq 0]}$$

$$= Ee^{-\lambda(\tau_0 - \sigma_1)} 1_{[\tau_0 \geq \sigma_1]}$$

$$= E\left(E(e^{-\lambda(\tau_0 - \sigma_1)} 1_{[\tau_0 \geq \sigma_1]} | \sigma_1)\right)$$

$$= E\left(e^{\lambda \sigma_1} \int_{\sigma_1}^\infty e^{-\lambda s} B(ds)\right)$$

$$= \int_0^\infty e^{\lambda y} \left(\int_y^\infty e^{-\lambda s} B(ds)\right) ae^{-ay} dy$$

$$= \int_{s=0}^\infty \left(\int_{y=0}^s ae^{-(a-\lambda)y} dy\right) e^{-\lambda s} B(ds)$$

$$= a(a-\lambda)^{-1} \int_0^\infty (1 - e^{-(a-\lambda)s}) e^{-\lambda s} B(ds)$$

$$= a(a-\lambda)^{-1} \int_0^\infty (e^{-\lambda s} - e^{-as}) B(ds)$$

$$= a(a-\lambda)^{-1}(\hat{B}(\lambda) - \hat{B}(a)).$$

We may now evaluate (7.8.3.2) using $\xi = \hat{B}(a)$ and $\mu a = \rho - 1$. We have

$$Ee^{-\lambda M_\infty} = Ee^{-\lambda W_\infty}$$

$$= \frac{-\mu a}{1 - a\lambda^{-1}(1 - \xi) - \lambda^{-1}(\lambda - a)\{a(a-\lambda)^{-1}(\hat{B}(\lambda) - \xi)\}}$$

$$= \frac{1 - \rho}{1 - a\lambda^{-1}(1 - \xi) + a\lambda^{-1}\hat{B}(\lambda) - \lambda^{-1}a\xi}$$

$$= \frac{1 - \rho}{1 - a\lambda^{-1} + a\lambda^{-1}\hat{B}(\lambda)}$$

$$= \frac{1 - \rho}{1 - \frac{a}{b}\left(\frac{1 - \hat{B}(\lambda)}{b^{-1}\lambda}\right)}.$$

Recall that $B_0(x) = b\int_0^x (1 - B(s)) ds$ has transform

$$\hat{B}_0(\lambda) = \frac{b(1 - \hat{B}(\lambda))}{\lambda},$$

so

$$Ee^{-\lambda W_\infty} = (1 - \rho)/1 - \rho\hat{B}_0(\lambda)$$

$$= \sum_{n=0}^\infty (1 - \rho)\rho^n \hat{B}_0^n(\lambda),$$

and, for $x \geq 0$,

$$P[W_\infty \leq x] = \sum_{n=0}^{\infty}(1 - \rho)\rho^n B_0^{n*}(x).$$

7.8.5. QUEUE LENGTHS.

Continue to suppose $\rho < 1$. Recall the notation

$$t_k = \sigma_1 + \cdots + \sigma_k, \quad k \geq 1,$$

which is the time of arrival of the kth customer, and

$$\nu(t) = 1 + \sup\{k : t_k \leq t\},$$

which is the number of arrivals in $[0, t]$. Let Q_n be the size of the queue left behind when the nth customer terminates service and leaves the system. Since customer number n arrives at t_n, enters service at $t_n + W_n$ and departs at $t_n + W_n + \tau_n$, we have for $n \geq 0$, $k \geq 1$

$$[Q_n < k] = [t_n + W_n + \tau_n < t_{n+k}]$$
$$= [W_n + \tau_n - \sum_{i=n+1}^{n+k} \sigma_i < 0].$$

For model building purists, the previous set identity can be used to define the non-negative integer valued random variable Q_n. Recall

$$W_n \in \sigma(\tau_0, \ldots, \tau_{n-1}, \sigma_1, \ldots, \sigma_n),$$

hence $W_n, \tau_n, \sum_{i=n+1}^{n+k} \sigma_i$ are all independent. From Proposition 7.8.5, we have that

$$W_n \stackrel{d}{=} M_n = \vee_{j=0}^{n} S_j \uparrow M_\infty \stackrel{d}{=} W_\infty.$$

Set $W(x) = P[W(\infty) \leq x]$. Let τ^*, t_k^* be independent and independent of the random walk with $\tau^* \stackrel{d}{=} \tau$ and $t_k^* \stackrel{d}{=} \sum_{i=n+1}^{n+k} \sigma_i$. Then

$$P[Q_n < k] = P[W_n + \tau_n - \sum_{i=n+1}^{n+k} \sigma_i < 0]$$
$$= P[M_n + \tau^* - t_k^* < 0],$$

as $n \to \infty$, which converges to

$$P[Q_\infty < k] := P[M_\infty + \tau^* - t_k^* < 0] = P[M_\infty + \tau^* < t_k^*],$$

where M_∞, τ^*, t_k^* are independent. Therefore, if we difference, we get for $k \geq 0$

$$
\begin{aligned}
q_k = P[Q_\infty = k] &= \lim_{n \to \infty} P[Q_n = k] \\
&= P[M_\infty + \tau^* < t_{k+1}^*] - P[M_\infty + \tau^* < t_k^*] \\
&= P[t_k^* \leq M_\infty + \tau^* \leq t_{k+1}^*] \\
&= P[\nu(M_\infty + \tau^*) = k + 1] \\
&= \int_0^\infty P[\nu(t) = k + 1] W * B(dt).
\end{aligned}
$$

In the case of the M/G/1 queue where $A(x) = 1 - e^{-ax}$, we have

$$
P[\nu(t) = k + 1] = e^{-at}(at)^k / k!
$$

and thus

(7.8.5.1) $$ q_k = \int_0^\infty \frac{e^{-at}(at)^k}{k!} W * B(dt). $$

Finally, consider the special case of the M/M/1 queue, where

$$
B(x) = 1 - e^{-bx}, \quad A(x) = 1 - e^{-ax}, \quad a > 0, b > 0, x > 0.
$$

Recall from (7.8.2.2) that in this case

$$
W(x) = P[M_\infty \leq x] = P[W_\infty \leq x] = 1 - \rho e^{-(b-a)x}, \quad x > 0,
$$

so that the distribution $W(\cdot)$ has an atom size $1 - \rho$ at zero. The distribution $W * B$ has density

$$
\int_{[0,y]} be^{-b(y-x)} W(dx)
$$

$$
= (1 - \rho)be^{-by} + \int_{[0,y]} be^{-b(y-x)} \rho(b - a)e^{-(b-a)x} dx
$$

(the first term comes from the atom at zero)

$$
\begin{aligned}
&= (1 - \rho)be^{-by} + b\rho(b - a)e^{-by} \int_0^y e^{ax} dx \\
&= (1 - \rho)be^{-by} + b\rho(b - a)e^{-by}(e^{ay} - 1)/a \\
&= (1 - \rho)be^{-by} + (b - a)(e^{-(b-a)y} - e^{-by})
\end{aligned}
$$

(recall that $ba^{-1}\rho = 1$)

$$= (b-a)e^{-by} - (b-a)e^{-by} + (b-a)e^{-(b-a)y}$$
$$= (b-a)e^{-(b-a)y}.$$

We may now evaluate (7.8.5.1) to get for $k \geq 0$

$$q_k = \int_0^\infty \frac{e^{-at}(at)^k}{k!}(b-a)e^{-(b-a)t}dt$$

$$= \frac{a^k(b-a)}{k!} \int_0^\infty t^k e^{-bt}dt$$

$$= \frac{a^k}{k!}(b-a)b^{-(k+1)} \int_0^\infty s^k e^{-s}ds$$

$$= (\frac{a}{b})^k(1-ab^{-1})\Gamma(k+1)/k!$$

$$= \rho^k(1-\rho),$$

yielding a geometric distribution.

EXERCISES

7.1. Here is a meta-theorem: If ξ and ξ' are two random sequences such that $\xi \overset{d}{=} \xi'$, then ξ and ξ' are probabilistically indistinguishable in the sense that they behave the same. Two examples are the following:

(a) If $\xi = (\xi_1, \xi_2, \dots)$ and $\xi' = (\xi_1', \xi_2', \dots)$, then if $\xi_n \to \xi$ a.s. then also there exists a random variable ξ' such that $\xi_n' \to \xi'$ a.s.

(b) If $\xi = (\xi_1, \xi_2, \dots)$ and $\xi' = (\xi_1', \xi_2', \dots)$, then if $\xi_n \overset{P}{\to} \xi$ then also there exists a random variable ξ' such that $\xi_n' \overset{P}{\to} \xi'$.

7.2. Use the definition of duality of stopping times to check the following statements:

(a) If both τ and τ' are dual to η, then $\tau = \tau'$;

(b) If τ is dual to both η and η', then $\eta = \eta'$ (Greenwood and Shaked, 1977).

7.3. If τ, η are dual stopping times, show

$$\sum_{n=0}^{\infty} u^n P[\tau > n] = \sum_{n=0}^{\infty} (Eu^\eta)^n,$$

that is,

$$\frac{(1 - Eu^\tau)}{1 - u} = \frac{1}{1 - Eu^\eta}.$$

Proceed directly from the definition.

7.4. In the M/G/1 queue, compute the Laplace transform of the limiting virtual waiting time distribution V. What if B is exponential?

7.5. Suppose $(\gamma, \bar{\gamma})$ and $(\eta, \bar{\eta})$ are two dual pairs of stopping times. Define

$$\tau = \min\{n : n \in \mathcal{M}_\gamma \cap \mathcal{M}_\eta\}$$
$$\bar{\tau} = \bar{\gamma} \wedge \bar{\eta}.$$

Show $(\tau, \bar{\tau})$ are dual (Greenwood and Shaked, 1977).

7.6. Set $b > 0$ and $\eta = \inf\{n : X_n > b\}$. Let

$$\tau = \begin{cases} 1, & \text{if } X_1 \leq b \\ \infty, & \text{if } X_1 > b. \end{cases}$$

Show (τ, η) are dual. Generate other dual pairs using Problem 7.5 (Greenwood and Shaked, 1977).

7.7. Let (τ, η) be dual. Suppose $P[\tau = \infty] > 0$, and define

$$L(\tau) = \max\{n \in \mathcal{M}_\tau\}.$$

Express the distribution of $S_{L(\tau)}$ in terms of S_τ and use duality to express this distribution in terms of η, S_η. If $\tau = N$, interpret $S_{L(N)}$.

7.8. Uniqueness of Wiener-Hopf Factors. Suppose for $n \geq 1$ there exist measures $H_n, \bar{H}_n$ concentrating on $(0, \infty)$ and $(-\infty, 0]$ respectively such that $\sum_{n=1}^{\infty} H_n(0, \infty) \leq 1, \sum_{n=0}^{\infty} \bar{H}_n(-\infty, 0] \leq 1$. Set

$$\chi(q, \zeta) = \sum_{n=1}^{\infty} q^n \int_{(0, \infty)} e^{i\zeta x} H_n(dx)$$
$$\bar{\chi}(q, \zeta) = \sum_{n=1}^{\infty} q^n \int_{(-\infty, 0]} e^{i\zeta x} \bar{H}_n(dx).$$

For a distribution F with $\hat{F}(\zeta) = \int_R e^{i\zeta x} F(dx)$, if

$$q - q\hat{F} = (1 - \chi)(1 - \bar{\chi}),$$

then

$$\chi(q, \zeta) = Eq^N e^{i\zeta S_N}, \quad \bar{\chi}(q, \zeta) = Eq^{\bar{N}} e^{i\zeta S_{\bar{N}}}.$$

7.9. If F is continuous and symmetric, derive the generating function of N and check that N is proper.

7.10. Give a direct proof that $\mu < 0$ implies

$$\sum_n \frac{1}{n} P[S_n > 0] < \infty$$

by showing that
   (a)
$$\sum_n P[|X_1| > n] < \infty, \quad \text{and}$$

   (b)
$$\sum_n \frac{1}{n^2} EX_1^2 1[|X_1| \le n] \infty$$

(Feller, 1971).

7.11. In a random walk, suppose $EX_1 > 0$. Set

$$N_x = \inf\{n : S_n > x\}, \quad x > 0,$$

$$R(x) = 1 + \sum_{j=1}^{\infty} P[S_1 > 0, \dots, S_j > 0, S_j \le x].$$

Show
$$EN_x = EN_0 R(x).$$

What is the relationship between $R(x)$ and the distribution of S_N?

7.12. Consider the following variant of the G/G/1 queue: Customers arrive at fixed intervals appearing at time points $n = 0, 1, 2, \dots$. Service times τ_n, $n \ge 0$, are integer valued with $P[\tau_0 = j] = p_j$, $j \ge 1$. Suppose $E\tau_0 < 1$ and $W_0 = k$ for $k \ge 1$. Compute the expected number of customers served in the initial busy period and the expected umber of customers served in subsequent busy periods (A. Lemoine, communication).

7.13. Consider the waiting time process $\{W_n\}$ associated with the random walk $\{S_n\}$.
   (a) Suppose $M_\infty = \vee_{n=0}^{\infty} S_n < \infty$. Show for any $x_0 \ge 0$ that

$$P[W_n \le x | W_0 = x_0] \to W(x)$$

as $n \to \infty$. Show W satisfies

(*)
$$W(x) = \int_{[0,\infty)} F(x - y) W(dy), \quad x \ge 0,$$

and that W is a probability distribution concentrating on $[0, \infty)$.

(b) For the G/M/1 queueing model with $EX_1 = 0$ show that

$$W(x) = \begin{cases} 0, & \text{if } x < 0, \\ 1 + \beta^{-1}x, & \text{if } x \geq 0, \end{cases}$$

is an unbounded solution of *.

(c) Let $J_{n+1} = \min\{0, X_{n+1} + W_n\}$. If $\mu = EX_1 \geq 0$, then $J_n \Rightarrow 0$, while if $\mu < 0$ then $J_n \Rightarrow J_\infty$ and $P[J_\infty \leq x] = J(x)$ is proper and satisfies

(**) $$J(x) = \int_{[0,\infty)} F(x - y)W(dy), \quad x < 0.$$

(Assume F is continuous. What if EX_1 does not exist?)

(d) To solve * consider the auxiliary equation ** for $x < 0$ in the unknowns W and J. (Assume as in (a) that $M_\infty < \infty$ a.s.) Then J is non-decreasing with $J(-\infty) = 0, J(0) < \infty$. Define

$$\hat{W}(\zeta) = \int_0^\infty e^{i\zeta x} W(dx),$$

$$\hat{J}(\zeta) = \int_{(-\infty,0]} e^{i\zeta x} J(dx),$$

and show

$$1 - \hat{F}(\zeta) = \frac{W(0) - \hat{J}(\zeta)}{\hat{W}(\zeta)}.$$

Using the Wiener-Hopf factorization ($q = 1$) prove (remember $M_\infty < \infty$)

$$\hat{W}(\zeta) = \frac{W(0)}{q - Ee^{i\zeta S_N}}$$

$$\hat{J}(\zeta) = W(0)Ee^{i\zeta S_N}.$$

(e) For the M/G/1 queueing model, show $J(x)$ is given by

$$J(x) = de^{ax}$$

for some constant d. Derive the Pollaczek-Khintchine formula for W.

Bibliography

Asmussen, S., *Applied Probability and Queues*, J. Wiley and Sons, New York, 1987.

Abramowitz, M. and Stegun, I., *Handbook of Mathematical Functions*, Dover, New York, 1965.

Athreya, K. and Ney, P., *Branching Processes*, Springer-Verlag, New York, 1972.

Bickel, P.J. and Doksum, K.A., *Mathematical Statistics: Basic Ideas and Selected Topics*, Holden Day, San Francisco, 1977

Billingsley, P., *Convergence of Probability Measures*, Wiley, New York, 1968.

Billingsley, P., *Weak Convergence of Measures: Applications in Probability*, Monograph No. 5, SIAM, Philadelphia, PA, 1971.

Billingsley, P., *Probability and Measure*, J. Wiley and Sons, New York, 1986.

Birnbaum, Z.W., *Numerical tabulation of the distribution of Kolmogorov's statistic ...*, J. Amer. Statist. Assoc. **47**(1952), 425–441.

Breiman, L., *Probability*, Addison Wesley, Reading, MA, 1968.

Browne, S., *Best choice problems and records in a linear birth process*, Preprint: Columbia University, New York, NY, 1991.

Bunge, J. and Nagaraja, H., *Exact distribution theory for some point process record models*, Adv. Applied Probability **24**(1992), 20–44.

Chung, K.L., *A Course in Probability Theory*, Academic Press, New York, NY, 1974.

Cinlar, E., *Introduction to Stochastic Processes*, Prentice-Hall, Englewood Cliffs, New Jersey, 1975.

Cinlar, E. and Jagers, P., *Two mean values which characterize the Poisson process*, J. Applied Probability **10**(1973), 678–681

Cohen, A., *An Elementary Treatise on Differential Equations*, Second Edition, D.C. Heath and Co., Boston, MA, 1933.

Cohen, J.W., *On Regenerative Processes in Queueing Theory* Lecture Notes in Economics and Mathematical Systems, **121** Springer-Verlag, New York, 1976.

Cox, D. and Lewis, P., *The Statistical Analysis of Series of Events*, Methuen and Co, LTD, London, UK, 1966.

Dantzig, D. van, *Sur la méthode des fonctions génératrices*, Colloques Internationaux du CNRS **13** (1948), 29–45.

Doob, J., *Stochastic Processes*, Wiley, New York, 1953.

Durrett, R. and Iglehart, D., *Functionals of Brownian meander and Brownian excursion*, Jour. Ann. Probability **5** (1977), 130–135.

Dwass, M., *Extremal processes*, Ann. Math. Statist. **35**(1964), 1718–1725.

Feigin, P., *On the characterization of point processes with the order statistic property*, J. Applied Probability **16**(1979), 297–304.

Feller, W., *An Introduction to Probability Theory and its Applications* Vol. I, Third edition, Wiley, New York, 1968.

Feller, W., *An Introduction to Probability Theory and its Applications*, Vol. II, second edition, Wiley, New York, 1971.

Freedman, D., *Brownian Motion and Diffusion*, Holden-Day, San Francisco, 1971.

Gnedenko, B.V., *Sur la distribution limité d'une série aléatoire*, Ann. Math. **44**(1943), 423–453.

Goldman, J., *Stochastic point processes: limit theorems*, Ann. Math. Statist. **38** (1967), 721–729.

Greenwood, P., *Wiener-Hopf decomposition of random walks and Levy processes*, Z. Wahrscheinlichkeitstheorie verw. Gebiete **34**(1976), 193–198.

Greenwood, P. and Shaked, M., *Fluctuations of random walk in R^d and storage systems*, Advances in Appl. Probability, **9**(1977), 566–587.

Greenwood, P. and Pitman, J., *Fluctuation identities for Levy processes and splitting at the maximum* , Adv. Applied Probability **12**(1980), 893–902.

Haan, L. de, *On Regular Variation and its Application to the Weak Convergence of Sample Extremes*, Mathematical Centre Tract 32 Mathematical Centre, Amsterdam, Holland, 1970.

Jacobs, P. and Lewis, P., *A mixed autoregressive moving average exponential sequence and point process (EARMA(1,1)*, Adv. Applied Probability **9**(1977), 87–104.

Jagers, P., *Branching Processes with Biological Applications*, Wiley, New York, 1975.

Kallenberg, O., *Random Measures*, Third edition, Akademie-Verlag, Berlin, 1983.

Karlin, S. and Taylor, H., *A First Course in Stochastic Processes*, Second edition, Academic Press, New York, 1975.

Karatzas, I. and Shreve, S., *Brownian Motion and Stochastic Calculus*, Springer-Verlag, New York, 1988.

Keiding, N., *Estimation in the birth process* Biometrika **61**(1974), 71–80.

Kelly, F., *Reversibility and Stochastic Networks*, J. Wiley and Sons, New York, 1979.

Kemeny, J.G. and Snell, J.L., *Finite Markov Chains*, Undergraduate Texts in Mathematics, Springer-Verlag, New York, 1976.

Kendell, D.G. *Branching processes since 1873*, J. London Math. Soc. 41(1966), 385–406.

Lawrance, A. and Lewis, P., *An exponential moving average sequence and point process (EMA 1)*, J. Applied Probability 14(1977), 98–113.

Leadbetter, M., Lindgren, G. and Rootzen, H., *Extremes and Related Properties of Random Sequences and Processes*, Springer-Verlag, New York, 1983.

Lemoine, A., *On two stationary distributions for the stable GI/G/1 queue*, J. Applied Probability 11(1974), 849–852.

Lindvall, T., *A probabilistic proof of Blackwell's renewal theorem*, Ann. Probability 5(1977), 482–485.

Lloyd, E.H., *Reservoirs with correlated inflows*, Technometrics 5(1963), 85–93.

Lucantoni, D., *New results on the single server queue with a batch Markovian arrival process*, Stochastic Models 7(1991), 1–46.

McFadden, J., *On the mixed Poisson process*, Sankhya Ser A 27(1968), 83–92.

McKean, H.P., Jr., *Stochastic Integrals*, Academic Press, New York, 1969.

Melamed, B., *Characterization of Poisson traffic streams in Jackson queueing networks*, Adv. Applied Probability 11(1979), 422–438.

Melamed, B., *On markov jump processes imbedded at jump epochs and their queueing-theoretic applications*, Math. Oper. Res. 7(1982), 233–244.

Miller, D.R., *Existence of limits in regenerative processes*, Ann. Math. Statist. 43(1972), 1275–1282.

Moran, P., *A non-Markovian quasi-Poisson process*, Studia Sci. Math. Hungar 2(1967), 425–429.

Neuts, M.F., *Probability*, Allyn and Bacon, Inc., Boston, Mass, 1973.

Neuts, M., *Matrix-Geometric Solutions in Stochastic Models*, Johns Hopkins University Press, Baltimore, 1981.

Neuts, M., *Structured Stochastic Matrices of M/G/1 Type and their Applications*, Marcel Dekker, New York, 1989.

Neuts, M., Liu, D. and Narayana, S., *Local poissonification of the Markovian arrival process*, Stochastic Models 8(1992), 87–130.

Neuts, M. and Resnick, S., *On the times of birth in a linear birth process*, J. Australian Math. Soc. 12(1971), 743–475.

Neveu, J., *Processus Ponctuels*, Ecole d'Eté de Probabilités de Saint-Flour VI., Lecture Notes in Mathematics, 598 (1976), Springer-Verlag, Berlin.

Pakes, A.J., *On dams with Markovian inputs*, J. Applied Probability 10 (1973), 317–329.

Pollard, D., *Convergence of Stochastic Processes*, Springer-Verlag, New York, 1984.

Prabhu, N.U., *Basic Queueing Theory*, Technical Report No. 478, School of ORIE, Cornell University, Ithaca, NY 14853, 1981.

Renyi, A., *On the theory of order statistics*, Acta Math. Acad. Sci. Hung. 4(1953), 191–231.

Renyi, A., *On two mathematical models of traffic on a divided highway*, J. Applied Probability 1 (1964), 311–320.

Resnick, Sidney, *Extreme Values, Regular Variation, and Point Processes*, Springer-Verlag, New York, 1987

Ross, S. *Introduction to Probability Models*, Third edition, Academic Press, Orlando, Florida, 1985.

Runnenburg, J. Th., *Probabilistic interpretation of some formulae in queueing theory*, Bull. Inst. Internat. Statist. 37(1958), 405–414.

Runnenburg, J. Th., *On the use of the method of collective marks in queueing theory*, Proc. Symp. on Congestion Theory, W.L. Smith and W.E. Wilkinson, editors, University of North Carolina Monographs, Chapel Hill, 1965, pp. 399–438.

Takacs, L., *Introduction to the Theory of Queues*, Oxford University Press, New York, 1962.

Taylor, H. and Karlin, S., *Introduction to Stochastic Modelling*, Academic Press, New York, 1984.

Thorisson, H., *The coupling of regenerative processes*, Adv. Appl. Probability 15(1983), 531–561.

Tin, P. and Phatarfod, R.M., *Some exact results for dams with Markovian inputs*, J. Applied Probability 13 (1976) 831–846.

Walrand, J., *An Introduction to Queueing Networks*, Prentice Hall, Englewood Cliffs, New Jersey, 1989.

Waugh, W.A. O'N., *Transformation of a birth process into a Poisson process*, J. R. Statist. Soc. B 32(1970), 418–431.

Wolff, R., *Stochastic Modeling and the Theory of Queues*, Prentice-Hall, Englewood Cliffs, New Jersey, 1989.

Index

Reviews of *Adventures in Stochastic Processes*

"Definitely the best textbook for a second course in probability now available. Written with excruciating lucidity, and with an excellent choice of exercises."
—Gian-Carlo Rota

"A very fine textbook which will become popular among students as well as among professors preparing an introductory course on stochastic processes."
—Int'le Math. Nachrichten

"A splendid book to bring home the value and importance of stochastic processes. Highly recommended."
—Choice

Also by the author

A Probability Path: **ISBN 0-8176-4055-X**

"This book is different from the classical textbooks on probability theory in that it treats the measure theoretic background not as a prerequisite but as an integral part of probability theory. The result is that the reader gets a thorough and well-structured framework needed to understand the deeper concepts of current day advanced probability as it is used in statistics, engineering, biology and finance... The pace of the book is quick and disciplined. Yet there are ample examples sprinkled over the entire book and each chapter finishes with a wealthy section of inspiring problems."
—Publications of the International Statistical Institute

"This book is a rather comfortable probability path leading from fundamental notions of measure theory and probability theory to martingale theory, since fundamental concepts and basic theorems are always illustrated by important examples and results of additional theoretical value."
—Statistics & Decision